Financial Derivatives
in Theory and Practice

Financial Derivatives in Theory and Practice

Revised Edition

P. J. HUNT
WestLB AG, London, UK

J. E. KENNEDY
University of Warwick, UK

John Wiley & Sons, Ltd

Other Wiley Editorial Offices

John Wiley & Sons Inc., 111 River Street, Hoboken, NJ 07030, USA

Jossey-Bass, 989 Market Street, San Francisco, CA 94103-1741, USA

Wiley-VCH Verlag GmbH, Boschstr. 12, D-69469 Weinheim, Germany

John Wiley & Sons Australia Ltd, 33 Park Road, Milton, Queensland 4064, Australia

John Wiley & Sons (Asia) Pte Ltd, 2 Clementi Loop #02-01, Jin Xing Distripark, Singapore
129809

John Wiley & Sons Canada Ltd, 22 Worcester Road, Etobicoke, Ontario, Canada M9W 1L1

Wiley also publishes its books in a variety of electronic formats. Some content that appears
in print may not be available in electronic books.

British Library Cataloguing in Publication Data

A catalogue record for this book is available from the British Library

ISBN 0-470-86358-7 (Cloth)
ISBN 0-470-86359-5 (Paper)

Typeset in 10/12pt Computer Modern by Laserwords Private Limited, Chennai, India

To Patrick.
Choose well.

Contents

Preface to revised edition

Since this book first appeared in 2000, it has been adopted by a number of universities as a standard text for a graduate course in finance. As a result we have produced this revised edition. The only differences in content between this text and its predecessor are the inclusion of an additional chapter of exercises with solutions, and the corrections of a number of errors.

Many of the exercise are variants of ones given to students of the M.Sc. in Mathematical Finance at the University of Warwick, so they have been tried and tested. Many provide drill in routine calculations in the interest-rate setting.

Since the book first appeared there has been further development in the modelling of interest rate derivatives. The modelling and approximation of market models has progressed further, as has that of Markov-functional models that are now used, in their multi-factor form, in a number of banks. A notable exclusion from this revised edition is any coverage of these advances. Those interested in this area can find some of this in Rebonato (2002) and Bennett and Kennedy (2004).

A few extra acknowledgements are due in this revised edition: to Noel Vaillant and Jørgen Aase Nielsen for pointing out a number of errors, and to Stuart Price for typing the exercise chapter and solutions.

Preface

The growth in the financial derivatives market over the last thirty years has been quite extraordinary. From virtually nothing in 1973, when Black, Merton and Scholes did their seminal work in the area, the total outstanding notional value of derivatives contracts today has grown to several trillion dollars. This phenomenal growth can be attributed to two factors.

The first, and most important, is the natural need that the products fulfil. Any organization or individual with sizeable assets is exposed to moves in the world markets. Manufacturers are susceptible to moves in commodity prices; multinationals are exposed to moves in exchange rates; pension funds are exposed to high inflation rates and low interest rates. Financial derivatives are products which allow all these entities to reduce their exposure to market moves which are beyond their control.

The second factor is the parallel development of the financial mathematics needed for banks to be able to price and hedge the products demanded by their customers. The breakthrough idea of Black, Merton and Scholes, that of pricing by arbitrage and replication arguments, was the start. But it is only because of work in the field of pure probability in the previous twenty years that the theory was able to advance so rapidly. Stochastic calculus and martingale theory were the perfect mathematical tools for the development of financial derivatives, and models based on Brownian motion turned out to be highly tractable and usable in practice.

Where this leaves us today is with a massive industry that is highly dependent on mathematics and mathematicians. These mathematicians need to be familiar with the underlying theory of mathematical finance and they need to know how to apply that theory in practice. The need for a text that addressed both these needs was the original motivation for this book. It is aimed at both the mathematical practitioner and the academic mathematician with an interest in the real-world problems associated with financial derivatives. That is to say, we have written the book that we would both like to have read when we first started to work in the area.

This book is divided into two distinct parts which, apart from the need to cross-reference for notation, can be read independently. Part I is devoted to the theory of mathematical finance. It is not exhaustive, notable omissions being a treatment of asset price processes which can exhibit jumps, equilibrium theory and the theory of optimal control (which underpins the pricing of American options). What we have included is the basic theory for continuous asset price processes with a particular emphasis on the martingale approach to arbitrage pricing.

The primary development of the finance theory is carried out in Chapters 1 and 7. The reader who is not already familiar with the subject but who has a solid grounding in stochastic calculus could learn most of the theory from these two chapters alone. The fundamental ideas are laid out in Chapter 1, in the simple setting of a single-period economy with only finitely many states. The full (and more technical) continuous time theory, is developed in Chapter 7. The treatment here is slightly non-standard in the emphasis it places on numeraires (which have recently become extremely important as a modelling tool) and the (consequential) development of the L^1 theory rather than the more common L^2 version. We also choose to work with the filtration generated by the assets in the economy rather than the more usual Brownian filtration. This approach is certainly more natural but, surprisingly, did not appear in the literature until the work of Babbs and Selby (1998).

An understanding of the continuous theory of Chapter 7 requires a knowledge of stochastic calculus, and we present the necessary background material in Chapters 2–6. We have gathered together in one place the results from this area which are relevant to financial mathematics. We have tried to give a full and yet readable account, and hope that these chapters will stand alone as an accessible introduction to this technical area. Our presentation has been very much influenced by the work of David Williams, who was a driving force in Cambridge when we were students. We also found the books of Chung and Williams (1990), Durrett (1996), Karatzas and Shreve (1991), Protter (1990), Revuz and Yor (1991) and Rogers and Williams (1987) to be very illuminating, and the informed reader will recognize their influence.

The reader who has read and absorbed the first seven chapters of the book will be well placed to read the financial literature. The last part of theory we present, in Chapter 8, is more specialized, to the field of interest rate models. The exposition here is partly novel but borrows heavily from the papers by Baxter (1997) and Jin and Glasserman (1997). We describe several different ways present in the literature for specifying an interest rate model and show how they relate to one another from a mathematical perspective. This includes a discussion of the celebrated work of Heath, Jarrow and Morton (1992) (and the less celebrated independent work of Babbs (1990)) which, for the first time, defined interest rate models directly in terms of forward rates.

Part II is very much about the practical side of building models for pricing derivatives. It, too, is far from exhaustive, covering only topics which directly reflect our experiences through our involvement in product development within the London interest rate derivative market. What we have tried to do in this part of the book, through the particular problems and products that we discuss, is to alert the reader to the issues involved in derivative pricing in practice and to give him a framework within which to make his own judgements and a platform from which to develop further models.

Chapter 9 sets the scene for the remainder of the book by identifying some basic issues a practitioner should be aware of when choosing and applying models. Chapters 10 and 11 then introduce the reader to the basic instruments and terminology and to the pricing of standard vanilla instruments using swaption measure. This pricing approach comes from fairly recent papers in the area which focus on the use of various assets as numeraires when defining models and doing calculations. This is actually an old idea dating back at least to Harrison and Pliska (1981) and which, starting with the work of Geman et al. (1995), has come to the fore over the past decade. Chapter 12 is on futures contracts. The treatment here draws largely on the work of Duffie and Stanton (1992), though we have attempted a cleaner presentation of this standard topic than those we have encountered elsewhere.

The remainder of Part II tackles pricing problems of increasing levels of complexity, beginning with single-currency European products and finishing with various Bermudan callable products. Chapters 13–16 are devoted to European derivatives. Chapters 13 and 15 present a new approach to this much-studied problem. These products are theoretically straightforward but the challenge for the practitioner is to ensure the model he employs in practice is well calibrated to market-implied distributions and is easy to implement. Chapter 14 discusses convexity corrections and the pricing of 'convexity-related' products using the ideas of earlier chapters. These are important products which have previously been studied by, amongst others, Coleman (1995) and Doust (1995). We provide explicit formulae for some commonly met products and then, in Chapter 16, generalize these to multi-currency products.

The last three chapters focus on the pricing of path-dependent and American derivatives. Short-rate models have traditionally been those favoured for this task because of their ease of implementation and because they are relatively easy to understand. There is a vast literature on these models, and Chapter 17 provides merely a brief introduction. The Vasicek-Hull-White model is singled out for more detailed discussion and an algorithm for its implementation is given which is based on a semi-analytic approach (taken from Gandhi and Hunt (1997)).

More recently attention has turned to the so-called market models, pioneered by Brace et al. (1997) and Miltersen et al. (1997), and extended by Jamshidian (1997). These models provided a breakthrough in tackling the issue of model calibration and, in the few years since they first appeared, a vast literature has developed around them. They, or variants of and approximations to them, are now starting to replace previous (short-rate) models within most

major investment banks. Chapter 18 provides a basic description of both the
LIBOR- and swap-based market models. We have not attempted to survey
the literature in this extremely important area and for this we refer the reader
instead to the article of Rutkowski (1999).

The book concludes, in Chapter 19, with a description of some of our own
recent work (jointly with Antoon Pelsser), work which builds on Hunt and
Kennedy (1998). We describe a class of models which can fit the observed
prices of liquid instruments in a similar fashion to market models but which
have the advantage that they can be efficiently implemented. These models,
which we call Markov-functional models, are especially useful for the pricing
of products with multi-callable exercise features, such as Bermudan swaptions
or Bermudan callable constant maturity swaps. The exposition here is similar
to the original working paper which was first circulated in late 1997 and
appeared on the Social Sciences Research Network in January 1998. A précis
of the main ideas appeared in *RISK Magazine* in March 1998. The final paper,
Hunt, Kennedy and Pelsser (2000), is to appear soon. Similar ideas have more
recently been presented by Balland and Hughston (2000).

We hope you enjoy reading the finished book. If you learn from this book
even one-tenth of what we learnt from writing it, then we will have succeeded
in our objectives.

Phil Hunt
Joanne Kennedy
31 December 1999

Acknowledgements

There is a long list of people to whom we are indebted and who, either directly or indirectly, have contributed to making this book what it is. Some have shaped our thoughts and ideas; others have provided support and encouragement; others have commented on drafts of the text. We shall not attempt to produce this list in its entirety but our sincere thanks go to all of you.

Our understanding of financial mathematics and our own ideas in this area owe much to two practitioners, Sunil Gandhi and Antoon Pelsser. PH had the good fortune to work with Sunil when they were both learning the subject at NatWest Markets, and with Antoon at ABN AMRO Bank. Both have had a marked impact on this book.

The area of statistics has much to teach about the art of modelling. Brian Ripley is a master of this art who showed us how to look at the subject through new eyes. Warmest and sincere thanks are also due to another colleague from JK's Oxford days, Peter Clifford, for the role he played when this book was taking shape.

We both learnt our probability while doing PhDs at the University of Cambridge. We were there at a particularly exciting time for probability, both pure and applied, largely due to the combined influences of Frank Kelly and David Williams. Their contributions to the field are already well known. Our thanks for teaching us some of what you know, and for teaching us how to use it.

Finally, our thanks go to our families, who provided us with the support, guidance and encouragement that allowed us to pursue our chosen careers. To Shirley Collins, Harry Collins and Elizabeth Kennedy: thank you for being there when it mattered. But most of all, to our parents: for the opportunities given to us at sacrifice to yourselves.

Part I
Theory

1

Single-Period Option Pricing

1.1 OPTION PRICING IN A NUTSHELL

To introduce the main ideas of option pricing we first develop this theory
in the case when asset prices can take on only a finite number of values
and when there is only one time-step. The continuous time theory which we
introduce in Chapter 7 is little more than a generalization of the basic ideas
introduced here.

The two key concepts in option pricing are *replication* and *arbitrage*. Option
pricing theory centres around the idea of replication. To price any derivative
we must find a portfolio of assets in the economy, or more generally a trading
strategy, which is guaranteed to pay out *in all circumstances* an amount
identical to the payout of the derivative product. If we can do this we have
exactly replicated the derivative.

This idea will be developed in more detail later. It is often obscured in
practice when calculations are performed 'to a formula' rather than from first
principles. It is important, however, to ensure that the derivative being priced
can be reproduced by trading in other assets in the economy.

An arbitrage is a trading strategy which generates profits from nothing
with no risk involved. Any economy we postulate must not allow arbitrage
opportunities to exist. This seems natural enough but it is essential for our
purposes, and Section 1.3.2 is devoted to establishing necessary and sufficient
conditions for this to hold.

Given the absence of arbitrage in the economy it follows immediately that
the value of a derivative is the value of a portfolio that replicates it. To
see this, suppose to the contrary that the derivative costs more than the
replicating portfolio (the converse can be treated similarly). Then we can sell
the derivative, use the proceeds to buy the replicating portfolio and still be
left with some free cash to do with as we wish. Then all we need do is use
the income from the portfolio to meet our obligations under the derivative
contract. This is an arbitrage opportunity – and we know they do not exist.

All that follows in this book is built on these simple ideas!

Financial Derivatives in Theory and Practice Revised Edition. P. J. Hunt and J. E. Kennedy
© 2004 John Wiley & Sons, Ltd ISBNs: 0-470-86358-7 (HB); 0-470-86359-5 (PB)

1.2 THE SIMPLEST SETTING

Throughout this section we consider the following simple situation. There are two assets, $A^{(1)}$ and $A^{(2)}$, with prices at time zero $A_0^{(1)}$ and $A_0^{(2)}$. At time 1 the economy is in one of two possible states which we denote by ω_1 and ω_2. We denote by $A_1^{(i)}(\omega_j)$ the price of asset i at time 1 if the economy is in state ω_j. Figure 1.1 shows figuratively the possibilities. We shall denote a portfolio of assets by $\phi = (\phi^{(1)}, \phi^{(2)})$ where $\phi^{(i)}$, the holding of asset i, could be negative.

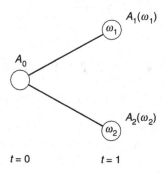

Figure 1.1 Possible states of the economy

We wish to price a derivative (of $A^{(1)}$ and $A^{(2)}$) which pays at time 1 the amount $X_1(\omega_j)$ if the economy is in state ω_j. We will do this by constructing a replicating portfolio. Once again, this is the *only* way that a derivative can be valued, although, as we shall see later, explicit knowledge of the replicating portfolio is not always necessary.

The first step is to check that the economy we have constructed is arbitrage-free, meaning we cannot find a way to generate money from nothing. More precisely, we must show that there is no portfolio $\phi = (\phi^{(1)}, \phi^{(2)})$ with one of the following (equivalent) conditions holding:

(i) $\phi^{(1)} A_0^{(1)} + \phi^{(2)} A_0^{(2)} < 0$, $\quad \phi^{(1)} A_1^{(1)}(\omega_j) + \phi^{(2)} A_1^{(2)}(\omega_j) \geq 0$, $\quad j = 1, 2$,

(ii) $\phi^{(1)} A_0^{(1)} + \phi^{(2)} A_0^{(2)} \leq 0$, $\quad \phi^{(1)} A_1^{(1)}(\omega_j) + \phi^{(2)} A_1^{(2)}(\omega_j) \geq 0$, $\quad j = 1, 2$,

where the second inequality is strict for some j.

Suppose there exist $\phi^{(1)}$ and $\phi^{(2)}$ such that

$$\phi^{(1)} A_1^{(1)}(\omega_j) + \phi^{(2)} A_1^{(2)}(\omega_j) = X_1(\omega_j), \quad j = 1, 2. \tag{1.1}$$

We say that ϕ is a replicating portfolio for X. If we hold this portfolio at time zero, at time 1 the value of the portfolio will be exactly the value of X, no matter which of the two states the economy is in. It therefore follows that a fair price for the derivative X (of $A^{(1)}$ and $A^{(2)}$) is precisely the cost of this portfolio, namely

$$X_0 = \phi^{(1)} A_0^{(1)} + \phi^{(2)} A_0^{(2)}.$$

Subtleties

The above approach is exactly the one we use in the more general situations later. There are, however, three potential problems:

(i) Equation (1.1) may have no solution.
(ii) Equation (1.1) may have (infinitely) many solutions yielding the same values for X_0.
(iii) Equation (1.1) may have (infinitely) many solutions yielding different values of X_0.

Each of these has its own consequences for the problem of derivative valuation.

(i) If (1.1) has no solution, we say the economy is *incomplete*, meaning that it is possible to introduce further assets into the economy which are not redundant and cannot be replicated. Any such asset is not a derivative of existing assets and cannot be priced by these methods.

(ii) If (1.1) has many solutions all yielding the same value X_0 then there exists a portfolio $\psi \neq 0$ such that

$$\psi \cdot A_0 = 0, \quad \psi \cdot A_1(\omega_j) = 0, \quad j = 1, 2.$$

This is not a problem and means our original assets are not all independent – one of these is a derivative of the others, so any further derivative can be replicated in many ways.

(iii) If (1.1) has many solutions yielding different values for X_0 we have a problem. There exist portfolios ϕ and ψ such that

$$(\phi - \psi) \cdot A_0 < 0$$
$$(\phi - \psi) \cdot A_1(\omega_j) = \big(X_1(\omega_j) - X_1(\omega_j)\big) = 0, \qquad j = 1, 2.$$

This is an arbitrage, a portfolio with strictly negative value at time zero and zero value at time 1.

In this final case our initial economy was poorly defined. Such situations can occur in practice but are not sustainable. From a derivative pricing viewpoint such situations must be excluded from our model. In the presence of arbitrage there is no unique fair price for a derivative, and in the absence of arbitrage the derivative value is given by the initial value of *any* replicating portfolio.

1.3 GENERAL ONE-PERIOD ECONOMY

We now develop in more detail all the concepts and ideas raised in the previous section. We restrict attention once again to a single-period economy for clarity, but introduce many assets and many states so that the essential structure and techniques used in the continuous time setting can emerge and be discussed.

We now consider an economy $\mathcal{E}$ comprising n assets with m possible states at time 1. Let Ω be the set of all possible states. We denote, as before, the individual states by $\omega_j, j = 1, 2, \ldots, m$, and the asset prices by $A_0^{(i)}$ and $A_1^{(i)}(\omega_j)$. We begin with some definitions.

Definition 1.1 *The economy $\mathcal{E}$ admits arbitrage if there exists a portfolio ϕ such that one of the following conditions (which are actually equivalent in this discrete setting) holds:*

(i) $\phi \cdot A_0 < 0$ and $\phi \cdot A_1(\omega_j) \geq 0$ for all j,
(ii) $\phi \cdot A_0 \leq 0$ and $\phi \cdot A_1(\omega_j) \geq 0$ for all j, with strict inequality for some j.

If there is no such ϕ then the economy is said to be arbitrage-free.

Definition 1.2 *A derivative X is said to be attainable if there exists some ϕ such that*

$$X_1(\omega_j) = \phi \cdot A_1(\omega_j) \quad \text{for all } j.$$

We have seen that an arbitrage-free economy is essential for derivative pricing theory. We will later derive conditions to check whether a given economy is indeed arbitrage-free. Here we will quickly move on to show, in the absence of arbitrage, how products can be priced. First we need one further definition.

Definition 1.3 *A pricing kernel Z is any strictly positive vector with the property that*

$$A_0 = \sum_j Z_j A_1(\omega_j). \tag{1.2}$$

The reason for this name is the role Z plays in derivative pricing as summarized by Theorem 1.4. It also plays an important role in determining whether or not an economy admits arbitrage, as described in Theorem 1.7.

1.3.1 Pricing

One way to price a derivative in an arbitrage-free economy is to construct a replicating portfolio and calculate its value. This is summarized in the first part of the following theorem. The second part of the theorem enables us to price a derivative without explicitly constructing a replicating portfolio, as long as we know one exists.

Theorem 1.4 *Suppose that the economy $\mathcal{E}$ is arbitrage-free and let X be an attainable contingent claim, i.e. a derivative which can be replicated with other assets. Then the fair value of X is given by*

$$X_0 = \phi \cdot A_0 \tag{1.3}$$

where ϕ solves

$$X_1(\omega_j) = \phi \cdot A_1(\omega_j) \quad \text{for all } j.$$

Furthermore, if Z is some pricing kernel for the economy then X_0 can also be represented as

$$X_0 = \sum_j Z_j X_1(\omega_j).$$

Proof: The first part of the result follows by the arbitrage arguments used previously. Moving on to the second statement, substituting the defining equation (1.2) for Z into (1.3) yields

$$X_0 = \phi \cdot A_0 = \sum_j Z_j \big(\phi \cdot A_1(\omega_j)\big) = \sum_j Z_j X_1(\omega_j).$$

$\square$

Remark 1.5: Theorem 1.4 shows how the pricing of a derivative can be reduced to calculating a pricing kernel Z. This is of limited value as it stands since we still need to show that the economy is arbitrage-free and that the derivative in question is attainable. This latter step can be done by either explicitly constructing the replicating portfolio or proving beforehand that *all* derivatives (or at least all within some suitable class) are attainable. If all derivatives are attainable then the economy is said to be *complete*. As we shall shortly see, both the problems of establishing no arbitrage and completeness for an economy are in themselves intimately related to the pricing kernel Z.

Remark 1.6: Theorem 1.4 is essentially a geometric result. It merely states that if X_0 and $X_1(\omega_j)$ are the projections of A_0 and $A_1(\omega_j)$ in some direction ϕ, and if A_0 is an affine combination of the vectors $A_1(\omega_j)$, then X_0 is the same affine combination of the vectors $X_1(\omega_j)$. Figure 1.2 illustrates this for the case $n = 2$, a two-asset economy.

1.3.2 Conditions for no arbitrage: existence of Z

The following result gives necessary and sufficient conditions for an economy to be arbitrage-free. The result is essentially a geometric one. However, its real importance as a tool for calculation becomes clear in the continuous time setting of Chapter 7 where we work in the language of the probabilist. We will see the result rephrased probabilistically later in this chapter.

Theorem 1.7 *The economy $\mathcal{E}$ is arbitrage-free if and only if there exists a pricing kernel, i.e. a strictly positive Z such that*

$$A_0 = \sum_j Z_j A_1(\omega_j).$$

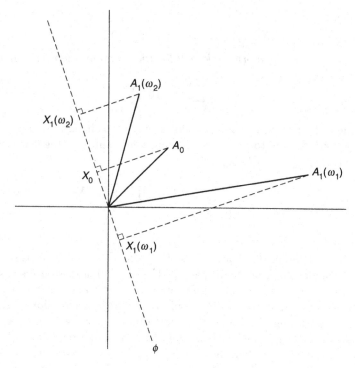

Figure 1.2 Geometry of option pricing

Proof: Suppose such a Z exists. Then for any portfolio ϕ,

$$\phi \cdot A_0 = \phi \cdot \left(\sum_j Z_j A_1(\omega_j) \right) = \sum_j Z_j \big(\phi \cdot A_1(\omega_j) \big). \qquad (1.4)$$

If ϕ is an arbitrage portfolio then the left-hand side of (1.4) is non-positive, and the right-hand side is non-negative. Hence (1.4) is identically zero. This contradicts ϕ being an arbitrage portfolio and so $\mathcal{E}$ is arbitrage-free.

Conversely, suppose no such Z exists. We will in this case construct an arbitrage portfolio. Let C be the convex cone constructed from $A_1(\cdot)$,

$$C = \Big\{ a : a = \sum_j Z_j A_1(\omega_j), Z \gg 0 \Big\}.$$

The set C is a non-empty convex set *not* containing A_0. Here $Z \gg 0$ means that all components of Z are greater than zero. Hence, by the separating hyperplane theorem there exists a hyperplane $H = \{x : \phi \cdot x = \beta\}$ that separates A_0 and C,

$$\phi \cdot A_0 \leq \beta \leq \phi \cdot a \quad \text{for all } a \in C.$$

The vector ϕ represents an arbitrage portfolio, as we now demonstrate.

First, observe that if $a \in \bar{C}$ (the closure of C) then $\mu a \in \bar{C}$, $\mu \geq 0$, and

$$\beta \leq \mu(\phi \cdot a).$$

Taking $\mu = 0$ yields $\beta \leq 0$, $\phi \cdot A_0 \leq 0$. Letting $\mu \uparrow \infty$ shows that $\phi \cdot a \geq 0$ for all $a \in \bar{C}$, in particular $\phi \cdot A_1(\omega_j) \geq 0$ for all j. So we have that

$$\phi \cdot A_0 \leq 0 \leq \phi \cdot A_1(\omega_j) \quad \text{for all } j, \tag{1.5}$$

and it only remains to show that (1.5) is not always identically zero. But in this case $\phi \cdot A_0 = \phi \cdot C = 0$ which violates the separating property for H. $\square$

Remark 1.8: Theorem 1.7 states that A_0 must be in the interior of the convex cone created by the vectors $A_1(\omega_j)$ for there to be no arbitrage. If this is not the case an arbitrage portfolio exists, as in Figure 1.3.

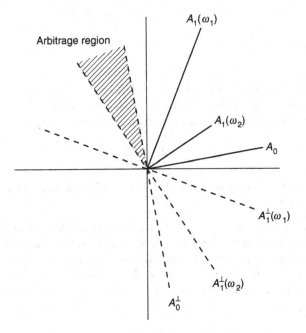

Figure 1.3 An arbitrage portfolio

1.3.3 Completeness: uniqueness of Z

If we are in a complete market all contingent claims can be replicated. As we have seen in Theorem 1.4, this enables us to price derivatives without having to explicitly calculate the replicating portfolio. This can be useful, particularly for the continuous time models introduced later. In our finite economy it is clear what conditions are required for completeness. There must be at least as many assets as states of the economy, i.e. $m \leq n$, and these must span an appropriate space.

Theorem 1.9 *The economy $\mathcal{E}$ is complete if and only if there exists a (generalized) left inverse A^{-1} for the matrix A where*

$$A_{ij} = A_1^{(i)}(\omega_j).$$

Equivalently, $\mathcal{E}$ is complete if and only if there exists no non-zero solution to the equation $Ax = 0$.

Proof: For the economy to be complete, given any contingent claim X, we must be able to solve

$$X_1(\omega_j) = \phi \cdot A_1(\omega_j) \quad \text{for all } j,$$

which can be written as

$$X_1 = A^T \phi. \tag{1.6}$$

The existence of a solution to (1.6) for all X_1 is exactly the statement that A^T has a right inverse (some matrix B such that $A^T B : \mathbb{R}^m \to \mathbb{R}^m$ is the identity matrix) and the replicating portfolio is then given by

$$\phi = (A^T)^{-1} X_1.$$

This is equivalent to the statement that A has full rank m and there being no non-zero solution to $Ax = 0$.   □

Remark 1.10: If there are more assets than states of the economy the hedge portfolio will not in general be uniquely specified. In this case one or more of the underlying assets is already completely specified by the others and could be regarded as a derivative of the others. When the number of assets matches the number of states of the economy A^{-1} is the usual matrix inverse and the hedge is unique.

Remark 1.11: Theorem 1.9 demonstrates clearly that completeness is a statement about the rank of the matrix A. Noting this, we see that there exist economies that admit arbitrage yet are complete. Consider A having rank $m < n$. Since $\dim(\text{Im}(A)^\perp) = n - m$ we can choose $A_0 \in \text{Im}(A)^\perp$, in which case there does not exist $Z \in \mathbb{R}^m$ such that $AZ = A_0$. By Theorem 1.7 the economy admits arbitrage, yet having rank m it is complete by Theorem 1.9.

In practice we will always require an economy to be arbitrage-free. Under this assumption we can state the condition for completeness in terms of the pricing kernel Z.

Theorem 1.12 *Suppose the economy $\mathcal{E}$ is arbitrage-free. Then it is complete if and only if there exists a unique pricing kernel Z.*

Proof: By Theorem 1.7, there exists some vector Z satisfying

$$A_0 = \sum_j Z_j A_1(\omega_j). \tag{1.7}$$

By Theorem 1.9, it suffices to show that Z being unique is equivalent to there being no solution to $Ax = 0$. If $\hat{Z}$ also solves (1.7) then $x = \hat{Z} - Z$ solves $Ax = 0$. Conversely, if Z solves (1.7) and $Ax = 0$ then $\hat{Z} = Z + \varepsilon x$ solves (1.7) for all $\varepsilon > 0$. Since $Z_j > 0$ for all j we can choose ε sufficiently small that $\hat{Z}_j > 0$ for all j, yielding a second pricing kernel. $\square$

The combination of Theorems 1.7 and 1.12 gives the well-known result that an economy is complete and arbitrage-free if and only if there exists a unique pricing kernel Z. This strong notion of completeness is not the primary one that we shall consider in the continuous time context of Chapter 7, where we shall say that an economy $\mathcal{E}$ is complete if it is $\mathcal{F}_T^A$-complete.

Definition 1.13 *Let $\mathcal{F}_1^A$ be the smallest σ-algebra with respect to which the map $A_1 : \Omega \to \mathbb{R}^m$ is measurable. We say the economy $\mathcal{E}$ is $\mathcal{F}_1^A$-complete if every $\mathcal{F}_1^A$-measurable contingent claim is attainable.*

The reason why $\mathcal{F}_1^A$ completeness is more natural to consider is as follows. Suppose there are two states in the economy, ω_i and ω_j, for which $A_1(\omega_i) = A_1(\omega_j)$. Then it is impossible, by observing only the prices A, to distinguish which state of the economy we are in, and in practice all we are interested in is derivative payoffs that can be determined by observing the process A. There is an analogue of Theorem 1.12 which covers this case.

Theorem 1.14 *Suppose the economy $\mathcal{E}$ is arbitrage-free. Then it is $\mathcal{F}_1^A$-complete if and only if all pricing kernels agree on $\mathcal{F}_1^A$, i.e. if $Z^{(1)}$ and $Z^{(2)}$ are two pricing kernels, then for every $F \in \mathcal{F}_1^A$,*

$$\mathbb{E}\left[Z^{(1)} 1_F\right] = \mathbb{E}\left[Z^{(2)} 1_F\right].$$

Proof: For a discrete economy, the statement that X_1 is $\mathcal{F}_1^A$-measurable is precisely the statement that $X_1(\omega_i) = X_1(\omega_j)$ whenever $A_1(\omega_i) = A_1(\omega_j)$. If this is the case we can identify any states ω_i and ω_j for which $A_1(\omega_i) = A_1(\omega_j)$. The question of $\mathcal{F}_1^A$-completeness becomes one of proving that this reduced economy is complete in the full sense. It is clearly arbitrage-free, a property it inherits from the original economy.

Suppose Z is a pricing kernel for the original economy. Then, as is easily verified,

$$\hat{Z} := \mathbb{E}[Z|\mathcal{F}_1^A] \qquad (1.8)$$

is a pricing kernel for the reduced economy. If the reduced economy is (arbitrage-free and) complete it has a unique pricing kernel $\hat{Z}$ by Theorem 1.12. Thus all pricing kernels for the original economy agree on $\mathcal{F}_1^A$ by (1.8). Conversely, if all pricing kernels agree on $\mathcal{F}_1^A$ then the reduced economy has a unique pricing kernel, by (1.8), thus is complete by Theorem 1.12. □

1.3.4 Probabilistic formulation

We have seen that pricing derivatives is about replication and the results we have met so far are essentially geometric in nature. However, it is standard in the modern finance literature to work in a probabilistic framework. This approach has two main motivations. The first is that probability gives a natural and convenient language for stating the results and ideas required, and it is also the most natural way to formulate models that will be a good reflection of reality. The second is that many of the sophisticated techniques needed to develop the theory of derivative pricing in continuous time are well developed in a probabilistic setting.

With this in mind we now reformulate our earlier results in a probabilistic context. In what follows let $\mathbb{P}$ be a probability measure on Ω such that $\mathbb{P}(\{\omega_j\}) > 0$ for all j. We begin with a preliminary restatement of Theorem 1.7.

Theorem 1.15 *The economy $\mathcal{E}$ is arbitrage-free if and only if there exists a strictly positive random variable Z such that*

$$A_0 = \mathbb{E}[ZA_1]. \qquad (1.9)$$

Extending Definition 1.3, we call Z a pricing kernel for the economy $\mathcal{E}$.

Suppose, further, that $\mathbb{P}(A_1^{(i)} \gg 0) = 1$, $A_0^{(i)} > 0$ for some i. Then the economy $\mathcal{E}$ is arbitrage-free if and only if there exists a strictly positive random variable κ with $\mathbb{E}[\kappa] = 1$ such that

$$\mathbb{E}\left[\kappa \frac{A_1}{A_1^{(i)}}\right] = \frac{A_0}{A_0^{(i)}}. \qquad (1.10)$$

Proof: The first result follows immediately from Theorem 1.7 by setting $Z(\omega_j) = Z_j/\mathbb{P}(\{\omega_j\})$. To prove the second part of the theorem we show that (1.9) and (1.10) are equivalent. This follows since, given either of Z or κ, we can define the other via

$$Z(\omega_j) = \kappa(\omega_j)\frac{A_0^{(i)}}{A_1^{(i)}(\omega_j)}.$$

□

Remark 1.16: Note that the random variable Z is simply a weighted version of the pricing kernel in Theorem 1.7, the weights being given by the probability measure $\mathbb{P}$. Although the measure $\mathbb{P}$ assigns probabilities to 'outcomes' ω_j, these probabilities are arbitrary and the role played by $\mathbb{P}$ here is to summarize which states may occur through the assignment of positive mass.

Remark 1.17: The random variable κ is also a reweighted version of the pricing kernel of Theorem 1.7 (and indeed of the one here). In addition to being positive, κ has expectation one, and so $\kappa_j := \kappa(\omega_j)/\mathbb{P}(\{\omega_j\})$ defines a probability measure. We see the importance of this shortly.

Remark 1.18: Equation (1.10) can be interpreted geometrically, as shown in Figure 1.4 for the two-asset case. No arbitrage is equivalent to $A_0 \in C$ where

$$C = \Big\{a : a = \sum_j Z_j A_1(\omega_j), Z \gg 0\Big\}.$$

Rescaling A_0 and each $A_1(\omega_j)$ does not change the convex cone C and whether or not A_0 is in C, it only changes the weights required to generate A_0 from the $A_1(\omega_j)$. Rescaling so that A_0 and the $A_1(\omega_j)$ all have the same (unit) component in the direction i ensures that the weights κ_j satisfy $\sum_j \kappa_j = 1$.

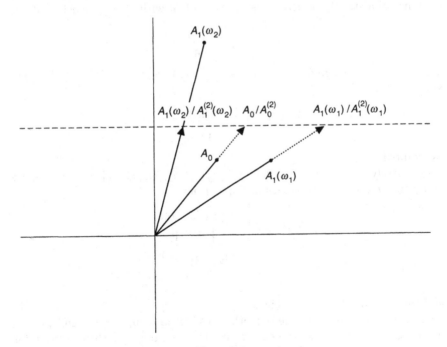

Figure 1.4 Rescaling asset prices

We have one final step to take to cast the results in the standard probabilistic format. This format replaces the problem of finding Z or κ by one of finding a probability measure with respect to which the *process A*, suitably rebased, is a *martingale*. Theorem 1.20 below is the precise statement of what we mean by this (equation (1.11)). We meet martingales again in Chapter 3.

Definition 1.19 *Two probability measures $\mathbb{P}$ and $\mathbb{Q}$ on the finite sample space Ω are said to be equivalent, written $\mathbb{P} \sim \mathbb{Q}$, if*

$$\mathbb{P}(F) = 0 \iff \mathbb{Q}(F) = 0,$$

for all $F \subseteq \Omega$. If $\mathbb{P} \sim \mathbb{Q}$ we can define the Radon–Nikodým derivative of $\mathbb{P}$ with respect to $\mathbb{Q}$, $\frac{d\mathbb{P}}{d\mathbb{Q}}$, by

$$\frac{d\mathbb{P}}{d\mathbb{Q}}(S) = \frac{\mathbb{P}(S)}{\mathbb{Q}(S)}.$$

Theorem 1.20 *Suppose $\mathbb{P}(A_1^{(i)} \gg 0) = 1$, $A_0^{(i)} > 0$ for some i. Then the economy $\mathcal{E}$ is arbitrage-free if and only if there exists a probability measure $\mathbb{Q}_i$ equivalent to $\mathbb{P}$ such that*

$$\mathbb{E}_{\mathbb{Q}_i}[A_1/A_1^{(i)}] = A_0/A_0^{(i)}. \tag{1.11}$$

The measure $\mathbb{Q}_i$ is said to be an equivalent martingale measure for $A^{(i)}$.

Proof: By Theorem 1.15 we must show that (1.11) is equivalent to the existence of a strictly positive random variable κ with $\mathbb{E}[\kappa] = 1$ such that

$$\mathbb{E}\left[\kappa \frac{A_1}{A_1^{(i)}}\right] = \frac{A_0}{A_0^{(i)}}.$$

Suppose $\mathcal{E}$ is arbitrage-free and such a κ exists. Define $\mathbb{Q}_i \sim \mathbb{P}$ by $\mathbb{Q}_i(\{\omega_j\}) = \kappa(\omega_j)\mathbb{P}(\{\omega_j\})$. Then

$$\mathbb{E}_{\mathbb{Q}_i}[A_1/A_1^{(i)}] = \mathbb{E}[\kappa A_1/A_1^{(i)}]$$
$$= A_0/A_0^{(i)}$$

as required.

Conversely, if (1.11) holds define $\kappa(\omega_j) = \mathbb{Q}_i(\{\omega_j\})/\mathbb{P}(\{\omega_j\})$. It follows easily that κ has unit expectation. Furthermore,

$$\mathbb{E}\left[\kappa \frac{A_1}{A_1^{(i)}}\right] = \mathbb{E}\left[\frac{d\mathbb{Q}_i}{d\mathbb{P}} \frac{A_1}{A_1^{(i)}}\right]$$
$$= \mathbb{E}_{\mathbb{Q}_i}[A_1/A_1^{(i)}]$$
$$= A_0/A_0^{(i)},$$

and there is no arbitrage by Theorem 1.15. $\square$

We are now able to restate our other results on completeness and pricing in this same probabilistic framework. In our new framework the condition for completeness can be stated as follows.

Theorem 1.21 *Suppose that no arbitrage exists and that* $\mathbb{P}(A_1^{(i)} \gg 0) = 1$, $A_0^{(i)} > 0$ *for some* i. *Then the economy* $\mathcal{E}$ *is complete if and only if there exists a unique equivalent martingale measure for the 'unit'* $A^{(i)}$.

Proof: Observe that in the proof of Theorem 1.20 we established a one-to-one correspondence between pricing kernels and equivalent martingale measures. The result now follows from Theorem 1.12 which establishes the equivalence of completeness to the existence of a unique pricing kernel. □

Our final result, concerning the pricing of a derivative, is left as an exercise for the reader.

Theorem 1.22 *Suppose* $\mathcal{E}$ *is arbitrage-free,* $\mathbb{P}(A_1^{(i)} \gg 0) = 1$, $A_0^{(i)} > 0$ *for some* i, *and that* X *is an attainable contingent claim. Then the fair value of* X *is given by*

$$X_0 = A_0^{(i)} \mathbb{E}_{\mathbb{Q}_i}[X_1/A_1^{(i)}],$$

where $\mathbb{Q}_i$ *is an equivalent martingale measure for the unit* $A^{(i)}$.

1.3.5 Units and numeraires

Throughout Section 1.3.4 we assumed that the price of one of the assets, asset i, was positive with probability one. This allowed us to use this asset price as a *unit*; the operation of dividing all other asset prices by the price of asset i can be viewed as merely recasting the economy in terms of this new unit.

There is no reason why, throughout Section 1.3.4, we need to restrict the unit to be one of the assets. All the results hold if the unit $A^{(i)}$ is replaced by some other unit U which is strictly positive with probability one and for which

$$U_0 = \mathbb{E}_{\mathbb{P}}(ZU_1) \tag{1.12}$$

where Z is some pricing kernel for the economy. Note, in particular, that we can always take $U_0 = 1$, $\mathbb{Q}(\{\omega_j\}) = 1/n$ and $U_1(\omega_j) = 1/(nZ_j\mathbb{P}(\{\omega_j\}))$.

Observe that (1.12) automatically holds (assuming a pricing kernel exists) when U is a derivative and thus is of the form $U = \phi \cdot A$. In this case we say that U is a *numeraire* and then we usually denote it by the symbol N in preference to U. In general there are more units than numeraires.

The ideas of numeraires, martingales and change of measure are central to the further development of derivative pricing theory.

1.4 A TWO-PERIOD EXAMPLE

We now briefly consider an example of a two-period economy. Inclusion of the extra time-step allows us to develop new ideas whilst still in a relatively

simple framework. In particular, to price a derivative in this richer setting we
shall need to define what is meant by a *trading strategy* which is *self-financing*.

For our two-period example we build on the simple set-up of Section 1.2.
Suppose that at the new time, time 2, the economy can be in one of four
states which we denote by ω'_j, $j = 1, \ldots, 4$, with the restriction that states
ω'_1 and ω'_2 can be reached only from ω_1 and states ω'_3 and ω'_4 can be reached
only from ω_2. Figure 1.5 summarizes the possibilities.

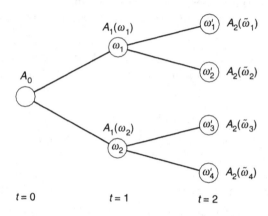

Figure 1.5 Possible states of a two-period economy

Let Ω now denote the set of all possible *paths*. That is, $\Omega = \{\tilde{\omega}_k, k = 1, \ldots, 4\}$ where $\tilde{\omega}_k = (\omega_1, \omega'_k)$ for $k = 1, 2$ and $\tilde{\omega}_k = (\omega_2, \omega'_k)$ for $k = 3, 4$. The
asset prices follow one of four paths with $A_t^{(i)}(\tilde{\omega}_k)$ denoting the price of asset
i at time $t = 1, 2$.

Consider the problem of pricing a derivative X which at time 2 pays an
amount $X_2(\tilde{\omega}_k)$. In order to price the derivative X we must be able to replicate
over all paths. We cannot do this by holding a static portfolio. Instead, the
portfolio we hold at time zero will in general need to be changed at time 1
according to which state ω_j the economy is in at this time. Thus, replicating
X in a two-period economy amounts to specifying a *process* ϕ which is non-
anticipative, i.e. a process which does not depend on knowledge of a future
state at any time. Such a process is referred to as a *trading strategy*. To find
the fair value of X we must be able to find a trading strategy which is *self-
financing*; that is, a strategy for which, apart from the initial capital at time
zero, no additional influx of funds is required in order to replicate X. Such a
trading strategy will be called *admissible*.

We calculate a suitable ϕ in two stages, working backwards in time. Suppose
we know the economy is in state ω_1 at time 1. Then we know from the one-
period example of Section 1.2 that we should hold a portfolio $\phi_1(\omega_1)$ of assets

satisfying

$$\phi_1(\omega_1) \cdot A_2(\omega_j') = X_2(\tilde{\omega}_j), \quad j = 1, 2.$$

Similarly, if the economy is in state ω_2 at time 1 we should hold a portfolio $\phi_1(\omega_2)$ satisfying

$$\phi_1(\omega_2) \cdot A_2(\omega_j') = X_2(\tilde{\omega}_j), \quad j = 3, 4.$$

Conditional on knowing the economy is in state ω_j at time 1, the fair value of the derivative X at time 1 is then $X_1(\omega_j) = \phi_1(\omega_j) \cdot A_1(\omega_j)$, the value of the replicating portfolio.

Once we have calculated ϕ_1 the problem of finding the fair value of X at time zero has been reduced to the one-period case, i.e. that of finding the fair value of an option paying $X_1(\omega_j)$ at time 1. If we can find ϕ_0 such that

$$\phi_0 \cdot A_1(\omega_j) = X_1(\omega_j), \quad j = 1, 2,$$

then the fair price of X at time zero is

$$X_0 = \phi_0 \cdot A_0.$$

Note that for the ϕ satisfying the above we have

$$\phi_0 \cdot A_1(\omega_j) = \phi_1(\omega_j) \cdot A_1(\omega_j), \quad j = 1, 2,$$

as must be the case for the strategy to be self-financing; the portfolio is merely rebalanced at time 1.

Though of mathematical interest, the multi-period case is not important in practice and when we again take up the story of derivative pricing in Chapter 7 we will work entirely in the continuous time setting. For a full treatment of the multi-period problem the reader is referred to Duffie (1996).

2

Brownian Motion

2.1 INTRODUCTION

Our objective in this book is to develop a theory of derivative pricing in continuous time, and before we can do this we must first have developed models for the underlying assets in the economy. In reality, asset prices are piecewise constant and undergo discrete jumps but it is convenient and a reasonable approximation to assume that asset prices follow a continuous process. This approach is often adopted in mathematical modelling and justified, at least in part, by results such as those in Section 2.3.1. Having made the decision to use continuous processes for asset prices, we must now provide a way to generate such processes. This we will do in Chapter 6 when we study stochastic differential equations. In this chapter we study the most fundamental and most important of all continuous stochastic processes, the process from which we can build all the other continuous time processes that we will consider, Brownian motion.

The physical process of Brownian motion was first observed in 1828 by the botanist Robert Brown for pollen particles suspended in a liquid. It was in 1900 that the mathematical process of Brownian motion was introduced by Bachelier as a model for the movement of stock prices, but this was not really taken up. It was only in 1905, when Einstein produced his work in the area, that the study of Brownian motion started in earnest, and not until 1923 that the existence of Brownian motion was actually established by Wiener. It was with the work of Samuelson in 1969 that Brownian motion reappeared and became firmly established as a modelling tool for finance.

In this chapter we introduce Brownian motion and derive a few of its more immediate and important properties. In so doing we hope to give the reader some insight into and intuition for how it behaves.

Financial Derivatives in Theory and Practice Revised Edition. P. J. Hunt and J. E. Kennedy
© 2004 John Wiley & Sons, Ltd ISBNs: 0-470-86358-7 (HB); 0-470-86359-5 (PB)

2.2 DEFINITION AND EXISTENCE

We begin with the definition.

Definition 2.1 *Brownian motion is a real-valued stochastic process with the following properties:*

(BM.i)   *Given any $t_0 < t_1 < \ldots < t_n$ the random variables $\{(W_{t_i} - W_{t_{i-1}}),$ $i = 1, 2, \ldots, n\}$ are independent.*
(BM.ii) *For any $0 \leq s \leq t, W_t - W_s \sim N(0, t - s)$.*
(BM.iii) W_t *is continuous in t almost surely (a.s.).*
(BM.iv) $W_0 = 0$ *a.s.*

Property (BM.iv) is sometimes omitted to allow arbitrary initial distributions, although it is usually included. We include it for definiteness and note that extending to the more general case is a trivial matter.

Implicit in the above definition is the fact that a process W exists with the stated properties and that it is unique. Uniqueness is a straightforward matter – any two processes satisfying (BM.i)–(BM.iv) have the same finite-dimensional distributions, and the finite-dimensional distributions determine the law of a process. In fact, Brownian motion can be characterized by a slightly weaker condition than (BM.ii):

(BM.ii′) For any $t \geq 0, h \geq 0$, the distribution of $W_{t+h} - W_t$ is independent of $t, \mathbb{E}[W_t] = 0$ and $\mathrm{var}[W_t] = t$.

It should not be too surprising that (BM.ii′) in place of (BM.ii) gives an equivalent definition. Given any $0 \leq s \leq t$, the interval $[s, t]$ can be subdivided into n equal intervals of length $(t - s)/n$ and

$$W_t - W_s = \sum_{i=1}^{n} W_{s+(t-s)i/n} - W_{s+(t-s)(i-1)/n},$$

i.e. as a sum of n independent, identically distributed random variables. A little care needs to be exercised, but it effectively follows from the central limit theorem and the continuity of W that $W_t - W_s$ must be Gaussian. Breiman (1992) provides all the details. The moment conditions in (BM.ii′) now force the process to be a standard Brownian motion (without them the conditions define the more general process $\mu t + \sigma W_t$).

The existence of Brownian motion is more difficult to prove. There are many excellent references which deal with this question, including those by Breiman (1992) and Durrett (1984). We will merely state this result. To do so, we must explicitly define a probability space $(\Omega, \mathcal{F}, \mathbb{P})$ and a process W on this space which is Brownian motion. Let $\Omega = \mathbb{C} \equiv \mathbb{C}(\mathbb{R}^+, \mathbb{R})$ be the set of continuous functions from $[0, \infty)$ to $\mathbb{R}$. Endow $\mathbb{C}$ with the metric of uniform convergence, i.e. for $x, y \in \mathbb{C}$

$$\rho(x, y) = \sum_{n=1}^{\infty} \left(\frac{1}{2}\right)^n \frac{\rho_n(x, y)}{1 + \rho_n(x, y)}$$

where

$$\rho_n(x, y) = \sup_{0 \leq t \leq n} |x(t) - y(t)|.$$

Now define $\mathcal{F} = \mathcal{C} \equiv \mathcal{B}(\mathbb{C})$, the Borel σ-algebra on $\mathbb{C}$ induced by the metric ρ.

Theorem 2.2 *For each $\omega \in \mathbb{C}, t \geq 0$, define $W_t(\omega) = \omega_t$. There exists a unique probability measure $\mathbb{W}$ (Wiener measure) on $(\mathbb{C}, \mathcal{C})$ such that the stochastic process $\{W_t, t \geq 0\}$ is Brownian motion (i.e. satisfies conditions (BM.i)–(BM.iv)).*

Remark 2.3: The set-up described above, in which the sample space Ω is the set of sample paths, is the *canonical set-up*. There are others, but the canonical set-up is the most direct and intuitive to consider. Note, in particular, that for this set-up *every* sample path is continuous.

In the above we have defined Brownian motion without reference to a filtration. Adding a filtration is a straightforward matter: it will often be taken to be $\{\mathcal{F}_t^W\}^\circ := \sigma(W_s, s \leq t)$, but it could also be more general. Brownian motion relative to a filtered probability space is defined as follows.

Definition 2.4 *The process W is Brownian motion with respect to the filtration $\{\mathcal{F}_t\}$ if:*
(i)   it is adapted to $\{\mathcal{F}_t\}$;
(ii) for all $0 \leq s \leq t, W_t - W_s$ is independent of $\mathcal{F}_s$;
(iii) it is a Brownian motion as defined in Definition 2.1.

2.3 BASIC PROPERTIES OF BROWNIAN MOTION

2.3.1 Limit of a random walk

If you have not encountered Brownian motion before it is important to develop an intuition for its behaviour. A good place to start is to compare it with a simple symmetric random walk on the integers, S_n. The following result roughly states that if we speed up a random walk and look at it from a distance (see Figure 2.1) it appears very much like Brownian motion.

Theorem 2.5 *Let $\{X_i, i \geq 1\}$ be independent and identically distributed random variables with $\mathbb{P}(X_i = 1) = \mathbb{P}(X_i = -1) = \frac{1}{2}$. Define the simple*

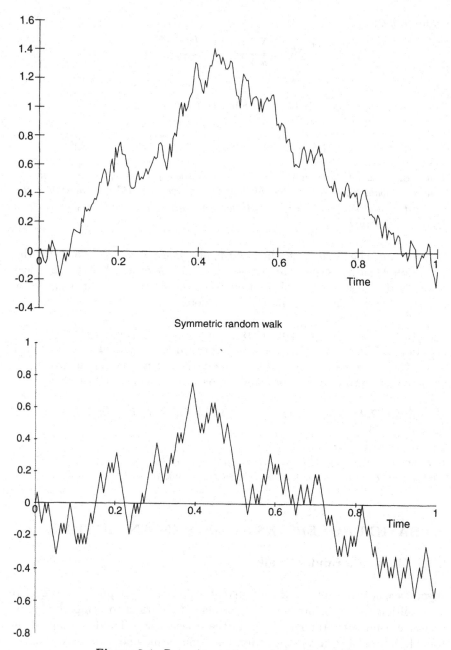

Figure 2.1. Brownian motion and a random walk

symmetric random walk $\{S_n, n \geq 1\}$ *as* $S_n = \sum_{i=1}^{n} X_i$, *and the rescaled random walk* $Z_n(t) := \frac{1}{\sqrt{n}} S_{\lfloor nt \rfloor}$ *where* $\lfloor x \rfloor$ *is the integer part of* x. *As* $n \to \infty$,

$$Z_n \Rightarrow W.$$

In the above, W is a Brownian motion and the convergence $\Rightarrow$ denotes weak convergence. In this setting this is equivalent to convergence of finite-dimensional distributions, meaning, here, that for any $t_1, t_2, \ldots, t_k$ and any $z \in \mathbb{R}^k$,

$$\mathbb{P}(Z_n(t_i) \leq z_i, \; i = 1, \ldots, k) \to \mathbb{P}(W_{t_i} \leq z_i, \; i = 1, \ldots, k)$$

as $n \to \infty$.

Proof: This follows immediately from the (multivariate) central limit theorem for the simple symmetric random walk. □

2.3.2 Deterministic transformations of Brownian motion

There is an extensive theory which studies what happens to a Brownian motion under various (random) transformations. One question people ask is what is the law of the process X defined by

$$X_t = W_{\tau(t)},$$

where the random clock τ is specified by

$$\tau(t) = \inf\{s \geq 0 : Y_s > t\},$$

for some process Y. In the case when Y, and consequently τ, is non-random the study is straightforward because the Gaussian structure of Brownian motion is retained. In particular, there are a well-known set of transformations of Brownian motion which produce another Brownian motion, and these results prove especially useful when studying properties of the Brownian sample path.

Theorem 2.6 *Suppose W is Brownian motion. Then the following transformations also produce Brownian motions:*

(i) $\widetilde{W}_t := cW_{t/c^2}$ *for any* $c \in \mathbb{R}\backslash\{0\}$;   *(scaling)*

(ii) $\widetilde{W}_t := tW_{1/t}$ *for* $t > 0, \widetilde{W}_0 := 0$;   *(time inversion)*

(iii) $\widetilde{W}_t := \{W_{t+s} - W_s : t \geq 0\}$ *for any* $s \in \mathbb{R}^+$.   *(time homogeneity)*

Proof: We must prove that (BM.i)–(BM.iv) hold. The proof in each case follows similar lines so we will only establish the time inversion result which is slightly more involved than the others.

Given any fixed $t_0 < t_1 < \ldots < t_n$ it follows from the Gaussian structure of W that $(\widetilde{W}_{t_0}, \ldots, \widetilde{W}_{t_n})$ also has a Gaussian distribution. A Gaussian distribution is completely determined by its mean and covariance and thus (BM.i) and (BM.ii) now follow if we can show that $\mathbb{E}[\widetilde{W}_t] = 0$ and $\mathbb{E}[\widetilde{W}_s \widetilde{W}_t] = s \wedge t$ for all s, t. But this follows easily:

$$\mathbb{E}[\widetilde{W}_t] = \mathbb{E}[t W_{1/t}] = t \mathbb{E}[W_{1/t}] = 0,$$
$$\mathbb{E}[\widetilde{W}_s \widetilde{W}_t] = \mathbb{E}[st W_{1/s} W_{1/t}] = st\big((1/t) \wedge (1/s)\big) = s \wedge t.$$

Condition (BM.iv), that $\widetilde{W}_0 = 0$, is part of the definition and, for any $t > 0$, the continuity of $\widetilde{W}$ at t follows from the continuity of W. It therefore only remains to prove that $\widetilde{W}$ is continuous at zero or, equivalently, that for all $n > 0$ there exists some $m > 0$ such that $\sup_{0<t\leq 1/m} |\widetilde{W}_t| < 1/n$. The continuity of $\widetilde{W}$ for $t > 0$ implies that

$$\sup_{0<t\leq 1/m} |\widetilde{W}_t| = \sup_{q \in \mathbb{Q} \cap (0,1/m]} |\widetilde{W}_q|,$$

and so

$$\mathbb{P}\Big(\lim_{t\to 0} \widetilde{W}_t = 0\Big) = \mathbb{P}\Big(\bigcap_n \bigcup_m \big\{ \sup_{q \in \mathbb{Q} \cap (0,1/m]} |\widetilde{W}_q| < 1/n \big\}\Big)$$
$$= \mathbb{P}\Big(\bigcap_n \bigcup_m \bigcap_{q \in \mathbb{Q} \cap (0,1/m]} \{ |\widetilde{W}_q| < 1/n \}\Big). \qquad (2.1)$$

There are a countable number of events in the last expression of equation (2.1) and each of these involves the process $\widetilde{W}$ at a single strictly positive time. The distributions of W and $\widetilde{W}$ agree for $t > 0$ and therefore this last probability is unaltered if the process $\widetilde{W}$ is replaced by the original Brownian motion W. But W is a.s. continuous at zero and so the probability in (2.1) is one and $\widetilde{W}$ is also a.s. continuous at zero. $\qquad \square$

2.3.3 Some basic sample path properties

The transformations just introduced have many useful applications. Two of them are given below.

Theorem 2.7 If W is a Brownian motion then, as $t \to \infty$,

$$\frac{W_t}{t} \to 0 \qquad a.s.$$

Proof: Writing $\widetilde{W}_t$ for the time inversion of W as defined in Theorem 2.6, $\mathbb{P}(\lim_{t\to\infty} \frac{W_t}{t} = 0) = \mathbb{P}(\lim_{s\to 0} \widetilde{W}_s = 0)$ and this last probability is one since $\widetilde{W}$ is a Brownian motion, continuous at zero. $\qquad \square$

Theorem 2.8 *Given a Brownian motion* W,

$$\mathbb{P}\left(\sup_{t\geq 0} W_t = +\infty, \inf_{t\geq 0} W_t = -\infty\right) = 1.$$

Proof: Consider first $\sup_{t\geq 0} W_t$. For any $a > 0$, the scaling property implies that

$$\mathbb{P}\left(\sup_{t\geq 0} W_t > a\right) = \mathbb{P}\left(\sup_{t\geq 0} cW_{t/c^2} > a\right)$$

$$= \mathbb{P}\left(\sup_{s\geq 0} W_s > a/c\right).$$

Hence the probability is independent of a and thus almost every sample path has a supremum of either 0 or ∞. In particular, for almost every sample path, $\sup_{t\geq 0} W_t = 0$ if and only if $W_1 \leq 0$ and $\sup_{t\geq 1}(W_t - W_1) = 0$. But then, defining $\widetilde{W}_t = W_{1+t} - W_1$,

$$p := \mathbb{P}\left(\sup_{t\geq 0} W_t = 0\right)$$

$$= \mathbb{P}\left(W_1 \leq 0, \sup_{t\geq 1}(W_t - W_1) = 0\right)$$

$$= \mathbb{P}(W_1 \leq 0)\mathbb{P}\left(\sup_{t\geq 0} \widetilde{W}_t = 0\right)$$

$$= \frac{1}{2}p.$$

We conclude that $p = 0$. A symmetric argument shows that $\mathbb{P}(\inf_{t\geq 0} W_t = -\infty) = 1$ and this completes the proof. $\square$

This result shows that the Brownian path will keep oscillating between positive and negative values over extended time intervals. The path also oscillates wildly over small time intervals. In particular, it is nowhere differentiable.

Theorem 2.9 *Brownian motion is nowhere differentiable with probability one.*

Proof: It suffices to prove the result on $[0, 1]$. Suppose $W.(\omega)$ is differentiable at some $t \in [0, 1]$ with derivative bounded in absolute value by some constant $N/2$. Then there exists some $n > 0$ such that

$$h \leq 4/n \quad \Rightarrow \quad |W_{t+h}(\omega) - W_t(\omega)| \leq Nh. \tag{2.2}$$

Denote the event that (2.2) holds for some $t \in [0, 1]$ by $A_{n,N}$. The event that Brownian motion is differentiable for some $t \in [0, 1]$ is a subset of the event $\bigcup_N \bigcup_n A_{n,N}$ and the result is proven if we can show that $\mathbb{P}(A_{n,N}) = 0$ for all n, N.

Fix n and N. Let $\Delta_n(k) = W_{(k+1)/n} - W_{k/n}$ and define $k_t^n = \inf\{k : k/n \geq t\}$. If (2.2) holds at t then it follows from the triangle inequality that $|\Delta_n(k_t^n + j)| \leq 7N/n$, for $j = 0, 1, 2$. Therefore

$$A_{n,N} \subseteq \bigcup_{k=0}^{n} \bigcap_{j=0}^{2} \{|\Delta_n(k+j)| \leq 7N/n\},$$

$$\mathbb{P}(A_{n,N}) \leq \mathbb{P}\left(\bigcup_{k=0}^{n} \bigcap_{j=0}^{2} \{|\Delta_n(k+j)| \leq 7N/n\}\right)$$

$$\leq (n+1)\mathbb{P}\left(\bigcap_{j=0}^{2} \{|\Delta_n(j)| \leq 7N/n\}\right)$$

$$= (n+1)\mathbb{P}\left(\{|\Delta_n(0)| \leq 7N/n\}\right)^3$$

$$\leq (n+1)\left(\frac{14N}{\sqrt{2\pi n}}\right)^3$$

$$\to 0 \qquad \text{as } n \to \infty,$$

the last equality following from the independence of the Brownian increments. If we now note that $A_{n,N}$ is increasing in n we can conclude that $\mathbb{P}(A_{n,N}) = 0$ for all N, n and we are done. □

The lack of differentiability of Brownian motion illustrates how irregular the Brownian path can be and implies immediately that the Brownian path is not of finite variation. This means it is not possible to use the Brownian path as an integrator in the classical sense. The following result is central to stochastic integration. A proof is provided in Chapter 3, Corollary 3.81.

Theorem 2.10 *For a Brownian motion W, define the doubly infinite sequence of (stopping) times T_k^n via*

$$T_0^n \equiv 0, \ T_{k+1}^n = \inf\{t > T_k^n : |W_t - W_{T_k^n}| > 2^{-n}\},$$

and let

$$[W]_t(\omega) := \lim_{n \to \infty} \sum_{k \geq 1} [W_{t \wedge T_k^n}(\omega) - W_{t \wedge T_{k-1}^n}(\omega)]^2.$$

The process $[W]$ is called the quadratic variation of W, and $[W]_t = t$ a.s.

2.4 STRONG MARKOV PROPERTY

Markov processes and strong Markov processes are of fundamental importance to mathematical modelling. Roughly speaking, a Markov process is one for which 'the future is independent of the past, given the present'; for any fixed $t \geq 0$, X is Markovian if the law of $\{X_s - X_t : s \geq t\}$ given X_t is independent of the law of $\{X_s : s \leq t\}$. This property obviously simplifies the study of Markov processes and often allows problems to be solved (perhaps numerically) which are not tractable for more general processes. We shall, in Chapter 6, give a definition of a (strong) Markov process which makes the above more precise. That Brownian motion is Markovian follows from its definition.

Theorem 2.11   *Given any $t \geq 0$, $\{W_s - W_t : s \geq t\}$ is independent of $\{W_s : s \leq t\}$ (and indeed $\mathcal{F}_t$ if the Brownian motion is defined on a filtered probability space).*

Proof: This is immediate from (BM.i).   □

The strong Markov property is a more stringent requirement of a stochastic process and is correspondingly more powerful and useful. To understand this idea we must here introduce the concept of a stopping time. We will reintroduce this in Chapter 3 when we discuss stopping times in more detail.

Definition 2.12   *The random variable T, taking values in $[0, \infty]$, is an $\{\mathcal{F}_t\}$ stopping time if*

$$\{T \leq t\} = \{\omega : T(\omega) \leq t\} \in \mathcal{F}_t,$$

for all $t \leq \infty$.

Definition 2.13   *For any $\{\mathcal{F}_t\}$ stopping time T, the pre-T σ-algebra $\mathcal{F}_T$ is defined via*

$$\mathcal{F}_T = \{F : \text{ for every } t \leq \infty, F \cap \{T \leq t\} \in \mathcal{F}_t\}.$$

Definition 2.12 makes precise the intuitive notion that a stopping time is one for which we know that it has occurred at the moment it occurs. Definition 2.13 formalizes the idea that the σ-algebra $\mathcal{F}_T$ contains all the information that is available up to and including the time T.

A strong Markov process, defined precisely in Chapter 6, is a process which possesses the Markov property, as described at the beginning of this section, but with the constant time t generalized to be an arbitrary stopping time T. Theorem 2.15 below shows that Brownian motion is also a strong Markov process. First we establish a weaker preliminary result.

Proposition 2.14   *Let W be a Brownian motion on the filtered probability space $(\Omega, \{\mathcal{F}_t\}, \mathcal{F}, \mathbb{P})$ and suppose that T is an a.s. finite stopping time taking on one of a countable number of possible values. Then*

$$\widetilde{W}_t := W_{t+T} - W_T$$

is a Brownian motion and $\{\widetilde{W}_t : t \geq 0\}$ is independent of $\mathcal{F}_T$ (in particular, it is independent of $\{W_s : s \leq T\}$).

Proof: Fix $t_1, \ldots, t_n \geq 0, F_1, \ldots, F_n \in \mathcal{B}(\mathbb{R}), F \in \mathcal{F}_T$ and let $\{\tau_j, j \geq 1\}$ denote the values that T can take. We have that

$$\mathbb{P}(\widetilde{W}_{t_i} \in F_i, i = 1, \ldots, n; F) = \sum_{j=1}^{\infty} \mathbb{P}(\widetilde{W}_{t_i} \in F_i, i = 1, \ldots, n; F; T = \tau_j)$$

$$= \sum_{j=1}^{\infty} \mathbb{P}(W_{t_i + \tau_j} - W_{\tau_j} \in F_i, i = 1, \ldots, n) \mathbb{P}(F; T = \tau_j)$$

$$= \mathbb{P}(W_{t_i} \in F_i, i = 1, \ldots, n) \mathbb{P}(F),$$

the last equality following from the Brownian shifting property and by performing the summation. This proves the result. $\qquad\square$

Theorem 2.15 *Let W be a Brownian motion on the filtered probability space $(\Omega, \{\mathcal{F}_t\}, \mathcal{F}, \mathbb{P})$ and suppose that T is an a.s. finite stopping time. Then*

$$\widetilde{W}_t := W_{t+T} - W_T$$

is a Brownian motion and $\{\widetilde{W}_t : t \geq 0\}$ is independent of $\mathcal{F}_T$ (in particular, it is independent of $\{W_s : s \leq T\}$).

Proof: As in the proof of Proposition 2.14, and adopting the notation introduced there, it suffices to prove that

$$\mathbb{P}(\widetilde{W}_{t_i} \in F_i, i = 1, \ldots, n; F) = \mathbb{P}(W_{t_i} \in F_i, i = 1, \ldots, n) \mathbb{P}(F).$$

Define the stopping times T_k via

$$T_k = \frac{q}{2^k} \quad \text{if} \quad \frac{q-1}{2^k} \leq T < \frac{q}{2^k}.$$

Each T_k can only take countably many values so, noting that $T_k \geq T$ and thus $F \in \mathcal{F}_{T_k}$, Proposition 2.14 applies to give

$$\mathbb{P}(\widetilde{W}_{t_i}^k \in F_i, i = 1, \ldots, n; F) = \mathbb{P}(W_{t_i}^k \in F_i, i = 1, \ldots, n) \mathbb{P}(F), \qquad 2.3$$

where $\widetilde{W}_t^k = W_{t+T_k} - W_{T_k}$. As $k \to \infty, T_k(\omega) \to T(\omega)$ and thus, by the almost sure continuity of Brownian motion, $\widetilde{W}_t^k \to \widetilde{W}_t$ for all t, a.s. The result now follows from (2.3) by dominated convergence. $\qquad\square$

2.4.1 Reflection principle

An immediate and powerful consequence of Theorem 2.15 is as follows.

Theorem 2.16 *If W is a Brownian motion on the space $(\Omega, \{\mathcal{F}_t\}, \mathcal{F}, \mathbb{P})$ and $T = \inf\{t > 0 : W_t \geq a\}$, for some a, then the process*

$$\widetilde{W}_t := \begin{cases} W_t, & t < T \\ 2a - W_t, & t \geq T \end{cases}$$

is also Brownian motion.

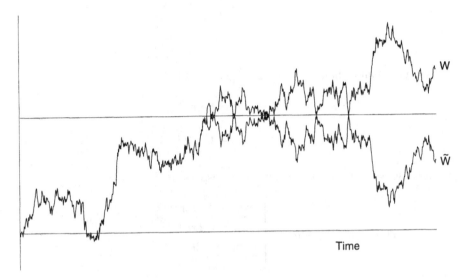

Figure 2.2 Reflected Brownian path

Proof: It is easy to see that T is a stopping time (look ahead to Theorem 3.40 if you want more details here), thus $W_t^* := W_{t+T} - W_T$ is, by the strong Markov property, a Brownian motion independent of $\mathcal{F}_T$, as is $-W_t^*$ (by the scaling property). Hence the following have the same distribution:

(i) $W_t \mathbb{1}_{\{t<T\}} + \left(a + W_{t-T}^*\right)\mathbb{1}_{\{t\geq T\}}$
(ii) $W_t \mathbb{1}_{\{t<T\}} + \left(a - W_{t-T}^*\right)\mathbb{1}_{\{t\geq T\}}.$

The first of these is just the original Brownian motion, the second is $\widetilde{W}$, so the result follows. $\square$

A 'typical' sample path for Brownian motion W and the reflected Brownian motion $\widetilde{W}$ is shown in Figure 2.2.

Example 2.17 The reflection principle can be used to derive the joint distribution of Brownian motion and its supremum, a result which is used for the analytic valuation of barrier options. Let W be a Brownian motion and $\widetilde{W}$ be the Brownian motion reflected about some $a > 0$. Defining $M_t = \sup_{s \leq t} W_s$

(and $\widetilde{M}$ similarly), it is clear that, for $x \leq a$,

$$\{\omega : W_t(\omega) \leq x, M_t(\omega) \geq a\} = \{\omega : \widetilde{W}_t(\omega) \geq 2a - x, \widetilde{M}_t(\omega) \geq a\},$$

and thus, denoting by Φ the normal distribution function,

$$\mathbb{P}(W_t \leq x, M_t \geq a) = \mathbb{P}(\widetilde{W}_t \geq 2a - x, \widetilde{M}_t \geq a)$$

$$= \mathbb{P}(\widetilde{W}_t \geq 2a - x)$$

$$= 1 - \Phi\left(\frac{2a - x}{\sqrt{t}}\right),$$

the second equality holding since $2a - x \geq a$. For $x > a$,

$$\mathbb{P}(W_t \geq x, M_t \geq a) = \mathbb{P}(W_t \geq x)$$

$$= 1 - \Phi\left(\frac{x}{\sqrt{t}}\right).$$

From these we can find the density with respect to x:

$$\mathbb{P}(W_t \in dx, M_t \geq a) = \begin{cases} \dfrac{\exp(-(2a - x)^2/2t)}{\sqrt{2\pi t}}, & x \leq a \\ \dfrac{\exp(-x^2/2t)}{\sqrt{2\pi t}}, & x > a. \end{cases}$$

The density with respect to a follows similarly.

3

Martingales

Martingales are amongst the most important tools in modern probability. They are also central to modern finance theory. The simplicity of the martingale definition, a process which has mean value at any future time, conditional on the present, equal to its present value, belies the range and power of the results which can be established. In this chapter we will introduce continuous time martingales and develop several important results. The value of some of these will be self-evident. Others may at first seem somewhat esoteric and removed from immediate application. Be assured, however, that we have included nothing which does not, in our view, aid understanding or which is not directly relevant to the development of the stochastic integral in Chapter 4, or the theory of continuous time finance as developed in Chapter 7.

A brief overview of this chapter and the relevance of each set of results is as follows. Section 3.1 below contains the definition, examples and basic properties of martingales. Throughout we will consider general continuous time martingales, although for the financial applications in this book we only need martingales that have continuous paths. In Section 3.2 we introduce and discuss three increasingly restrictive classes of martingales. As we impose more restrictions, so stronger results can be proved. The most important class, which is of particular relevance to finance, is the class of uniformly integrable martingales. Roughly speaking, any uniformly integrable martingale M, a stochastic process, can be summarized by M_∞, a random variable: given M we know M_∞, given M_∞ we can recover M. This reduces the study of these martingales to the study of random variables. Furthermore, this class is precisely the one for which the powerful and extremely important optional sampling theorem applies, as we show in Section 3.3. A second important space of martingales is also discussed in Section 3.2, square-integrable martingales. It is not obvious in this chapter why these are important and the results may seem a little abstract. If you want motivation, glance ahead to Chapter 4, otherwise take our word for it that they are important.

Section 3.3 contains a discussion of stopping times and, as mentioned

Financial Derivatives in Theory and Practice Revised Edition. P. J. Hunt and J. E. Kennedy
© 2004 John Wiley & Sons, Ltd ISBNs: 0-470-86358-7 (HB); 0-470-86359-5 (PB)

above, the optional sampling theorem ('if a martingale is uniformly integrable then the mean value of the martingale at any stopping time, conditional on the present, is its present value'). The quadratic variation for continuous martingales is introduced and studied in Section 3.4. The results which we develop are interesting in their own right, but the motivation for the discussion of quadratic variation is its vital role in the development of the stochastic integral. The chapter concludes with two sections on processes more general than martingales. Section 3.5 introduces the idea of localization and defines *local* martingales and *semimartingales*. Then, in Section 3.6, *supermartingales* are considered and the important Doob–Meyer decomposition theorem is presented. We will call on the latter result in Chapter 8 when we study term structure models.

3.1 DEFINITION AND BASIC PROPERTIES

Definition 3.1   *Let $(\Omega, \mathcal{F}, \mathbb{P})$ be a probability triple and $\{\mathcal{F}_t\}$ be a filtration on $\mathcal{F}$. A stochastic process M is an $\{\mathcal{F}_t\}$ martingale (or just martingale when the filtration is clear) if:*

(M.i) M is adapted to $\{\mathcal{F}_t\}$;
(M.ii) $\mathbb{E}[|M_t|] < \infty$ for all $t \geq 0$;
(M.iii) $\mathbb{E}[M_t|\mathcal{F}_s] = M_s$ a.s., for all $0 \leq s \leq t$.

Remark 3.2: Property (M.iii) can also be written as $\mathbb{E}[(M_t - M_s)\mathbf{1}_F] = 0$ for all $s \leq t$, for all $F \in \mathcal{F}_s$. We shall often use this representation in proofs.

Example 3.3   Brownian motion is a rich source of example martingales. Let W be a Brownian motion and $\{\mathcal{F}_t\}$ be the filtration generated by W. Then it is easy to verify directly that each of the following is a martingale:

   (i) $\{W_t, t \geq 0\}$;
   (ii) $\{W_t^2 - t, t \geq 0\}$;
   (iii) $\{\exp(\lambda W_t - \frac{1}{2}\lambda^2 t), t \geq 0\}$ for any $\lambda \in \mathbb{R}$ (Wald's martingale).

Taking Wald's martingale for illustration, property (M.i) follows from the definition of $\{\mathcal{F}_t\}$, and property (M.ii) follows from property (M.iii) by setting $s = 0$. To prove property (M.iii), note that for $t \geq 0$,

$$\mathbb{E}[M_t] = \int_{-\infty}^{\infty} \exp(\lambda u - \tfrac{1}{2}\lambda^2 t) \frac{\exp(-u^2/2t)}{\sqrt{2\pi t}}\, du$$
$$= \int_{-\infty}^{\infty} \frac{\exp(-(u - \lambda t)^2/2t)}{\sqrt{2\pi t}}\, du$$
$$= 1.$$

Thus, appealing also to the independence of Brownian increments, $\mathbb{E}[M_t - M_s | \mathcal{F}_s] = \mathbb{E}[M_t - M_s] = 0$, which establishes property (M.iii).

We will meet these martingales again later.

Example 3.4   Most martingales explicitly encountered in practice are Markov processes, but they need not be, as this example demonstrates. First define a *discrete time martingale* M via

$$M_0 = 0$$

$$M_n = M_{n-1} + \alpha_n, \qquad n \geq 1,$$

where the α_n are independent, identically distributed random variables taking the values ± 1 with probability $1/2$. Viewed as a discrete time process, M is Markovian. Now define $M_t^c = M_{\lfloor t \rfloor}$ where $\lfloor t \rfloor$ is the integer part of t. This is a continuous time martingale which is not Markovian. The process (M_t, t) is Markovian but it is not a martingale.

Example 3.5   Not all martingales behave as one might expect. Consider the following process M. Let T be a random exponentially distributed time, $\mathbb{P}(T > t) = \exp(-t)$, and define M via

$$M_t = \begin{cases} 1 & \text{if } t - T \in \mathbb{Q}^+, \\ 0 & \text{otherwise,} \end{cases}$$

$\mathbb{Q}^+$ being the positive rationals. Conditions (M.i) and (M.ii) of Definition 3.1 are clearly satisfied for the filtration $\{\mathcal{F}_t\}$ generated by the process M. Further, for any $t \geq s$ and any $F \in \mathcal{F}_s$,

$$\mathbb{E}[M_t \mathbf{1}_F] \leq \mathbb{E}[\mathbf{1}_{\{t-T \in \mathbb{Q}^+\}}] = 0 = \mathbb{E}[M_s \mathbf{1}_F].$$

Thus $\mathbb{E}[M_t | \mathcal{F}_s] = M_s$ a.s., which is condition (M.iii), and M is a martingale.

Example 3.5 seems counter-intuitive and we would like to eliminate from consideration martingales with behaviour such as this. We will do this by imposing a (right-) continuity constraint on the paths of martingales that we will consider henceforth. We shall see, in Theorem 3.8, that this restriction is not unnecessarily restrictive.

Definition 3.6   *A function x is said to be càdlàg (continu à droite, limites à gauche) if it is right-continuous with left limits,*

$$\lim_{h \downarrow 0} x_{t+h} = x_t$$

$$\lim_{h \downarrow 0} x_{t-h} \text{ exists.}$$

In this case we define $x_{t-} := \lim_{h \downarrow 0} x_{t-h}$ (which is left-continuous with right limits).

We say that a stochastic process X is càdlàg if, for almost every $\omega, X.(\omega)$ is a càdlàg function. If $X.(\omega)$ is càdlàg for every ω we say that X is entirely càdlàg.

The martingale in Example 3.5 is clearly not càdlàg, but it has a càdlàg *modification* (the process $M \equiv 0$).

Definition 3.7 *Two stochastic processes X and Y are modifications (of each other) if, for all t,*

$$\mathbb{P}(X_t = Y_t) = 1.$$

We say X and Y are indistinguishable if

$$\mathbb{P}(X_t = Y_t, \text{ for all } t) = 1.$$

Theorem 3.8 *Let M be a martingale with respect to the right-continuous and complete filtration $\{\mathcal{F}_t\}$. Then there exists a unique (up to indistinguishability) modification M^* of M which is càdlàg and adapted to $\{\mathcal{F}_t\}$ (hence is an $\{\mathcal{F}_t\}$ martingale).*

Remark 3.9: Note that it is not necessarily the case that every sample path is càdlàg, but there will be a set (having probability one) of sample paths, in which set every sample path will be càdlàg. This is important to bear in mind when, for example, proving several results about stopping times which are results about filtrations and not about probabilities (see Theorems 3.37 and 3.40).

Remark 3.10: The restriction that $\{\mathcal{F}_t\}$ be right-continuous and complete is required to ensure that the modification is adapted to $\{\mathcal{F}_t\}$. It is for this reason that the concept of completeness of a filtration is required in the study of stochastic processes.

A proof of Theorem 3.8 can be found, for example, in Karatzas and Shreve (1991). Càdlàg processes (or at least piecewise continuous ones) are natural ones to consider in practice and Theorem 3.8 shows that this restriction is exactly what it says and no more – the càdlàg restriction does not in any way limit the finite-dimensional distributions that a martingale can exhibit. The 'up to indistinguishability' qualifier is, of course, necessary since we can always modify any process on a null set. Henceforth we will restrict attention to càdlàg processes. The following result will often prove useful.

Theorem 3.11 *Let X and Y be two càdlàg stochastic processes such that, for all t,*

$$\mathbb{P}(X_t = Y_t) = 1,$$

i.e. they are modifications. Then X and Y are indistinguishable.

Proof: By right-continuity,

$$\{X_t \neq Y_t, \text{ some } t\} = \bigcup_{q \in \mathbb{Q}} \{X_q \neq Y_q\},$$

and thus

$$\mathbb{P}(X_t \neq Y_t, \text{ some } t) \leq \sum_{q \in \mathbb{Q}} \mathbb{P}(X_q \neq Y_q) = 0 \,.$$

□

So we can safely restrict to càdlàg processes, and henceforth whenever we talk about a martingale we will mean its càdlàg version. The finite-dimensional distributions of a martingale (indeed, of any stochastic process) determine its law, and this in turn uniquely determines the càdlàg version of the martingale. Thus questions about the sample paths of a martingale are now (implicitly) reduced to questions about the finite-dimensional distributions. Consequently it now makes sense to ask questions about the sample paths of continuous time martingales as these will be measurable events *on the understanding that any such question is about the càdlàg version of the martingale.*

3.2 CLASSES OF MARTINGALES

We now introduce three subcategories of martingales which are progressively more restrictive. These are $\mathcal{L}^1$-bounded martingales, uniformly integrable martingales and square-integrable martingales. The concept of uniform integrability is closely tied to conditional expectation and hence to the martingale property (see Remark 3.23). Consequently the set of uniformly integrable martingales is often the largest and most natural class for which many important results hold. Note that Brownian motion is not included in any of the classes below.

3.2.1 Martingales bounded in $\mathcal{L}^1$

First a more general definition which includes $\mathcal{L}^1$.

Definition 3.12 *We define $\mathcal{L}^p$ to be the space of random variables Y such that*

$$\mathbb{E}[|Y|^p] < \infty \,.$$

A stochastic process X is said to be bounded in $\mathcal{L}^p$ (or in $\mathcal{L}^p$) if

$$\sup_{t \geq 0} \mathbb{E}[|X_t|^p] < \infty \,.$$

If the process X is bounded in $\mathcal{L}^2$ we say X is square-integrable.

We will often also use $\mathcal{L}^p$ to denote the space of stochastic processes bounded in $\mathcal{L}^p$.

Here we are interested in $\mathcal{L}^1$-bounded martingales. Note that if M is a martingale then $\mathbb{E}[|M_t|] < \infty$ for all t. Further, it follows from the conditional form of Jensen's inequality that, for $s \leq t$,

$$\mathbb{E}[|M_t|] = \mathbb{E}\mathbb{E}[|M_t||\mathcal{F}_s] \geq \mathbb{E}|\mathbb{E}[M_t|\mathcal{F}_s]| = \mathbb{E}[|M_s|].$$

Hence $\sup_{t\geq 0}\mathbb{E}[|M_t|] = \lim_{t\to\infty}\mathbb{E}[|M_t|]$ and a martingale being bounded in $\mathcal{L}^1$ is a statement about the martingale as $t \to \infty$.

The restriction of $\mathcal{L}^1$-boundedness is sufficient to prove one important convergence theorem. We omit the proof which, although not difficult, would require us to introduce a few extra ideas. A proof can be found, for example, in Revuz and Yor (1991).

Theorem 3.13 (Doob's martingale convergence theorem) *Let M be a càdlàg martingale bounded in $\mathcal{L}^1$. Then $M_\infty(\omega) := \lim_{t\to\infty} M_t(\omega)$ exists and is finite almost surely.*

Remark 3.14: It follows from Fatou's lemma that $\mathbb{E}[|M_\infty|] \leq \liminf \mathbb{E}[|M_t|] < \infty$. One might be tempted to conclude that

(i) $M_t \to M_\infty$ in $\mathcal{L}^1$, meaning $\mathbb{E}[|M_t - M_\infty|] \to 0$ as $t \to \infty$,
and
(ii) $M_t = \mathbb{E}[M_\infty|\mathcal{F}_t]$.

Neither of these is true in general. If they are to hold we need to work with the more restrictive class of uniformly integrable martingales.

3.2.2 Uniformly integrable martingales

The two properties described in Remark 3.14 are very useful when they hold. We shall use the representation $M_t = \mathbb{E}[M_\infty|\mathcal{F}_t]$ in Chapter 4 when we develop the stochastic integral. The point is that this representation allows us to identify a martingale M which is a stochastic process, with a random variable, M_∞. To this end we introduce the idea of a uniformly integrable martingale.

Definition 3.15 *A family $\mathcal{C}$ of (real-valued) random variables is uniformly integrable (UI) if, given any $\varepsilon > 0$, there exists some $K_\varepsilon < \infty$ such that*

$$\mathbb{E}[|X|\mathbb{1}_{\{|X|>K_\varepsilon\}}] < \varepsilon$$

for all $X \in \mathcal{C}$.

A martingale M is UI if the family of random variables $\{M_t, t \geq 0\}$ is UI.

Remark 3.16: The effect of imposing the uniform integrability assumption is to control the tail behaviour of the martingale. It is just sufficient for Theorem 3.18, the convergence result described in Remark 3.14, to hold.

Clearly $\sup_{t\geq 0}\mathbb{E}\big[|M_t|\big] < 1+K_1$ for a UI martingale, so uniform integrability is indeed a stronger requirement than $\mathcal{L}^1$-boundedness. Note the following result, however, which shows that, in one sense at least, uniform integrability is only just stronger than $\mathcal{L}^1$-boundedness.

Theorem 3.17 *Suppose M is a martingale bounded in $\mathcal{L}^p$ for some $p > 1$. Then M is uniformly integrable.*

Proof: Let $A = \sup_{t\geq 0}\mathbb{E}\big[|M_t|^p\big]$. Defining $K_\varepsilon = {}^{p-1}\!\!\sqrt{A/\varepsilon}$, we have

$$\sup_{t\geq 0}\mathbb{E}[|M_t|1_{\{|M_t|>K_\varepsilon\}}] \leq \sup_{t\geq 0}\mathbb{E}\left[|M_t|\left|\frac{M_t}{K_\varepsilon}\right|^{p-1}1_{\{|M_t|>K_\varepsilon\}}\right]$$

$$\leq \sup_{t\geq 0}\mathbb{E}\left[\frac{|M_t|^p}{K_\varepsilon^{p-1}}\right] = \frac{A}{K_\varepsilon^{p-1}} = \varepsilon.$$

$\square$

And now for the theorem that shows UI martingales to be exactly the right concept.

Theorem 3.18 *For a càdlàg martingale M the following are equivalent:*

(i) $M_t \to M_\infty$ *in $\mathcal{L}^1$ (for some random variable M_∞).*
(ii) *There exists some $M_\infty \in \mathcal{L}^1$ such that $M_t = \mathbb{E}[M_\infty|\mathcal{F}_t]$ a.s. for all t.*
(iii) *M is uniformly integrable.*

Proof: (i) $\Rightarrow$ (ii): Given $t \geq 0$ and $F \in \mathcal{F}_t$, for all $s \geq t$,

$$\mathbb{E}[(M_\infty - M_t)1_F] = \mathbb{E}[(M_s - M_t)1_F] + \mathbb{E}[(M_\infty - M_s)1_F]$$
$$= \mathbb{E}[(M_\infty - M_s)1_F]$$
$$\leq \mathbb{E}[|M_\infty - M_s|] \to 0$$

as $s \to \infty$. This holds for all $F \in \mathcal{F}_t$, and M_t is $\mathcal{F}_t$-measurable, thus we conclude that $\mathbb{E}[M_\infty|\mathcal{F}_t] = M_t$.

(ii) $\Rightarrow$ (iii): Note the following three results.
(a) $\mathbb{E}[|M_t|1_{\{|M_t|>K\}}] \leq \mathbb{E}[|M_\infty|1_{\{|M_t|>K\}}]$ (by Jensen's inequality and the tower property).
(b) $K\mathbb{P}(|M_t| > K) \leq \mathbb{E}[|M_t|] \leq \mathbb{E}[|M_\infty|]$ (by truncating the expectation and Jensen's inequality, respectively).
(c) If $X \in \mathcal{L}^1$, then for any $\varepsilon > 0$, there exists some $\delta > 0$ such that

$$\mathbb{P}(F) < \delta \Rightarrow \mathbb{E}[|X|1_F] < \varepsilon.$$

Given any $\varepsilon > 0$, choose δ such that (c) holds for $X = M_\infty$, and choose K_ε in (b) so that $\mathbb{P}(|M_t| > K_\varepsilon) < \delta$. It follows from (a) and (c) that

$$\mathbb{E}\big[|M_t|\mathbb{1}_{\{|M_t|>K\}}\big] < \varepsilon$$

for all t, hence M is uniformly integrable.

(iii) $\Rightarrow$ (i): Since M is uniformly integrable, it is bounded in $\mathcal{L}^1$ (Remark 3.16), hence $M_t \to M_\infty$ a.s. for some $M_\infty \in \mathcal{L}^1$ (Theorem 3.13). Almost sure convergence of a UI process implies $\mathcal{L}^1$ convergence, so (i) holds. $\square$

Exercise 3.19: Establish condition (c) in the above proof: if $X \in \mathcal{L}^1$, then for any $\varepsilon > 0$, there exists some $\delta > 0$ such that

$$\mathbb{P}(F) < \delta \Rightarrow \mathbb{E}\big[|X|\mathbb{1}_F\big] < \varepsilon.$$

Exercise 3.20: Establish the convergence result used to prove (iii) $\Rightarrow$ (i) above: suppose $X_t \in \mathcal{L}^1$ for each $t \geq 0$ and $X \in \mathcal{L}^1$. Then $X_t \to X$ in $\mathcal{L}^1$ if and only if the following hold:
(i) $X_t \to X$ in probability;
(ii) the family $\{X_t, t \geq 0\}$ is UI.

Exercise 3.21: Prove the following which we shall need later: if $\{X_n, n \geq 1\}$ is a UI family of random variables such that $X_n \to X$ a.s. as $n \to \infty$, then $X \in \mathcal{L}^1$ (and thus $X_n \to X$ in $\mathcal{L}^1$ by Exercise 3.20).

A proof of these standard results can be found in Rogers and Williams (1994).

Remark 3.22: Given any uniformly integrable martingale, we can identify it with an $\mathcal{L}^1$ random variable $X = M_\infty$. Conversely, given any random variable $X \in \mathcal{L}^1$ we can define a UI martingale via

$$M_t := \mathbb{E}[X|\mathcal{F}_t]. \tag{3.1}$$

For each t, the random variable M_t in (3.1) is not unique, but if M_t and M_t^* both satisfy (3.1) then $M_t = M_t^*$ a.s. That is, if M and M^* are two martingales satisfying (3.1) for all t, they are modifications of each other. By Theorem 3.8, if $\{\mathcal{F}_t\}$ is right-continuous and complete, and we will insist on this in practice, then there is a unique (up to indistinguishability) martingale M satisfying (3.1) that is càdlàg.

Remark 3.23: A slight modification of the above proof of (ii)$\Rightarrow$(iii) establishes the following result: for $X \in \mathcal{L}^1$ the class

$$\{\mathbb{E}[X|\mathcal{G}] : \mathcal{G} \text{ a sub-}\sigma\text{-algebra of } \mathcal{F}\}$$

is uniformly integrable.

Two important examples of UI martingales are bounded martingales (obvious from the definition) and martingales of the form $\widetilde{M}_t := M_{t \wedge n}$ where M is a martingale and n is some constant (using property (ii) of Theorem 3.18).

3.2.3 Square-integrable martingales

We are working towards the ultimate goal, in Chapter 4, of defining the stochastic integral, and it is for this reason that we introduce the space of square-integrable martingales. The motivation for this is as follows. In constructing the stochastic integral we will need to consider the limit M of a sequence of martingales $M^{(n)}$. We require this limit also to be a martingale, so we need to introduce a space of martingales that is complete. If we restrict to a subclass of UI integrable martingales we are able to identify each martingale M with its limit random variable M_∞ and the problem of finding a complete (normed) space of martingales reduces to finding a complete (normed) space of random variables. Such a space is well known, the space L^2 of (equivalence classes of) square-integrable random variables. Those not familiar with square-integrable random variables and their properties can find all the important results in Appendix 2.

Definition 3.24 *The martingale spaces $\mathcal{M}^2, \mathcal{M}_0^2, c\mathcal{M}^2$ and $c\mathcal{M}_0^2$ are defined as follows.*

$$\mathcal{M}^2 = \{\text{martingales } M : M_\infty \in \mathcal{L}^2\}$$
$$\mathcal{M}_0^2 = \{M \in \mathcal{M}^2 : M_0 = 0 \text{ a.s.}\}$$
$$c\mathcal{M}^2 = \{M \in \mathcal{M}^2 : M \text{ is continuous a.s.}\}$$
$$c\mathcal{M}_0^2 = \{M \in \mathcal{M}^2 : M_0 = 0, M \text{ is continuous a.s.}\}.$$

We will need the following result in the special case $p = 2$. It can be found in any standard text on martingales.

Theorem 3.25 (Doob's $\mathcal{L}^p$ inequality) *Let M be a càdlàg martingale, bounded in $\mathcal{L}^p$, for some $p > 1$, and define*

$$M^* := \sup_{t \geq 0} |M_t|.$$

Then $M^* \in \mathcal{L}^p$ and

$$\mathbb{E}\left[(M^*)^p\right] \leq \left(\frac{p}{p-1}\right)^p \sup_{t \geq 0} \mathbb{E}\left[(M_t)^p\right].$$

Corollary 3.26 *Suppose M is a càdlàg martingale bounded in $\mathcal{L}^p$, for some $p > 1$. Then*

$$M_t \to M_\infty \text{ a.s. and in } \mathcal{L}^p$$

and

$$\sup_{t \geq 0} \mathbb{E}\left[(M_t)^p\right] = \lim_{t \to \infty} \mathbb{E}\left[(M_t)^p\right] = \mathbb{E}\left[(M_\infty)^p\right].$$

Proof: It is immediate that $M \in \mathcal{L}^1$, hence $M_t \to M_\infty$ a.s. by Theorem 3.13. Further, applying Doob's $\mathcal{L}^p$ inequality,

$$|M_t - M_\infty|^p \leq (2M^*)^p \in \mathcal{L}^1$$

so $M_t \to M_\infty$ in $\mathcal{L}^p$ by the dominated convergence theorem. Now, for all $s \leq t$, the conditional form of Jensen's inequality implies that

$$\mathbb{E}\left[|M_s|^p\right] = \mathbb{E}\left[|\mathbb{E}[M_t|\mathcal{F}_s]|^p\right] \leq \mathbb{E}\left[|M_t|^p\right]$$

so $\mathbb{E}\left[|M_t|^p\right]$ is increasing in t. This establishes the first equality. The second equality follows immediately from $\mathcal{L}^p$ convergence. □

Theorem 3.27 *The spaces $\mathcal{M}^2, \mathcal{M}_0^2, c\mathcal{M}^2$ and $c\mathcal{M}_0^2$ are Hilbert spaces for the norm*

$$\|M\|_2 := \left(\mathbb{E}[M_\infty^2]\right)^{\frac{1}{2}}.$$

(So the corresponding inner product on this space is $\langle M, N \rangle = \mathbb{E}[M_\infty N_\infty]$.)

Remark 3.28: Recall that a Hilbert space is a complete inner product space. We shall use the inner product structure in Chapter 5.

Remark 3.29: To be strictly correct we need to be a little more careful in our definitions before Theorem 3.27 can hold. Let $\|\cdot\|$ denote the classical L^2 norm $\|X\| = (\mathbb{E}[X^2])^{1/2}$. The problem is that if X and Y are two random variables which are a.s. equal, then $\|X - Y\| = 0$ but, in general, $X \neq Y$. Hence $\|\cdot\|$ is not a norm on the space $\mathcal{L}^2$. For this to be true we must identify any two random variables which are almost surely equal. Similarly here, we must identify any two martingales which are modifications of each other. In common with most other authors, we will do this without further comment whenever we need to. Appendix 2 contains a fuller discussion of the space L^2.

Proof: In each case the trick is to identify a martingale with its limit random variable $M_\infty \in \mathcal{L}^2$, which we can do by Theorems 3.17 and 3.18. All the properties of a Hilbert space are inherited trivially from those of L^2, with the exception of completeness. We prove this for $c\mathcal{M}_0^2$.

To prove $c\mathcal{M}_0^2$ is complete we must show that any Cauchy sequence $M^{(n)}$ converges to some limit $M \in c\mathcal{M}_0^2$. So let $M^{(n)}$ be some Cauchy sequence with respect to $\|\cdot\|_2$ in $c\mathcal{M}_0^2$ and let (recalling Remark 3.29),

$$M_\infty := \lim_{n\to\infty} M_\infty^{(n)} \quad \text{in } \mathcal{L}^2.$$

$$M_t := \mathbb{E}[M_\infty | \mathcal{F}_t],$$

where the limit exists since L^2 is complete and each $M_\infty^{(n)} \in \mathcal{L}^2$. It is immediate from the definition of the norm $\|\cdot\|$ on L^2 that $\lim_{n\to\infty} \|M^{(n)} - M\|_2 = 0$, so it remains only to prove that $M \in c\mathcal{M}_0^2$. That M is a martingale is immediate from its definition, so it suffices to prove that it is continuous and null at zero. But, for each n, $M^{(n)} - M$ is bounded in $\mathcal{L}^2$, so Corollary 3.26 implies that

$$\sup_{t\geq 0} \mathbb{E}[(M_t^{(n)} - M_t)^2] = \|M_\infty^{(n)} - M_\infty\|_2 \qquad (3.2)$$

which converges to zero as $n \to \infty$, thus $M_0 = 0$. Further, it follows from (3.2) and Doob's $\mathcal{L}^2$ inequality that there exists some subsequence n_k on which

$$\sup_{t\geq 0} |M_t^{(n_k)}(\omega) - M_t(\omega)| \to 0$$

for almost all ω. But the uniform limit of a continuous function is continuous, hence M is continuous almost surely. $\qquad\square$

Exercise 3.30: Convince yourself of the final step in this proof; if $X_n \to X$ in $\mathcal{L}^p$ then $X_n \to X$ in probability, and if $X_n \to X$ in probability then there exists some subsequence X_{n_k} such that $X_{n_k} \to X$ a.s.

The space $c\mathcal{M}_0^2$ will be of particular relevance for the development of the stochastic integral. It is convenient in that development to start with martingale integrators which start at zero, hence the reason for not working in the more general space $c\mathcal{M}^2$.

3.3 STOPPING TIMES AND THE OPTIONAL SAMPLING THEOREM

3.3.1 Stopping times

Stopping times perform a very important role in both the theory of stochastic processes and finance. The theory of stopping times is a detailed and deep one, but for our purposes we can avoid most of it. Roughly speaking, a stopping time is one for which 'you know when you have got there'. Here is the formal definition of a stopping time.

Definition 3.31   *The random variable* T, *taking values in* $[0, \infty]$, *is an* $\{\mathcal{F}_t\}$ *stopping time if*

$$\{T \leq t\} = \{\omega : T(\omega) \leq t\} \in \mathcal{F}_t$$

for all $t \leq \infty$.

Definition 3.32   *For any* $\{\mathcal{F}_t\}$ *stopping time* T, *the pre-T σ-algebra* $\mathcal{F}_T$ *is defined via*

$$\mathcal{F}_T = \{F \in \mathcal{F} : for\ every\ t \leq \infty, F \cap \{T \leq t\} \in \mathcal{F}_t\}.$$

Technical Remark 3.33: Definition 3.31 is as we would expect. Note that the inequalities are not strict in either definition; if they were the time T would be an *optional* time. If the filtration $\{\mathcal{F}_t\}$ is right-continuous, optional and stopping times agree. See Revuz and Yor (1991) for more on this.

Remark 3.34: The pre-T σ-algebra is the information that we will have available at the random time T. In particular, if T is the constant time t, $\mathcal{F}_T = \mathcal{F}_t$.

Exercise 3.35:   Show that the following are stopping times:
(i)   constant times;
(ii) $S \wedge T, S \vee T$ and $S + T$ for stopping times S and T;
(iii) $\sup_n\{T_n\}$ for a sequence of stopping times $\{T_n\}$;
(iv) $\inf_n\{T_n\}$ when $\{\mathcal{F}_t\}$ is right-continuous.

Exercise 3.36:   Show that if S and T are stopping times, then $\mathcal{F}_{S \wedge T} = \mathcal{F}_S \cap \mathcal{F}_T$. Also, if $\{\mathcal{F}_t\}$ is right-continuous and $T_n \downarrow T$ are stopping times then $\mathcal{F}_T = \bigcap_n \mathcal{F}_{T_n}$.

The most important stopping times for the development of the stochastic integral will be hitting times of sets by a stochastic process X, that is stopping times of the form

$$H_\Gamma(\omega) := \inf\{t > 0 : X_t(\omega) \in \Gamma\}. \tag{3.3}$$

Note the strict inequality in this definition. The story is different if the inequality is softened (see Rogers and Williams (1987), for example). Times of the form (3.3) are not, in general, stopping times. The following results are sufficient for our purposes and give conditions on either the process X or the filtration $\{\mathcal{F}_t\}$ under which H_Γ is a stopping time. Indeed, for our applications X will always be a.s. continuous and the filtration will always be complete.

Theorem 3.37 *Let Γ be an open set and suppose X is $\{\mathcal{F}_t\}$ adapted; if X is entirely càdlàg, or if X is càdlàg and $\{\mathcal{F}_t\}$ is complete, then H_Γ is an $\{\mathcal{F}_{t+}\}$ stopping time.*

Proof: It suffices to prove that $\{H_\Gamma < t\} \in \mathcal{F}_t$ for all t since then $\{H_\Gamma \leq t\} = \bigcap_{q \in (t, t+1/n] \cap \mathbb{Q}} \{H_\Gamma < q\} \in \mathcal{F}_{t+1/n}$ for all $n > 0$. Taking the intersection over all $n > 0$ shows that $\{H_\Gamma \leq t\} \in \mathcal{F}_{t+}$.

Denote by Ω^* the set of ω for which $X(\omega)$ is càdlàg and by A_t the set

$$A_t := \bigcup_{q \in (0,t) \cap \mathbb{Q}} \{\omega : X_q(\omega) \in \Gamma\}$$

$$= \bigcup_{q \in (0,t) \cap \mathbb{Q}} X_q^{-1}(\Gamma).$$

Note that $A_t \in \mathcal{F}_t$ since Γ is open, and on Ω^* the events $\{H_\Gamma < t\}$ and A_t coincide. Thus we can write

$$\{H_\Gamma < t\} = \left(\Omega^* \cap \{H_\Gamma < t\}\right) \cup \left((\Omega^*)^c \cap \{H_\Gamma < t\}\right)$$

$$= \left(\Omega^* \cap A_t\right) \cup \left((\Omega^*)^c \cap \{H_\Gamma < t\}\right). \tag{3.4}$$

When X is entirely càdlàg, $\Omega^* = \Omega$ and $\{H_\Gamma < t\} = A_t \in \mathcal{F}_t$. When $\{\mathcal{F}_t\}$ is complete both Ω^* and the second event in (3.4), which has probability zero, are in $\mathcal{F}_t$, as is A_t, hence $\{H_\Gamma < t\} \in \mathcal{F}_t$ as required. $\qquad \square$

Corollary 3.38 *Suppose, in addition to the hypotheses of Theorem 3.37, that $\{\mathcal{F}_t\}$ is right-continuous. Then H_Γ is an $\{\mathcal{F}_t\}$ stopping time.*

Exercise 3.39: Let Γ be an open set and let X be adapted to $\{\mathcal{F}_t\}$. Suppose that X is entirely continuous, or that X is a.s. continuous and $\{\mathcal{F}_t\}$ is complete. Show that H_Γ need not be an $\{\mathcal{F}_t\}$ stopping time.

Theorem 3.40 *Let Γ be a closed set and suppose X is $\{\mathcal{F}_t\}$ adapted. If X is entirely càdlàg, or if X is càdlàg and $\{\mathcal{F}_t\}$ is complete, then*

$$H_\Gamma^* := \inf\{t > 0 : X_t(\omega) \in \Gamma \text{ or } X_{t-}(\omega) \in \Gamma\}$$

is an $\{\mathcal{F}_t\}$ stopping time.

Proof: We prove the entirely càdlàg version of the result. The simple argument used in Theorem 3.37 extends this result to the càdlàg case. Let $\rho(x, \Gamma) :=$ $\inf\{|x - y| : y \in \Gamma\}$. Defining the open set $\Gamma_n = \{x : \rho(x, \Gamma) < n^{-1}\}$, those points within distance n^{-1} of Γ, we have $\Gamma = \bigcap_n \Gamma_n$, and so

$$\{H_\Gamma^* \leq t\} = \{X_t \in \Gamma\} \cup \{X_{t-} \in \Gamma\} \cup \left(\bigcap_n \bigcup_{q \in (0,t) \cap \mathbb{Q}} \{X_q \in \Gamma_n\} \right) \in \mathcal{F}_t. \qquad \square$$

Corollary 3.41 *If X is entirely continuous (or if X is continuous and $\{\mathcal{F}_t\}$ is complete), $\{\mathcal{F}_t\}$ adapted, Γ a closed set, then H_Γ is an $\{\mathcal{F}_t\}$ stopping time.*

Example 3.42 Consider a Brownian motion W on its canonical set-up (Theorem 2.2), and let

$$T = \inf\{t > 0 : W_t + t \notin [-a, b]\}$$

for some $a, b > 0$. The set $(-\infty, a) \cup (b, \infty)$ is an open subset of $\mathbb{R}$ and W is entirely continuous, so T is an $\{\mathcal{F}_t\}$ stopping time by Corollary 3.39 above.

Example 3.43 Now let W be a two-dimensional Brownian motion on a complete space $(\Omega, \{\mathcal{F}_t\}, \mathcal{F}, \mathbb{P})$ and let

$$T = \inf\{t > 0 : W_t \notin D\}$$

where

$$D = \{(x, y) : x^2 + y^2 < K\}.$$

Now, since Brownian motion is a.s. continuous and $\{\mathcal{F}_t\}$ is complete, we can apply Corollary 3.41 to deduce that T is an $\{\mathcal{F}_t\}$ stopping time.

Given the intuitive interpretation of $\mathcal{F}_T$ as the amount of information available at the stopping time T, we would also hope that $X_T \in \mathcal{F}_T$ if X is a process adapted to $\{\mathcal{F}_t\}$. It is true in certain circumstances.

Theorem 3.44 *Let X be adapted to $\{\mathcal{F}_t\}$ and let T be some $\{\mathcal{F}_t\}$ stopping time. If X is entirely càdlàg, or if X is càdlàg and $\{\mathcal{F}_t\}$ is complete, then X_T is $\mathcal{F}_T$-measurable.*

Remark 3.45: Theorem 3.44 holds more generally when X is *progressively measurable*, as defined later in Theorem 3.60. This is a joint measurability condition which holds when X is càdlàg and adapted and $\{\mathcal{F}_t\}$ is complete.

Remark 3.46: If $T(\omega) = \infty$ and $X_\infty(\omega)$ exists, we define $X_T(\omega) = X_\infty(\omega)$. If $T(\omega) = \infty$ but X_∞ does not exist, we take $X_T(\omega) = 0$ by convention.

An important idea for the results to come is that of a stopped process.

Definition 3.47 *Given a stochastic process X and a stopping time T, the stopped process X^T is defined by*

$$X_t^T(\omega) := X_{t \wedge T}(\omega).$$

There is a natural and immediate extension of Theorem 3.44 to processes as follows.

Corollary 3.48 *If T is an $\{\mathcal{F}_t\}$ stopping time and if X is $\{\mathcal{F}_t\}$-adapted and entirely càdlàg (or càdlàg and $\{\mathcal{F}_t\}$ is complete), then the process X^T is $\{\mathcal{F}_{t \wedge T}\}$ adapted (and therefore also $\{\mathcal{F}_t\}$-adapted).*

3.3.2 Optional sampling theorem

Now that we have met the concept of a martingale and a stopping time, it is time to put the two together and ask the question: does the martingale property

$$M_t = \mathbb{E}[M_T | \mathcal{F}_t]$$

hold if T is a stopping time? In general the answer is no, as can be seen by taking M to be Brownian motion and $T = \inf\{t > 0 : M_t \geq 1\}$. There is an important situation when it is true, however – when M is a UI martingale. Indeed, this property is a characterizing property of UI martingales, as Theorems 3.50 and 3.53 demonstrate.

We must first prove a restricted version of the result.

Theorem 3.49 *Suppose $S \leq T$ are two stopping times taking at most a finite number of values (possibly including ∞), and let M be some UI, càdlàg martingale. Then*

$$M_S = \mathbb{E}[M_T | \mathcal{F}_S] = \mathbb{E}[M_\infty | \mathcal{F}_S] \quad \text{a.s.}$$

Proof: It suffices to prove the first equality since the second then follows by applying it to the stopping time $T = \infty$. Let $0 = t_0 \leq \ldots \leq t_n \leq \infty$ be a set of values containing all those that S and T can take. For any $F \in \mathcal{F}_S$ we have

$$(M_T - M_S)\mathbf{1}_F = \left(\sum_{j=1}^n (M_{t_j} - M_{t_{j-1}})\mathbf{1}_{\{S < t_j \leq T\}} \right)\mathbf{1}_F$$

and thus

$$
\begin{aligned}
\mathbb{E}\Big[(M_T - M_S)\mathbb{1}_F\Big] &= \sum_{j=1}^n \mathbb{E}\Big[(M_{t_j} - M_{t_{j-1}})\mathbb{1}_{\{(S<t_j\leq T)\}}\mathbb{1}_F\Big] \\
&= \sum_{j=1}^n \mathbb{E}\mathbb{E}\Big[(M_{t_j} - M_{t_{j-1}})\mathbb{1}_{\{S<t_j\leq T\}}\mathbb{1}_F\,|\mathcal{F}_{t_{j-1}}\Big] \\
&= \sum_{j=1}^n \mathbb{E}\Big[\mathbb{1}_{\{S<t_j\leq T\}}\mathbb{1}_F\mathbb{E}\Big[(M_{t_j} - M_{t_{j-1}})|\mathcal{F}_{t_{j-1}}\Big]\Big] \\
&= 0.
\end{aligned}
$$

This holds for all $F \in \mathcal{F}_S$, and M_S is $\mathcal{F}_S$-measurable; hence

$$
\mathbb{E}[M_T|\mathcal{F}_S] = \mathbb{E}[M_S|\mathcal{F}_S] = M_S \qquad \text{a.s.}
$$

$\square$

It is now a short step to prove the more general result.

Theorem 3.50 *If M is a uniformly integrable, càdlàg martingale and $S \leq T$ are two stopping times, then*

$$
M_S = \mathbb{E}[M_T|\mathcal{F}_S] = \mathbb{E}[M_\infty|\mathcal{F}_S] \qquad \text{a.s.}
$$

Proof: If suffices to prove that, for any stopping time S,

$$
M_S = \mathbb{E}[M_\infty|\mathcal{F}_S], \tag{3.5}
$$

since then

$$
\mathbb{E}[M_T|\mathcal{F}_S] = \mathbb{E}\Big[\mathbb{E}[M_\infty|\mathcal{F}_T]|\mathcal{F}_S\Big] = \mathbb{E}[M_\infty|\mathcal{F}_S] = M_S.
$$

To establish (3.5), we need to show that, for all $F \in \mathcal{F}_S$, $\mathbb{E}[M_S\mathbb{1}_F] = \mathbb{E}[M_\infty\mathbb{1}_F]$. Define the sequence of $\{\mathcal{F}_t\}$ stopping times $S_k \downarrow S$ as $k \to \infty$ via

$$
S_k = \begin{cases} +\infty & \text{if} & S \geq k, \\ \frac{q}{2^k} & \text{if} & \frac{q-1}{2^k} \leq S < \frac{q}{2^k} < k. \end{cases}
$$

Since M is UI, M_∞ exists and, for each k, the pair of stopping times $\{S_k, \infty\}$ satisfy the conditions of Theorem 3.49. Thus,

$$
M_{S_k} = \mathbb{E}[M_\infty|\mathcal{F}_{S_k}] \qquad \text{a.s.} \tag{3.6}
$$

Conditioning on $\mathcal{F}_S$ ($\subseteq \mathcal{F}_{S_k}$), we deduce that

$$
\mathbb{E}[M_{S_k}|\mathcal{F}_S] = \mathbb{E}[M_\infty|\mathcal{F}_S],
$$

i.e. $\mathbb{E}[M_{S_k}\mathbb{1}_F] = \mathbb{E}[M_\infty\mathbb{1}_F]$ for all $F \in \mathcal{F}_S$. The result now follows if we can show that, for all $F \in \mathcal{F}_S$, $\mathbb{E}[M_{S_k}\mathbb{1}_F] \to \mathbb{E}[M_S\mathbb{1}_F]$ as $k \to \infty$, i.e. that $M_{S_k} \to M_S$ in an $\mathcal{L}^1$ sense. But the representation in (3.6) and Remark 3.23 show that the family $\{M_{S_k}\}$ is UI, and the right-continuity of M ensures that $M_{S_k} \to M_S$ a.s. These two conditions combined ensure the $\mathcal{L}^1$ convergence that we require (Exercise 3.21). $\square$

The following result was established as part of this last proof and will be needed shortly.

Corollary 3.51 *If M is a uniformly integrable, càdlàg martingale and S is a stopping time, then $\mathbb{E}[|M_S|] < \infty$.*

Remark 3.52: There are several special cases of the optional sampling theorem which are often stated in slightly different form. For example, we can drop the UI requirement on M if the stopping time T is bounded by some n since M^n is a UI martingale. Also, if M is bounded up until the stopping time T the result also holds since M^T is UI (and we shall shortly prove that it is also a martingale).

There is a very useful partial converse to the optional sampling theorem which characterizes UI martingales as precisely those for which the optional sampling theorem holds.

Theorem 3.53 *Let M be some càdlàg and adapted process such that, for every stopping time T,*

$$\mathbb{E}[|M_T|] < \infty \quad \text{and} \quad \mathbb{E}[M_T] = 0.$$

Then M is a uniformly integrable martingale.

Proof: As per our earlier convention, define $M_\infty(\omega) = \lim_{t \to \infty} M_t(\omega)$ if the limit exists, $M_\infty(\omega) = 0$ otherwise. Note that M_∞ thus defined is in $\mathcal{L}^1$ (taking $T = \infty$ in the hypothesis of the theorem). By Theorem 3.18, it suffices to show that $M_t = \mathbb{E}[M_\infty|\mathcal{F}_t]$, i.e. for all $t \geq 0$, all $F \in \mathcal{F}_t$,

$$\mathbb{E}[M_t \mathbb{1}_F] = \mathbb{E}[M_\infty \mathbb{1}_F]. \tag{3.7}$$

Define the stopping time T via

$$T(\omega) = \begin{cases} t & \text{if } \omega \in F \\ \infty & \text{if } \omega \in F^c. \end{cases}$$

Applying the hypotheses to stopping times T and ∞ yields

$$\mathbb{E}[M_\infty] = \mathbb{E}[M_\infty \mathbb{1}_F] + \mathbb{E}[M_\infty \mathbb{1}_{F^c}] = 0,$$
$$\mathbb{E}[M_T] = \mathbb{E}[M_t \mathbb{1}_F] + \mathbb{E}[M_\infty \mathbb{1}_{F^c}] = 0.$$

Combining these two yields (3.7) as required. $\square$

This provides the final result needed to prove the following important proposition.

Proposition 3.54 *Let M be an $\{\mathcal{F}_t\}$ martingale and T be an $\{\mathcal{F}_t\}$ stopping time. If M is entirely càdlàg, or if M is càdlàg and $\{\mathcal{F}_t\}$ is complete, then M^T is also an $\{\mathcal{F}_t\}$ martingale. That is, the class of martingales is stable under stopping.*

Proof: We must verify properties (M.i)–(M.iii) of Definition 3.1. The first, that M^T is adapted, is an immediate consequence of Corollary 3.48. To establish property (M.ii), note that M^t is a UI martingale (since $M_s^t = \mathbb{E}[M_\infty^t|\mathcal{F}_s]$ for all s), and thus Corollary 3.51 implies that $\mathbb{E}[|M_T^t|] = \mathbb{E}[|M_t^T|] < \infty$ as required. It remains to show that M^T has the martingale property, (M.iii). This is clearly equivalent to showing that, for all $0 \le s \le t$,

$$\mathbb{E}[M_t^{T \wedge t}|\mathcal{F}_s] = M_s^{T \wedge t} \qquad ,$$

and we do this by showing that $M^{T \wedge t}$ is a UI martingale. We appeal to Theorem 3.53. Since M^t is UI, the optional sampling theorem implies that, for all stopping times S,

$$M_0 = \mathbb{E}[M_{T \wedge S}^t] = \mathbb{E}[M_S^{T \wedge t}] \, ,$$

and Corollary 3.51 implies that $\mathbb{E}[|M_S^{T \wedge t}|] = \mathbb{E}[|M_{T \wedge S}^t|] < \infty$. We are done. $\square$

Example 3.55 Let W be a Brownian motion and define $T = \inf\{t > 0 : W_t \notin [-a, b]\}$, $a, b > 0$. We will show that:

(i) $T < \infty$ a.s.;
(ii) $\mathbb{P}(W_T = b) = a/(a + b)$;
(iii) $\mathbb{E}[T] = ab$.

To do this we will consider stopped versions of each of the martingales introduced in Example 3.3. Consider first the martingale $M_t^\lambda = \exp\left(\lambda W_{t \wedge T} - \frac{1}{2}\lambda^2(t \wedge T)\right)$ which is bounded by zero below and by $\exp(\lambda(a + b))$ above, so is certainly UI. Applying the optional sampling theorem at the stopping time T yields

$$\begin{aligned}
1 = \mathbb{E}[M_0^\lambda] &= \mathbb{E}[\exp(\lambda W_T - \tfrac{1}{2}\lambda^2 T)] \\
&= \mathbb{E}[\exp(\lambda W_T - \tfrac{1}{2}\lambda^2 T)\mathbb{1}_{\{T<\infty\}}] \\
&\to \mathbb{E}[\mathbb{1}_{\{T<\infty\}}]
\end{aligned}$$

as $\lambda \to 0$ by the dominated convergence theorem. This proves (i) (which also follows from Theorem 2.8). Now applying the optional sampling theorem to the UI martingale W^T yields

$$\begin{aligned}
0 = \mathbb{E}[W_0] &= \mathbb{E}[W_T^T] \\
&= \mathbb{E}[b\mathbb{1}_{\{W_T=b\}} - a\mathbb{1}_{\{W_T=-a\}}] \\
&= b\mathbb{P}(W_T = b) - a(1 - \mathbb{P}(W_T = b)),
\end{aligned}$$

which is (ii). Finally, apply the optional sampling theorem to the UI martingale $M_t^n := W_{t \wedge n}^2 - (t \wedge n)$ at the stopping time T to obtain

$$0 = \mathbb{E}[W_{T \wedge n}^2 - (T \wedge n)].$$

As $n \to \infty$, $T \wedge n \uparrow T$ a.s., and $W_{T \wedge n}^2 \le (a+b)^2$, so applications of the monotone and dominated convergence theorems yield

$$\mathbb{E}[T] = \mathbb{E}[W_T^2] = \frac{ab^2 + ba^2}{a+b} = ab.$$

3.4 VARIATION, QUADRATIC VARIATION AND INTEGRATION

We are approaching the point where we can start to define the stochastic integral. Before we can do so, there is one more idea which we must introduce, that of *quadratic variation*. First we will introduce the more classical concept of *total variation* and review its role in integration.

3.4.1 Total variation and Stieltjes integration

Definition 3.56 *Let* $\{X_t, t \ge 0\}$ *be some stochastic process. The total variation process* V_X *of* X *is defined, for each* $t \ge 0$, *as*

$$V_X(t, \omega) = \sup_{\Delta} \sum_{i=1}^{n(\Delta)} |X(t_i, \omega) - X(t_{i-1}, \omega)|,$$

where $\Delta = \{t_0 = 0 < t_1 < \ldots < t_{n(\Delta)} = t\}$ *is a partition of* $[0, t]$.

Remark 3.57: Clearly this is a sample path definition and, for each ω, it agrees with the classical definition of total variation for functions.

Proposition 3.58 *If* X *is* $\{\mathcal{F}_t\}$-*adapted and entirely càdlàg (or càdlàg and* $\{\mathcal{F}_t\}$ *is complete) then* V_X *is also entirely càdlàg (respectively càdlàg) and* $\{\mathcal{F}_t\}$-*adapted.*

Proof: We prove the result when X is entirely càdlàg, the càdlàg result then following along the lines of the proof of Theorem 3.37. Given any ω and t we have, since $X(\omega)$ is càdlàg, that

$$V_X(t, \omega) = \sup_n \sum_{i=1}^{\lfloor 2^n t \rfloor + 1} \left| X(t \wedge (i/2^n), \omega) - X(t \wedge ((i-1)/2^n), \omega) \right|.$$

This is straightforward to verify and is left as an exercise for the reader. For each n the approximating sum is clearly $\mathcal{F}_t$-measurable, and the supremum of a countable number of measurable functions is itself measurable, thus V_X is $\{\mathcal{F}_t\}$-adapted. The càdlàg property is immediate. $\qquad\square$

Definition 3.59 *The process X is said to be of finite variation if the following conditions hold:*

(i) $V_X(t,\omega) < +\infty$ *for all t, a.s.;*

(ii) *X is càdlàg;*

(iii) *X is $\{\mathcal{F}_t\}$-adapted.*

The first property is the main defining characteristic of a finite variation process. The other two are conditions we have imposed and will continue to impose on all the processes that will interest us.

The advantage of a process being of finite variation is that it is easy to define a stochastic integral with respect to a finite variation process. This is done as for Stieltjes integration, as summarized by the following theorem.

Theorem 3.60 *Let X be some finite variation process. Then there exist increasing, adapted and càdlàg processes X^+ and X^- such that*

$$X = X^+ - X^- \quad \text{a.s.}$$

Furthermore, since both X^+ and X^- are càdlàg, we can define the stochastic integral $H \bullet X$ for any bounded $\mathcal{B}(\mathbb{R}^+) \otimes \mathcal{F}$-measurable integrand H to be

$$(H \bullet X)_t(\omega) = \int_0^t H_u(\omega)dX_u^+(\omega) - \int_0^t H_u(\omega)dX_u^-(\omega) \qquad (3.8)$$

(set it to zero if $X(\omega)$ is not càdlàg) where the integrals in (3.8) are Lebesgue–Stieltjes integrals. The stochastic integral $(H \bullet X)$ satisfies conditions (i) and (ii) of a finite variation process.

A process H is called progressively measurable with respect to $\{\mathcal{F}_t\}$ if, for each t, the map $(s,\omega) \to H_s(\omega)$ from $[0,t] \times \Omega \to (\mathbb{R}, \mathcal{B}(\mathbb{R}))$ is $\mathcal{B}([0,t]) \otimes \{\mathcal{F}_t\}$-measurable. If H is bounded and progressively measurable then $H \bullet X$ is also $\{\mathcal{F}_t\}$-adapted, hence $H \bullet X$ is of finite variation.

Proof: The functions $X^+ = \frac{1}{2}(X + V_X)$, $X^- = \frac{1}{2}(X - V_X)$ have the desired properties, as is easily verified. It is then standard integration theory that (3.8) is well defined for any bounded H. That $V_{H \bullet X}(t) < \infty$ follows from the boundedness of H over $[0,t]$ and the finite total variation of X^+ and X^- over $[0,t]$. That $H \bullet X$ is càdlàg and (when H is progressively measurable) adapted is straightforward.   □

Remark 3.61: Note in the above proof that the definition of the integral is not dependent on the particular decomposition of X (which is not unique).

What we have succeeded in doing is to define the stochastic integral of a bounded process with respect to a finite variation integrator, and we have done this on a pathwise basis using no more than standard integration theory. The boundedness condition on the integrand is easily weakened, but we shall defer this until Chapter 4. The question now is, which processes are of finite variation? Does this technique allow us to define $H \bullet W$ for a Brownian motion W?

Theorem 3.62 *The only continuous martingales of finite variation are a.s. constant.*

Proof: Let M be some continuous finite variation martingale and, without loss, suppose $M_0 = 0$. We further suppose, by augmentation if necessary, that the filtration is complete (note that M is still a martingale for this larger filtration). We show that $\widehat{M} \equiv 0$ where $\widehat{M} = M^T$ and

$$T = \inf\{t > 0 : V_M(t) > K\},$$

for some arbitrary K. This is clearly sufficient.

Since M is continuous, to prove that $\widehat{M} \equiv 0$ a.s. it suffices to prove that $\widehat{M}_q = 0$ a.s. for all $q \in \mathbb{Q}^+$. This follows if we can establish that $\mathbb{E}[\widehat{M}_t^2] = 0$ for all $t \geq 0$. But for any partition Δ of $[0, t]$ we have

$$\widehat{M}_t^2 = \sum_{j=1}^{n(\Delta)} (\widehat{M}_{t_j}^2 - \widehat{M}_{t_{j-1}}^2)$$

$$= \sum_{j=1}^{n(\Delta)} (\widehat{M}_{t_j} - \widehat{M}_{t_{j-1}})^2 + 2 \sum_{j=1}^{n(\Delta)} (\widehat{M}_{t_j} - \widehat{M}_{t_{j-1}})\widehat{M}_{t_{j-1}}. \qquad (3.9)$$

Since T is a stopping time (Corollary 3.39), M^T is a martingale (Proposition 3.54) and so the second term in (3.9) has expectation zero. Thus

$$\mathbb{E}[\widehat{M}_t^2] = \mathbb{E}\left[\sum_{j=1}^{n(\Delta)} (\widehat{M}_{t_j} - \widehat{M}_{t_{j-1}})^2\right]$$

$$\leq E\left[V_{\widehat{M}}(t) \sup_{j \leq n(\Delta)} |\widehat{M}_{t_j} - \widehat{M}_{t_{j-1}}|\right]$$

$$\leq K\mathbb{E}\left[\sup_{j \leq n(\Delta)} |\widehat{M}_{t_j} - \widehat{M}_{t_{j-1}}|\right].$$

Taking a successively finer submesh, this supremum decreases a.s. to zero by the continuity of $\widehat{M}$, hence, by dominated convergence, the expectation also decreases to zero as required. □

So if we want to define a stochastic integral relative to a martingale integrator we will have to work harder. All non-trivial continuous martingales are of infinite variation. However, they are all of finite *quadratic variation*, and it is this property that will allow us to define the stochastic integral for continuous martingale integrators in Chapter 4.

3.4.2 Quadratic variation

The fact that the only continuous martingales of finite variation are constant means we will not be able to define the stochastic integral relative to martingales as in the previous section. We will also need to replace the variation process with something, and this is quadratic variation.

Definition 3.63 *Let X be some continuous process null at zero. Define the doubly infinite sequence of times T_k^n (or $T_k^n(X)$ when we need to be explicit) via*

$$T_0^n \equiv 0, \ T_{k+1}^n = \inf\{t > T_k^n : |X_t - X_{T_k^n}| > 2^{-n}\}.$$

The quadratic variation of the process X, written $[X]$, is defined to be

$$[X]_t(\omega) := \lim_{n \to \infty} \sum_{k \geq 1} [X_{t \wedge T_k^n}(\omega) - X_{t \wedge T_{k-1}^n}(\omega)]^2,$$

or $+\infty$ when the limit does not exist.

Remark 3.64: We have in this definition been very specific about the sequence of meshes to be used to define the quadratic variation. In particular, they are all nested. It is also possible to define quadratic variation using other nested meshes and similar results hold to the ones that we prove in this and the next chapter. The reason for our particular choice is that we will be able to prove several almost sure results, whereas when using other definitions the results are only true in probability. It makes no essential difference to the theory, but almost sure results are often easier to appreciate.

Remark 3.65: The restriction in the definition of quadratic variation that the meshes be nested differs from the classical definition (for a function) under which the quadratic variation of any non-trivial continuous martingale is infinity. This contrasts with our next result.

Theorem 3.66 *For any continuous martingale M the quadratic variation process $[M]$ is adapted and a.s. finite, continuous and increasing.*

Remark 3.67: As usual, continuous here means entirely continuous or a.s. continuous on a complete filtered space. We need this to ensure that the $T_k^n(M)$ are stopping times in the proof below.

Proof: We will for now only prove the result for the case when M is bounded (and thus, in particular, $M \in c\mathcal{M}_0^2$). The general result, and more, follows easily once we have met the idea of localization in Section 3.5, by the inclusion of the phrase 'By localization we may assume M is bounded' at the start of this proof.

We make the following definitions:

$$t_k^n = t \wedge T_k^n(M),$$
$$s_k^n = \sup_j\{t_j^{n-1} : t_j^{n-1} \leq t_k^n\},$$
$$\Delta_k^n = M_{t_k^n} - M_{t_{k-1}^n},$$
$$H_k^n = M_{s_k^n} - M_{t_k^n},$$

and note that

$$|H_k^n| \leq 2^{-(n-1)}. \tag{3.10}$$

The times s_k^n and t_k^n are illustrated in Figure 3.1.

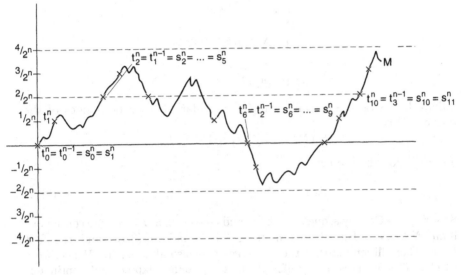

Figure 3.1 The times s_k^n and t_k^n

Defining

$$A_t^n = \sum_{k \geq 1} \left(M_{t_k^n} - M_{t_{k-1}^n} \right)^2,$$

it is readily seen that

$$A_t^n = M_t^2 - 2N_t^n \tag{3.11}$$

where

$$N_t^n = \sum_{k \geq 1} M_{t_{k-1}^n} \left(M_{t_k^n} - M_{t_{k-1}^n} \right)$$
$$= \sum_{k \geq 1} M_{s_{k-1}^{n+1}} \left(M_{t_k^{n+1}} - M_{t_{k-1}^{n+1}} \right).$$

Invoking the notation above, $N_t^n - N_t^{n+1} = \sum_{k \geq 1} H_{k-1}^{n+1} \Delta_k^{n+1}$, whence

$$\mathbb{E}[(N_t^n - N_t^{n+1})^2] = \mathbb{E}[\sum_{j,k \geq 1} H_{j-1}^{n+1} H_{k-1}^{n+1} \Delta_j^{n+1} \Delta_k^{n+1}]$$

$$= \sum_{j,k \geq 1} \mathbb{E}\mathbb{E}[H_{j-1}^{n+1} H_{k-1}^{n+1} \Delta_j^{n+1} \Delta_k^{n+1} | \mathcal{F}_{t_{j-1}^{n+1} \vee t_{k-1}^{n+1}}]$$

$$= \sum_{k \geq 1} \mathbb{E}[(H_{k-1}^{n+1})^2 (\Delta_k^{n+1})^2]$$

$$\leq 4^{-n} \sum_{k \geq 1} \mathbb{E}[(\Delta_k^{n+1})^2]$$

$$\leq 4^{-n} \mathbb{E}[M_t^2].$$

Taking limits as $t \to \infty$, and applying Corollary 3.26 to both sides of the above, yields

$$\mathbb{E}[(N_\infty^n - N_\infty^{n+1})^2] \leq 4^{-n} \mathbb{E}[M_\infty^2],$$

and thus, for all $m \geq n$,

$$\mathbb{E}[(N_\infty^n - N_\infty^m)^2] \leq \frac{1}{4^{n-1}} \mathbb{E}[M_\infty^2]. \tag{3.12}$$

So $\{N^n\}$ is a Cauchy sequence in $c\mathcal{M}_0^2$ and converges in $\mathcal{L}^2$ to some continuous limit $N \in c\mathcal{M}_0^2$. But it follows from the bound in (3.12) and the first Borel–Cantelli lemma that the convergence is also almost sure. Hence, from (3.11), A^n converges a.s., uniformly in t, to some adapted and continuous process $[M]$.

It remains to prove that $[M]$ is increasing. For each ω and any fixed $s \leq t$, define $k(m, \omega) = \max\{k : T_k^m(\omega) \leq s\}, \hat{k}(m, \omega) = \min\{k : T_k^m(\omega) \geq t\}$. Note that, as $m \to \infty, T_{k(m,\omega)}^m \uparrow s^* \leq s, T_{\hat{k}(m,\omega)}^m \downarrow t^* \geq t$, and that

$$[M]_t(\omega) - [M]_s(\omega) = [M]_{t^*}(\omega) - [M]_{s^*}(\omega)$$

$$= \lim_{n \to \infty} \lim_{m \to \infty} A^n(T_{\hat{k}(m,\omega)}^m) - A^n(T_{k(m,\omega)}^m) \geq 0.$$

$\square$

We will need the following corollary in the next chapter.

Corollary 3.68 *Let M be some continuous martingale null at zero. If S_k^n is any doubly infinite sequence of stopping times, increasing in k, such that, for each n,*

(i) $\{T_k^n, k \geq 0\} \subseteq \{S_k^n, k \geq 0\}$,

(ii) $\{S_k^n, k \geq 0\} \subseteq \{S_k^{n+1}, k \geq 0\}$,

where $\{T_k^n\}$ is as in Definition 3.63, then, for almost every ω,

$$[M]_t(\omega) := \lim_{n \to \infty} \sum_{k \geq 1} [M_{t \wedge S_k^n}(\omega) - M_{t \wedge S_{k-1}^n}(\omega)]^2. \tag{3.13}$$

Proof: The proof of Theorem 3.66 goes through, with S_k^n replacing T_k^n, unaltered except for (3.10), which becomes

$$|H_k^n| \leq 2^{-n},$$

and subsequent consequential changes to future inequalities in the proof. We thus conclude that the right-hand side of (3.13) has the properties of $[M]$ as stated in Theorem 3.66. But $[M]$ is the only process with these properties, as we show in Meyer's theorem below. $\square$

Remark 3.69: Note that quadratic variation is a sample path property. If we change the probability measure on our space from $\mathbb{P}$ to some equivalent probability measure $\mathbb{Q}$ (so $\mathbb{P}(F) = 0 \Leftrightarrow \mathbb{Q}(F) = 0$), the quadratic variation of M will not change and remains a.s. finite.

We now give a very important result which follows almost immediately from what we have done above. This is a special case of the Doob–Meyer theorem, the full version of which we meet in Section 3.6.

Theorem 3.70 (Meyer) *The quadratic variation $[M]$ of a continuous square-integrable martingale M is the unique increasing, continuous process such that $M^2 - [M]$ is a UI martingale.*

Proof: We saw in the proof above that for $M \in c\mathcal{M}_0^2$, $[M]$ has the required properties, so it remains only to prove the uniqueness. Suppose A is another process with these properties. Both $[M]$ and A are increasing, hence $A - [M]$ is of finite variation. Further, $A - [M] = (M^2 - [M]) - (M^2 - A)$ is a continuous martingale, null at zero. By Theorem 3.62 it must therefore be identically zero. $\square$

3.4.3 Quadratic covariation

The definition of the quadratic covariation of two processes is an obvious extension of the quadratic variation for a single process.

Definition 3.71 *Let X and Y be continuous processes null at zero and let the doubly infinite sequence of times $T_k^n(X)$ and $T_k^n(Y)$ be as defined in Definition 3.63. Now define the doubly infinite sequence S_k^n to be the ordered union of the $T_k^n(X)$ and $T_k^n(Y)$,*

$$S_0^n = 0,$$

$$S_k^n = \inf\left\{t \in \{T_k^n(X) : k \geq 0\} \cup \{T_k^n(Y) : k \geq 0\} : t > S_{k-1}^n\right\}, \quad k > 0.$$

The quadratic covariation of the processes X and Y, written $[X, Y]$, is defined to be

$$[X, Y]_t(\omega) := \lim_{n \to \infty} \sum_{k \geq 1} [X_{t \wedge S_k^n}(\omega) - X_{t \wedge S_{k-1}^n}(\omega)][Y_{t \wedge S_k^n}(\omega) - Y_{t \wedge S_{k-1}^n}(\omega)],$$

or $+\infty$ when the limit does not exist.

Theorem 3.72 (Polarization identity) *For almost every ω,*

$$[M, N](\omega) = \tfrac{1}{2}\big([M + N] - [M] - [N]\big)(\omega).$$

Proof: This is an easy consequence of Corollary 3.68 and is left as an exercise.
□

The polarization identity reduces most results about quadratic covariation to the analogous results for the quadratic variation. An important example is the following generalization of Theorem 3.70. The proof is left as an exercise.

Corollary 3.73 *Let M and N be continuous square-integrable martingales. The quadratic covariation $[M, N]$ of M and N is the unique finite variation, continuous process such that $MN - [M, N]$ is a UI martingale. Thus, the process MN is a UI martingale if and only if $[M, N] = 0$.*

3.5 LOCAL MARTINGALES AND SEMIMARTINGALES

Many of the processes that we meet in finance are martingales, but this class of processes is certainly too restrictive. In this section we extend this class, eventually to *semimartingales*. All the processes that we shall encounter will be semimartingales. First we shall meet *local martingales* and the idea of localization, a simple yet, as we shall see repeatedly, very powerful technique.

3.5.1 The space $c\mathcal{M}_{\text{loc}}$

We have seen that the class of (continuous) martingales is stable under stopping (Proposition 3.54). We now define the class of processes which are *reduced* to $\mathcal{M}_0$, the class of martingales null at zero, by an increasing sequence of stopping times. This technique, of taking a stochastic process and using an increasing sequence of stopping times to generate a sequence of processes with some desirable property, is what we mean by *localization*.

Definition 3.74 *Let M be an adapted process null at 0. Then M is called a local martingale null at 0, and we write $M \in \mathcal{M}_{0,\text{loc}}$, if there exists an increasing sequence $\{T_n\}$ of stopping times with $T_n \uparrow \infty$ a.s. such that each stopped process M^{T_n} is a martingale (null at 0). If M is also continuous we write $M \in c\mathcal{M}_{0,\text{loc}}$. The sequence $\{T_n\}$ is referred to as a reducing sequence for M (into $\mathcal{M}_0$).*

A process M is called a (continuous) local martingale, written $M \in \mathcal{M}_{\text{loc}}$ (respectively $M \in c\mathcal{M}_{\text{loc}}$), if M_0 is $\mathcal{F}_0$-measurable and $M - M_0 \in \mathcal{M}_{0,\text{loc}}$ (respectively $M - M_0 \in c\mathcal{M}_{0,\text{loc}}$).

Remark 3.75: Observe that no integrability condition is imposed on the random variable M_0.

Remark 3.76: A martingale is a local martingale but the converse is not true. It is a common misconception to believe that if we have integrability (i.e. $\mathbb{E}\big[|M_t|\big] < \infty$ for all $t \geq 0$) then this will ensure that a local martingale is a martingale. Indeed a UI local martingale need not be a martingale. A counter-example to this and other related examples can be found in Revuz and Yor (1991, p. 182).

Remark 3.77: In contrast to Remark 3.76, uniform integrability can sometimes help establishing that a local martingale is a true martingale. If M is a local martingale and for each $t > 0$ the family $\{M_t^{T_n}\}$ is UI then the dominated convergence theorem can be applied to show that $\mathbb{E}[M_t|\mathcal{F}_s] = M_s$, so M is a martingale.

In Chapter 4, we shall define the stochastic integral for integrators in $c\mathcal{M}_0^2$ and then use localization to extend the theory to more general continuous integrators. Should we not, therefore, consider processes which localize to this smaller class? The answer lies in the following simple lemma. This lemma is an example of the simplification that comes from restricting to continuous integrators. Let $c\mathcal{M}_{0,\text{loc}}^2$ be the space of local martingales which can be reduced to martingales in $c\mathcal{M}_0^2$.

Lemma 3.78 *Let* $M \in c\mathcal{M}_{0,\text{loc}}$ *and, for each* $n \in \mathbb{N}$, *define*

$$S_n := \inf\{t > 0 : |M_t| > n\}.$$

Then for each n, M^{S_n} *is an a.s. bounded martingale, whence*

$$c\mathcal{M}_{0,\text{loc}} = c\mathcal{M}_{0,\text{loc}}^2.$$

Proof: Suppose $\{T_k\}$ is a reducing sequence for M and that we stop the martingale M^{T_k} at the stopping time S_n. Recalling that the space $c\mathcal{M}_0$ is stable under stopping, the stopped process $(M^{T_k})^{S_n} = M^{T_k \wedge S_n}$ is also a martingale (null at 0) and for $s < t$ we have

$$M_{T_k \wedge S_n \wedge s} = \mathbb{E}\big[M_{T_k \wedge S_n \wedge t}|\mathcal{F}_s\big].$$

As $k \uparrow \infty$, $M_{T_k \wedge S_n \wedge t} \to M_{S_n \wedge t}$ a.s. and, since M is continuous,

$$|M_{T_k \wedge S_n \wedge t}| \leq n$$

almost surely for all k. It now follows from the dominated convergence theorem that

$$M_{S_n \wedge s} = \mathbb{E}[M_{S_n \wedge t}|\mathcal{F}_s],$$

and we are done. $\qquad\square$

Noting that $S_n \uparrow \infty$, it is clear from the lemma that we can always use the sequence $\{S_n\}$ to check if a given continuous adapted process is in $c\mathcal{M}_{0,\text{loc}}$.

This sequence can also be used to complete the proofs of Theorems 3.62 and 3.66. Indeed, these results can both be extended to local martingales, as summarized in the following two theorems.

Theorem 3.79 *For any $M \in c\mathcal{M}_{0,\text{loc}}$, the quadratic variation process $[M]$ is adapted and a.s. finite, continuous and increasing. Furthermore, $[M]$ is the unique adapted, continuous, increasing process such that*

$$M^2 - [M] \in c\mathcal{M}_{0,\text{loc}}. \tag{3.14}$$

Proof: For each $n \in \mathbb{N}$, define S_n as in Lemma 3.78. Each M^{S_n} is bounded, hence Theorem 3.66 holds and shows that $[M^{S_n}]$ is finite, adapted, continuous and increasing. That $[M]$ is adapted now follows since, given any $t \geq 0$, $[M]_t(\omega) = \sup_n [M^{S_n}]_t(\omega)$, and the supremum of a countable measurable family is itself measurable.

Let E_t be the event that $[M]$ is continuous and increasing over the interval $[0, t]$, and let E_t^n be the corresponding event for $[M^{S_n}]$. Theorem 3.66 shows that $\mathbb{P}(E_t^n) = 1$ for all n and t. Since $M_t(\omega) = \lim_{n \uparrow \infty} M_t^{S_n}(\omega)$ for almost all ω, $\mathbf{1}_{E_t^n} \downarrow \mathbf{1}_{E_t}$ and thus

$$\mathbb{P}(E_t) = \lim_{n \uparrow \infty} \mathbb{P}(E_t^n) = 1,$$

as required.

Property (3.14) follows by a similar argument. The uniqueness claim for $[M]$ is immediate from the next result, which extends Theorem 3.62 (and completes that proof). $\qquad\square$

Theorem 3.80 *Suppose $M \in c\mathcal{M}_{0,\text{loc}}$. If M has paths of finite variation then $M \equiv 0$ a.s.*

Proof: Let V_M denote the total variation process of M. Stopping M at the stopping time $T_n = \inf\{t > 0 : V_M(t) > n\}$ reduces the claim here to the case covered by Theorem 3.66, i.e. when M is a bounded continuous martingale with paths of bounded variation. $\qquad\square$

A simple corollary identifies the quadratic variation of Brownian motion.

Corollary 3.81 *If W is a Brownian motion then $[W]_t = t$.*

Proof: Example 3.3 established the fact that $W_t^2 - t$ is a martingale. The result thus follows from Theorem 3.79. $\qquad\square$

We conclude this section with the extension of Theorem 3.79 to the corresponding result for the quadratic variation of two local martingales. The proof follows from the polarization identity.

Corollary 3.82 *Let $M, N \in c\mathcal{M}_{0,\text{loc}}$. The quadratic covariation $[M, N]$ of M and N is the unique finite variation, continuous process such that $MN - [M, N] \in c\mathcal{M}_{0,\text{loc}}$. Thus, the process MN is a continuous local martingale if and only if $[M, N] = 0$.*

3.5.2 Semimartingales

The class of semimartingales is a relatively minor extension of the class of local martingales but it is, in fact, of fundamental importance. It is a very general class of processes and it is, in a sense, the largest class of integrators for which the stochastic integral can be defined (Theorem 4.27).

Definition 3.83 *A process X is called a semimartingale (relative to the set-up $(\Omega, \{\mathcal{F}_t\}, \mathcal{F}, \mathbb{P}))$ if X is an adapted process which can be written in the form*

$$X = X_0 + M + A, \tag{3.15}$$

where X_0 is an $\mathcal{F}_0$-measurable random variable, M is a local martingale null at zero and A is an adapted càdlàg process, also null at zero, having paths of finite variation.

We denote by $\mathcal{S}$ the space of semimartingales and by $c\mathcal{S}$ the subspace of continuous semimartingales.

We shall usually be considering only *continuous* semimartingales (certainly when we define the stochastic integral). The continuity assumption simplifies the theory. It turns out that when X is continuous, the processes M and A in the decomposition (3.15) can also be taken to be continuous. It then follows from Theorem 3.80 that this continuous decomposition is unique up to indistinguishability (but see Remark 3.85). This allows us to make the following definition.

Definition 3.84 *If X is a continuous semimartingale and the decomposition (3.15) is taken to be such that M and A are continuous, we call M the continuous local martingale part of X and denote it by $M = X^{\mathrm{loc}}$.*

Remark 3.85: Note that the decomposition (3.15) is not unique (even when X is continuous) if we allow M and A to have discontinuities since there are non-trivial discontinuous (local) martingales which are of finite variation.

To complete our study of semimartingales we now consider the quadratic variation of a continuous semimartingale which is, of course, defined via Definition 3.63. The following lemma allows us to show that, if X is a continuous semimartingale then $[X] = [X^{\mathrm{loc}}]$, so we already know all about $[X]$. We omit the proof of the lemma. The result is an obvious extension of the equivalent result for a martingale given in Corollary 3.68 once one has seen the proof of the integration by parts result given in Section 4.6. The reader who wants to prove this now should look ahead to equation (4.22).

Lemma 3.86 *Let X be a continuous semimartingale and let $\{T_k^n\}$ be as in Definition 3.63. If S_k^n is any doubly infinite sequence of stopping times, increasing in k, such that, for each n,*
  (i) $\{T_k^n : k \geq 0\} \subseteq \{S_k^n : k \geq 0\}$,

(ii) $\{S_k^n : k \geq 0\} \subseteq \{S_k^{n+1} : k \geq 0\}$,

then, for almost all ω,

$$[X]_t(\omega) = \lim_{n \to \infty} \sum_{k \geq 1} \left[X_{t \wedge S_k^n}(\omega) - X_{t \wedge S_{k-1}^n}(\omega)\right]^2. \qquad (3.16)$$

Theorem 3.87 *For any continuous semimartingale $X \in cS$,*

$$[X] = [X^{\mathrm{loc}}]$$

and, if Y is another continuous semimartingale,

$$[X, Y] = [X^{\mathrm{loc}}, Y^{\mathrm{loc}}].$$

Proof: The second equality follows by applying polarization to the first.

By stopping X at the time $T = \inf\{t > 0 : V_A(t) > K\}$, where A is the finite variation part of X, we may assume that $V_A \leq K$; the general result then follows by localization. Suppose X has decomposition (3.15) and consider the doubly infinite sequence of times S_k^n defined, for each n, via

$$S_0^n = 0,$$

$$S_k^n = \inf\left\{t \in \{T_k^n(X) : k \geq 0\} \cup \{T_k^n(M) : k \geq 0\} : t > S_{k-1}^n\right\}, \quad k > 0,$$

where $\{T_k^n(X)\}$ (respectively $\{T_k^n(M)\}$) are the defining stopping times in the definition of $[X]$ (respectively $[M]$). Letting $s_k^n := S_k^n \wedge t$, we now have by Lemma 3.86,

$$\begin{aligned}
[X]_t(\omega) &= \lim_{n \to \infty} \sum_{k \geq 1} \left(X_{s_k^n}(\omega) - X_{s_{k-1}^n}(\omega)\right)^2 \\
&= \lim_{n \to \infty} \left[\sum_{k \geq 1} \left(M_{s_k^n}(\omega) - M_{s_{k-1}^n}(\omega)\right)^2 \right. \\
&\quad + \sum_{k \geq 1} \left(2\left(M_{s_k^n}(\omega) - M_{s_{k-1}^n}(\omega)\right) + \left(A_{s_k^n}(\omega) - A_{s_{k-1}^n}(\omega)\right)\right) \\
&\quad \left. \times \left(A_{s_k^n}(\omega) - A_{s_{k-1}^n}(\omega)\right)\right],
\end{aligned}$$

for almost all ω. For each ω, the first term on the right-hand side converges to $[M]_t$ by Corollary 3.68. The remaining term is, for each n, bounded by $3 \times 2^{-n} V_A(t) \leq 3 \times 2^{-n} K \to 0$ as $n \to \infty$. This completes the proof. $\square$

Remark 3.88 (Change of filtrations and measure): The definition of quadratic variation is a sample path definition and is in no way dependent on the probability measure $\mathbb{P}$ or the filtration $\{\mathcal{F}_t\}$. Clearly the statement that the quadratic variation is a.s. finite and increasing does depend on the probability measure, but if $[X]$ is a.s. finite and increasing under $\mathbb{P}$ it will also be a.s. finite and increasing for any $\mathbb{Q} \sim \mathbb{P}$.

Remark 3.89: Suppose X is a continuous semimartingale on $(\Omega, \{\mathcal{F}_t\}, \mathcal{F}, \mathbb{P})$. What happens to X if we change $\{\mathcal{F}_t\}$ or $\mathbb{P}$? We shall see in Chapter 5 that if $\mathbb{Q} \sim \mathbb{P}$ then X is also a semimartingale for $\mathbb{Q}$, but the decomposition (3.15) will, in general, change.

Clearly X will not necessarily remain a continuous semimartingale if we change the filtration $\{\mathcal{F}_t\}$, to a finer filtration for example.

3.6 SUPERMARTINGALES AND THE DOOB–MEYER DECOMPOSITION

Supermartingales are an obvious generalization of martingales and will play an important role in the development of term structure models in Chapter 8.

Definition 3.90 *Let $(\Omega, \mathcal{F}, \mathbb{P})$ be a probability triple and $\{\mathcal{F}_t\}$ be a filtration on $\mathcal{F}$. A stochastic process X is an $\{\mathcal{F}_t\}$ supermartingale if:*

(i) X *is adapted to* $\{\mathcal{F}_t\}$;
(ii) $\mathbb{E}\big[|X_t|\big] < \infty$ *for all* $t \geq 0$;
(iii) $\mathbb{E}[X_t|\mathcal{F}_s] \leq X_s$ *a.s., for all* $0 \leq s \leq t$.

If $-X$ *is a supermartingale then we say that X is a submartingale.*

Remark 3.91: A supermartingale is a process which 'drifts down on average'. Clearly any martingale is also a supermartingale (indeed a process is a martingale if and only if it is both a supermartingale and a submartingale). It is also an immediate consequence of Fatou's lemma that if X is a local martingale bounded below by some constant then X is also a supermartingale.

Many of the results stated in this chapter for martingales also hold more generally for supermartingales. We shall not explicitly need most of these more general results. Note, in particular, however, that Theorem 3.8 holds for a supermartingale X (it has a càdlàg modification if $\{\mathcal{F}_t\}$ is right-continuous and complete), as does Doob's (super)martingale convergence theorem.

We finish this chapter with a result of great importance within probability theory, the Doob–Meyer decomposition theorem (we have already met a special case of this decomposition in Theorem 3.70). There are many different statements of this result and the one given here is particularly suited to an application in Chapter 8. A proof can be found in Elliott (1982, p. 83).

Theorem 3.92 (Doob–Meyer) *Suppose $\{X_t\}$, $t \in [0, \infty]$, is a càdlàg supermartingale. Then X has a unique decomposition of the form*

$$X_t = M_t - A_t. \tag{3.17}$$

Here A is an increasing predictable process null at zero a.s., and there is an increasing sequence $\{T_n\}$ of stopping times such that $T_n \uparrow \infty$ a.s. and each stopped process M^{T_n} is a uniformly integrable martingale.

Remark 3.93: We will define the concept of a *predictable process* in Definition 4.2. In particular, 'predictable' implies 'adapted' which is enough for our uses later.

Corollary 3.94 *If X is an a.s. positive càdlàg local martingale integrable at zero then the conclusions of Theorem 3.92 hold.*

Proof: Being a positive local martingale integrable at zero, X is a supermartingale. Therefore the hypotheses of Theorem 3.92, and consequently the conclusions, hold.   □

Remark 3.95: Note that equation (3.17) ensures that the supermartingale X is also a semimartingale.

4

Stochastic Integration

In Chapter 3 we often needed a filtration to be complete and right-continuous. We shall throughout the remainder of this book, unless otherwise stated, assume this to be the case. We note that a filtration can always be augmented so that these properties hold as described in Appendix 1.

4.1 OUTLINE

It is easy to get lost in the details of any account of stochastic integration, so before we begin we will take this opportunity to step back and put all the results to follow into perspective. At the end of this chapter we will have assigned a meaning to expressions of the form

$$I_t = (H \bullet X)_t \equiv \int_0^t H_u \, dX_u \qquad (4.1)$$

for a broad class of integrand processes H and integrators X. All the ideas presented in this outline are introduced with more detailed discussion in later sections.

The integrators: continuous semimartingales

Throughout we will restrict attention to integrators which are continuous. A deeper theory is required if we wish to drop this restriction and it is totally unnecessary to do so for our purposes – remember the integrators will be asset price processes and we will insist that they are continuous.

Recall that by a semimartingale we mean a process of the form $X = X_0 + M + A$ where X_0 is a random variable, M is a continuous *local martingale* (null at zero) and A is a finite variation process (also null at zero). It turns out that the class of semimartingales is, in a sense made precise in Theorem 4.27, the largest class of integrators for which the stochastic integral can be defined.

Financial Derivatives in Theory and Practice Revised Edition. P. J. Hunt and J. E. Kennedy
© 2004 John Wiley & Sons, Ltd   ISBNs: 0-470-86358-7 (HB); 0-470-86359-5 (PB)

The integrands: predictable processes

Suppose we wish to interpret the stochastic integral (4.1) as the gain from trading in the asset X, and that H_t is the holding at t of that asset. At the very least we want (4.1) to have a meaning when H is a simple process, one for which there are only a finite number of time points at which we change the amount of asset X that we hold. Clearly we must also insist that H does not look into the future otherwise there will be arbitrage, so H must at the very least be adapted. In continuous time, even being adapted is not (in general) enough to give (4.1) a reasonable interpretation. The problem is that in continuous time the future and the past meet at the present (see Example 4.1) so we must be more careful.

Consider for a moment the simple integrands H. If we insist that H is left-continuous we have already decided just before t what the integrand will be at t. An immediate consequence of this is that, if H is simple and left-continuous and if X is a martingale, then $H \bullet X$ is also a martingale. This is exactly what we want because it rules out arbitrage in the economy. We shall in Section 4.3.1 give a more precise definition of a simple process and shall include left-continuity as a requirement. Of course, simple integrands are not enough for a complete theory of stochastic integration so we must extend the class. If we do so, in a natural way, we end up with exactly the class of *predictable* processes. This class is, in a sense we make explicit in Section 4.2, the smallest reasonable class of integrands that contains all simple left-continuous integrands. All left-continuous processes are predictable. We shall see (from the martingale representation theorem in Chapter 5) that this class of integrands is large enough to encompass all the processes that might be of interest.

The predictability restriction is what prevents the integrand from anticipating the future. We must also ensure that the stochastic integral does not explode, just as we must when defining a classical Lebesgue–Stieltjes integral. We will discuss two ways of doing this. The first is to make H *locally bounded*, meaning there is a sequence of stopping times $T_n \uparrow \infty$ such that H^{T_n} is bounded. The second, which only works for a local martingale integrator M, is to insist that $H \in \Pi(M)$ (see Section 4.5.1). More on this later.

Constructing the integral

The construction of the stochastic integral is done in two stages. The first step is to define the integral for the case when both H and X are bounded. This is the majority of the work, and we will outline it shortly. The second step, which is more straightforward, is to extend from bounded H and X to the general case. This is achieved using the idea of localization which was introduced in Section 3.5. The argument goes as follows. Suppose H and $X = X_0 + M + A$ are general and let $\{T_n\}$ be an increasing sequence of stopping times, $T_n \uparrow \infty$, such that $H1(0, T_n]$ and M^{T_n} are bounded. We define the general stochastic integral $H \bullet X$ via

$$\int_0^t H_u(\omega)\, dX_u(\omega) = \int_0^t H_u(\omega)\, dM_u(\omega) + \int_0^t H_u(\omega)\, dA_u(\omega)$$

$$:= \lim_{n\uparrow\infty} \int_0^{t \wedge T_n(\omega)} H_u(\omega)\, dM_u(\omega) + \int_0^{t \wedge T_n(\omega)} H_u(\omega)\, dA_u(\omega)$$

$$= \lim_{n\uparrow\infty} \int_0^t H_u(\omega) 1(0, T_n(\omega)]\, dM_u^{T_n(\omega)}(\omega)$$

$$+ \int_0^t H_u(\omega) 1(0, T_n(\omega)]\, dA_u(\omega).$$

The final term on the right-hand side is a classical Lebesgue–Stieltjes integral, so we know what that means (Theorem 3.60 has already covered this). All that remains to complete the definition is to define $H \bullet M$ for bounded predictable H and bounded local martingales M, i.e. bounded martingales M. This is the crux of the construction and the topic of Section 4.3.

Construction of the stochastic integral for integrators in $c\mathcal{M}_0^2$

We need only construct the integral for bounded $M \in c\mathcal{M}_0^2$, but the technique we use works for any $M \in c\mathcal{M}_0^2$.

The construction, which has many features in common with the construction of the classical integral, is completed as follows.

(1) Define the integral $H \bullet M$ in an 'obvious' way for a class of simple integrands which are piecewise constant (between a finite number of stopping times). We denote by $\mathcal{U}$ the space of all such integrands.

(2) Define the integral for a general $H \in \bar{\mathcal{U}}$ (the closure of $\mathcal{U}$) by continuity,

$$H \bullet M := \lim_{n\to\infty} H^n \bullet M, \qquad (4.2)$$

where $H^n \in \mathcal{U}$. What we mean by the closure of $\mathcal{U}$ and the limit in (4.2) is made precise in Section 4.3. The construction is complete once we observe that $\bar{\mathcal{U}}$ contains all bounded predictable integrands.

4.2 PREDICTABLE PROCESSES

A theory of stochastic integration must include integrands which are constant until some stopping time T and are then zero. But should we allow integrands H of the form

$$H = 1(0, T] \qquad (4.3)$$

or of the form

$$H = 1(0, T),$$

or both? Both these processes are adapted, but the following example, taken from Durrett (1996), shows that (for a general theory) we should restrict to the former.

Example 4.1 Let $(\Omega, \mathcal{F}, \mathbb{P})$ be a probability space supporting two random variables T and ξ defined by

$$\mathbb{P}(T \leq t) = t \wedge 1,$$
$$\mathbb{P}(\xi = 1) = \mathbb{P}(\xi = -1) = \tfrac{1}{2}.$$

Define the càdlàg process X via

$$X = \xi 1[T, \infty],$$

and define the filtration $\{\mathcal{F}_t\}$ by $\mathcal{F}_t = \sigma(X_u, u \leq t)$. Note that X is an $\{\mathcal{F}_t\}$ martingale and consider the 'stochastic integral' $I_t = \int_0^t X_u \, dX_u$. The integral I can, in this example, be defined pathwise as a Lebesgue–Stieltjes integral, so the stochastic integral must (surely!) be defined to agree with this,

$$I_t(\omega) = \int_0^t X_u(\omega) \, dX_u(\omega) = \xi^2 1_{\{t \geq T\}} = 1_{\{t \geq T\}}.$$

The integral I is not a martingale and the 'trading strategy' $H = X$ has generated a guaranteed unit profit by time one.

The problem in this example is that the integrand is right-continuous and thus able to take advantage of the jump in the integrator. Of course, this problem cannot occur when the integrator is continuous, which is the case for which we present the theory, but the point is well made. See Remark 4.4 below for more on this point.

We will develop a theory that excludes right-continuous integrands and contains integrands of the type (4.3) and finite linear combinations thereof. To do this we view a general integrand as a map

$$H : (0, \infty) \times \Omega \to \mathbb{R}$$
$$(t, \omega) \mapsto H_t(\omega).$$

We will define a σ-algebra on the product space $(0, \infty) \times \Omega$ and insist that all integrands are *jointly* measurable with respect to this σ-algebra. In so doing we are reducing a stochastic process (a collection of measurable maps indexed by t) to a single measurable map from a larger space, and thus we are able to view the integrands as members of a classical (L^2) Hilbert space.

This identification is important in extending the definition of the integral from simple integrands to more general bounded integrands.

The above joint measurability is strictly stronger than the requirement that H be adapted. There is no unique choice for the σ-algebra to impose on the product space. The one we shall work with, as given by the following definition, is the smallest σ-algebra which is sufficient for the development of the general theory. See Remark 4.4 for further comment.

Definition 4.2 *The predictable σ-algebra $\mathcal{P}$ on $(0, \infty) \times \Omega$ is the smallest σ-algebra on $(0, \infty) \times \Omega$ such that every adapted process having paths which are left-continuous with limits from the right is $\mathcal{P}$-measurable.*

A process H with time parameter set $(0, \infty)$ is called predictable if it is $\mathcal{P}$-measurable as a map from $(0, \infty) \times \Omega$ to $(\mathbb{R}, \mathcal{B}(\mathbb{R}))$, and we write $H \in \mathcal{P}$. We say $H \in b\mathcal{P}$ if H is also bounded.

Technical Remark 4.3: The astute reader will notice that we have not defined a predictable process at zero. We do not need the integrand there and excluding it avoids troublesome technicalities.

In fact, in the definition of the predictable σ-algebra we can replace left-continuous processes by bounded linear combinations of processes of the simple form (4.3). We will build up the definition of the stochastic integral starting with this most basic type of integrands.

Technical Remark 4.4: For continuous integrators it is possible to define a theory of stochastic integration for a wider class of integrands. In the extreme, the integrand can be merely adapted if the integrator satisfies extra regularity conditions (Karatzas and Shreve (1991)), and this includes the case when Brownian motion is the integrator. Without these restrictions, continuity of X is sufficient to develop a theory for integrands which are *progressively measurable* as defined in Theorem 3.60. This is the approach of Revuz and Yor (1991). However, as they point out, extending the class of integrands in this way is of no real advantage since the stochastic integral of a progressively measurable integrand is indistinguishable from the stochastic integral of a corresponding predictable integrand.

4.3 STOCHASTIC INTEGRALS: THE L^2 THEORY

We are now in a position to develop the stochastic integral with respect to an integrator $M \in c\mathcal{M}_0^2$. This is the heart of the definition of the general stochastic integral; after this section the rest is plain sailing.

Throughout we will restrict attention to a fixed integrator $M \in c\mathcal{M}_0^2$ (recall $c\mathcal{M}_0^2$ is the space of continuous square-integrable martingales null at zero). As noted in Section 4.1, at this stage we need to define the stochastic integral with respect to M at least for all integrands H which are *bounded* and predictable. In so doing we shall actually define the integral for more general integrands.

The first stage in the construction is to define the stochastic integral for a class $\mathcal{U}$ of *simple* integrands for which the desired definition is obvious and can be done pathwise. The next step is to extend this definition by a continuity argument to $\bar{\mathcal{U}}$, the 'closure' of $\mathcal{U}$. However, before we can talk about the closure of $\mathcal{U}$ we need to define a norm (or at least a topology) on the space of integrands, and to talk about continuity we further need a norm (topology) on the space of resultant stochastic integrals.

For the space of stochastic integrals the norm is easily defined. We will show that for any $H \in \mathcal{U}$, the square-integrability property of M ensures that $H \bullet M \in c\mathcal{M}_0^2$, and for $c\mathcal{M}_0^2$ we have already defined the norm $\| \cdot \|_2$ in Section 3.2.3. The existence of the increasing process $[M]$ is what allows us to define a norm $\| \cdot \|_M$ for $H \in \mathcal{P}$, and this we do in Section 4.3.2. With these definitions we shall see that the continuous linear map

$$I : \mathcal{U} \to c\mathcal{M}_0^2$$

$$H \mapsto H \bullet M$$

is an isometry and thus can be extended uniquely to a map $I : \bar{\mathcal{U}} \to c\mathcal{M}_0^2$. Clearly the space $\bar{\mathcal{U}}$ depends on M via the norm $\| \cdot \|_M$. As Theorem 4.10 states, this is precisely the Hilbert space $L^2(M)$ which we introduce shortly.

4.3.1 The simplest integral

Fix $M \in c\mathcal{M}_0^2$. For H of the form (4.3) we make the obvious pathwise definition for the stochastic integral $H \bullet M := \int H dM$ as

$$(H \bullet M)_t(\omega) = \left(\int_0^t H dM \right)(\omega) := (M_{t \wedge T} - M_0)(\omega).$$

That is,

$$H \bullet M = M^T \quad \text{(since } M_0 = 0\text{)}.$$

Note that clearly $H \bullet M \in c\mathcal{M}_0^2$.

We now extend the definition of the stochastic integral by linearity to the space of finite linear combinations of $1(0, T]$-type processes.

Definition 4.5 *For two stopping times S and T, denote by $1(S, T]$ the process*

$$1(S, T](t, \omega) = \begin{cases} 1 & \text{if } S(\omega) < t \leq T(\omega) \\ 0 & \text{otherwise,} \end{cases}$$

and let $\mathcal{U}$ be the space of simple processes H of the form

$$H(t, \omega) = \sum_{i=1}^{n} c_{i-1} 1(T_{i-1}, T_i](t, \omega)$$

where $\{T_k\}$ is a finite sequence of stopping times with $0 \leq T_0 \leq T_1 \leq \ldots \leq T_n$ and $\{c_k\}$ is a finite sequence of constants.

For any $H \in \mathcal{U}$, we define $H \bullet M$ (by the obvious extension of (4.4)) as

$$(H \bullet M)_t(\omega) = \sum_{i=1}^{n} c_{i-1} \left[M_{T_i \wedge t}(\omega) - M_{T_{i-1} \wedge t}(\omega) \right]. \tag{4.5}$$

It is easy to check that the space $\mathcal{U}$ is indeed the space of linear combinations of $1(0, T]$-type processes, and furthermore the definition (4.5) is independent of the particular representation chosen for any $H \in \mathcal{U}$. The following obvious result will prove useful shortly.

Lemma 4.6 Fix $M \in c\mathcal{M}_0^2$ and let $H \in \mathcal{U}$. Then $H \bullet M \in c\mathcal{M}_0^2$ and

$$\|H \bullet M\|_2^2 = \mathbb{E}\left[(H \bullet M)_\infty^2\right] = \mathbb{E}\left[\sum_{i=1}^{n} c_{i-1}^2 (M_{T_i}^2 - M_{T_{i-1}}^2)\right].$$

Remark 4.7: Note that it is vital at this point that $\{\mathcal{F}_t\}$ be complete for $H \bullet M$ to be adapted to $\{\mathcal{F}_t\}$ (Corollary 3.48). This is the only point in the construction of the stochastic integral at which we need to appeal to completeness.

4.3.2 The Hilbert space $L^2(M)$

We proved in Theorem 3.66 that the quadratic variation $[M]$ is an increasing process, and this property gives us a candidate norm $\| \cdot \|_M$ on the space of predictable processes $\mathcal{P}$, namely

$$\|H\|_M = \left(\mathbb{E} \int_0^\infty H_s^2 d[M]_s \right)^{\frac{1}{2}}.$$

There is one problem with this, however, one we have met before, which is the fact that any H which is indistinguishable from zero, yet is not strictly zero, will have $\|H\|_M = 0$. Thus $\| \cdot \|_M$ is not strictly a norm on $\mathcal{P}$. The problem also arises more generally if $[M]$ is not strictly increasing. To get around this we must identify any processes H and K for which $\|H - K\|_M = 0$. We will do this whenever necessary without further comment. The only concession we shall make is to stick to the convention, as best we can, that we refer to spaces of equivalence classes using Roman font and actual processes using script font. We make the following definitions.

Definition 4.8 The spaces $\mathcal{L}^2(M)$ and $L^2(M)$ are defined to be

$$\mathcal{L}^2(M) := \{H \in \mathcal{P} \text{ such that } \|H\|_M < \infty\},$$
$$L^2(M) := \{\text{equivalence classes of } H \in \mathcal{P} \text{ such that } \|H\|_M < \infty\}.$$

Technical Remark 4.9: With this definition $\| \cdot \|_M$ is a true norm for the space $L^2(M)$, and under this norm $L^2(M)$ is actually a Hilbert space. To see this, we view $L^2(M)$ as the space $L^2((0, \infty) \times \Omega, \mathcal{P}, \mu_m)$ where μ_m is the unique measure on $((0, \infty) \times \Omega, \mathcal{P})$ such that

$$\mu_m([s, t] \times F) = \mathbb{E}\big[([M]_t - [M]_s)1_F\big]$$

for $s < t, F \in \mathcal{F}_s$. See Appendix 2 for more detail on L^2.

Equipped with a norm we can now identify the closure of $\mathcal{U}$, the set of integrands to which we can hope to extend the stochastic integral by considering sequences of simple integrands. Standard results from analysis allow us to state the following result. A proof can be found in Rogers and Williams (1987).

Theorem 4.10   *Viewed as a subspace of $L^2(M)$, the closure of $\mathcal{U}$ under the norm $\| \cdot \|_M$ is precisely $\mathcal{L}^2(M)$.*

Remark 4.11: Note that we have the relationships $\mathcal{U} \subset b\mathcal{P} \subset \bar{\mathcal{U}} = \mathcal{L}^2(M)$ (interpreting $\bar{\mathcal{U}}$ as a set of processes again). Thus $\mathcal{L}^2(M)$ is the smallest 'closed' space, for the norm $\| \cdot \|_M$, containing all bounded predictable processes. Note, however, that if we change M then $\mathcal{L}^2(M)$ also changes.

4.3.3 The L^2 integral

Given $M \in c\mathcal{M}_0^2$, how do we extend the definition of the stochastic integral $H \bullet M$ to include as integrands all H in $\mathcal{L}^2(M)$? The following extension of Lemma 4.6 is crucial.

Lemma 4.12   *Fix $M \in c\mathcal{M}_0^2$. For any $H \in \mathcal{U}, H \bullet M \in c\mathcal{M}_0^2$ and*

$$\|H \bullet M\|_2 = \|H\|_M, \tag{4.6}$$

i.e. the map $I_M : (\mathcal{U}, \| \cdot \|_M) \to (c\mathcal{M}_0^2, \| \cdot \|_2)$ is an isometry. In particular, if $\{H^n\}$ is a Cauchy sequence in $\mathcal{U}$, then $\{H^n \bullet M\}$ is Cauchy in $c\mathcal{M}_0^2$.

Proof: From Lemma 4.6 we have that $H \bullet M \in c\mathcal{M}_0^2$ and that

$$\|H \bullet M\|_2^2 = \mathbb{E}\left[\sum_{i=1}^{n} c_{i-1}^2 (M_{T_i}^2 - M_{T_{i-1}}^2)\right].$$

But $M^2 - [M]$ is a uniformly integrable martingale (Theorem 3.70) and so for each $i \in \{1, \ldots, n\}$ we have

$$\mathbb{E}\big[M_{T_i}^2 - M_{T_{i-1}}^2 | \mathcal{F}_{T_{i-1}}\big] = \mathbb{E}\big[[M]_{T_i} - [M]_{T_{i-1}} | \mathcal{F}_{T_{i-1}}\big].$$

Thus

$$\|H \bullet M\|_2^2 = \mathbb{E}\left[\sum_{i=1}^n c_{i-1}^2 \left([M]_{T_i} - [M]_{T_{i-1}}\right)\right] = \mathbb{E}\left[\int_0^\infty H_u^2 \, d[M]_u\right]$$

as required.

The Cauchy property is immediate from (4.6).   □

We will shortly apply this result to complete the definition of the L^2 stochastic integral. Before we do so we take a brief pause for thought. To define the integral in general, ideally we would like to do this on a pathwise basis as follows:

(i) For each $H \in \mathcal{L}^2(M)$ identify an approximating sequence $\{H^n\} \subset \mathcal{U}$ for which $\|H^n - H\|_M \to 0$ as $n \to \infty$.
(ii) Show that the corresponding sequence of stochastic integrals $H^n \bullet M$ converges a.s. to some limit $H \bullet M$. On the null set where the sequence does not converge, define $H \bullet M \equiv 0$.

Given any Cauchy sequence $H^n \in \mathcal{L}^2(M)$ it follows from Lemma 4.12 that the sequence $H^n \bullet M$ converges in $\mathcal{L}^2$, hence converges a.s. along a fast subsequence, so step (ii) presents no problem. Furthermore, the *existence* of a suitable sequence for step (i) is not in doubt – it follows from the fact that $L^2(M)$ is complete (Appendix 2). However, there are many possible approximating sequences which could be used in step (i) and the remaining problem is to identify one in particular which will be used for the definition. In the special case when H is left-continuous there is an obvious choice of approximating sequence, as we show in the next section, and this completes the construction. However, for general $H \in \mathcal{L}^2(M)$ there is no constructive way to define the approximating sequence H^n. This would not be a problem if the limit were independent of the approximating sequence, but given any two approximating sequences $\{H^n\}$ and $\{K^n\}$, the limits (along a fast subsequence) $H \bullet M$ and $K \bullet M$ will in general disagree on a null set. Thus our inability to explicitly and uniquely identify an approximating sequence of integrands procludes us from *defining* the stochastic integral as a pathwise limit (but see Section 4.3.4 for what is true pathwise).

These observations force us to be less ambitious in constructing the stochastic integral. The following result essentially completes that definition. It defines the stochastic integral but only up to an equivalence class. When we need to consider the stochastic integral as a true stochastic process and not just as an equivalence class, as we do when considering sample path properties, we choose some (appropriate) member of the equivalence class.

Recall that we can identify $c\mathcal{M}_0^2$ with a classical L^2 space by imposing the norm $\|\cdot\|_2$ and taking equivalence classes. For the remainder of this section we denote this space of equivalence classes by $L^2(c\mathcal{M}_0^2)$.

Theorem 4.13 *Given $M \in c\mathcal{M}_0^2$ and $H \in \mathcal{L}^2(M)$, there exists a sequence $\{H^n\} \subset \mathcal{U}$ and $N \in c\mathcal{M}_0^2$ such that*

$$\|H^n - H\|_M \to 0 \tag{4.7}$$
$$\|H^n \bullet M - N\|_2 \to 0 \tag{4.8}$$

as $n \to \infty$. Furthermore, if $\hat{M}$ and $\hat{N}$ lie in the same equivalence class of $c\mathcal{M}_0^2$ as M and N respectively, and if $\hat{H}$ and H lie in the same equivalence class of $\mathcal{L}^2(M)$, then (4.7) and (4.8) hold for $\hat{M}$, $\hat{N}$ and $\hat{H}$.

Viewing M and H as elements of $L^2(c\mathcal{M}_0^2)$ and $L^2(M)$ respectively, we define the stochastic integral $H \bullet M \in L^2(c\mathcal{M}_0^2)$ to be the unique $N \in L^2(c\mathcal{M}_0^2)$ for which (4.7) and (4.8) hold. Viewing M and H as elements of $c\mathcal{M}_0^2$ and $\mathcal{L}^2(M)$, we define the stochastic integral $H \bullet M \in c\mathcal{M}_0^2$ to be some $N \in c\mathcal{M}_0^2$ for which (4.7) and (4.8) hold.

Proof: By Theorem 4.10, given any $H \in L^2(M) \equiv \bar{\mathcal{U}}$, we can find a sequence $\{H^n\} \subset \mathcal{U}$ such that $H^n \to H$ in $L^2(M)$. The sequence $\{H^n\}$ is Cauchy hence, by Lemma 4.12, $\{H^n \bullet M\}$ is a Cauchy sequence in $L^2(c\mathcal{M}_0^2)$ which, since $L^2(c\mathcal{M}_0^2)$ is complete, converges to a limit in $L^2(c\mathcal{M}_0^2)$. We define $H \bullet M$ to be this limit. That $H \bullet M$ is well defined (as an equivalence class), i.e. independent of the approximating sequence $\{H^n\}$, follows from (4.6) for $H^n \in \mathcal{U}$. This completes the construction at the equivalence class level. The remainder of the theorem is obvious. □

Corollary 4.14 *For any $M \in c\mathcal{M}_0^2$ and $H \in \mathcal{L}^2(M)$, $H \bullet M \in c\mathcal{M}_0^2$ and*

$$\mathbb{E}\big[(H \bullet M)_\infty^2\big] := \|H \bullet M\|_2^2 = \|H\|_M^2 =: \mathbb{E}\left(\int_0^\infty H_u^2 \, d[M]_u\right).$$

4.3.4 Modes of convergence to $H \bullet M$

It is worth spending a little more time understanding in more detail the way in which the integrals $H^n \bullet M$, $H^n \in \mathcal{U}$, converge to $H \bullet M$. Theorem 4.13 states that, given any $H \in \mathcal{L}^2(M)$, there exists some continuous martingale $H \bullet M \in c\mathcal{M}_0^2$ such that

$$\mathbb{E}[(H^n \bullet M - H \bullet M)_\infty^2] \to 0 \tag{4.9}$$

whenever

$$\|H^n - H\|_M \to 0.$$

Suppose we consider all the equivalence classes of $c\mathcal{M}_0^2$ and select one element within each class. We define this process to be the stochastic integral $H \bullet M$ of all H for which (4.7) and (4.8) hold. The convergence in (4.9) is $\mathcal{L}^2$

convergence. An immediate consequence of this is convergence in probability, so we can replace (4.9) by

$$\mathbb{P}\left(\sup_{t \geq 0} |(H^n \bullet M - H \bullet M)_t| > \varepsilon\right) \to 0$$

as $n \to \infty$ for any $\varepsilon > 0$.

As mentioned above, what one would like ideally is almost sure convergence. But convergence in probability implies almost sure convergence *along a subsequence*, so we can conclude, given any $H \in \mathcal{L}^2(M)$, that there exists a sequence $\{H^{n'}\} \subset \mathcal{U}$ such that, for almost every ω,

$$(H^{n'} \bullet M)_t(\omega) \to (H \bullet M)_t(\omega).$$

Thus, at least along a fast subsequence, once we have defined the stochastic integral, any approximating sequence converges a.s. to this stochastic integral. However, any two convergent subsequences will misbehave on different null sets, and it is this that prevents us from *defining* the stochastic integral as a pathwise limit.

There is a special case when the stochastic integral can be defined pathwise. When H is bounded and left-continuous we have the following result which explicitly identifies the stochastic integral, on a pathwise basis, as the limit of a Riemann sum.

Theorem 4.15 *Let $M \in c\mathcal{M}_0^2$ and $H \in b\mathcal{P}$ be left-continuous. Define the doubly infinite sequence of stopping times $\{T_k^n\}$ via*

$$T_0^n \equiv 0, \; T_{k+1}^n = \inf\{t > T_k^n : |H_t - H_{T_k^n}| > 2^{-n}\}.$$

Then, for almost every ω,

$$\left(\int_0^t H_u \, dM_u\right)(\omega) = \lim_{n \to \infty} \sum_{k \geq 1} H_{t \wedge T_{k-1}^n}(\omega)\left[M_{t \wedge T_k^n}(\omega) - M_{t \wedge T_{k-1}^n}(\omega)\right].$$

Proof: The method of proof is similar to that used in Theorem 3.66. We leave this as an exercise for the reader. $\qquad\Box$

Remark 4.16: We shall shortly define the stochastic integral for a wider class of integrands and integrators; all the comments in this section hold in this greater generality. In particular, Theorem 4.15 holds if M is a general continuous semimartingale and H is locally bounded and left-continuous.

4.4 PROPERTIES OF THE STOCHASTIC INTEGRAL

We now review some basic properties of the stochastic integral. These extend
to the more general integral which we define shortly, and indeed some of them
are needed to be able to carry out that extension.

The first result shows how the stochastic integral behaves with respect
to stopping and plays a key role in its extension by localization in the
next section. The second identifies the quadratic variation of the L^2
stochastic integral. One property we expect is that integration be associative,
$H \bullet (K \bullet M) = (HK) \bullet M$. This is the fourth result we prove, and to do so
we need first to establish the Kunita–Watanabe characterization, the third
theorem.

The first result proves that stopping the stochastic integral is the same as
stopping either the integrand or the integrator.

Theorem 4.17 *Fix $M \in c\mathcal{M}_0^2$. Then for any $H \in \mathcal{L}^2(M)$ and any stopping
time T,*
$$(H \bullet M)^T = H1(0,T] \bullet M = H \bullet M^T.$$

Proof: For general $H \in \mathcal{L}^2(M)$, let $\{H^n\} \subset \mathcal{U}$ be such that $\|H^n - H\|_M \to 0$
as $n \to \infty$. For $H \in \mathcal{U}$ the result is straightforward, thus in particular
$$(H^n \bullet M)^T = H^n 1(0,T] \bullet M = H^n \bullet M^T.$$

The general result is now immediate from the following observations,

(i) Using the optional sampling theorem and Jensen's inequality, it is clear
that $c\mathcal{M}_0^2$ is stable under stopping (that is, if $M \in c\mathcal{M}_0^2$ and T is a
stopping time, then $M^T \in c\mathcal{M}_0^2$). Thus,

$$\begin{aligned}
\|(H^n \bullet M)^T - (H \bullet M)^T\|_2^2 &= \mathbb{E}\Big[(H^n \bullet M - H \bullet M)_T^2\Big] \\
&= \mathbb{E}\Big[\Big(\mathbb{E}[(H^n \bullet M - H \bullet M)_\infty | \mathcal{F}_T]\Big)^2\Big] \\
&\leq \mathbb{E}\big[(H^n \bullet M - H \bullet M)_\infty^2\big] \\
&= \|H^n \bullet M - H \bullet M\|_2^2 \to 0
\end{aligned}$$

(the second equality is the optional sampling theorem, the inequality
is Jensen and the convergence is from the definition of the stochastic
integral);

(ii) $\|H^n 1(0,T] - H1(0,T]\|_M \leq \|H^n - H\|_M \to 0$, thus $\|H^n 1(0,T] \bullet M - H1(0,T] \bullet M\|_2 \to 0$;

(iii) $M^T \in c\mathcal{M}_0^2$, $\|H^n - H\|_M \to 0$ thus $\|H^n \bullet M^T - H \bullet M^T\|_2 \to 0$. $\square$

For integrands $H \in \mathcal{L}^2(M)$ we have seen that $H \bullet M \in c\mathcal{M}_0^2$. The following
result identifies the quadratic variation process of $H \bullet M$ and the covariation
of any two stochastic integrals in terms of certain Lebesgue–Stieltjes integrals.

Theorem 4.18 (Quadratic variation, quadratic covariation) *Fix* $M \in c\mathcal{M}_0^2$. *For any* $H \in L^2(M)$,

$$(H \bullet M)_t^2 - \int_0^t H_u^2 \, d[M]_u$$

is a continuous UI martingale and thus

$$[H \bullet M]_t = \int_0^t H_u^2 \, d[M]_u \, . \tag{4.10}$$

Further, if $N \in c\mathcal{M}_0^2$ *and* $K \in L^2(N)$, *then*

$$[H \bullet M, K \bullet N]_t = \int_0^t H_u K_u \, d[M, N]_u \, . \tag{4.11}$$

Proof: Let

$$X_t := (H \bullet M)_t^2 - \int_0^t H_u^2 \, d[M]_u \, .$$

Clearly X is continuous. To prove X is a UI martingale we use Theorem 3.53. Let T be a stopping time. Since $H \bullet M$ is UI (being in $c\mathcal{M}_0^2$), the optional sampling theorem combined with Jensen's inequality shows that

$$\mathbb{E}[(H \bullet M)_T^2] \leq \mathbb{E}[(H \bullet M)_\infty^2] \, .$$

Further, $\int_0^T H_u^2 \, d[M]_u \leq \int_0^\infty H_u^2 \, d[M]_u$ which is integrable by Corollary 4.14, and thus we can conclude that the random variable $\int_0^T H_u^2 \, d[M]_u$ is integrable and so

$$\mathbb{E}[|X_T|] \leq \mathbb{E}[(H \bullet M)_T^2] + \mathbb{E}\left[\int_0^T H_u^2 \, d[M]_u\right] < \infty \, .$$

Next observe, by Corollary 4.14 and Theorem 4.17, that

$$\mathbb{E}\left[\int_0^T H_u^2 \, d[M]_u\right] = \mathbb{E}\left[\int_0^\infty (H1(0,T])_u^2 \, d[M]_u\right]$$
$$= \mathbb{E}[(H1(0,T] \bullet M)_\infty^2]$$
$$= \mathbb{E}[[(H \bullet M)_\infty^T]^2]$$
$$= \mathbb{E}[(H \bullet M)_T^2] \, .$$

Thus $\mathbb{E}[X_T] = 0$ and X is indeed a UI martingale by Theorem 3.53.

The identity (4.10) follows from Theorem 3.70. To establish (4.11) it is sufficient to show

$$[H \bullet M, N]_t = \int_0^t H_u \, d[M, N]_u \, . \tag{4.12}$$

Equation (4.12) can be proved by again using Theorem 3.53 to show that

$$(H \bullet M)N - H \bullet [M, N]$$

is a continuous UI martingale. The details are left to the reader as an exercise.
□

As our next result indicates, Lebesgue–Stieltjes integration can be used to form the basis of an alternative characterization of the stochastic integral. This is, in fact, how the L^2 stochastic integral was originally defined by Kunita and Watanabe (1967).

Theorem 4.19 Let $M \in c\mathcal{M}_0^2$ and $H \in \mathcal{L}^2(M)$. The stochastic integral $H \bullet M$ is the unique (up to indistinguishability) martingale in $c\mathcal{M}_0^2$ such that

$$[H \bullet M, N] = H \bullet [M, N] \tag{4.13}$$

for every $N \in c\mathcal{M}_0^2$.

Proof: We already know from (4.12) above that $H \bullet M$ satisfies (4.13). To show uniqueness, suppose X is some other martingale in $c\mathcal{M}_0^2$ such that

$$[X, N] = H \bullet [M, N] \tag{4.14}$$

for every $N \in c\mathcal{M}_0^2$. Subtracting (4.13) from (4.14) we have, for each $t \geq 0$, that

$$[X - H \bullet M, N]_t = 0 \quad \text{a.s.}$$

In particular, taking $N = X - H \bullet M$, we have that $[X - H \bullet M, X - H \bullet M] = 0$ and so $\mathbb{E}[(X - H \bullet M)^2] = 0$ by Theorem 3.70. Hence $X = H \bullet M$ a.s. □

Corollary 4.20 (Associativity) Fix $M \in c\mathcal{M}_0^2$. If $K \in \mathcal{L}^2(M)$ and $H \in \mathcal{L}^2(K \bullet M)$, then $HK \in \mathcal{L}^2(M)$ and

$$(HK) \bullet M = H \bullet (K \bullet M).$$

Proof: Since, by Theorem 4.18, $[K \bullet M] = \int_0^t K_u^2 \, d[M]_u$ for all $t \geq 0$, we have

$$\mathbb{E}\left[\int_0^t H_u^2 K_u^2 \, d[M]_u \right] = \mathbb{E}\left[\int_0^t H_u^2 \, d[K \bullet M]_u \right]$$

and so $HK \in \mathcal{L}^2(M)$.

By Theorem 4.19, for any $N \in c\mathcal{M}_0^2$ we know that

$$[K \bullet M, N] = \int_0^t K_u \, d[M, N]_u$$

and so, appealing to the associativity of Stieltjes integrals,

$$[(HK) \bullet M, N] = \int_0^t H_u K_u d[M, N]_u$$

$$= \int_0^t H_u d[K \bullet M, N]_u$$

$$= [H \bullet (K \bullet M), N].$$

The result now follows from the uniqueness in Theorem 4.19. $\qquad\square$

4.5 EXTENSIONS VIA LOCALIZATION

The theory we have developed so far for the stochastic integral, though elegant and self-contained, is too restrictive. For example, the class of integrators $c\mathcal{M}_0^2$ does not include Brownian motion and we certainly need to extend the theory to include this for our applications. Fortunately localization allows us to extend the stochastic integral without much difficulty.

Our first use of localization is to extend the class of integrators to continuous local martingales null at zero, $c\mathcal{M}_{0,\text{loc}}$. This is not the final class of integrators for our theory; for that we must wait until Section 4.5.2.

4.5.1 Continuous local martingales as integrators

We are now ready to extend the definition of the stochastic integral to the class of integrators $c\mathcal{M}_{0,\text{loc}}$. The class of integrands we shall use again depends on the integrator M and we denote it by $\Pi(M)$. This class is defined as follows:

$$\Pi(M) := \left\{ H \in \mathcal{P} \text{ such that } \int_0^t H_u^2 d[M]_u < \infty \text{ a.s., for all } t \right\}. \quad (4.15)$$

Observe that if $M \in c\mathcal{M}_0^2$, in which case $\mathcal{L}^2(M)$ is defined, we have

$$\mathcal{L}^2(M) \subseteq \Pi(M).$$

Thus the definition of the stochastic integral given below broadens the class of integrands as well as the class of integrators.

Fix $M \in c\mathcal{M}_{0,\text{loc}}$ and let $H \in \Pi(M)$. For $n \in \mathbb{N}$, let

$$S_n := \inf\{t > 0 : |M_t| > n\},$$

$$R_n := \inf\{t > 0 : \int_0^t H_u^2 d[M]_u > n\}, \quad (4.16)$$

$$T_n := R_n \wedge S_n.$$

For each n, T_n is a stopping time and $T_n \uparrow \infty$ a.s. Further, note that

$$M^{T_n} \in c\mathcal{M}_0^2,$$

$$H1(0, T_n] \in L^2(M^{T_n}),$$

and so, using Theorem 4.13, for each n we may define the stochastic integral

$$(H1(0, T_n]) \bullet M^{T_n} \in c\mathcal{M}_0^2.$$

It is immediate from Theorem 4.17 that, for $1 \le n \le m$, we have

$$\left((H1(0, T_m]) \bullet M^{T_m} \right)^{T_n} = (H1(0, T_n]) \bullet M^{T_n}. \tag{4.17}$$

Equation (4.17) tells us we can use localization to provide a consistent definition of the stochastic integral $H \bullet M$.

Definition 4.21 *Let $M \in c\mathcal{M}_{0,\text{loc}}$ and $H \in \Pi(M)$. We define the stochastic integral $H \bullet M$ to be the process in $c\mathcal{M}_{0,\text{loc}}$ defined by*

$$H \bullet M(\omega) = \lim_{n \to \infty} (H1(0, T_n])(\omega) \bullet M^{T_n}(\omega),$$

where $\{T_n\}$ is defined by (4.16).

Remark 4.22: Note the statement included in this theorem that the stochastic integral with respect to a local martingale integrator is itself a local martingale. This fact is used repeatedly in calculations and applications (see Example 4.37 for one example).

Given $M \in c\mathcal{M}_{0,\text{loc}}$, $\Pi(M)$ is in fact the largest possible class of integrands we can take. The interested reader is referred to Rogers and Williams (1987) or Karatzas and Shreve (1991) for an explanation of why this is the case.

With the obvious modifications, the properties of the L^2 stochastic integral established in Section 4.4 carry over to this more general case.

4.5.2 Semimartingales as integrators

The final step in our construction of the stochastic integral is to extend the class of integrators to continuous semimartingales. In order to discuss integration with respect to a semimartingale we replace the obvious choice, $\Pi(X)$ (as defined in (4.15)), by a slightly smaller class of integrands that does not depend on X^{loc}, the space of *locally bounded predictable processes*. We need to do this because $\Pi(X)$ does not depend on the finite variation part of X and so, in general, denoting the finite variation part of X by A, the integral $H \bullet A$ will not exist for $H \in \Pi(X)$. As will be clear from the definition, the new space we introduce is independent of X and as such can be used as part of a more general *stochastic calculus*.

Definition 4.23 *A process H is called* locally bounded and predictable, *and we write $H \in \ell b\mathcal{P}$, if it is predictable and if there exists an increasing sequence $\{T_n\}$ of stopping times with $T_n \uparrow \infty$ a.s. and a sequence $\{c_n\}$ of constants such that, for all ω,*

$$|H1(0, T_n]| < c_n. \tag{4.18}$$

The sequence $\{T_n\}$ is referred to as a reducing sequence *as $H1(0, T_n] \in b\mathcal{P}$.*

Remark 4.24: Since $\{\mathcal{F}_t\}$ is assumed to be complete it is sufficient that (4.18) hold for almost all ω. To see this, let Ω^* be the set of paths for which (4.18) holds. Now define the stopping times $\{\hat{T}_n\}$ by

$$\hat{T}_n(\omega) = \begin{cases} T_n(\omega) & \omega \in \Omega^* \\ 0 & \text{otherwise.} \end{cases}$$

This is a reducing sequence of stopping times for H.

All càdlàg processes H such that $\limsup_{t \downarrow 0} |H_t| < \infty$ a.s. are seen to be in $\ell b\mathcal{P}$ by taking

$$T_n = \inf\{t > 0 : |H_t| > n\}, \tag{4.19}$$

which ensures that (4.18) holds a.s. Furthermore, we have the following important result alluded to above.

Lemma 4.25 *For any $M \in c\mathcal{M}_{0,\text{loc}}$ (or more generally $X \in cS$),*

$$\ell b\mathcal{P} \subset \Pi(M).$$

Proof: Given $H \in \ell b\mathcal{P}$, take the reducing sequence (4.19) and observe that

$$\int_0^{T_n} H_u^2 \, d[M]_u \leq n^2 [M]_{T_n} < \infty \text{ a.s.}$$

$\square$

The lemma means that, for $H \in \ell b\mathcal{P}$, $M \in c\mathcal{M}_{0,\text{loc}}$, we can define $H \bullet M$ as in Section 4.5.1. Clearly, for $H \in \ell b\mathcal{P}$ and A a continuous adapted finite variation process, we can define $H \bullet A$ as the pathwise Lebesgue–Stieltjes integral (as we did in Chapter 3 for bounded H)

$$(H \bullet A)_t(\omega) = \int_0^t H_u(\omega) \, dA_u(\omega).$$

Noting that $H \in \mathcal{P}$ implies H is progressively measurable, Theorem 3.60 ensures that $H \bullet A$ is again continuous adapted and of finite variation. Thus we have the following extension of the stochastic integral.

Definition 4.26 *For X a continuous semimartingale as at (3.15) and for $H \in \ell b \mathcal{P}$ we define the stochastic integral $H \bullet X$ to be the continuous semimartingale (null at zero) given by*

$$\int H_u \, dX_u \equiv H \bullet X := H \bullet M + H \bullet A.$$

4.5.3 The end of the road!

The concept of a semimartingale was originally arrived at by *ad hoc* means. However, it can be argued that any reasonable integrator is a semimartingale as follows. Recall that we began the development of the stochastic integral by taking as integrand the class $\mathcal{U}$ of simple predictable processes of the form

$$H = \sum_{i=1}^{n} c_{i-1} 1(T_{i-1}, T_i].$$

For any continuous adapted process X and any $H \in \mathcal{U}$, we could define a 'stochastic integral' $H \bullet X$ in the obvious way via

$$(H \bullet X)_t = \sum_{i=1}^{n} c_{i-1}(X_{T_i \wedge t} - X_{T_{i-1} \wedge t}).$$

Let H and the sequence $\{H^n\}$ be in $\mathcal{U}$. For a general X, even if we have uniform convergence of $\{H^n\}$ to H (a very strict form of convergence) this is not enough to ensure the weak requirement that we have convergence in probability of $(H^n \bullet X)_t$ to $(H \bullet X)_t$. The following result tells us that it is precisely the class of semimartingales for which the map $H \to H \bullet X$ satisfies this mildest continuity condition, and thus for which a sensible extension of this map is possible. A proof of this result can be found in Dellacherie and Meyer (1980).

Theorem 4.27 *Suppose X is a continuous adapted process such that, whenever $\{H^n\}$ is a sequence in $\mathcal{U}$ with*

$$\sup_{t,\omega} |H^n(t,\omega)| \to 0,$$

as $n \to \infty$, then for each t

$$(H^n \bullet X)_t \to 0$$

in probability. Then X is a continuous semimartingale.
    The converse result is also true.

4.6 STOCHASTIC CALCULUS: ITÔ'S FORMULA

4.6.1 Integration by parts and Itô's formula

Having defined the stochastic integral we now need to develop the rules by which it can be manipulated, i.e. stochastic calculus. The main tool is a change of variable formula for stochastic integrals known as Itô's formula. We begin by proving a special case of the result, the integration by parts formula, which shows that the equivalent classical result must be amended when dealing with stochastic integrals to include a correction term.

Theorem 4.28 (Integration by parts formula) *Let X and Y be continuous semimartingales (with respect to $(\Omega, \{\mathcal{F}_t\}, \mathcal{F}, \mathbb{P})$). Then, a.s., for all $t \geq 0$,*

$$X_t Y_t = X_0 Y_0 + \int_0^t X_u \, dY_u + \int_0^t Y_u \, dX_u + [X, Y]_t. \qquad (4.20)$$

In particular,

$$X_t^2 = X_0^2 + 2 \int_0^t X_u \, dX_u + [X]_t. \qquad (4.21)$$

Proof: We first prove (4.21) from which we can deduce (4.20) by polarization,

$$X_t Y_t = \tfrac{1}{2} \left((X_t + Y_t)^2 - X_t^2 - Y_t^2 \right).$$

Suppose X has decomposition (3.15). It suffices, by use of the reducing sequence $S_n = \inf\{t > 0 : |X_t| + |M_t| > n\}$, $n \geq 1$, to assume that both X and M are bounded. Using the notation of Theorem 3.66, we can write

$$
\begin{aligned}
X_t^2 - X_0^2 &= \sum_{k \geq 1} (X_{t_k^n}^2 - X_{t_{k-1}^n}^2) + 2 \sum_{k \geq 1} X_{t_{k-1}^n} (X_{t_k^n} - X_{t_{k-1}^n}) \\
&= \sum_{k \geq 1} (X_{t_k^n}^2 - X_{t_{k-1}^n}^2) + 2 \sum_{k \geq 1} X_{t_{k-1}^n} (M_{t_k^n} - M_{t_{k-1}^n}) \\
&\qquad + 2 \sum_{k \geq 1} X_{t_{k-1}^n} (A_{t_k^n} - A_{t_{k-1}^n}). \qquad (4.22)
\end{aligned}
$$

For each ω, the first term on the right-hand side converges to $[X]_t$ by definition, and the second converges a.s. to $X \bullet M$ by Theorem 4.15 and Remark 4.16. The final term is, for each ω, merely a discrete approximation to the Lebesgue–Stieltjes integral $\int X_u \, dA_u$ (which is also continuous), thus converges to $X \bullet A$ by the classical (Lebesgue) dominated convergence theorem (recall X is bounded). $\qquad\square$

Remark 4.29: Observe that if X or Y is of finite variation then (4.20) reduces to the ordinary integration by parts formula for Lebesgue-Stieltjes integrals since in that case $[X, Y]_t \equiv 0$.

Remark 4.30: From the particular case (4.21), if M is a local martingale we now have an expression for the local martingale $M^2 - [M]$ (see Theorem 3.79) in terms of a stochastic integral,

$$M_t^2 - [M]_t = M_0^2 + 2 \int_0^t M_u \, dM_u \, .$$

Equation (4.21) shows that if X_t is any continuous semimartingale and $f(x) = x^2$ then $f(X_t)$ is also a continuous semimartingale. Itô's formula tells us that this is the case for any function f which is C^2 and provides the decomposition for the continuous semimartingale $f(X_t)$.

Theorem 4.31 (Itô's formula)   Let $f : \mathbb{R} \to \mathbb{R}$ be C^2 and let X be a *continuous semimartingale. Then, a.s., for all $t \geq 0$,*

$$f(X_t) = f(X_0) + \int_0^t f'(X_u) \, dX_u + \tfrac{1}{2} \int_0^t f''(X_u) \, d[X]_u \, . \qquad (4.23)$$

In particular, if X has the decomposition $X = X_0 + M + A$ then $f(X_t)$ has the decomposition

$$f(X_t) = f(X_0) + \int_0^t f'(X_u) \, dM_u + \left\{ \int_0^t f'(X_u) \, dA_u + \tfrac{1}{2} \int_0^t f''(X_u) \, d[M]_u \right\}, \qquad (4.24)$$

and is thus a continuous semimartingale.

Remark 4.32: Note that $f'(X_t)$ is a continuous adapted process in $\ell b \mathcal{P}$ and so the stochastic integral is well defined. The second integral in (4.23) is to be understood in the Lebesgue–Stieltjes sense.

Proof: The second result is a simple consequence of the first. To prove the first, note that (4.23) trivially holds for $f(x) \equiv 1$ and $f(x) = x$. Further, if (4.23) holds for f and g, then it also holds for $f + g$ and, by Theorem 4.28, for fg. We can conclude that the result holds for all polynomial functions.

We now prove the result for general f. By localization we may assume that X, $[M]$ and V_A are bounded, by K say. By the continuity of (4.24) in t, we need only prove that, for all $t \geq 0$ and all $\varepsilon > 0$,

$$\mathbb{P}\Big(|(\mathcal{G}f)_t| > \varepsilon \Big) = 0 \, ,$$

where $(\mathcal{G}f)_t$ is the difference between the left- and the right-hand sides of (4.23). We prove this by polynomial approximation.

Fix $\varepsilon > 0$, and let p_n be a polynomial such that

$$\sup_{|x| \leq K} |p_n(x) - f(x)| + |p_n'(x) - f'(x)| + |p_n''(x) - f''(x)| < n^{-1} \, . \qquad (4.25)$$

We have just shown that $(\mathcal{G}p_n)_t = 0$, a.s., so $(\mathcal{G}f)_t = (\mathcal{G}(f - p_n))_t$ a.s. It follows from (4.25) and the bounds on M and V_A that

$$|(\mathcal{G}(f - p_n))_t| \le \frac{2 + \frac{3}{2}K}{n} + \left|\int_0^t (p_n' - f')\,dM_u\right|.$$

For n sufficiently large $(> 2(2 + \frac{3}{2}K)/\varepsilon)$,

$$\mathbb{P}\big(|(\mathcal{G}f)_t| > \varepsilon\big) = \mathbb{P}\big(\mathcal{G}(f - p_n)_t > \varepsilon\big)$$

$$\le \mathbb{P}\left(\left|\int_0^t (p_n' - f')\,dM_u\right| > \varepsilon/2\right)$$

$$\le \frac{4\mathbb{E}\left[|\int_0^t (p_n' - f')\,dM_u|^2\right]}{\varepsilon^2} \qquad \text{(Chebyshev)}$$

$$= \frac{4\mathbb{E}\left[|\int_0^t (p_n' - f')^2\,d[M]_u|^2\right]}{\varepsilon^2}$$

$$\le \frac{4K}{n^2\varepsilon^2},$$

the last equality following from the isometry property, Corollary 4.14. Taking n sufficiently large, we may make this as small as we wish. $\qquad\square$

4.6.2 Differential notation

A shorthand for Itô's formula is to write it in *differential notation* as

$$df(X) = f'(X)dX + \tfrac{1}{2}f''(X)d[X].$$

To facilitate computations using Itô's formula (especially in the multidimensional case to follow) we make the following definition. For continuous semimartingales X and Y, let

(i) $$dX\,dY := d[X, Y] \quad (= dY\,dX).$$

Since

(ii) $$dX\,dA = 0 \text{ if } A \text{ is of finite variation,}$$

if Z is also a continuous semimartingale, we have

(iii) $$(dX\,dY)dZ = dX(dY\,dZ) = 0,$$

and similarly for higher-order multiples. If, in addition, $H, K \in \ell b\mathcal{P}$, recalling from Theorem 4.18 that

$$d[H \bullet X, K \bullet Y] = HK\,d[X, Y],$$

we have

(iv) $$(HdX)(KdY) = HKdXdY.$$

Properties (ii)–(iv) allow us to perform calculations easily, treating each term 'as if it were a scalar', but cancelling terms involving three or more differentials, or two differentials if at least one is of finite variation.

A simple example relates the quadratic variation of a continuous semimartingale to that of its local martingale part. If X has the decomposition $X = X_0 + M + A$, we have

$$(dX)^2 = (dM + dA)^2 = (dM)^2 + 2dMdA + (dA)^2 = (dM)^2.$$

We shall make free use of this type of differential notation in our calculations.

Using this formalism we can rewrite Itô's formula as

$$df(X) = f'(X)dX + \tfrac{1}{2}f''(X)(dX)^2.$$

This suggests one could base an alternative direct proof for Itô's formula on Taylor's theorem. The interested reader is referred to Karatzas and Shreve (1991) for a treatment along these lines.

Example 4.33 Let W be a standard Brownian motion relative to $(\Omega, \{\mathcal{F}_t\}, \mathcal{F}, \mathbb{P})$ and let $H \in \Pi(W)$. Then, for $t \geq 0$, we can define

$$\zeta_t := \int_0^t H_u \, dW_u - \tfrac{1}{2} \int_0^t H_u^2 \, ds.$$

Consider the process

$$Z_t := \exp(\zeta_t).$$

Applying Itô's formula to the semimartingale ζ, we have, for $f \in C^2$,

$$df(\zeta_t) = f'(\zeta_t)d\zeta_t + \tfrac{1}{2}f''(\zeta_t)(d\zeta_t)^2$$

where

$$(d\zeta_t)^2 = (H_t dW_t - \tfrac{1}{2}H_t^2 dt)^2 = H_t^2(dW_t)^2 = H_t^2 dt.$$

Taking $f(x) = \exp(x)$, we obtain

$$dZ_t = Z_t(H_t dW_t - \tfrac{1}{2}H_t^2 dt) + \tfrac{1}{2}Z_t H_t^2 dt$$
$$= Z_t H_t dW_t.$$

Noting $Z_0 = 1$ and writing this in integral form, we have shown that, for all $t \geq 0$, Z_t satisfies

$$Z_t = 1 + \int_0^t Z_u H_u \, dW_u. \tag{4.26}$$

In fact Z is the unique solution to (4.26). We prove this as follows. Suppose X is some other solution to (4.26) and define $Y_t := Z_t^{-1} = \exp(-\zeta_t)$. Apply Itô's formula to obtain

$$dY_t = -Y_t H_t dW_t + Y_t H_t^2 dt$$

and consider the process XY. From the integration by parts formula we have

$$d(X_t Y_t) = X_t dY_t + Y_t dX_t + d[X,Y]_t = X_t Y_t H_t^2 dt + d[X,Y]_t .$$

Further,
$$d[X,Y]_t = (X_t H_t dW_t)(-Y_t H_t dW_t) = -X_t Y_t H_t^2 dt ,$$

and so $d(X_t Y_t) = 0$,
$$X_t Y_t = X_0 Y_0 = 1 .$$

Thus, $X_t = Z_t$ a.s. for all $t \geq 0$.

Finally, observe from (4.26) that $Z_t = \exp(\zeta_t)$ is a local martingale. A natural question to ask is for which processes $H \in \Pi(W)$ is Z a true martingale? We will return to this question in connection with Girsanov's Theorem in the next chapter.

4.6.3 Multidimensional version of Itô's formula

We now state a version of Itô's formula for functions of several semimartingles. We will need this multidimensional version for later applications.

Definition 4.34   *A process* $X = (X^{(1)}, \dots, X^{(n)})$ *defined relative to* $(\Omega, \{\mathcal{F}_t\}, \mathcal{F}, \mathbb{P})$ *with values in* $\mathbb{R}^n$ *is called a continuous semimartingale if each coordinate process* $X^{(i)}$ *is a continuous semimartingale.*

Theorem 4.35 (Itô's formula)   *Let* $f : \mathbb{R}^n \to \mathbb{R}$ *be* $C^2(\mathbb{R}^n)$ *and let* $X = (X^{(1)}, \dots, X^{(n)})$ *be a continuous semimartingale in* $\mathbb{R}^n$. *Then, a.s.,* $f(X_t)$ *is a continuous semimartingale and*

$$f(X_t) = f(X_0) + \sum_{i=1}^{n} \int_0^t \frac{\partial f}{\partial x_i}(X_u) \, dX_u^{(i)}$$

$$+ \frac{1}{2} \sum_{i,j=1}^{n} \int_0^t \frac{\partial^2 f}{\partial x_i \partial x_j}(X_u) \, d[X^{(i)}, X^{(j)}]_u .$$

The proof goes through as for the one-dimensional case with only the notation being more difficult. We mention two extensions of Itô's formula which are not difficult to prove.

Extension 1: If some of the $X^{(i)}$ are of finite variation then the function f need only be of class C^1 in the corresponding coordinates. In particular, if X is a continuous semimartingale in $\mathbb{R}^1$, if A is a continuous process of finite variation, also in $\mathbb{R}^1$, and f is $C^{2,1}(\mathbb{R}, \mathbb{R})$ then

$$f(X_t, A_t) = f(X_0, A_0) + \int_0^t \frac{\partial f}{\partial x}(X_u, A_u)\, dX_u$$

$$+ \int_0^t \frac{\partial f}{\partial a}(X_u, A_u)\, dA_u + \frac{1}{2}\int_0^t \frac{\partial^2 f}{\partial x^2}(X_u, A_u)\, d[X]_u\,.$$

Example 4.36 (Doléan's exponential) Let X be a continuous semimartingale with $X_0 = 0$ and define

$$\mathcal{E}(X)_t = \exp(X_t - \tfrac{1}{2}[X]_t)\,.$$

Then $\mathcal{E}(X)_t$ is the unique (continuous) semimartingale Z such that

$$Z_t = 1 + \int_0^t Z_u\, dX_u\,. \tag{4.27}$$

Note that we proved a special case of this result in Example 4.33 where we took $X_t = (H \bullet W)_t$. This time, apply Itô's formula to the two-dimensional semimartingale $(X, [X])$ with $f(x, y) = \exp(x - \tfrac{1}{2}y)$ to obtain

$$d\mathcal{E}(X)_t = \mathcal{E}(X)_t dX_t - \tfrac{1}{2}\mathcal{E}(X)_t d[X]_t + \tfrac{1}{2}\mathcal{E}(X)_t d[X]_t$$

$$= \mathcal{E}(X)_t dX_t\,.$$

The uniqueness can be proved as in Example 4.33. The process $\mathcal{E}(X)$ is called the exponential of X and (4.27) is known as an exponential stochastic differential equation.

Extension 2: Itô's formula still holds if the function f is defined and in C^2 only on an open set and if X takes its values a.s. in this set. This means that we can apply Itô's formula to $\log(X_t)$, for example, if X_t is a strictly positive process.

Example 4.37 Let (X, Y) be a two-dimensional Brownian motion started at $(X_0, Y_0) \neq (0, 0)$, and set

$$V_t = X_t^2 + Y_t^2\,.$$

By Itô's formula

$$dV_t = 2(X_t dX_t + Y_t dY_t) + 2dt\,.$$

and so

$$d[V]_t = 4(X_t^2 + Y_t^2)dt = 4V_t dt.$$

Next define $T_n := \inf\{t > 0 : V_t < n^{-2}\}$ where we assume $V_0 > n^{-2}$, and consider the stopped process $M_t = \log(V_t^{T_n})$. For the stopped process V^{T_n}, which is a.s. strictly positive, we have

$$dV_t^{T_n} = 1(0, T_n]dV_t$$

and so, by Itô's formula applied to M,

$$
\begin{aligned}
dM_t &= 1(0, T_n]\left(M_t^{-1}dV_t - \tfrac{1}{2}M_t^{-2}d[V]_t\right) \\
&= 1(0, T_n]2V_t^{-1}(X_t dX_t + Y_t dY_t).
\end{aligned} \tag{4.28}
$$

From (4.28) we see that $M = \log(V^{T_n})$ defines a local martingale (Remark 4.22). We can use this to establish the following facts:

(i) two-dimensional Brownian motion hits any open set containing the origin with probability one;

(ii) two-dimensional Brownian motion hits the origin itself with probability zero.

To show this, define

$$S_N = \inf\{t > 0 : V_t > N^2\}$$

where N is chosen large enough so that $V_0 < N^2$. Now M^{S_N} is a bounded local martingale, hence a UI martingale, and so by the optional sampling theorem we conclude

$$\mathbb{E}\left[\log(V_{T_n \wedge S_N})\right] = \log(V_0). \tag{4.29}$$

We know from Theorem 2.8 that $S_N < \infty$ a.s., so (4.29) yields

$$\log(N^2)p_{n,N} + \log(n^{-2})(1 - p_{n,N}) = \log(V_0),$$

where $p_{n,N} = \mathbb{P}(S_N < T_n)$, and thus

$$p_{n,N} = \frac{\log(V_0) + 2\log(n)}{2\log(N) + 2\log(n)}.$$

Given any open set A containing the origin, the open ball $\{(x, y) : x^2 + y^2 < n^{-2}\}$ lies strictly within A for some n, hence by the continuity of Brownian motion the probability of hitting A is greater than $p_{n,N}$. Further,

$$\{T_n < \infty\} = \bigcup_N \{T_n < N\},$$

so the monotone convergence theorem implies that

$$\mathbb{P}(T_n < \infty) = \lim_{N \uparrow \infty} p_{n,N} = 1,$$

which establishes (i).

To prove (ii), note that, for $N > \sqrt{V_0}$, $n > 1/\sqrt{V_0}$,

$$\mathbb{P}(H_0 < \infty) = \lim_{N \uparrow \infty} \mathbb{P}(H_0 < S_N)$$
$$= \lim_{N \uparrow \infty} \lim_{n \uparrow \infty} \mathbb{P}(T_n < S_N) = 0.$$

4.6.4 Lévy's theorem

Lévy's theorem is a powerful and extremely useful result which shows that if X is a continuous local martingale with quadratic variation $[X]_t = t$, then X must be Brownian motion. Here we present the elegant proof of this result provided by Kunita and Watanabe (1967), a simple application of Itô's formula.

Theorem 4.38 Let X be some continuous d-dimensional local martingale adapted to the filtration $\{\mathcal{F}_t\}$. Then X is an $\{\mathcal{F}_t\}$ Brownian motion if and only if

$$[X^{(i)}, X^{(j)}]_t = \delta_{ij} t \qquad \text{a.s.}$$

for all i, j and t.

Proof: If X is a Brownian motion the result is merely a d-dimensional version of Corollary 3.81. The converse is a little more involved. Given an arbitrary $\lambda \in \mathbb{R}^d$, define the (complex-valued) continuous semimartingale $f(X_t, t)$ to be

$$f(X_t, t) := \exp(i\lambda \cdot X_t + \tfrac{1}{2}|\lambda|^2 t)$$
$$= \exp(\tfrac{1}{2}|\lambda|^2 t)\big(\cos(\lambda \cdot X_t) + i\sin(\lambda \cdot X_t)\big).$$

By Itô's formula,

$$df(X_t, t) = \sum_{k=1}^{d} \frac{\partial f(X_t, t)}{\partial X_t^{(k)}} dX_t^{(k)} + \frac{\partial f(X_t, t)}{\partial t} dt$$

$$+ \tfrac{1}{2} \sum_{k,\ell=1}^{d} \frac{\partial^2 f(X_t, t)}{\partial X_t^{(k)} \partial X_t^{(\ell)}} dX_t^{(k)} dX_t^{(\ell)}$$

$$= \sum_{k=1}^{d} i\lambda_k f(X_t, t) dX_t^{(k)} + \tfrac{1}{2}|\lambda|^2 f(X_t, t)\, dt$$

$$- \tfrac{1}{2} \sum_{k,\ell=1}^{d} \lambda_k \lambda_\ell \delta_{k\ell} f(X_t, t)\, dt$$

$$= \sum_{k=1}^{d} i\lambda_k f(X_t, t) dX_t^{(k)} .$$

Thus f is a local martingale. But, given any $T \geq 0$, f is bounded (in absolute value and thus in each component) on $[0, T]$ by $\exp(\tfrac{1}{2}|\lambda|^2 T)$ (which is its modulus at time T) so f is, in fact, a true martingale. Therefore, given any $s \leq t$,

$$\mathbb{E}[f(X_t, t)|\mathcal{F}_s] = f(X_s, s) \quad \text{a.s.}$$

which can be rewritten as

$$\mathbb{E}[\exp(i\lambda \cdot (X_t - X_s))|\mathcal{F}_s] = \exp(-\tfrac{1}{2}|\lambda|^2(t - s)) \quad \text{a.s.} \qquad (4.30)$$

Equation (4.30), which holds for all $\lambda \in \mathbb{R}^d$, shows that $X_t - X_s$ is independent of $\mathcal{F}_s$. Taking expectations in (4.30) yields

$$\mathbb{E}[\exp(i\lambda \cdot (X_t - X_s))] = \exp(-\tfrac{1}{2}|\lambda|^2(t - s)),$$

which is the characteristic function of a $N(0, t - s)$ random variable. Thus we have established the remaining requirement of Definition 2.4 for X to be a Brownian motion. $\qquad \square$

5

Girsanov and Martingale Representation

This chapter is devoted to two important results in the study of financial models, Girsanov's theorem and the martingale representation theorem. The common theme to these results is the study of change of probability measure.

Girsanov's theorem shows how the law of a semimartingale changes when the original probability measure $\mathbb{P}$ is replaced by some *equivalent* probability measure $\mathbb{Q}$. It turns out that a semimartingale remains a semimartingale and Girsanov's theorem relates the semimartingale decomposition under each of the measures $\mathbb{P}$ and $\mathbb{Q}$ to the *Radon–Nikodým derivative of $\mathbb{Q}$ with respect to $\mathbb{P}$*. This is useful when pricing derivatives because, as we saw in Chapter 1, it is convenient to work in some probability measure where the assets of the economy are martingales.

The martingale representation theorem gives necessary and sufficient conditions for a family of (local) martingales to be representable as stochastic integrals with respect to some other (finite) set of (local) martingales. This result is important in a financial context because it provides conditions under which a model is complete.

5.1 EQUIVALENT PROBABILITY MEASURES AND THE RADON-NIKODÝM DERIVATIVE

In this section we collect together some basic results concerning equivalent probability measures which will be used extensively in later sections, particularly the discussion of Girsanov's theorem. Recall that we met the idea of equivalent measures in the discrete setting of Chapter 1.

5.1.1 Basic results and properties

Let us begin our discussion with a formal definition.

Financial Derivatives in Theory and Practice Revised Edition. P. J. Hunt and J. E. Kennedy
© 2004 John Wiley & Sons, Ltd   ISBNs: 0-470-86358-7 (HB); 0-470-86359-5 (PB)

Definition 5.1 *Let $\mathbb{P}$ and $\mathbb{Q}$ be two probability measures on the measurable space $(\Omega, \mathcal{F})$. We say that $\mathbb{Q}$ is* absolutely continuous *with respect to $\mathbb{P}$ (with respect to the σ-algebra $\mathcal{F}$), written $\mathbb{Q} \ll \mathbb{P}$, if for all $F \in \mathcal{F}$,*

$$\mathbb{P}(F) = 0 \;\Rightarrow\; \mathbb{Q}(F) = 0.$$

If both $\mathbb{P} \ll \mathbb{Q}$ and $\mathbb{Q} \ll \mathbb{P}$ then $\mathbb{P}$ and $\mathbb{Q}$ are said to be equivalent *(with respect to $\mathcal{F}$), written $\mathbb{P} \sim \mathbb{Q}$.*

Remark 5.2: The measures $\mathbb{P}$ and $\mathbb{Q}$ could, and often will, be defined on some σ-algebra $\mathcal{G}$ strictly larger than $\mathcal{F}$. Equivalence of $\mathbb{P}$ and $\mathbb{Q}$ with respect to $\mathcal{F} \subset \mathcal{G}$ does not imply equivalence with respect to $\mathcal{G}$, but the converse is trivially true.

If two measures $\mathbb{P}$ and $\mathbb{Q}$ are equivalent we can freely talk about results holding almost surely without specifying which of the two measures we are working with, and this we shall do throughout. Many of the results which we are about to present for equivalent measures also hold when the measures are not equivalent but when one is absolutely continuous with respect to the other. In this latter case it is important to qualify an almost sure statement to include the measure under which the result is true. You are referred to Protter (1990) and Revuz and Yor (1991) for the more general statement of those results.

Equivalent probability measures can be related to each other via a positive random variable, and this is the content of the Radon–Nikodým theorem. This is a classical result in measure theory and we state it here without proof (which can be found, for example, in Billingsley (1986)).

Theorem 5.3 *Suppose $\mathbb{P}$ and $\mathbb{Q}$ are equivalent probability measures on the space $(\Omega, \mathcal{F})$. Then there exists a strictly positive random variable $\rho \in \mathcal{L}^1(\Omega, \mathcal{F}, \mathbb{P})$ which is a.s. unique and is such that, for all $F \in \mathcal{F}$,*

$$\mathbb{Q}(F) = \int_F \rho \, d\mathbb{P} = \mathbb{E}_{\mathbb{P}}[\rho \mathbb{1}_F]. \tag{5.1}$$

Further, $\rho^{-1} \in \mathcal{L}^1(\Omega, \mathcal{F}, \mathbb{Q})$ and

$$\mathbb{P}(F) = \int_F \rho^{-1} \, d\mathbb{Q} = \mathbb{E}_{\mathbb{Q}}[\rho^{-1} \mathbb{1}_F]. \tag{5.2}$$

Conversely, given any strictly positive random variable $\rho \in \mathcal{L}^1(\Omega, \mathcal{F}, \mathbb{P})$ such that $\mathbb{E}_{\mathbb{P}}[\rho] = 1$, there exists some unique measure $\mathbb{Q} \sim \mathbb{P}$ for which equations (5.1) and (5.2) hold.

This result allows us to make the following definition.

Definition 5.4 *The random variable ρ of Theorem 5.3 is referred to as (a version of) the Radon–Nikodým derivative of $\mathbb{Q}$ relative to $\mathbb{P}$ on $(\Omega, \mathcal{F})$. Noting that any two versions agree a.s., we write*

$$\frac{d\mathbb{Q}}{d\mathbb{P}}\bigg|_{\mathcal{F}} = \rho \quad a.s.$$

Similarly, we have

$$\frac{d\mathbb{P}}{d\mathbb{Q}}\bigg|_{\mathcal{F}} = \rho^{-1} \quad a.s.$$

Remark 5.5: These last two results take on a very concrete form in the case when Ω is a finite set, $\{\omega_i : i = 1, 2, \ldots, n\}$. In this case we may as well assume (the general case being a trivial extension) that $\mathcal{F}$ comprises all subsets of Ω. Whenever $\mathbb{P}(\omega_i) > 0$ the random variable ρ is now just

$$\rho(\omega_i) = \frac{\mathbb{Q}(\omega_i)}{\mathbb{P}(\omega_i)},$$

which is finite and strictly positive. For all other ω_i we can set $\rho(\omega_i)$ to be any value, zero for example. These latter ω_i have probability zero (under both $\mathbb{P}$ and $\mathbb{Q}$) so the ambiguity in this case is immaterial.

Remark 5.6: Suppose we only know that $\mathbb{Q}$ is absolutely continuous with respect to $\mathbb{P}$. In this case the Radon–Nikodým theorem states that there exists a positive random variable $\rho \in \mathcal{L}^1(\Omega, \mathcal{F}, \mathbb{P})$ such that (5.1) holds ($\mathbb{P}$-a.s.). Since we now only know that $\rho \geq 0$ ($\mathbb{P}$-a.s.) we cannot necessarily recover $\mathbb{P}$ from $\mathbb{Q}$ and ρ, i.e. (5.2) does not hold in general. To see this, consider $F = \{\omega \in \Omega : \rho(\omega) = 0\}$. From (5.1)

$$\mathbb{Q}(F) = \mathbb{E}_{\mathbb{P}}[\rho \mathbf{1}_F] = 0,$$

and so the random variable ρ is strictly positive $\mathbb{Q}$-a.s. and ρ^{-1} is well-defined under $\mathbb{Q}$. But if $\mathbb{Q}$ is absolutely continuous with respect to $\mathbb{P}$ but not equivalent then $\mathbb{P}(F) > 0$. If, in addition, (5.2) were to hold then we would have

$$\mathbb{P}(F) = \mathbb{E}_{\mathbb{Q}}[\rho^{-1} \mathbf{1}_F] = 0,$$

a contradiction.

A straightforward application of (the usual) monotone class arguments (see, for example, Rogers and Williams (1994)) extends (5.1) and (5.2) to corresponding statements for random variables.

Proposition 5.7 *Let X be some random variable defined on $(\Omega, \mathcal{F})$ and suppose that $\mathbb{P} \sim \mathbb{Q}$ with respect to $\mathcal{F}$. Then, in each case provided the expectation exists, expectations under $\mathbb{P}$ and $\mathbb{Q}$ are related via*

$$\mathbb{E}_{\mathbb{Q}}[X] = \mathbb{E}_{\mathbb{P}}[\rho X]$$

and

$$\mathbb{E}_{\mathbb{P}}[X] = \mathbb{E}_{\mathbb{Q}}[\rho^{-1} X],$$

where

$$\rho = \left.\frac{d\mathbb{Q}}{d\mathbb{P}}\right|_{\mathcal{F}}.$$

Our intention to study equivalent measures on filtered probability spaces leads us to ask what happens when we start with two measures $\mathbb{P} \sim \mathbb{Q}$ with respect to some σ-algebra $\mathcal{F}$ and consider the σ-algebra $\mathcal{G} \subset \mathcal{F}$. We find the following result which extends Theorem 5.3 and Proposition 5.7.

Theorem 5.8 *Let $\mathbb{P}$ and $\mathbb{Q}$ be two probability measures on the space $(\Omega, \mathcal{F})$ and suppose $\mathcal{G} \subset \mathcal{F}$ is some σ-algebra. If $\mathbb{P} \sim \mathbb{Q}$ with respect to $\mathcal{F}$ then $\mathbb{P} \sim \mathbb{Q}$ with respect to $\mathcal{G}$ and*

$$\left.\frac{d\mathbb{Q}}{d\mathbb{P}}\right|_{\mathcal{G}} = \mathbb{E}_{\mathbb{P}}[\rho | \mathcal{G}] \quad \text{a.s.}$$

$$\left.\frac{d\mathbb{P}}{d\mathbb{Q}}\right|_{\mathcal{G}} = \mathbb{E}_{\mathbb{Q}}[\rho^{-1} | \mathcal{G}] \quad \text{a.s.}$$

where

$$\rho = \left.\frac{d\mathbb{Q}}{d\mathbb{P}}\right|_{\mathcal{F}} \quad \text{a.s.}$$

Furthermore, given any $X \in \mathcal{L}^1(\Omega, \mathcal{G}, \mathbb{Q})$, $Y \in \mathcal{L}^1(\Omega, \mathcal{G}, \mathbb{P})$,

$$\mathbb{E}_{\mathbb{Q}}[X] = \mathbb{E}_{\mathbb{P}}[\hat{\rho} X]$$
$$\mathbb{E}_{\mathbb{P}}[Y] = \mathbb{E}_{\mathbb{Q}}[\hat{\rho}^{-1} Y]$$

where

$$\hat{\rho} = \left.\frac{d\mathbb{Q}}{d\mathbb{P}}\right|_{\mathcal{G}}.$$

Proof: That $\mathbb{P}$ and $\mathbb{Q}$ are equivalent with respect to $\mathcal{G}$ is immediate. From (5.1) and the tower property we have, for all $G \in \mathcal{G} \subset \mathcal{F}$,

$$\mathbb{Q}(G) = \mathbb{E}_{\mathbb{P}}[\rho 1_G] = \mathbb{E}_{\mathbb{P}}\big[\mathbb{E}_{\mathbb{P}}[\rho \mid \mathcal{G}] 1_G\big].$$

Since $\mathbb{E}_{\mathbb{P}}[\rho | \mathcal{G}]$ is a $\mathcal{G}$-measurable random variable we conclude from Theorem 5.3 (applied to $\mathcal{G}$) that

$$\left.\frac{d\mathbb{Q}}{d\mathbb{P}}\right|_{\mathcal{G}} = \mathbb{E}_{\mathbb{P}}[\rho | \mathcal{G}] \quad \text{a.s.}$$

The proof of the rest of the theorem follows similarly. $\qquad\square$

Corollary 5.9 *Let* $\mathbb{P}, \mathbb{Q}$ *and* ρ *be as in Theorem 5.8. Given any* $X \in \mathcal{L}^1(\Omega, \mathcal{F}, \mathbb{Q})$, *and a* σ*-algebra* $\mathcal{G} \subset \mathcal{F}$,

$$\mathbb{E}_{\mathbb{Q}}[X|\mathcal{G}] = \frac{\mathbb{E}_{\mathbb{P}}[\rho X|\mathcal{G}]}{\mathbb{E}_{\mathbb{P}}[\rho|\mathcal{G}]} \quad a.s. \tag{5.3}$$

Proof: Given any $G \in \mathcal{G}$, since the right-hand side of (5.3) is $\mathcal{G}$-measurable, it follows from Theorem 5.8 that

$$\mathbb{E}_{\mathbb{Q}}\left[1_G \frac{\mathbb{E}_{\mathbb{P}}[\rho X|\mathcal{G}]}{\mathbb{E}_{\mathbb{P}}[\rho|\mathcal{G}]}\right] = \mathbb{E}_{\mathbb{P}}\left[1_G \frac{\mathbb{E}_{\mathbb{P}}[\rho X|\mathcal{G}]}{\mathbb{E}_{\mathbb{P}}[\rho|\mathcal{G}]}\left(\frac{d\mathbb{Q}}{d\mathbb{P}}\bigg|_{\mathcal{G}}\right)\right]$$
$$= \mathbb{E}_{\mathbb{P}}\left[1_G \mathbb{E}_{\mathbb{P}}[\rho X|\mathcal{G}]\right]$$
$$= \mathbb{E}_{\mathbb{P}}\left[1_G \rho X\right]$$
$$= \mathbb{E}_{\mathbb{Q}}\left[1_G X\right].$$

Thus, by the definition of conditional expectation, the right-hand side of (5.3) is indeed $\mathbb{E}_{\mathbb{Q}}[X|\mathcal{G}]$. $\square$

5.1.2 Equivalent and locally equivalent measures on a filtered space

It is interesting and, as we shall see shortly, very important to consider how Theorem 5.3 and Proposition 5.7 can be refined when the space $(\Omega, \mathcal{F})$ is further endowed with a filtration $\{\mathcal{F}_t\}$. We have already provided the basis for this extension with Theorem 5.8 and Corollary 5.9. Our first result extends Theorem 5.3 and motivates the rest of this section.

Theorem 5.10 *Let* $\mathbb{P}$ *and* $\mathbb{Q}$ *be probability measures on the filtered space* $(\Omega, \{\mathcal{F}_t\}, \mathcal{F})$, *and suppose that* $\mathbb{Q} \sim \mathbb{P}$ *with respect to* $\mathcal{F}$. *Suppose further that the filtration* $\{\mathcal{F}_t\}$ *satisfies the usual conditions. Then, for all* $t \in [0, \infty]$,

$$\rho_t := \frac{d\mathbb{Q}}{d\mathbb{P}}\bigg|_{\mathcal{F}_t}$$

defines an a.s. strictly positive UI $\{\mathcal{F}_t\}$ *martingale under* $\mathbb{P}$.

Proof: Since $\mathbb{P}$ and $\mathbb{Q}$ are equivalent with respect to $\mathcal{F}$, there exists a strictly positive random variable $\rho \in \mathcal{L}^1(\Omega, \mathcal{F}, \mathbb{P})$ such that $\rho = \frac{d\mathbb{Q}}{d\mathbb{P}}\big|_{\mathcal{F}}$. Define the uniformly integrable martingale $\rho_t := \mathbb{E}_{\mathbb{P}}[\rho|\mathcal{F}_t]$, $t \in [0, \infty]$. It follows from Theorem 5.8 that $\rho_t = \frac{d\mathbb{Q}}{d\mathbb{P}}\big|_{\mathcal{F}_t}$ a.s., so all that remains is to prove that the martingale ρ is a.s. strictly positive.

Since $\{\mathcal{F}_t\}$ satisfies the usual conditions ρ has a càdlàg version (Theorem 3.8). Given any fixed $t \geq 0$ it is clear that ρ_t is strictly positive, being the conditional expectation of a strictly positive random variable. Combining these two observations, we see that

$$\mathbb{P}(\rho_t > 0, \text{ all } t \geq 0) = \mathbb{P}\left(\bigcup_{q \in \mathbb{Q}^+} \rho_q > 0\right) = 1.$$

    $\square$

Remark 5.11: It is clear from the above proof that when the usual conditions do not apply, ρ_t is a.s. strictly positive for any given $t \geq 0$. However, $\rho(\omega)$ will not necessarily be a.s. strictly positive for all t simultaneously.

The martingale ρ plays a crucial role in Girsanov's theorem, Theorem 5.20, a result which extends Proposition 5.7 to processes and explains how semimartingales behave under a change of probability measure. This will be discussed in Section 5.2.

The remainder of this section will be devoted to proving various converse statements to Theorem 5.10. This is important because it will enable us to start with some probability measure $\mathbb{P}$ and construct some other equivalent measure $\mathbb{Q}$ by using a $\mathbb{P}$ martingale.

The converse to Theorem 5.10 is not as straightforward as one might at first hope. Indeed it is immediate that if $\mathcal{F}$ is richer than $\mathcal{F}_\infty$ it is not possible to recover $\frac{d\mathbb{Q}}{d\mathbb{P}}\big|_{\mathcal{F}}$ from the martingale ρ. The nearest thing to a converse for Theorem 5.10 is the following result, a consequence of Theorems 5.3 and 5.10.

Theorem 5.12 *Let $\mathbb{P}$ be a probability measure on $(\Omega, \{\mathcal{F}_t\}, \mathcal{F})$. Given any a.s. strictly positive UI $(\{\mathcal{F}_t\}, \mathbb{P})$ martingale ρ with $\mathbb{E}_\mathbb{P}[\rho_\infty] = 1$,*

$$\frac{d\mathbb{Q}}{d\mathbb{P}}\bigg|_{\mathcal{F}_\infty} := \rho_\infty$$

defines (via equation (5.1)) a measure $\mathbb{Q} \sim \mathbb{P}$ with respect to $\mathcal{F}_\infty$.

So given a strictly positive UI $(\{\mathcal{F}_t\}, \mathbb{P})$ martingale it is possible to use this to define some other measure $\mathbb{Q} \sim \mathbb{P}$ with respect to $\mathcal{F}_\infty$. But what if ρ is a strictly positive martingale but not a UI martingale (which is a much more restrictive requirement)? In this case the situation is somewhat more subtle, as the following results demonstrate. Note, however, that if we are working only on some finite time horizon $[0, T]$ then any càdlàg martingale is UI so the complications below (with regard to completing filtrations in particular) do not arise.

Definition 5.13 *Two probability measures $\mathbb{P}$ and $\mathbb{Q}$ defined on the filtered space $(\Omega, \{\mathcal{F}_t\}, \mathcal{F})$ are said to be locally equivalent (with respect to the filtration $\{\mathcal{F}_t\}$) if, for all $t \geq 0$, $\mathbb{P} \sim \mathbb{Q}$ with respect to $\mathcal{F}_t$, i.e. for all $F \in \bigcup_{t \in [0,\infty)} \mathcal{F}_t$,*

$$\mathbb{P}(F) = 0 \iff \mathbb{Q}(F) = 0.$$

Theorem 5.14 *Let $\mathbb{P}$ and $\mathbb{Q}$ be probability measures on $(\Omega, \{\mathcal{F}_t\}, \mathcal{F})$, and suppose that $\mathbb{Q}$ is locally equivalent to $\mathbb{P}$ with respect to $\{\mathcal{F}_t\}$. Then*

$$\rho_t := \frac{d\mathbb{Q}}{d\mathbb{P}}\bigg|_{\mathcal{F}_t}$$

defines an $\{\mathcal{F}_t\}$ martingale under $\mathbb{P}$ which is a.s. strictly positive for each $t \geq 0$. If, in addition, $\{\mathcal{F}_t\}$ satisfies the usual conditions under $\mathbb{P}$ then ρ has an a.s. strictly positive version.

Conversely, if ρ is a strictly positive martingale on the probability space $(\Omega, \{\mathcal{F}_t\}, \mathcal{F}, \mathbb{P})$, then for each $T \geq 0$,

$$\left.\frac{d\mathbb{Q}^T}{d\mathbb{P}}\right|_{\mathcal{F}_T} = \rho_T$$

defines a probability measure $\mathbb{Q}^T \sim \mathbb{P}$ on $(\Omega, \mathcal{F}_T)$. Furthermore, the family $\{\mathbb{Q}^T : T \geq 0\}$ is consistent, meaning that if $T \geq S$ and $F \in \mathcal{F}_S$, then $\mathbb{Q}^T(F) = \mathbb{Q}^S(F)$.

Proof: This result is an immediate consequence of Theorems 5.10 and 5.3. $\square$

The natural question to ask now, and the final one we shall address in this section, is: given a strictly positive $(\{\mathcal{F}_t\}, \mathbb{P})$ martingale, as in Theorem 5.14, does there exist some $\mathbb{Q}$ defined on $\mathcal{F}_\infty$ such that $\mathbb{Q}(F) = \mathbb{Q}^T(F)$ for all $F \in \mathcal{F}_T$, all $T > 0$, i.e. does ρ define some probability measure $\mathbb{Q}$ on $\mathcal{F}_\infty$ which is locally equivalent to $\mathbb{P}$? In general the answer to this question is no, as Example 5.28 below demonstrates. It does, however, hold in one important special case. This result follows from Theorem 5.14 and the Daniel–Kolmogorov consistency theorem. For a detailed proof see Karatzas and Shreve (1991, p. 192).

Theorem 5.15 Suppose ρ is a strictly positive martingale on the probability space $(\Omega, \{\mathcal{F}_t\}, \mathcal{F}, \mathbb{P})$ and that $\mathcal{F}_t^\circ = \sigma(X_u : u \leq t)$ for some process X (note crucially that the filtration is not complete). Then there exists some probability measure $\mathbb{Q}$ defined on $\mathcal{F}_\infty^\circ$ which is locally equivalent to $\mathbb{P}$ with respect to $\{\mathcal{F}_t^\circ\}$ and is such that

$$\left.\frac{d\mathbb{Q}}{d\mathbb{P}}\right|_{\mathcal{F}_t^\circ} = \rho_t \quad a.s.$$

5.1.3 Novikov's condition

We will see, in Section 5.2, how the law of a $\mathbb{P}$-continuous semimartingale changes under a locally equivalent measure $\mathbb{Q}$. Often in practice we find ourselves in some measure $\mathbb{P}$ with a given continuous semimartingale and we want to find some other measure $\mathbb{Q}$ under which this continuous semimartingale has particular properties. For example, in a financial context it is important to find a measure under which a continuous semimartingale becomes a (local) martingale. In such situations Girsanov's theorem usually gives enough information to identify a candidate for the Radon–Nikodým process ρ which connects $\mathbb{P}$ and $\mathbb{Q}$. It is then obvious that the candidate ρ is a *local* martingale under $\mathbb{P}$ but not that it is a true martingale. A non-negative local martingale is always a supermartingale (a consequence of Fatou's lemma) and thus it is immediate that it is a martingale if and only if

$$\mathbb{E}_{\mathbb{P}}[\rho_t] = 1, \tag{5.4}$$

for all $t \geq 0$. It is useful, therefore, to have criteria which are sufficient to guarantee that (5.4) holds. One such sufficient condition is known as Novikov's condition. For a proof and a discussion of an alternative criterion the reader is referred to Chapter VIII of Revuz and Yor (1991).

Theorem 5.16 (Novikov) *Let $(\Omega, \{\mathcal{F}_t\}, \mathcal{F}, \mathbb{P})$ be a filtered probability space supporting a continuous local martingale X and a strictly positive random variable $\rho_0 \in \mathcal{L}^1(\Omega, \mathcal{F}_0, \mathbb{P})$, and define $\rho = \rho_0 \mathcal{E}(X)$. If, for all $t \geq 0$,*

$$\mathbb{E}\big[\exp(\tfrac{1}{2}[X]_t)\big] < \infty,$$

then $\mathbb{E}[\rho_t] = \mathbb{E}[\rho_0]$ for all $t \geq 0$ and thus ρ is a martingale.

This result only applies to martingales ρ of the form $\rho = \rho_0 \mathcal{E}(X)$, for some X. But this class includes all strictly positive continuous martingales, as the following lemma (to which we also appeal later) summarizes.

Lemma 5.17 *Suppose $\mathbb{Q}$ is locally equivalent to $\mathbb{P}$ with respect to $\{\mathcal{F}_t\}$ and that the strictly positive $\mathbb{P}$ martingale $\rho_t = \frac{d\mathbb{Q}}{d\mathbb{P}}\big|_{\mathcal{F}_t}$ has a continuous version. Then*

$$X_t := \int_0^t \rho_u^{-1} d\rho_u$$

is the unique continuous local martingale such that

$$\rho_t = \rho_0 \mathcal{E}(X_t) := \rho_0 \exp(X_t - \tfrac{1}{2}[X]_t). \tag{5.5}$$

Moreover,

$$\frac{d\mathbb{P}}{d\mathbb{Q}}\bigg|_{\mathcal{F}_t} = \rho_0^{-1} \mathcal{E}(-X_t)$$

Proof: The result follows by applying Itô's formula to (5.5). $\qquad\square$

Example 5.18 Suppose W is a d-dimensional Brownian motion on the filtered probability space $(\Omega, \{\mathcal{F}_t\}, \mathcal{F}, \mathbb{P})$ which satisfies the usual conditions and let $C \in \Pi(W)$ (the d-dimensional version of $\Pi(W)$ introduced in Chapter 4). Then $X = C \bullet W$ defines an $(\{\mathcal{F}_t\}, \mathbb{P})$ local martingale. If, in addition, $\mathbb{E}_{\mathbb{P}}[\frac{1}{2} \int_0^T |C_u|^2 \, du] < \infty$ then

$$\rho_t := \mathcal{E}\left(\int_0^t C_u \cdot dW_u\right)$$

defines an $(\{\mathcal{F}_t\}_{t \leq T}, \mathbb{P})$ martingale.

5.2 GIRSANOV'S THEOREM

We have previously stated that a semimartingale remains a semimartingale under an equivalent change of probability measure. This result is known as Girsanov's theorem, a result which also establishes explicitly the link between the decomposition of an arbitrary semimartingale X under the two equivalent measures $\mathbb{P}$ and $\mathbb{Q}$ and the Radon–Nikodým derivative process $\rho_t = \frac{d\mathbb{Q}}{d\mathbb{P}}\big|_{\mathcal{F}_t}$. Girsanov's theorem and its proof are the subject of this section.

We have in this book only developed results for continuous semimartingales and so here we study Girsanov's theorem in a restricted context. In particular, we will need to assume that the Radon–Nikodým process connecting the two measures is itself continuous. Nevertheless the results we obtain cover many of the most important applications and certainly everything needed for this book. As in Section 5.1, we will state all the results here for (locally) equivalent measures, although they do generalize to the case when we only have (local) absolute continuity of one measure with respect to the other.

The technique of studying problems using change of measure is unique to stochastic calculus, having no counterpart in the classical theory of integration, and plays a surprisingly important role in further developments of the theory. The reader will briefly be introduced to one significant use of the idea when we discuss stochastic differential equations driven by Brownian motion in the Chapter 6.

5.2.1 Girsanov's theorem for continuous semimartingales

Let $\mathbb{Q}$ be a probability measure on $(\Omega, \mathcal{F})$ locally equivalent to $\mathbb{P}$ with respect to $\{\mathcal{F}_t\}$. Clearly a process of finite variation under $\mathbb{P}$ is also a process of finite variation under $\mathbb{Q}$. However, local martingales may not have the martingale property under a change of measure and the first part of Girsanov's theorem gives the decomposition of a continuous local martingale under $\mathbb{P}$ as a continuous semimartingale under $\mathbb{Q}$. The general decomposition then follows.

We shall need the following lemma which is a step in the right direction.

Lemma 5.19 *Suppose $\mathbb{Q}$ is locally equivalent to $\mathbb{P}$ with respect to the filtration $\{\mathcal{F}_t\}$ and let*

$$\rho_t = \frac{d\mathbb{Q}}{d\mathbb{P}}\bigg|_{\mathcal{F}_t}.$$

Then:

(i) *M is an $(\{\mathcal{F}_t\}, \mathbb{Q})$ martingale $\Leftrightarrow \rho M$ is an $(\{\mathcal{F}_t\}, \mathbb{P})$ martingale;*

(ii) *M is an $(\{\mathcal{F}_t\}, \mathbb{Q})$ local martingale $\Leftrightarrow \rho M$ is an $(\{\mathcal{F}_t\}, \mathbb{P})$ local martingale.*

Proof: To establish (i), we must check that conditions (M.i)–(M.iii) of Definition 3.1 are satisfied. The adaptedness property (M.i) is immediate. Conditions (M.ii) and (M.iii) follow from Corollary 5.9.

Part (ii) is now an exercise in localization of part (i) and is left to the reader.
□

Theorem 5.20 (Girsanov's theorem) *Let $\mathbb{Q}$ be locally equivalent to $\mathbb{P}$ with respect to $\{\mathcal{F}_t\}$ and suppose the martingale ρ satisfying $\rho_t := \frac{d\mathbb{Q}}{d\mathbb{P}}\big|_{\mathcal{F}_t}$ has a continuous version. If M is a continuous local martingale under $\mathbb{P}$, then*

$$\tilde{M}_t := M_t - \int_0^t \rho_u^{-1} d[\rho, M]_u$$

is a continuous local martingale under $\mathbb{Q}$ and

$$[M]_t = [\tilde{M}]_t.$$

More generally, if Y is a continuous semimartingale under $\mathbb{P}$ with canonical decomposition

$$Y_t = Y_0 + M_t + A_t$$

where M is a $\mathbb{P}$-local martingale and the process A is of finite variation, then

$$Y_t = Y_0 + \left(M_t - \int_0^t \rho_u^{-1} d[\rho, M]_u \right) + \left(\int_0^t \rho_u^{-1} d[\rho, M]_u + A_t \right)$$

is the canonical decomposition of Y under $\mathbb{Q}$.

Proof: The second statement is an immediate consequence of the first, which we now prove. Under $\mathbb{Q}$ and, since $\mathbb{P}$ is locally equivalent to $\mathbb{Q}$, also under $\mathbb{P}$, ρ is a.s. strictly positive and continuous (the strict positivity follows as in the proof of Theorem 5.10). Thus the process ρ^{-1} is locally bounded and predictable and the integral $\int_0^t \rho_u^{-1} d[\rho, M]_u$ exists and is a well-defined, continuous finite variation process under both $\mathbb{P}$ and $\mathbb{Q}$. Clearly $\tilde{M}_t$ is a continuous semimartingale under $\mathbb{P}$.

Appealing to Lemma 5.19, to show $\tilde{M}_t$ is a $\mathbb{Q}$ local martingale we need only prove that $\rho_t \tilde{M}_t$ is a $\mathbb{P}$ local martingale. First note that, since M and $\tilde{M}$ differ only by a finite variation process, the quadratic variation processes for M and $\tilde{M}$ agree. Further, since the covariation of a process of bounded variation with a semimartingale is identically zero, we have

$$[\rho, M]_t = [\rho, \tilde{M}]_t.$$

Since $\rho_t \tilde{M}_t$ is the product of two continuous $\mathbb{P}$-semimartingales, we can apply integration by parts (Theorem 4.28) to obtain

$$\rho_t \tilde{M}_t = \rho_0 \tilde{M}_0 + \int_0^t \tilde{M}_u d\rho_u + \int_0^t \rho_u d\tilde{M}_u + [\rho, \tilde{M}]_t$$

$$= \rho_0 M_0 + \int_0^t \tilde{M}_u d\rho_u + \int_0^t \rho_u dM_u - [\rho, M]_t + [\rho, \tilde{M}]_t$$

$$= \rho_0 M_0 + \int_0^t \tilde{M}_u d\rho_u + \int_0^t \rho_u dM_u.$$

Thus $\rho \tilde{M}$ is a sum of terms each of which is a stochastic integral with respect to a $\mathbb{P}$ local martingale and so itself is a $\mathbb{P}$ local martingale, and we are done. $\square$

Remark 5.21: The hypothesis that $\mathbb{Q}$ is locally equivalent to $\mathbb{P}$ in Girsanov's theorem can be relaxed. If we replace it by the requirement that $\mathbb{Q}$ be locally absolutely continuous with respect to $\mathbb{P}$, the above results remain true as stated.

Because the Radon–Nikodým process is assumed to be continuous, Lemma 5.17 shows that it can be written in exponential form. This yields the following alternative presentation of Girsanov's theorem in this case.

Corollary 5.22 *Suppose $\mathbb{Q}$ is locally equivalent to $\mathbb{P}$ with respect to $\{\mathcal{F}_t\}$ and that the strictly positive $\mathbb{P}$ martingale $\rho_t = \frac{d\mathbb{Q}}{d\mathbb{P}}\big|_{\mathcal{F}_t}$ has a continuous version. Let*

$$X_t := \int_0^t \rho_u^{-1} d\rho_u$$

as in Lemma 5.17. If Y is a continuous semimartingale under $\mathbb{P}$ with canonical decomposition $Y = Y_0 + M + A$, then Y has the canonical decomposition

$$Y_t = Y_0 + \left(M_t - [M, X]_t \right) + \left([M, X]_t + A_t \right)$$

under $\mathbb{Q}$.

5.2.2 Girsanov's theorem for Brownian motion

The building block for most continuous state-space models, and all those in this book, is Brownian motion. It is valuable to study Girsanov's theorem in the context of a Brownian motion where the extra structure means that much more can be said, beginning with the following simple corollary of Corollary 5.22.

Corollary 5.23 *Under the assumptions of Corollary 5.22, if W is an $(\{\mathcal{F}_t\}, \mathbb{P})$ Brownian motion then*

$$\widetilde{W} := W - [W, X]$$

is an $(\{\mathcal{F}_t\}, \mathbb{Q})$ Brownian motion.

Proof: From Corollary 5.22, $\widetilde{W}$ is a local martingale under $\mathbb{Q}$. That $\widetilde{W}$ is a Brownian motion now follows from Lévy's theorem (Theorem 4.38) since $[W]_t = [\widetilde{W}]_t = t$. $\square$

An essential ingredient of the above corollary is the assumption that the Radon–Nikodým process ρ be continuous. Subject to this assumption, we have identified a simple but significant fact and one we shall make repeated use of in later chapters: under a locally equivalent change of measure a Brownian motion remains a Brownian motion apart from the addition of a finite variation process.

In the context of Brownian motion, the assumption that ρ is continuous is not as restrictive as it may at first appear. This a consequence of Theorem 5.49, the martingale representation theorem for Brownian motion, which is discussed in Section 5.3.2. An abridged version of that result is as follows.

Theorem   Let $(\Omega, \{\mathcal{F}_t\}, \mathcal{F}, \mathbb{P})$ be a probability space supporting a d-dimensional Brownian motion W and let $\{\mathcal{F}_t^W\}$ be the augmented natural filtration generated by W. Then any local martingale N adapted to $\{\mathcal{F}_t^W\}$ can be written in the form

$$N_t = N_0 + \int_0^t H_u \cdot dW_u,$$

for some $\{\mathcal{F}_t^W\}$-predictable $H \in \Pi(W)$.

As a consequence of this theorem, if $\{\mathcal{F}_t\} = \{\mathcal{F}_t^W\}$ then any local martingale, in particular the Radon–Nikodým process, automatically has a continuous version (since the stochastic integral above is continuous) and thus satisfies the hypothesis of Corollary 5.23.

In the case when we have local equivalence between $\mathbb{P}$ and $\mathbb{Q}$ and we are again working with the augmented Brownian filtration, we can apply the martingale representation theorem to be more explicit about the form of the Radon–Nikodým process and the finite variation process appearing in Corollary 5.23.

Theorem 5.24   Let W be a d-dimensional $(\{\mathcal{F}_t\}, \mathbb{P})$ Brownian motion and let $\{\mathcal{F}_t^W\}$ be the filtration generated by W, augmented to satisfy the usual conditions. Suppose $\mathbb{Q}$ is a probability measure locally equivalent to $\mathbb{P}$ with respect to $\{\mathcal{F}_t^W\}$. Then there exists an $\{\mathcal{F}_t^W\}$ predictable $\mathbb{R}^d$-valued process C such that

$$\rho_t := \left.\frac{d\mathbb{Q}}{d\mathbb{P}}\right|_{\mathcal{F}_t^W} = \exp\left(\int_0^t C_u \cdot dW_u - \tfrac{1}{2}\int_0^t |C_u|^2 du\right). \qquad (5.6)$$

Conversely, if ρ is a strictly positive $\{\mathcal{F}_t^W : 0 \le t \le T\}$ martingale, for some $T \in [0, \infty]$, with $\mathbb{E}_\mathbb{P}[\rho_T] = 1$, then ρ has the representation in (5.6) and defines a measure $\mathbb{Q} \equiv \mathbb{Q}^T \sim \mathbb{P}$ with respect to $\mathcal{F}_T^W$.

In either of the above cases, under $\mathbb{Q}$,

$$\widetilde{W}_t := W_t - \int_0^t C_u du$$

is an $(\{\mathcal{F}_t^W\}, \mathbb{Q})$ Brownian motion (with the time horizon being restricted to $[0, T]$ in the latter case).

Remark 5.25: In the converse statement above we have imposed the usual conditions on $\{\mathcal{F}_t^W\}$ in order to ensure that the martingale representation theorem holds. It is then not true, in general, that there exists some measure $\mathbb{Q}$ defined on $\mathcal{F}_\infty^W$ which is locally equivalent to $\mathbb{P}$ with respect to $\{\mathcal{F}_t^W\}$, as Example 5.28 below demonstrates. For this reason the converse statement in Theorem 5.24 requires either a finite time horizon $(T < \infty)$ or that ρ be a UI martingale $(T = \infty)$.

Proof: Theorem 5.14 implies that the Radon–Nikodým process ρ is a strictly positive $(\{\mathcal{F}_t^W\}, \mathbb{P})$ martingale and it follows from the martingale representation theorem that it has a continuous version. Further, since $\mathcal{F}_0^W$ is trivial and $\mathbb{E}[\rho_0] = 1$ we conclude that $\rho_0 \equiv 1$ $\mathbb{P}$-a.s. By Lemma 5.17,

$$X_t := \int_0^t \rho_u^{-1} d\rho_u$$

is the unique continuous $(\{\mathcal{F}_t^W\}, \mathbb{P})$ local martingale such that $\rho = \mathcal{E}(X)$. Again appealing to the martingale representation theorem, X has an integral representation

$$X_t = \int_0^t C_u \cdot dW_u$$

for some $\{\mathcal{F}_t^W\}$ predictable $C \in \Pi(W)$. This proves (5.6).

Conversely, the fact that a strictly positive $(\{\mathcal{F}_t^W\}, \mathbb{P})$ martingale defines a measure $\mathbb{Q}^T$ which is equivalent to $\mathbb{P}$ with respect to $\mathcal{F}_T^W$ follows from Theorem 5.14. The exponential representation of ρ in Equation (5.6) follows exactly as above.

Noting that $[W, X]_t = \int_0^t C_u du$, the final statement of the theorem follows from Corollary 5.23. □

This leads to the following corollary which pulls together several of the above results in a form useful for application. The proof is left as a simple exercise.

Corollary 5.26 *Under the assumptions and using the notation of the first part of Theorem 5.24, any $(\{\mathcal{F}_t^W\}, \mathbb{P})$ semimartingale Y is also an $(\{\mathcal{F}_t^W\}, \mathbb{Q})$ semimartingale and has the canonical decompositions under $\mathbb{P}$ and $\mathbb{Q}$ respectively*

$$Y_t = Y_0 + \int_0^t \sigma_u \cdot dW_u + A_t$$

$$= Y_0 + \int_0^t \sigma_u \cdot d\widetilde{W}_u + \left(\int_0^t C_u \cdot \sigma_u du + A_t \right)$$

for some $\sigma \in \Pi(W)$ and some finite variation process A.

Remark 5.27: Note that the extra finite variation process introduced above as a result of the change of measure is absolutely continuous with respect to Lebesgue measure.

Example 5.28 Now the promised example (see comments preceding Theorem 5.15) which illustrates the difficulties caused by the usual conditions when the time horizon is infinite. Let $(\Omega, \{\mathcal{F}_t\}, \mathcal{F}, \mathbb{P})$ be a probability space supporting a one-dimensional Brownian motion W, and let μ be a non-zero constant. Then it is easily checked (for example using Novikov's condition, or by direct calculation as we did in Example 3.3) that

$$\rho_t := \exp(\mu W_t - \tfrac{1}{2}\mu^2 t), \quad t \geq 0,$$

is an $(\{(\mathcal{F}_t^W)^\circ\}, \mathbb{P})$ martingale and so can be used to define a measure $\mathbb{Q}$ on $(\Omega, (\mathcal{F}_\infty^W)^\circ)$ which is locally equivalent to $\mathbb{P}$ with respect to $\{(\mathcal{F}_t^W)^\circ\}$ (Theorem 5.15). By Corollary 5.23, the process $\widetilde{W}_t := W_t - \mu t$ is an $(\{(\mathcal{F}_t^W)^\circ\}, \mathbb{Q})$ Brownian motion.

Although the measures $\mathbb{P}$ and $\mathbb{Q}$ are locally equivalent with respect to $\{(\mathcal{F}_t^W)^\circ\}$ they are not equivalent with respect to $(\mathcal{F}_\infty^W)^\circ$. Nor are they locally equivalent with respect to $\{\mathcal{F}_t^W\}$, the $\mathbb{P}$-augmentation of $\{(\mathcal{F}_t^W)^\circ\}$ with respect to $(\mathcal{F}_\infty^W)^\circ$. To see this, consider the event

$$\Lambda := \{\omega \in \Omega : \lim_{t \to \infty} t^{-1} W_t = \mu\} \in (\mathcal{F}_\infty^W)^\circ.$$

Clearly $\mathbb{Q}(\Lambda) = 1$ but $\mathbb{P}(\Lambda) = 0$ (Theorem 2.7) and so $\mathbb{Q}$ is not absolutely continuous with respect to $\mathbb{P}$ on the σ-algebra $(\mathcal{F}_\infty^W)^\circ$. Furthermore, $\Lambda \in \mathcal{F}_t^W$ for all $t \geq 0$ since it has probability zero under $\mathbb{P}$, and so $\mathbb{Q}$ is not locally equivalent to $\mathbb{P}$ with respect to $\{\mathcal{F}_t^W\}$.

We can say more. The process W is an $(\{(\mathcal{F}_t^W)^\circ\}, \mathbb{P})$ Brownian motion, and it is also an $(\{(\mathcal{F}_t^W)\}, \mathbb{P})$ Brownian motion since $\mathbb{P}$-augmentation has no material effect. Girsanov's theorem proves that $\widetilde{W}$ is an $(\{(\mathcal{F}_t^W)^\circ\}, \mathbb{Q})$ Brownian motion but, given *any* event $A \in \mathcal{F}_\infty^W$,

$$\mathbb{P}(A \cap \Lambda) = 0,$$

so $A \cap \Lambda \in \mathcal{F}_0^W$. Under the measure $\mathbb{P}$ this has little effect, adding only null sets, but under $\mathbb{Q}$,

$$\mathbb{Q}(A \cap \Lambda) = \mathbb{Q}(A) \neq 0,$$

in general. Thus, given any 'interesting' event under the measure $\mathbb{Q}$ it is effectively included in $\mathcal{F}_0^W$. Thus $\widetilde{W}$ cannot be an $(\{\mathcal{F}_t^W\}, \mathbb{Q})$ Brownian motion because $\widetilde{W}_{t+s} - \widetilde{W}_t$ is not independent of $\mathcal{F}_t^W$ for any t, s.

A consequence of this example is that it is not possible to work with the completed Brownian filtration over an infinite horizon when constructing a locally equivalent measure from a martingale (recall that there is no problem if the martingale is UI). It is possible to derive an analogue of Theorem 5.24 when the filtration is not complete, but to do so it is necessary to work on the canonical set-up (as introduced in Section 2.2 and discussed further in Section 6.3). This additional restriction is required because it is not, in general, possible to define the stochastic integral if the filtration is not complete, *except when working on the canonical set-up*. For more on this see, for example, Karatzas and Shreve (1991).

Of course, none of these complications arise when considering only a finite time horizon T and completing, for example, with respect to $(\mathcal{F}_T^W)^\circ$.

5.3 MARTINGALE REPRESENTATION THEOREM

An important question in finance is: *which contingent claims (i.e. payoffs) can be replicated by trading in the assets of an economy?* It is only derivatives with these payoffs that can be priced using arbitrage ideas. A related and equally important question in probability theory is the following: *given a (vector) martingale M, under what conditions can all (local) martingales N be represented in the form*

$$N_t = N_0 + \int_0^t H_u \cdot dM_u,$$

for some $H \in \Pi(M)$.

An answer to this second question is provided by the martingale representation theorem, and this section is devoted to providing a proof. In keeping with the rest of this book, we shall only prove the martingale representation theorem in the case when the martingale M is continuous. A more general treatment can be found in Jacod (1979).

If your objective in reading this section is only to obtain a precise statement of the results that are needed in Chapter 7 where we discuss completeness, you need only understand the definition of an equivalent martingale measure (Definition 5.40) and the martingale representation theorem itself, Corollary 5.47 and Theorem 5.50. Combine these with Theorem 6.38 of Chapter 6 and you have all you need. If, on the other hand, you wish to understand why the martingale representation theorem holds, then read on.

The proof of the martingale representation theorem follows a familiar path. First we prove a version of the result for the representation of martingales in $\mathcal{M}_0^2$ and then we extend to all local martingales by localization arguments. As usual, all the hard work is done in proving the former result. We do this in two steps. First, we show that any $N \in \mathcal{M}_0^2$ can be written as the sum of a term we can represent as a stochastic integral plus a term orthogonal to it. The second step is to provide necessary and sufficient conditions for the orthogonal term to be zero. For this we shall introduce the idea of an equivalent martingale measure, a measure equivalent to the original probability measure $\mathbb{P}$ and under which M remains a martingale. The connection between equivalent martingale measures and martingale representation may seem surprising at first, but remember that the Radon–Nikodým process connecting two equivalent probability measures is a UI martingale, and if the martingale representation theorem holds then we must be able to represent this martingale as a stochastic integral. The connection is made precise in the proofs to follow.

Throughout this section we will be working on a filtered probability space $(\Omega, \{\mathcal{F}_t\}, \mathcal{F}, \mathbb{P})$ satisfying the usual conditions. The usual conditions are needed for Proposition 5.45 and all dependent results to hold.

5.3.1 The space $\mathcal{I}^2(M)$ and its orthogonal complement

We begin with the definition of $\mathcal{I}^2(M)$ and one of its important properties, that of *stability*.

Definition 5.29 *For a continuous (vector) martingale M we define the space $\mathcal{I}^2(M)$ to be*

$$\mathcal{I}^2(M) = \{N \in c\mathcal{M}_0^2 : N = H \bullet M, \text{ for some } H \in \Pi(M)\}.$$

That is, $\mathcal{I}^2(M)$ is the set of square-integrable martingales which can be represented as a stochastic integral with respect to M.

Remark 5.30: When M is a vector martingale we will use the notation $H \bullet M$ to mean $\sum_i H^{(i)} \bullet M^{(i)}$.

Remark 5.31: Of course, in Definition 5.29 $H \bullet M$ is automatically continuous, being a stochastic integral for a continuous integrator M.

Proposition 5.32 *The space $\mathcal{I}^2(M)$ is a stable subspace of $\mathcal{M}_0^2$, meaning it satisfies the following two conditions:*

(S.i) It is closed (under the norm $\| \cdot \|_2$).
(S.ii) It is stable under stopping, meaning that if $N \in \mathcal{I}^2(M)$ and T is a stopping time, then $N^T \in \mathcal{I}^2(M)$.

Remark 5.33: Notice in Proposition 5.32 that we are considering the space $\mathcal{M}_0^2$ rather than the smaller space $c\mathcal{M}_0^2$ in which all martingales are continuous. Here, and repeatedly in what follows, we could have chosen to work with $c\mathcal{M}_0^2$. The reason why the precise choice is irrelevant is that when the martingale representation theorem holds (with respect to some continuous martingale M) *all* martingales are continuous and so $\mathcal{M}_0^2 = c\mathcal{M}_0^2$.

Remark 5.34: We remind the reader once more of a point made repeatedly in Chapter 4. Condition (S.i) makes no sense when the set $\mathcal{I}^2(M)$ is considered as a set of stochastic processes since then $\|\cdot\|_2$ is not a norm. It is only strictly correct when we identify any two processes which are indistinguishable and work with equivalence classes of processes. As we did in Chapter 4, we shall make this identification whenever required in what follows without further comment. The same is also true for the various other $\mathcal{L}^2$ spaces that we encounter.

Proof: If $N = H \bullet M \in \mathcal{I}^2(M)$ then

$$N^T = (H \bullet M)^T = H1(0, T] \bullet M \in \mathcal{I}^2(M)$$

by Theorem 4.17, so (S.ii) holds.

The proof of (S.i) is an adaptation of the usual isometry argument. Let $H^n \bullet M$ be a Cauchy sequence (using the norm $\|\cdot\|_2$) in $c\mathcal{M}_0^2$ with limit $N \in c\mathcal{M}_0^2$ ($c\mathcal{M}_0^2$ is complete). By the stochastic integral isometry (Corollary 4.14),

$$\|H^n - H^m\|_M = \|H^n \bullet M - H^m \bullet M\|_2,$$

so $\{H^n\}$ is also Cauchy in the space $L^2(M)$, which is also complete. It therefore has a limit H in $L^2(M)$ and $N = H \bullet M \in \mathcal{I}^2(M)$. Thus $\mathcal{I}^2(M)$ is closed. $\square$

The space $\mathcal{I}^2(M)$ is precisely those martingales in $\mathcal{M}_0^2$ which can be represented as a stochastic integral with respect to M. We shall see in Theorem 5.37 that being able to represent any element of $\mathcal{M}_0^2$ as a stochastic integral is equivalent to the orthogonal complement of $\mathcal{I}^2(M)$ in $\mathcal{M}_0^2$ being zero.

Definition 5.35 *Two square-integrable martingales M and N are said to be orthogonal, written $M \perp N$, if $\mathbb{E}[M_\infty N_\infty] = 0$. Given any $\mathcal{N} \subset \mathcal{M}_0^2$, we define the orthogonal complement of $\mathcal{N}$, $\mathcal{N}^\perp$, to be*

$$\mathcal{N}^\perp = \{L \in \mathcal{M}_0^2 : L \perp N \text{ for all } N \in \mathcal{N}\}.$$

Remark 5.36: Recall from Chapter 3 that we consider the space $\mathcal{M}_0^2$ to be a Hilbert space with norm $\|M\|_2^2 = \mathbb{E}[M_\infty^2]$. The inner product associated with the norm $\|\cdot\|_2$ is then just $\langle M, N \rangle = \mathbb{E}[M_\infty N_\infty]$, so orthogonality of martingales in $\mathcal{M}_0^2$ as defined above is the usual concept of orthogonality for a Hilbert space.

The following result is now an immediate consequence of standard linear analysis and Remark 5.36.

Theorem 5.37 $\mathcal{M}_0^2 = \mathcal{I}^2(M) \oplus \left(\mathcal{I}^2(M)\right)^\perp$. *That is, any $M \in \mathcal{M}_0^2$ has a unique decomposition $M = N + L$ where $N \in \mathcal{I}^2(M)$ and $L \in \left(\mathcal{I}^2(M)\right)^\perp$.*

In the results to follow we will be working with elements of $\mathcal{I}^2(M)$ and $\left(\mathcal{I}^2(M)\right)^{\perp}$. We will need to consider elements of each space stopped at some stopping time T. We have seen (Proposition 5.32) that $\mathcal{I}^2(M)$ is stable, so stopping an element in $\mathcal{I}^2(M)$ produces another element of $\mathcal{I}^2(M)$. The same is true of $\left(\mathcal{I}^2(M)\right)^{\perp}$, a fact we shall also need.

Proposition 5.38 *The space $\left(\mathcal{I}^2(M)\right)^{\perp}$ is a stable subspace of $\mathcal{M}_0^2$.*

Proof: It is a standard result from linear analysis that $\left(\mathcal{I}^2(M)\right)^{\perp}$, the orthogonal complement of a closed space, is closed, so condition (S.i) is straightforward. To prove that $\left(\mathcal{I}^2(M)\right)^{\perp}$ is closed under stopping (condition (S.ii)), let $N \in \mathcal{I}^2(M)$, $L \in \left(\mathcal{I}^2(M)\right)^{\perp}$ and let T be some stopping time. It follows from the optional sampling theorem and the tower property, respectively, that

$$\mathbb{E}[N_T L_T] = \mathbb{E}\big[N_T \mathbb{E}[L_\infty | \mathcal{F}_T]\big] = \mathbb{E}[N_T L_\infty] = \mathbb{E}[N_\infty^T L_\infty], \qquad (5.7)$$

and

$$\mathbb{E}[N_T L_T] = \mathbb{E}\big[\mathbb{E}[N_\infty | \mathcal{F}_T] L_T\big] = \mathbb{E}[N_\infty L_T] = \mathbb{E}[N_\infty L_\infty^T]. \qquad (5.8)$$

But $\mathcal{I}^2(M)$ is stable, so N^T is orthogonal to L and thus (5.7) equates to zero. This is identically equal to (5.8), thus N is orthogonal to L^T. Finally, since N was arbitrary we conclude that $L^T \in \left(\mathcal{I}^2(M)\right)^{\perp}$ and so $\left(\mathcal{I}^2(M)\right)^{\perp}$ is stable under stopping.   □

This proposition allows us to establish the following characterization of the continuous martingales within the space $\left(\mathcal{I}^2(M)\right)^{\perp}$, one we use below when establishing the martingale representation theorem.

Lemma 5.39 *Let M be a continuous martingale.*
(i) *Suppose that $N \in c\mathcal{M}_0^2$. Then $N \in (\mathcal{I}^2(M))^{\perp}$ if and only if NM is a local martingale.*
(ii) *Suppose that N is a bounded martingale. Then $N \in (\mathcal{I}^2(M))^{\perp}$ if and only if NM is a martingale.*

Proof: Throughout this proof we will, without further comment, make repeated use of Corollary 3.82 which states that if N and M are continuous local martingales then NM is a local martingale if and only if $[N, M] = 0$.
Proof of (i): Suppose $N \in \left(\mathcal{I}^2(M)\right)^{\perp}$ and let T be some arbitrary stopping time. For each n, define $S_n = \inf\{t > 0 : |M_t - M_0| > n\}$ and note that $M^{S_n} - M_0 = 1(0, S_n] \bullet M \in \mathcal{I}^2(M)$. By the stability of $\mathcal{I}^2(M)$ and $\left(\mathcal{I}^2(M)\right)^{\perp}$ respectively, $N^T \perp (M^{S_n} - M_0)^T$ and so

$$\mathbb{E}[N_T(M_T^{S_n} - M_0)] = 0.$$

Furthermore, $\mathbb{E}[|N_T(M_T^{S_n} - M_0)|] \le n\mathbb{E}[|N_T|] < \infty$ since N is a UI martingale, and $N(M^{S_n} - M_0)$ is continuous, so Theorem 3.53 implies that $N(M^{S_n} - M_0)$

is a UI martingale. It now follows that $[N, M^{S_n}] = 0$ and, letting $n \to \infty$, $[N, M] = 0$ so NM is a local martingale.

Conversely, suppose that NM is a local martingale, hence $[N, M] = 0$. Let $L = H \bullet M \in \mathcal{I}^2(M)$. By (the local martingale extension of) Theorem 4.18, $[N, L] = [N, H \bullet M] = H \bullet [N, M] = 0$, so NL is a local martingale. But then

$$NL = NL - [N, L]$$
$$= \tfrac{1}{2}\Big(\big((N+L)^2 - [N+L]\big) - \big(N^2 - [N]\big) - \big(L^2 - [L]\big)\Big)$$

expresses NL as a sum of UI martingales, hence it is itself a UI martingale and $\mathbb{E}[N_\infty L_\infty] = 0$. This proves the result.

Proof of (ii): As above, we can conclude that if $N \in (\mathcal{I}^2(M))^\perp$ then NM is a local martingale. Note that the continuous process NM is a martingale if and only if the stopped process $(NM)^t$ is a UI martingale for all $t \geq 0$. Thus, by Theorem 3.53, to prove that NM is a martingale it suffices to prove that $\mathbb{E}[|(NM)_T^t|] < \infty$ and $\mathbb{E}[(NM)_T^t] = (NM)_0$ for all stopping times T.

To prove that $\mathbb{E}[|(NM)_T^t|] < \infty$, note that M^t is a UI martingale so it follows from Corollary 3.51 that $\mathbb{E}[|(NM)_T^t|] \leq K\mathbb{E}[|M_T^t|] < \infty$, K being some constant which bounds N. It remains to prove that $\mathbb{E}[(NM)_T^t] = (NM)_0$. To see this, let $X^n := N^{t \wedge S_n} M^t$ where S_n is as defined above. Then

$$\mathbb{E}[X_T^n] = \mathbb{E}\mathbb{E}[X_T^n | \mathcal{F}_{T \wedge S_n}]$$
$$= \mathbb{E}\mathbb{E}[N_{T \wedge S_n}^t M_T^t | \mathcal{F}_{T \wedge S_n}]$$
$$= \mathbb{E}[N_{T \wedge S_n}^t M_{T \wedge S_n}^t]$$
$$= (NM)_0\,,$$

the third equality following from the optional sampling theorem applied to the UI martingale M^t, the final equality following since $(NM)^{S_n}$ is a bounded local martingale, thus a UI martingale. Now let $n \to \infty$. In so doing, we see that $X_T^n \to (NM)_T$ a.s., and, since $|X_T^n| \leq K|M_T^t|$ which is bounded in $\mathcal{L}^1$, the dominated convergence theorem implies that

$$\mathbb{E}[(NM)_T^t] = \mathbb{E}[\lim_{n \to \infty} X_T^n]$$
$$= \lim_{n \to \infty} \mathbb{E}[X_T^n]$$
$$= (NM)_0^t\,,$$

which completes the proof.

The converse result follows exactly as in (i). $\qquad\square$

5.3.2 Martingale measures and the martingale representation theorem

The martingale representation property is concerned with the ability, on some given probability space $(\Omega, \{\mathcal{F}_t\}, \mathcal{F}, \mathbb{P})$, to represent an arbitrary local martingale as a stochastic integral with respect to some other given martingale. The martingale representation theorem relates this property to the *extremality* of $\mathbb{P}$ within the set of *equivalent martingale measures for M*. To procede we therefore need the following definitions.

Definition 5.40 *Let $(\Omega, \{\mathcal{F}_t\}, \mathcal{F}, \mathbb{P})$ be a probability space satisfying the usual conditions and let M be a continuous (possibly vector) martingale under $\mathbb{P}$. The set of equivalent martingale measures for M, written $\mathrm{M}(M)$, is the set of probability measures $\mathbb{Q}$ satisfying each of the following conditions:*

(i) $\mathbb{Q} \sim \mathbb{P}$ with respect to $\mathcal{F}$;
(ii) $\mathbb{Q}$ and $\mathbb{P}$ agree on $\mathcal{F}_0$;
(iii) M is an $\{\mathcal{F}_t\}$ martingale under $\mathbb{Q}$.

Remark 5.41: Although it is not explicit in the notation, note that $\mathrm{M}(M)$ is dependent on the whole probability space $(\Omega, \{\mathcal{F}_t\}, \mathcal{F}, \mathbb{P})$.

Remark 5.42: We have insisted in the definition that $(\Omega, \{\mathcal{F}_t\}, \mathcal{F}, \mathbb{P})$ satisfies the usual conditions. This is essential for Proposition 5.45 and all subsequent results to hold. Note that when this is true condition (i) is redundant since it follows from condition (ii).

The following lemma is a trivial consequence of the above definition.

Lemma 5.43 *The set $\mathrm{M}(M)$ is convex.*

Definition 5.44 *The measure $\mathbb{P}$ is said to be extremal for $\mathrm{M}(M)$ if, whenever $\mathbb{Q}, \mathbb{R} \in \mathrm{M}(M)$ and $\mathbb{P} = \lambda\mathbb{Q} + (1-\lambda)\mathbb{R}$, $0 < \lambda < 1$, then $\mathbb{P} = \mathbb{Q} = \mathbb{R}$.*

We shall need the following result which we state without proof. A probabilistic proof can be found in Protter (1990), an analytical one in Revuz and Yor (1991). This is a vital step in the proof of the martingale representation theorem but we would have to include several new ideas to provide a proof.

Proposition 5.45 *If $\mathbb{P}$ is extremal in $\mathrm{M}(M)$ then every $(\{\mathcal{F}_t\}, \mathbb{P})$ local martingale is continuous.*

This proposition enables us to state and prove the main result of this section, the martingale representation theorem for square-integrable martingale.

Theorem 5.46 *Let $(\Omega, \{\mathcal{F}_t\}, \mathcal{F}, \mathbb{P})$ satisfy the usual conditions and let M be a continuous martingale under $\mathbb{P}$. The following are equivalent:*

(i) $\mathbb{P}$ is extremal for $\mathrm{M}(M)$;
(ii) $\mathcal{I}^2(M) = \mathcal{M}_0^2$.

Proof: (i) $\Rightarrow$ (ii): We already know that $\mathcal{M}_0^2 = \mathcal{I}^2(M) \oplus \mathcal{I}^2(M)^\perp$ by Theorem 5.37. It suffices to prove that $\mathcal{I}^2(M)^\perp = \{0\}$.

Let $N \in \mathcal{I}^2(M)^\perp$. Since $\mathbb{P}$ is extremal, N is continuous by Proposition 5.45 whence, defining $T = \inf\{t > 0 : |N_t| > \frac{1}{2}\}$, N^T is bounded by $\frac{1}{2}$. Thus we can use N^T to define two new measures $\mathbb{Q}$ and $\mathbb{R}$ via

$$\frac{d\mathbb{Q}}{d\mathbb{P}}\bigg|_{\mathcal{F}} := (1 - N_\infty^T) = (1 - N_T), \qquad \frac{d\mathbb{R}}{d\mathbb{P}}\bigg|_{\mathcal{F}} = (1 + N_T).$$

Clearly $\mathbb{P} = \frac{1}{2}\mathbb{Q} + \frac{1}{2}\mathbb{R}$, $\mathbb{P} \sim \mathbb{Q} \sim \mathbb{R}$, and they agree on $\mathcal{F}_0$ (since $N_0 = 0$). Furthermore, noting that $N^T \in \left(\mathcal{I}^2(M)\right)^\perp$ and is bounded, we have that $N^T M$ is a martingale under $\mathbb{P}$ by Lemma 5.39. Thus, subtracting this from the $(\{\mathcal{F}_t\}, \mathbb{P})$ martingale M we see that $M(1 - N^T)$ is also a martingale under $\mathbb{P}$, and it then follows from Lemma 5.19 that M is an $(\{\mathcal{F}_t\}, \mathbb{Q})$ martingale. But $\mathbb{P}$ is extremal by hypothesis, thus $\mathbb{P} = \mathbb{Q} = \mathbb{R}$ and $N \equiv 0$.

(ii) $\Rightarrow$ (i): Let $\mathbb{Q}, \mathbb{R} \in \mathbb{M}(M)$ and $0 < \lambda < 1$ be such that $\mathbb{P} = \lambda\mathbb{Q} + (1 - \lambda)\mathbb{R}$. Define the UI martingale ρ via

$$\rho_t = \mathbb{E}\left[\frac{d\mathbb{Q}}{d\mathbb{P}}\bigg|\mathcal{F}_t\right], \qquad t \in [0, \infty].$$

Observe that ρ is bounded by λ^{-1}, a fact which follows since

$$1 = \mathbb{E}_\mathbb{P}\left[\frac{d\mathbb{P}}{d\mathbb{P}}\bigg|\mathcal{F}_\infty\right] = \lambda\rho_\infty + (1 - \lambda)\mathbb{E}_\mathbb{P}\left[\frac{d\mathbb{R}}{d\mathbb{P}}\bigg|\mathcal{F}_\infty\right] \geq \lambda\rho_\infty,$$

and $M\rho$ is an $(\{\mathcal{F}_t\}, \mathbb{P})$ martingale by Lemma 5.19. Now, define the martingale $N := \rho - \rho_0$. The martingale property of M and ρM under $\mathbb{P}$ now implies that NM is a martingale under $\mathbb{P}$. Furthermore, the measures $\mathbb{P}$ and $\mathbb{Q}$ agree on $\mathcal{F}_0$, so $\rho_0 = 1$ a.s., and $N = \rho - 1$ is thus bounded by $\lambda^{-1} + 1$. Lemma 5.39 now implies that $N \in \mathcal{I}^2(M)^\perp$ and thus, from (ii), $N \equiv 0$ and so $\mathbb{P} = \mathbb{Q} = \mathbb{R}$ and $\mathbb{P}$ is extremal. $\qquad\square$

5.3.3 Extensions and the Brownian case

Theorem 5.46 is the core result in martingale representation theory for continuous martingale integrators. However, as it stands it is restrictive and it is not yet obvious that the result is of any practical significance. In this section we begin to rectify these shortcomings.

First, for a given continuous martingale M, we extend the class of processes we can represent to include all local martingales. This follows easily by localization.

Corollary 5.47 *Let* $(\Omega, \{\mathcal{F}_t\}, \mathcal{F}, \mathbb{P})$ *satisfy the usual conditions and let* M *be a continuous martingale under* $\mathbb{P}$. *Define*

$$\mathcal{I}(M) = \{N \in \mathcal{M}_{0,\text{loc}} : N = H \bullet M, \text{ for some } H \in \Pi(M)\}.$$

The following are equivalent:

(i) $\mathbb{P}$ *is extremal for* $\mathbb{M}(M)$;
(ii) $\mathcal{I}(M) = \mathcal{M}_{0,\text{loc}}$.

Proof: (i) $\Rightarrow$ (ii): If $\mathbb{P}$ is extremal then $N \in \mathcal{M}_{0,\text{loc}}$ is continuous by Proposition 5.45 and so can be localized using $S_n = \inf\{t > 0 : |M_t| > n\}$. Theorem 5.46 now implies that there exists some $H^n \in \Pi(M)$ such that

$$N_t^{S_n} = \int_0^t H_u^n \cdot dM_u .$$

Defining $H = \sum_{n=1}^{\infty} 1(S_{n-1}, S_n] H^n \in \Pi(M)$ yields $N = H \bullet M$.

(ii) $\Rightarrow$ (i): If $\mathcal{I}(M) = \mathcal{M}_{0,\text{loc}}$ then $c\mathcal{M}_0^2 = \mathcal{I}^2(M)$ so the result follows from Theorem 5.46. □

Given any probability space $(\Omega, \{\mathcal{F}_t\}, \mathcal{F}, \mathbb{P})$ one often wants to be able to represent (local) martingales which are adapted to a smaller filtration $\mathcal{G}_t \subset \mathcal{F}_t$, $\mathcal{G} \subset \mathcal{F}$. But, of course, $\mathbb{P}$ induces a probability measure on $\mathcal{G}$ and we could throughout have chosen to work on the smaller space $(\Omega, \{\mathcal{G}_t\}, \mathcal{G}, \mathbb{P})$. As long as M is $\{\mathcal{G}_t\}$-adapted and we modify Definition 5.40 for an equivalent martingale measure (replacing $(\{\mathcal{F}_t\}, \mathcal{F})$ with $(\{\mathcal{G}_t\}, \mathcal{G})$ throughout), all the results above remain valid.

Replication of derivative payoffs is more a question of representing random variables than of representing (local) martingales. However, the two ideas are intimately related and it is possible to move from one to the other. The following result shows how random variable representation follows from our results so far.

Corollary 5.48 *Let* $(\Omega, \{\mathcal{F}_t\}, \mathcal{F}, \mathbb{P})$ *be a probability space supporting a continuous martingale* $M \in c\mathcal{M}_0$ *and let* $\{\mathcal{F}_t^M\}$ *be the filtration generated by* M, *augmented to satisfy the usual conditions. If* $\mathbb{P}$ *is extremal for* $\mathbb{M}(M)$, *defined relative to* $(\Omega, \{\mathcal{F}_t^M\}, \mathcal{F}_\infty^M, \mathbb{P})$, *then for any* $Y \in \mathcal{L}^1(\mathcal{F}_\infty^M)$ *there exists some* $\{\mathcal{F}_t^M\}$-*predictable* $H \in \Pi(M)$ *such that*

$$Y = \mathbb{E}[Y] + \int_0^\infty H_u \cdot dM_u .$$

Proof: Defining $Y_t = \mathbb{E}[Y | \mathcal{F}_t^M]$, Corollary 5.47 yields

$$Y_t = Y_0 + \int_0^t H_u \cdot dM_u$$

for some $\{\mathcal{F}_t^M\}$-predictable $H \in \Pi(M)$. Since $M_0 = 0$, $\mathcal{F}_0^M$ is trivial, Y_0 is a constant a.s., and $Y_0 := \mathbb{E}[Y | \mathcal{F}_0^M] = \mathbb{E}[Y]$. The result follows. □

As an example of this corollary we consider the case when the martingale M is actually a Brownian motion. In this special case it is possible to establish the extremality of the measure $\mathbb{P}$ in the set $\mathbb{M}(M)$ by showing that $\mathbb{M}(M) = \{\mathbb{P}\}$.

Theorem 5.49   *Let $(\Omega, \{\mathcal{F}_t\}, \mathcal{F}, \mathbb{P})$ be a probability space supporting a Brownian motion W and let $\{\mathcal{F}_t^W\}$ be the augmented natural filtration generated by W. Then any local martingale N adapted to $\{\mathcal{F}_t^W\}$ can be written in the form*

$$N_t = N_0 + \int_0^t H_u \cdot dW_u,$$

for some $\{\mathcal{F}_t^W\}$-predictable $H \in \Pi(W)$. In particular, any $Y \in \mathcal{L}^1(\mathcal{F}_\infty^W)$ has the form

$$Y = \mathbb{E}[Y] + \int_0^\infty H_u \cdot dW_u.$$

Proof: We work on the space $(\Omega, \{\mathcal{F}_t^W\}, \mathcal{F}_\infty^W, \mathbb{P})$. The results follow from Corollaries 5.47 and 5.48 if we can show that $\mathbb{P}$ is extremal for $\mathbb{M}(W)$ (on this set-up). We show that $\mathbb{P}$ is the unique element of $\mathbb{M}(W)$.

Suppose $\mathbb{Q} \in \mathbb{M}(W)$. The quadratic variation $[W]$ is independent of probability measure and, since $\mathbb{Q} \sim \mathbb{P}$ (with respect to $\{\mathcal{F}_t^W\}$), $[W]_t = t$ a.s. under both $\mathbb{P}$ and $\mathbb{Q}$. It follows from Lévy's theorem that W is also a Brownian motion under $\mathbb{Q}$. Therefore all finite-dimensional distributions for W agree under $\mathbb{P}$ and $\mathbb{Q}$, and thus $\mathbb{P} = \mathbb{Q}$ on $\{(\mathcal{F}_t^W)^\circ\}$. Completing with respect to $\mathbb{P}$ (or $\mathbb{Q}$ which is equivalent) gives agreement on $\{\mathcal{F}_t^W\}$ as required.
□

Our main interest in these results is to be able to represent a contingent claim (the random variable Y) as a stochastic integral with respect to asset prices (the martingale M). Asset price processes will later be specified as the solution to some stochastic differential equation (SDE). We will see (Theorem 6.38) that the martingale representation property can be related to the SDE which defines the asset price processes. We will also provide sufficient conditions on the coefficients of the SDE to ensure that the martingale representation theorem holds.

In order to establish Theorem 6.38 we will need the following extension of Corollaries 5.47 and 5.48. The extension replaces the martingale integrator M with a more general local martingale. Of course, in this context the relevant set of measures to consider is not $\mathbb{M}(M)$ but $\mathbb{L}(M)$, the set of equivalent *local martingale measures for M*. The representation results are then identical. A proof, which is effectively a localization argument based on the results established so far, can be found in Revuz and Yor (1991).

Theorem 5.50   *Let $\mathbb{L}(M)$ be the set of equivalent local martingale measures for M (defined exactly as in Definition 5.40 with 'martingale' replaced by 'local martingale' throughout). If the martingale M in Corollaries 5.47 and 5.48 is replaced by a local martingale and the set $\mathbb{M}(M)$ is replaced by $\mathbb{L}(M)$ then the conclusions of the corollaries remain valid.*

6

Stochastic Differential Equations

6.1 INTRODUCTION

An *ordinary differential equation* (ODE) is a way to define (and generate) a function x by specifying its local behaviour in terms of its present value (and more generally past evolution):

$$\frac{dx}{dt} = b(x_t, t),$$

$$x_0 = \xi.$$

In general, the functional form $x(t)$ will not be known explicitly but, under suitable regularity conditions on the function b, various existence and uniqueness results can be established.

A *stochastic differential equation* (SDE) is the stochastic analogue of the ODE and its use is similar (although more wide-ranging). In general an SDE may take the form

$$dX_t = \sigma_t \, dZ_t$$

for suitable σ and some semimartingale Z. Similar questions of existence and uniqueness then arise as in the case of ODEs. We have already met one important example of an SDE such as this, the exponential SDE of Example 4.36,

$$dX_t = X_t \, dZ_t,$$

for a general semimartingale Z, and we established there that the solution was unique (more precisely we established *pathwise uniqueness*, a concept that we will meet shortly).

In this chapter we shall discuss the special case of SDEs of the form

$$dX_t = b(X_t)dt + \sigma(X_t)dW_t \tag{6.1}$$

Financial Derivatives in Theory and Practice Revised Edition. P. J. Hunt and J. E. Kennedy
© 2004 John Wiley & Sons, Ltd   ISBNs: 0-470-86358-7 (HB); 0-470-86359-5 (PB)

where W is a d-dimensional Brownian motion. Here the process X is n-dimensional and t may be one of its components (so (6.1) automatically includes the time-inhomogeneous case). This is by far the most studied case and is more than adequate for our purposes. For more general treatments see Revuz and Yor (1991), Protter (1990) and Jacod (1979) (in increasing generality).

By driving the SDE with Brownian motion we are building up more general processes from a process we understand (a little!); by making (σ, b) depend only on X_t we create Markovian models. It is an easy exercise to show that if σ and b are bounded and X satisfies (6.1) then

$$\lim_{h \to 0} \frac{1}{h} \mathbb{E}[X_{t+h} - X_t | \mathcal{F}_t^X] = b(X_t)$$

$$\lim_{h \to 0} \frac{1}{h} \mathbb{E}[(X_{t+h} - X_t)(X_{t+h} - X_t)^T | \mathcal{F}_t^X] = \sigma(X_t)\sigma(X_t)^T := a(X_t).$$

Consequently b is often refered to as the *drift* of the process X, and the matrix a is its *diffusion matrix*. Processes such as this are often known collectively as *diffusion processes*.

In Section 6.2 we shall discuss precisely what is meant by an SDE, and in Section 6.4 (after an aside on the canonical set-up in Section 6.3) what is meant by a solution to an SDE such as (6.1). Section 6.4 also discusses the ideas of existence and uniqueness of a solution to an SDE. There are two types of solution to an SDE, *strong* and *weak*, and corresponding to each there are notions of existence and uniqueness. These are each discussed, along with the relationships between them. All these result are of a general nature. In Section 6.5 we establish Itô's result regarding (strong) existence and uniqueness when the SDE (σ, b) obeys (local) Lipschitz conditions. These are the first results which allow us to decide, for a specific SDE, whether existence and uniqueness hold and as such they are of considerable importance in practice. More general results can be established in this area but we shall not have need of them. The chapter concludes with a proof of the strong Markov property for a solution to a (locally) Lipschitz SDE in Section 6.6, and a result relating (weak) uniqueness for an SDE to the martingale representation theorem in Section 6.7. It is the latter result that we shall appeal to when discussing completeness in Chapter 7.

6.2 FORMAL DEFINITION OF AN SDE

By the SDE (σ, b) we shall mean an equation of the form

$$dX_t = b(X_t)dt + \sigma(X_t)dW_t,$$

or, in integral form,

$$X_t = X_0 + \int_0^t b(X_u)du + \int_0^t \sigma(X_u)dW_u, \qquad (6.2)$$

where W is a d-dimensional Brownian motion and

the maps $b : \mathbb{R}^n \to \mathbb{R}^n$ and $\sigma : \mathbb{R}^n \to \mathbb{R}^n \times \mathbb{R}^d$ are Borel measurable.

The SDE is defined by the pair (σ, b) so we shall refer to it by this label. The measurability restrictions on (σ, b) ensure that for any continuous adapted X the processes $b(X.)$ and $\sigma(X.)$ are predictable so, subject to boundedness restrictions discussed in Remark 6.6 below, the right-hand side of (6.2) is a stochastic integral as defined in Chapter 4.

Note that the underlying probability space $(\Omega, \{\mathcal{F}_t\}, \mathcal{F}, \mathbb{P})$ is *not* specified as part of the SDE, nor is the distribution of X_0 – they are part of the solution.

For future reference we shall denote by $| \cdot |$ the Euclidean metric on the spaces $\mathbb{R}^n$ and $\mathbb{R}^n \times \mathbb{R}^d$ as appropriate.

6.3 AN ASIDE ON THE CANONICAL SET-UP

We met the canonical set-up for a continuous process in Chapter 2 when discussing the existence of Brownian motion. It also arises in the study of SDEs, so we now give a reminder and extension of the discussion there.

Let $\mathbb{C} = \mathbb{C}(\mathbb{R}^+, \mathbb{R})$ be the set of continuous functions $\omega : [0, \infty) \to \mathbb{R}$ and let us endow it with the metric of uniform convergence, i.e. for $\omega, \tilde{\omega} \in \mathbb{C}$

$$\rho(\omega, \tilde{\omega}) = \sum_{j=1}^{\infty} \left(\frac{1}{2}\right)^j \frac{\rho_j(\omega, \tilde{\omega})}{1 + \rho_j(\omega, \tilde{\omega})}$$

where

$$\rho_j(\omega, \tilde{\omega}) = \sup_{0 \leq t \leq j} |\omega(t) - \tilde{\omega}(t)|.$$

Now we define $\mathcal{C} := \mathcal{B}(\mathbb{C})$, the Borel σ-algebra on $\mathbb{C}$ induced by the metric ρ. Furthermore, for all $t \geq 0$ we can define the projection map $\pi_t : \mathbb{C} \to \mathbb{R}$ via

$$\pi_t(\omega) = \omega_t,$$

for all $\omega \in \mathbb{C}$. Then the family $\{\mathcal{C}_t\}$ defined by

$$\mathcal{C}_t := \sigma(\pi_s : s \leq t)$$

defines a filtration on $\mathbb{C}$. (In fact $\mathcal{C}_\infty = \mathcal{C}$, as shown in Williams (1991).) We let $(\mathbb{C}^n, \{\mathcal{C}_t^n\}, \mathcal{C}^n)^\circ$ denote the (obvious) n-dimensional version of this space, *n-dimensional path space.* The $(\dots)^\circ$ is used to signify the fact that this space has not been augmented.

Now suppose that X is some a.s. continuous n-dimensional process on the complete probability space $(\Omega, \{\mathcal{F}_t\}, \mathcal{F}, \mathbb{P})$. Then X induces a probability measure, $\tilde{\mathbb{P}}$, on $(\mathbb{C}^n, \{\mathcal{C}_t^n\}, \mathcal{C}^n)^\circ$ given by, for all $A \in \mathcal{C}^n$,

$$\tilde{\mathbb{P}}(A) = \mathbb{P}(X_. \in A).$$

The stochastic process $\tilde{X}$ defined on the probability space $(\mathbb{C}^n, \mathcal{C}^n, \{\mathcal{C}_t^n\}, \tilde{\mathbb{P}})^\circ$ via

$$\tilde{X}_t(\omega) := \omega_t$$

has the same law as X. We denote by $(\mathbb{C}^n, \{\mathcal{C}_t^n\}, \mathcal{C}^n, \tilde{\mathbb{P}})$ the probability space $(\mathbb{C}^n, \{\mathcal{C}_t^n\}, \mathcal{C}^n)^\circ$ augmented with respect to $(\mathcal{C}, \tilde{\mathbb{P}})$, and refer to it as the *canonical set-up corresponding to* X.

In a similar fashion to the above, if ξ is a random variable defined on $(\Omega, \mathcal{F}, \mathbb{P})$ and taking values in $\mathbb{R}^m$, the canonical set-up corresponding to ξ is the space $(\mathbb{R}^m, \mathcal{B}, \mathbb{P}^*)$ where, for all $A \in \mathcal{B}(\mathbb{R}^m)$,

$$\mathbb{P}^*(A) = \mathbb{P}(\xi \in A),$$

and $\mathcal{B}$ is $\mathcal{B}(\mathbb{R}^m)$ completed with respect to $\mathbb{P}^*$.

Combining these two ideas we obtain the following generalization.

Definition 6.1 *Let* $(\Omega, \{\mathcal{F}_t\}, \mathcal{F}, \mathbb{P})$ *be a complete filtered probability space supporting a random variable* $\xi \in \mathbb{R}^m$ *and some continuous process* X *taking values in* $\mathbb{R}^n$. *Define the filtration* $\{\mathcal{G}_t^\circ\}$ *and the* σ-*algebra* $\mathcal{G}^\circ$ *on* $\mathbb{R}^m \times \mathbb{C}^n$ *via*

$$\mathcal{G}_t^\circ := \mathcal{B}(\mathbb{R}^m) \otimes \mathcal{C}_t^n$$

$$\mathcal{G}^\circ := \mathcal{G}_\infty.$$

Then the canonical set-up corresponding to (ξ, X) *is the probability space* $(\mathbb{R}^m \times \mathbb{C}^n, \{\mathcal{G}_t\}, \mathcal{G}, \tilde{\mathbb{P}})$, *where*

(i) the measure $\tilde{\mathbb{P}}$ *is given, for each* $A \in \mathcal{G}^\circ$, *by*

$$\tilde{\mathbb{P}}(A) = \mathbb{P}((\xi, X_.) \in A),$$

(ii) $\mathcal{G}_t$ *and* $\mathcal{G}$ *are, respectively,* $\mathcal{G}_t^\circ$ *and* $\mathcal{G}^\circ$ *augmented with respect to* $(\mathcal{G}^\circ, \tilde{\mathbb{P}})$.

Furthermore, $(\tilde{\xi}, \tilde{X})$, *defined, for each* $(r, x) \in \mathbb{R}^m \times \mathbb{C}^n$, *by* $\tilde{\xi}(r, x) = r$, $\tilde{X}(r, x) = x$, *has the same law under* $\tilde{\mathbb{P}}$ *as* (ξ, X) *does under* $\mathbb{P}$.

6.4 WEAK AND STRONG SOLUTIONS

In this section we discuss what is meant by a solution to an SDE. There are two basic types of solution, weak and strong, with strong being a special case of weak. We describe each of these, the associated concepts of existence and uniqueness for an SDE and the relationships between them.

6.4.1 Weak solutions

The broadest notion of a solution is defined as follows. The (redundant) qualifier weak is always included to contrast this general situation with the special case of a strong solution.

Definition 6.2 *A weak solution of the SDE* (σ, b) *with initial distribution* μ *is a sextuple* $(\Omega, \{\mathcal{F}_t\}, \mathcal{F}, \mathbb{P}, W, X)$ *such that:*

(i) $(\Omega, \{\mathcal{F}_t\}, \mathcal{F}, \mathbb{P})$ *is a filtered probability space satisfying the usual conditions;*

(ii) W *is a d-dimensional Brownian motion with respect to the filtration* $\{\mathcal{F}_t\}$, *and* X *is a continuous n-dimensional process adapted to* $\{\mathcal{F}_t\}$;

(iii) X_0 *has distribution* μ;

(iv) *for all* $t \geq 0$,

$$\int_0^t |b(X_u)| + |\sigma(X_u)|^2 du < \infty \quad \text{a.s.;} \tag{6.3}$$

(v) *for all* $t \geq 0$,

$$X_t = X_0 + \int_0^t b(X_u) du + \int_0^t \sigma(X_u) dW_u \quad \text{a.s.}$$

For reasons which shall become clear later (related to strong solutions) the sextuple $(\Omega, \{\mathcal{F}_t\}, \mathcal{F}, \mathbb{P}, W, X_0)$ *is referred to as the initial data of the solution. The process* X *is refered to as the solution process.*

Remark 6.3: Note that an SDE is specified only by the pair (σ, b), so a solution comprises the probability space as well as the processes W and X on that space.

Remark 6.4: The filtration $\{\mathcal{F}_t\}$ need not be the augmentation of $\{\sigma(X_0) \vee \mathcal{F}_t^W\}$. A general filtration $\{\mathcal{F}_t\}$ can always be replaced by $\{\mathcal{F}_t^{(W,X)}\}$ but not in general by $\{\sigma(X_0) \vee \mathcal{F}_t^W\}$. Solutions satisfying the latter property are the subject of the next section.

Remark 6.5: Note that, because W is a Brownian motion relative to the filtration $\{\mathcal{F}_t\}$ and X_0 is $\mathcal{F}_0$-measurable, it follows immediately that X_0 must be independent of W.

Remark 6.6: The finiteness condition (6.3) is required to ensure the integrals (6.2) exist. The condition on σ ensures that $\sigma(X) \in \Pi(W)$ (the obvious generalization of $\Pi(W)$ defined in Section 4.5.1 for the case $d = 1$) and the condition on b ensures that the first integral in (6.2) exists as a (pathwise) classical integral.

Remark 6.7: Note that the continuity of X means that conditions (iv) and (v), which as specified hold separately at each t, also hold for all t simultaneously.

What we mean by weak existence for an SDE is the following.

Definition 6.8 *We say that weak existence holds for the SDE (σ, b) if, for every probability measure μ on $(\mathbb{R}^n, \mathcal{B}(\mathbb{R}^n))$, there exists a weak solution to (σ, b) with initial law μ.*

There are two natural notions of uniqueness for weak solutions to an SDE.

Definition 6.9 *We say that weak uniqueness, or uniqueness in law, holds for the SDE (σ, b) if, given any two weak solutions $(\Omega, \{\mathcal{F}_t\}, \mathcal{F}, \mathbb{P}, W, X)$ and $(\tilde{\Omega}, \{\tilde{\mathcal{F}}_t\}, \tilde{\mathcal{F}}, \tilde{\mathbb{P}}, \tilde{W}, \tilde{X})$ with the same initial distribution μ, the laws of X and $\tilde{X}$ agree.*

Definition 6.10 *Pathwise uniqueness holds for the SDE (σ, b) if, given any two (weak) solutions $(\Omega, \{\mathcal{F}_t\}, \mathcal{F}, \mathbb{P}, W, X)$ and $(\Omega, \{\mathcal{F}_t\}, \mathcal{F}, \mathbb{P}, W, \tilde{X})$, with the same initial data $(\Omega, \mathcal{F}, \{\mathcal{F}_t\}, \mathbb{P}, W, X_0)$, the processes X and $\tilde{X}$ are indistinguishable,*

$$\mathbb{P}(X_t = \tilde{X}_t, \text{ for all } t) = 1.$$

Remark 6.11: If the filtrations for the two solutions in the above definition are allowed to differ, we say that *strict pathwise uniqueness* holds. It can be shown that this reduces to the apparently weaker notion of pathwise uniqueness as defined above.

Uniqueness in law is the most obvious notion of uniqueness for weak solutions of an SDE because the law is the only thing which can be compared in general, since the probability spaces may be different. It is the relevant notion of uniqueness we shall need when we revisit the martingale representation theorem in Section 6.7. Pathwise uniqueness is only a statement about two solutions on the same set-up, but is then a more stringent requirement.

The relationship between these two notions of uniqueness is not immediately obvious. However, the following is true.

Theorem 6.12 *Pathwise uniqueness implies weak uniqueness.*

This fact was first proved by Yamada and Watanabe (1971). Given any two weak solutions it is clear from the discussion in Section 6.3 that, for either of the two solutions, a corresponding weak solution can be defined on the canonical set-up (for the Brownian motion and the solution process jointly) $(\mathbb{R}^{d+n}, \mathbb{C}^{d+n}, \{\mathcal{C}_t^{d+n}\}, \mathcal{C}^{n+d})$. The proof of Yamada and Watanabe rests on the fact that any two weak solutions can be put on the *same* canonical set-up $(\mathbb{C}^{d+2n}, \{\mathcal{C}_t^{d+2n}\}, \mathcal{C}^{d+2n})$ by using regular conditional probabilities. Roughly speaking, the idea is as follows. Starting from the two weak solution processes X and $\tilde{X}$, and the corresponding Brownian motions W and $\tilde{W}$, construct the regular conditional probabilities $\mathbb{P}(X.|W.)$ and $\tilde{\mathbb{P}}(\tilde{X}.|\tilde{W}.)$. Now define a probability measure on the space $\mathbb{C}^{d+2n}$ so that the first d components have the law of a Brownian motion (using Wiener measure), the next n components have the law of X (using the probabilities $\mathbb{P}(X.|W.)$) and the final n components have the law of $\tilde{X}$ (using the probabilities $\tilde{\mathbb{P}}(\tilde{X}.|\tilde{W}.)$). In so doing we have succeeded in putting both solutions on the same space and so pathwise uniqueness now implies that they must be indistinguishable, hence must have the same law.

Remark 6.13: Note in the above discussion that when discussing a weak solution the process (W, X) can be transferred to the canonical space $(\mathbb{C}^{d+n}, \{\mathcal{C}_t^{d+n}\}, \mathcal{C}^{d+n})$ but not necessarily $(\mathbb{C}^d, \{\mathcal{C}_t^d\}, \mathcal{C}^d)$ which is canonical for the Brownian motion W. This contrasts with the discussion below for strong solutions to an SDE.

6.4.2 Strong solutions

For any weak solution to the SDE (σ, b), the process (W, X) is adapted to the filtration $\{\mathcal{F}_t^{(W,X)}\}$. In many situations an SDE is used as a way to specify a model for some physical quantity which is subject to uncertainty, in our case asset prices, and for this a general weak solution is sufficient. For some applications, however, both the driving Brownian motion W and the process X have independent significance. In signal processing, for example, the Brownian motion will be an *input* noise and X will be the *output*. In such situations it is important that, at the very least, X be adapted to $\{\mathcal{F}_t^{(X_0,W)}\}$ since knowledge of W and X_0 should be all that is required to have complete knowledge of X. This leads to the concept of a strong solution introduced below. An even more direct link between the input W and the output X would be the existence of a pathwise relationship between them,

$$X. = F(X_0, W.) \tag{6.4}$$

for some mapping F. We shall soon provide conditions under which such a map exists.

Definition 6.14 *A strong solution of the SDE (σ, b) is a (weak) solution $(\Omega, \{\mathcal{F}_t\}, \mathcal{F}, \mathbb{P}, W, X)$ with the additional property that the solution process X*

is adapted to the filtration $\{\mathcal{F}_t^{(X_0,W)}\}$, *the filtration* $\{\sigma(X_0) \vee \mathcal{F}_t^W\}$ *augmented to include all* $(\mathcal{F}, \mathbb{P})$ *null sets.*

Remark 6.15: Our definition of strong is identical to that of Revuz and Yor (1991) but different from that of Rogers and Williams (1987) who require more of their strong solutions. Where we differ from Revuz and Yor (1991) is in the definition of weak; they define a weak solution to be one which is weak in our sense but not strong.

It may at first appear surprising that situations arise where the solution process X is not adapted to the filtration $\{\mathcal{F}^{(X_0,W)}\}$. The following famous example of Tanaka shows that they do.

Example 6.16   (Tanaka)   Consider the SDE

$$dX_t = \operatorname{sgn}(X_t)dW_t. \tag{6.5}$$

To see that weak existence holds for this SDE take any probability space $(\Omega, \{\mathcal{F}_t\}, \mathcal{F}, \mathbb{P})$ supporting a Brownian motion B and an independent random variable ξ having law μ. Now define

$$X_t = \xi + B_t,$$

$$W_t = \int_0^t \operatorname{sgn}(X_t)dX_t. \tag{6.6}$$

Equation (6.6) defines a local martingale with $(dW_t)^2 = (dB_t)^2 = dt$, so W is a Brownian motion by Lévy's theorem (Theorem 4.38). Furthermore, X clearly satisfies (6.5) so weak existence holds.

Weak uniqueness also holds since any (weak) solution to (6.5) is Brownian motion, again by Lévy's theorem, but pathwise uniqueness does not hold since $X_t = \xi - B_t$ is another weak solution. Finally, note the plausible fact (from (6.6)), proved in Rogers and Williams (1987), that $\mathcal{F}_t^W = \mathcal{F}_t^{|X|}$, so X is not adapted to $\{\mathcal{F}_t^W\}$ and the solution is not strong. (To prove this we would need to study the local time of X, which is beyond the scope of this book.)

Motivated by the fact that for a strong solution the 'input' process W is often specified, as is the probability space, the notion of strong existence for an SDE is stricter than the obvious counterpart of Definition 6.8.

Definition 6.17 *We say strong existence holds for the SDE* (σ, b) *if, for all initial data* $(\Omega, \{\mathcal{F}_t\}, \mathcal{F}, \mathbb{P}, W, \xi)$, *there exists some* X *such that* $(\Omega, \{\mathcal{F}_t\}, \mathcal{F}, \mathbb{P}, W, X)$ *is a strong solution to* (σ, b) *(with* $X_0 = \xi$ *a.s. of course).*

Remark 6.18: It is, of course, the use of $(\Omega, \{\mathcal{F}_t\}, \mathcal{F}, \mathbb{P}, W, \xi)$ as an input that motivated the label 'initial data', one which is less natural in the context of a weak solution.

Definition 6.19 *Strong uniqueness holds for (σ, b) if, given any initial data $(\Omega, \{\mathcal{F}_t\}, \mathcal{F}, \mathbb{P}, W, \xi)$ and any two strong solutions with this initial data, $(\Omega, \{\mathcal{F}_t\}, \mathcal{F}, \mathbb{P}, W, X)$ and $(\Omega, \{\mathcal{F}_t\}, \mathcal{F}, \mathbb{P}, W, \tilde{X})$, the processes X and $\tilde{X}$ are indistinguishable,*

$$\mathbb{P}(X_t = \tilde{X}_t, \text{ for all } t) = 1.$$

Returning to the question of existence or otherwise of the function F in (6.4), this is not true in general. However, when strong existence holds the story is altogether clearer, essentially because a solution must exist on the canonical set-up corresponding to the input Brownian motion and the initial condition.

Theorem 6.20 *Suppose strong existence holds for the SDE (σ, b). Fix some probability distribution μ on $\mathbb{R}^n$ and consider the initial data $(\tilde{\Omega}, \{\mathcal{G}_t\}, \mathcal{G}, \tilde{\mathbb{P}}, \tilde{W}, \tilde{\xi})$ where $\{\mathcal{G}_t\}$ and $\mathcal{G}$ are as in Definition 6.1 and*

$$\tilde{\Omega} = \mathbb{R}^n \times \mathbb{C}^d,$$
$$\tilde{\mathbb{P}} = \mu \times \mathbb{Q}^d,$$
$$\tilde{\xi}(\omega) := \tilde{\xi}(r, x) = r,$$
$$\tilde{W}(\omega) := \tilde{W}(r, x) = w,$$

where $\mathbb{Q}^d$ is Wiener measure on $(\mathbb{C}^d, \mathcal{C}^d)$. By strong existence, for this initial data there exists some solution process $\tilde{X}$ such that

$$\tilde{X}_t = \tilde{\xi} + \int_0^t b(s, \tilde{X}.) \, ds + \int_0^t \sigma(s, \tilde{X}.) \, dW_s$$

satisfying the finiteness condition (6.3).
    It now follows that there exists a function

$$F_\mu : \mathbb{R}^n \times \mathbb{C}^d \to \mathbb{C}^n$$

such that, for all $A \in \mathcal{C}_t^n$,

$$F_\mu^{-1}(A) \in \mathcal{G}_t$$

and such that

$$\tilde{X}.(\omega) = F_\mu(\tilde{\xi}(\omega), \tilde{W}.(\omega))$$

for almost all ω.
    Furthermore, given any initial data $(\Omega, \{\mathcal{F}_t\}, \mathcal{F}, \mathbb{P}, W, \xi)$ with ξ having distribution μ, the process

$$X.(\omega) = F_\mu(\xi(\omega), W.(\omega))$$

satisfies

$$X_t = \xi + \int_0^t b(s, X.) \, ds + \int_0^t \sigma(s, X.) \, dW_s.$$

Remark 6.21: It is the existence of a solution on the canonical set-up for (ξ, W) (compare with Remark 6.13) which ensures existence of the map F_μ, and this in turn ensures there exists a solution on any set-up. Thus strong existence holds if and only if there is a solution on the canonical set-up for (ξ, W) for all random variables ξ.

A proof of this can be found in Rogers and Williams (1987). Note, of course, that the function F_μ is not uniquely defined because changing its definition on a $\mathbb{P}$-null set makes no difference. If strong uniqueness holds for the SDE then any two such functions satisfying the conditions of the theorem will have

$$\mathbb{P}\left(F_\mu^{(1)}(X_0(\omega), W(\omega)) = F_\mu^{(2)}(X_0(\omega), W(\omega)) \right) = 1.$$

Remark 6.22: In general the function F_μ depends on the initial distribution μ. Rogers and Williams (1987) discuss this point further in the context of stochastic flows.

6.4.3 Tying together strong and weak

We have introduced two types of solution to an SDE, two notions of existence and three of uniqueness. In this section we catalogue the relationship between them.

Clearly any strong solution is also a weak solution and strong existence certainly implies weak existence. On the uniqueness question we have already seen that pathwise uniqueness implies uniqueness in law and it is immediate that pathwise uniqueness also implies strong uniqueness. There are no other general relationships between any two of the concepts already introduced. However, when these are combined we get the following.

Theorem 6.23 (Yamada and Watanabe) *For an SDE (σ, b), weak existence plus pathwise uniqueness implies strong existence plus strong uniqueness.*

This is a powerful and surprising result. Weak existence states only that for each initial distribution μ there is *some* probability space $(\Omega, \{\mathcal{F}_t\}, \mathcal{F}, \mathbb{P})$ on which (W, X) can be defined, whereas strong existence states that there exists a solution for *all* $(\Omega, \{\mathcal{F}_t\}, \mathcal{F}, \mathbb{P})$ which support a Brownian motion W and a random variable ξ with distribution μ.

Remark 6.24: We pointed out in Remark 6.15 that our concept of a strong solution is different from that of Rogers and Williams (1987). In fact, their definition is slightly stronger and with their definition Theorem 6.23 can be strengthened to an equivalence, i.e. strong existence and strong uniqueness also implies weak existence and pathwise uniqueness. Rogers and Williams (1987) contains the details.

6.5 ESTABLISHING EXISTENCE AND UNIQUENESS: ITÔ THEORY

The general theory in the previous section defines and relates the different concepts of existence and uniqueness for solutions to an SDE but gives no method for establishing whether or not these properties hold for a given SDE. In this section we reproduce Itô's theory for strong solutions to (locally) Lipschitz SDEs. This theory shows that if the SDE (σ, b) is (locally) Lipschitz and satisfies a linear growth constraint then strong existence and pathwise uniqueness hold. This was the first work done on SDEs and is still the most important technique for generating strong solutions.

In finance applications the existence of a strong solution is not necessary since in this context the use of an SDE is to generate a model for the asset prices and there is no interpretation given to the driving Brownian motion as an input. It is convenient, however, to work with a locally Lipschitz SDE because existence and uniqueness results are relatively easy to prove, as is the strong Markov property. In this case, since strong existence and pathwise uniqueness hold, *all* the results of Section 6.4 apply. In particular, the solution is also unique in law (Theorem 6.12) and this yields the martingale representation result (Theorem 6.38) which we shall need when we discuss continuous time finance theory in Chapter 7.

Two topics that we shall not discuss in this book are the questions of when weak existence and the strong Markov property hold for an SDE. As mentioned above, both hold for (locally) Lipschitz SDEs, but this is a very particular and restrictive case.

The primary tool for the study of weak existence is Girsanov's theorem. The basic idea of this approach is to start with some (relatively simple) SDE (σ, b) for which we know there is a weak solution $(\Omega, \{\mathcal{F}_t\}, \mathcal{F}, \mathbb{P}, W, X)$. If we now change measure to some other measure $\mathbb{Q} \sim \mathbb{P}$ then under $\mathbb{Q}$ the sextuple $(\Omega, \{\mathcal{F}_t\}, \mathcal{F}, \mathbb{Q}, \widetilde{W}, X)$, $\widetilde{W}$ here being the Brownian motion of Corollary 5.23, will be a solution to a different SDE $(\tilde{\sigma}, \tilde{b})$. In practice, of course, the idea is used in reverse. Given some SDE $(\tilde{\sigma}, \tilde{b})$, it is reduced to a simpler one using a change of measure and the solution to this simpler SDE then provides a solution to $(\tilde{\sigma}, \tilde{b})$.

Another technique for establishing weak existence, one which is also important when studying the strong Markov property, is the *martingale problem*. This is a very important topic but one which we shall not consider further. The interested reader is referred to Revuz and Yor (1991), Rogers and Williams (1987) or Strook and Varadhan (1979).

6.5.1 Picard–Lindelöf iteration and ODEs

In the theory of ODEs it is known that, for any constant $\xi \in \mathbb{R}^n$, the equation

$$
\begin{aligned}
dx_t &= b(x_t)dt, \\
x_0 &= \xi,
\end{aligned}
\tag{6.7}
$$

has a unique solution if the function b is globally Lipschitz, i.e. if there exists some $K > 0$ such that, for all $x, y \in \mathbb{R}^n$,

$$
|b(x) - b(y)| < K|x - y|.
\tag{6.8}
$$

Without this condition (or some relaxation of it) it is difficult to say much in general. The way this result is proved is by use of a Picard–Lindelöf iteration as follows. Define a sequence of functions $\{x^{(k)}\}$ by

$$
\begin{aligned}
x_t^{(0)} &= \xi, \quad \text{for all } t \\
x^{(k+1)} &= \mathcal{G}x^{(k)},
\end{aligned}
$$

where

$$
(\mathcal{G}x)_t = \xi + \int_0^t b(x_u)du.
\tag{6.9}
$$

We fix some T and prove existence and uniqueness over $t \in [0, T]$ which is clearly enough. We begin by establishing an inequality which we shall use to prove that the sequence $\{x^{(k)}\}$ converges to a unique limit satisfying (6.7). The inequality we establish is an L^2 inequality, whereas it is more usual for ODEs to use a simpler L^1 inequality. We use the L^2 result because this can be extended to a stochastic equivalent which is needed when considering SDEs. Defining $z_t^* := \sup_{0 \le u \le t} |z_u|$, it follows from (6.9), the Cauchy–Schwarz inequality and (6.8) respectively that, for all $0 \le t \le T$,

$$
\begin{aligned}
(\mathcal{G}x - \mathcal{G}y)_t^{*2} &= \left(\sup_{0 \le s \le t} \left| \int_0^s (b(x_u) - b(y_u))du \right| \right)^2 \\
&\le \left(\int_0^t |b(x_u) - b(y_u)|du \right)^2 \\
&\le t \int_0^t |b(x_u) - b(y_u)|^2 du \\
&\le K^2 T \int_0^t (x - y)_u^{*2} du.
\end{aligned}
$$

By induction, therefore, for bounded, continuous x, y,

$$
(\mathcal{G}^k x - \mathcal{G}^k y)_t^{*2} \le (K^2 T)^k \frac{t^{k-1}}{(k-1)!} \int_0^T (x - y)_u^{*2} du \le \frac{(KT)^{2k}}{(k-1)!} (x - y)_T^{*2}.
\tag{6.10}
$$

We now establish existence of a solution to (6.7). Given any $m, k > 0$,

$$\left(x^{(m+k)} - x^{(k)}\right)_T^{*2} = \sum_{j=0}^{m-1} \left(x^{(k+j+1)} - x^{(k+j)}\right)_T^{*2}$$

$$= \sum_{j=0}^{m-1} \left(\mathcal{G}^{k+j} x^{(1)} - \mathcal{G}^{k+j} x^{(0)}\right)_T^{*2}$$

$$\leq \sum_{j=0}^{m-1} \frac{(KT)^{2(k+j)}}{(k+j-1)!} \left(x^{(1)} - x^{(0)}\right)_T^{*2}$$

$$\leq \sum_{j=0}^{\infty} \frac{(KT)^{2(k+j)}}{(-1)!j!} \left(x^{(1)} - x^{(0)}\right)_T^{*2}$$

$$\leq \exp\left((KT)^2\right) \frac{(KT)^{2k}}{(k-1)!} \left(x^{(1)} - x^{(0)}\right)_T^{*2}$$

$$\to 0,$$

as $k \to \infty$, so the sequence of functions $\{x^{(k)}\}$ is Cauchy for the supremum norm on $[0, T]$ and thus converges to some continuous limit function x which satisfies

$$\lim_{k \to \infty} \left(x - x^{(k)}\right)_T^{*2} = 0. \tag{6.11}$$

To show that this limit function x solves (6.7), observe that (6.10) implies that

$$\lim_{k \to \infty} \left(\mathcal{G}x - x^{(k)}\right)_T^{*2} = \lim_{k \to \infty} \left(\mathcal{G}x - \mathcal{G}x^{(k-1)}\right)_T^{*2}$$

$$\leq (KT)^2 \left(x - x^{(k-1)}\right)_T^{*2} \to 0,$$

so $x = \mathcal{G}x$ as required.

To establish uniqueness, note that any solution to (6.7) satisfies $x = \mathcal{G}^k x$ for all $k > 0$. Thus it follows from (6.10) that if x and y are two solutions, $(x - y)_T^{*2} = 0$, i.e. $x \equiv y$ and any solution to (6.7) is unique.

6.5.2 A technical lemma

Itô's idea was to apply a similar Picard–Lindelöf procedure to an SDE to obtain a pathwise unique solution. The key to the proof for the ODE was inequality (6.10) for which a stochastic analogue must be found. This is the content of this section.

Lemma 6.25 *Let* $(\Omega, \{\mathcal{F}_t\}, \mathcal{F}, \mathbb{P})$ *be a filtered probability space supporting a d-dimensional Brownian motion W and suppose the n-dimensional process Z is defined by*

$$Z_t = \int_0^t b_u du + \int_0^t \sigma_u dW_u$$

where b and σ are $\{\mathcal{F}_t\}$-predictable processes. Then, for all $t \geq 0$,

$$\mathbb{E}[Z_t^{*2}] \leq 2(4+t)\mathbb{E}\left[\int_0^t (|b_u|^2 + |\sigma_u|^2)du\right]. \qquad (6.12)$$

Now let ξ be an $\mathcal{F}_0$-measurable random variable and define the operator $\mathcal{G}$ (or $\mathcal{G}^\xi$ when we wish to emphasize the role of ξ) by

$$(\mathcal{G}X)_t = \xi + \int_0^t b(X_u)du + \int_0^t \sigma(X_u)dW_u,$$

where now σ and b are Borel measurable functions on $\mathbb{R}^n$ satisfying the Lipschitz conditions

$$\begin{aligned} |b(x) - b(y)| &\leq K|x-y|, \\ |\sigma(x) - \sigma(y)| &\leq K|x-y|, \end{aligned} \qquad (6.13)$$

and X is a continuous adapted process taking values in $\mathbb{R}^n$. Then, given any fixed $T > 0$ and for all $0 \leq t \leq T$, $k > 0$ and any continuous adapted processes X and Y,

$$\begin{aligned} \mathbb{E}[(\mathcal{G}^k X - \mathcal{G}^k Y)_t^{*2}] &\leq C_T^k \frac{t^{k-1}}{(k-1)!}\mathbb{E}\left[\int_0^T (X-Y)_u^{*2}du\right] \\ &\leq C_T^k \frac{T^k}{(k-1)!}\mathbb{E}[(X-Y)_T^{*2}], \end{aligned} \qquad (6.14)$$

where $C_T = 4K^2(4+T)$.

Proof: Write $Z_t = M_t + A_t$ where

$$M_t = \int_0^t \sigma_u dW_u,$$

$$A_t = \int_0^t b_u du.$$

Trivially, for each $t \geq 0$,

$$|Z_t|^2 \leq 2(|M_t|^2 + |A_t|^2),$$

and further

$$Z_t^{*2} \leq 2(M_t^{*2} + A_t^{*2}),$$

whence

$$E[Z_t^{*2}] \leq 2E[M_t^{*2}] + 2\mathbb{E}[A_t^{*2}].$$

From Doob's $\mathcal{L}^2$ inequality for a continuous local martingale in n dimensions (Exercise 6.26) we have

$$\mathbb{E}[M_t^{*2}] \leq 4\mathbb{E}\left[\left(\int_0^t \sigma_u dW_u\right)^2\right] = 4\mathbb{E}\left[\int_0^t |\sigma_u|^2 du\right],$$

and by Jensen's inequality,

$$\mathbb{E}[A_t^{*2}] \leq \mathbb{E}\left[\left(\int_0^t |b_u| du\right)^2\right] \leq \mathbb{E}\left[t\int_0^t |b_u|^2 du\right].$$

Inequality (6.12) now follows immediately.

We now prove (6.14). The second inequality is trivial. We prove the first by induction on k as follows. Apply (6.12) to $Z_t = (\mathcal{G}U - \mathcal{G}V)_t$, where U and V are continuous adapted processes, to obtain

$$\mathbb{E}\left[(\mathcal{G}U - \mathcal{G}V)_t^{*2}\right] \leq 2(4+t)\left(\mathbb{E}\left[\int_0^t |b(U_u) - b(V_u)|^2 du\right]\right.$$
$$\left. + \mathbb{E}\left[\int_0^t |\sigma(U_u) - \sigma(V_u)|^2 du\right]\right)$$
$$\leq C_T \mathbb{E}\left[\int_0^t |U_u - V_u|^2 du\right]$$
$$= C_T \int_0^t \mathbb{E}\left[|U_u - V_u|^2\right] du, \tag{6.15}$$

the last inequality following from the Lipschitz conditions (6.13). Setting $U = X, V = Y$ in (6.15) yields (6.14) when $k = 1$. For general k, set $U = \mathcal{G}^k X$, $V = \mathcal{G}^k Y$ and use the inductive hypothesis to obtain

$$\mathbb{E}\left[(\mathcal{G}^{k+1}X - \mathcal{G}^{k+1}Y)_t^{*2}\right] \leq C_T \int_0^t \mathbb{E}\left[|\mathcal{G}^k X_u - \mathcal{G}^k Y_u|^2\right] du$$
$$\leq C_T \int_0^t \mathbb{E}\left[(\mathcal{G}^k X - \mathcal{G}^k Y)_u^{*2}\right] du$$
$$\leq C_T \int_0^t C_T^k \frac{v^{k-1}}{(k-1)!}\mathbb{E}\left[\int_0^T (X - Y)_u^{*2} du\right] dv$$
$$= C_T^{k+1}\frac{t^k}{k!}\mathbb{E}\left[\int_0^T (X - Y)_u^{*2} du\right],$$

and we are done. $\qquad\qquad\square$

Exercise 6.26: Prove that Doob's $\mathcal{L}^2$ inequality for one-dimensional càdlàg martingales generalizes to *continuous* n-dimensional local martingales.

6.5.3 Existence and uniqueness for Lipschitz coefficients

Armed with Lemma 6.25, we are now ready to prove Itô's existence and uniqueness results for SDEs. The first theorem requires global Lipschitz restrictions on the coefficients of the SDE. We relax these later to local Lipschitz conditions in the presence of a linear growth constraint.

Theorem 6.27 *Suppose the (Borel measurable) functions b and σ satisfy the global Lipschitz growth conditions*

$$|b(x) - b(y)| \leq K|x - y|$$
$$|\sigma(x) - \sigma(y)| \leq K|x - y|$$

for all $x, y \in \mathbb{R}^n$ and some $K > 0$. Then strong existence and pathwise uniqueness hold for the SDE (σ, b).

Proof: Let $(\Omega, \{\mathcal{F}_t\}, \mathcal{F}, \mathbb{P})$ be a probability space supporting a d-dimensional Brownian motion W and some $\mathcal{F}_0$-measurable random variable ξ. We must show there is a strong solution to (σ, b) (adapted to the filtration $\mathcal{F}_t^{(\xi, W)}$) with initial condition ξ. We construct such a solution in the case when ξ is bounded by using a stochastic anologue of Picard–Lindelöf iteration. This is sufficient since the general result then follows from this special case by truncation.

Let $\mathcal{G}$ be the operator of Lemma 6.25 and define

$$X_t^{(0)} = \xi, \quad \text{for all } t$$
$$X^{(k+1)} = \mathcal{G}X^{(k)} = \mathcal{G}^{k+1}X^{(0)}.$$

We show that for any fixed $T > 0$, $X^{(k)}$ converges a.s., uniformly on the interval $[0, T]$, to a continuous $\{\mathcal{F}_t^{(\xi, W)}\}$-adapted process X which satisfies $X = \mathcal{G}X$. Since T is arbitrary this is sufficient to establish that X is a strong solution to (σ, b).

So fix $T > 0$ and note that, for $t \in [0, T]$,

$$(X^{(1)} - X^{(0)})_t = \int_0^t b(\xi)du + \int_0^t \sigma(\xi)dW_u = tb(\xi) + \sigma(\xi)(W_t - W_0),$$

and thus, noting that ξ and W are independent,

$$\mathbb{E}\big[(X^{(1)} - X^{(0)})_t^{*2}\big] \leq 2t^2\mathbb{E}[b^2(\xi)] + 2\mathbb{E}[\sigma^2(\xi)]\mathbb{E}[W_t^{*2}]$$
$$\leq 2T^2\mathbb{E}[b^2(\xi)] + 8T\mathbb{E}[\sigma^2(\xi)]$$
$$=: B < \infty.$$

Now apply Lemma 6.25 to obtain

$$\mathbb{E}\big[(X^{(k+1)} - X^{(k)})_T^{*2}\big] = \mathbb{E}\big[(\mathcal{G}^k X^{(1)} - \mathcal{G}^k X^{(0)})_T^{*2}\big]$$
$$\leq C_T^k \frac{T^k}{(k-1)!}B,$$

and thus, by Chebyshev's inequality,

$$\mathbb{P}\left(\sup_{0 \le t \le T} (X_t^{(k+1)} - X_t^{(k)}) > 2^{-k}\right) \le 4^k C_T^k \frac{T^k}{(k-1)!} B. \tag{6.16}$$

The first Borel–Cantelli lemma can now be applied to (6.16) to conclude that

$$\mathbb{P}\left(\sup_{0 \le t \le T} (X_t^{(k+1)} - X_t^{(k)}) > 2^{-k} \text{ infinitely often}\right) = 0.$$

Let Ω^* be the set, having probability one, given by

$$\Omega^* := \left\{\omega : \sup_{0 \le t \le T} |X_t^{(k+1)}(\omega) - X_t^{(k)}(\omega)| > 2^{-k} \text{ finitely often}\right\}.$$

For $\omega \in \Omega^*$, we see that $\{X^{(k)}(\omega)\}$ is a Cauchy sequence in the space $\mathbb{C}([0,T], \mathbb{R}^n)$ (continuous functions from $[0, T]$ to $\mathbb{R}^n$) endowed with the supremum norm, $x_T^* := \sup_{0 \le t \le T} |x_t|$, and hence converges to a limit $X_\cdot(\omega)$. That is, $X^{(k)}$ converges a.s. uniformly on $[0, T]$ to the continuous limit process X. This limit process is $\{\mathcal{F}_t^{(\xi, W)}\}$-adapted since each $X_t^{(k)}$ is $\mathcal{F}_t^{(\xi, W)}$-measurable and the a.s. limit of a measurable random variable is measurable. It can now also be concluded, using (6.14) and an argument similar to the one used to establish (6.11), that

$$\lim_{k \to \infty} \mathbb{E}[(X - X^{(k)})_T^{*2}] = 0.$$

That X solves $X = \mathcal{G}X$ follows since, again from Lemma 6.25,

$$\begin{aligned}
\mathbb{E}[(X - \mathcal{G}X)_T^{*2}] &\le 2\mathbb{E}[(X - X^{(k)})_T^{*2}] + 2\mathbb{E}[(X^{(k)} - \mathcal{G}X)_T^{*2}] \\
&\le 2\mathbb{E}[(X - X^{(k)})_T^{*2}] + 2C_T\mathbb{E}[(X^{(k-1)} - X)_T^{*2}] \\
&\to 0
\end{aligned}$$

as $k \to \infty$, so $X^{(k)}$ also converges to $\mathcal{G}X$.

We conclude by establishing pathwise uniqueness. Let X and Y be two solution processes having the same initial data. Clearly, for any stopping time τ, $(\mathcal{G}^k X^\tau)_{T \wedge \tau} = (\mathcal{G}^k X)_{T \wedge \tau}$, and similarly for Y, so Lemma 6.25 applied to X^τ and Y^τ yields

$$\begin{aligned}
\mathbb{E}[(X^\tau - Y^\tau)_T^{*2}] &= \mathbb{E}[(X - Y)_{T \wedge \tau}^{*2}] \\
&= \mathbb{E}[(\mathcal{G}^k X - \mathcal{G}^k Y)_{T \wedge \tau}^{*2}] \\
&= \mathbb{E}[(\mathcal{G}^k X^\tau - \mathcal{G}^k Y^\tau)_{T \wedge \tau}^{*2}] \\
&\le \mathbb{E}[(\mathcal{G}^k X^\tau - \mathcal{G}^k Y^\tau)_T^{*2}] \\
&\le C_T^k \frac{T^k}{(k-1)!} \mathbb{E}[(X^\tau - Y^\tau)_T^{*2}]. \tag{6.17}
\end{aligned}$$

Taking $\tau = T_k := \inf\{t > 0 : |X_t| + |Y_t| > k\}$ in (6.17), the right-hand side of (6.17) is bounded, hence converges to zero as $k \to \infty$. Thus $X^{T_k} = Y^{T_k}$ a.s. on $[0, T]$. Letting $k \to \infty$, and noting that $T_k \uparrow T$ (since both X and Y are a.s. finite), now establishes that $X = Y$ a.s. on $[0, T]$. Since T is arbitrary, pathwise uniqueness holds. □

Corollary 6.28 *Suppose the (Borel measurable) functions b and σ are locally Lipschitz, meaning that for each $N > 0$ there exists some constant $K_N > 0$ such that*

$$|b(x) - b(y)| \leq K_N |x - y|,$$
$$|\sigma(x) - \sigma(y)| \leq K_N |x - y|, \tag{6.18}$$

whenever $|x|, |y| \leq N$. Suppose further that, for some $K > 0$, the following linear growth conditions also hold:

$$|b(x)| \leq K(1 + |x|),$$
$$|\sigma(x)| \leq K(1 + |x|). \tag{6.19}$$

Then strong existence and pathwise uniqueness hold for the SDE (σ, b).

Remark 6.29: As is the case also for ODEs, it is not possible to relax the statement of this theorem by removing the linear growth constraints.

Proof: Let

$$b^N(x) = b(x)\mathbb{1}_{\{|x| \leq N\}} + b\left(N\frac{x}{|x|}\right)\mathbb{1}_{\{|x| > N\}},$$

$$\sigma^N(x) = \sigma(x)\mathbb{1}_{\{|x| \leq N\}} + \sigma\left(N\frac{x}{|x|}\right)\mathbb{1}_{\{|x| > N\}}.$$

For each N the SDE (σ^N, b^N) is globally Lipschitz so admits a pathwise unique strong solution X^N by Theorem 6.27. Now define

$$\tau_N := \inf\{t > 0 : |X_t^N| > N\},$$
$$X_t := \lim_{N \to \infty} X_{t \wedge \tau_N}^N. \tag{6.20}$$

Note that, for all $M \geq N$, $(X_{t \wedge \tau_N}^M - X_{t \wedge \tau_N}^N) = 0$ since (σ^N, b^N) and (σ^M, b^M) agree for $|x| \leq N$. Hence the limit in (6.20) will exist a.s. if $\sup_N \tau_N = +\infty$ a.s., and will satisfy the SDE (σ, b). It will also be the unique such process since each X^N is unique.

We suppose that $|X_0|$ is bounded by some constant, A say, and consider only a finite interval $[0, T]$. The general case of unbounded X_0 follows by the usual truncation argument. Defining $E_t^N := \mathbb{E}[(X^N - X_0)_t^{*2}]$, it is clear that E^N is increasing and finite for all $t \in [0, T]$, thus differentiable Lebesgue almost everywhere. Furthermore, applying (6.12) and (6.19),

$$E_t^N \leq 4(4 + t)\mathbb{E}\left[K^2 \int_0^t (1 + |X_u^N|)^2 du\right]$$

$$\leq 8(4 + t)\mathbb{E}\left[K^2 \int_0^t (1 + |X_0|)^2 + (X^N - X_0)_u^{*2} du\right]$$

$$\leq 8(4 + T)K^2\left((A + 1)^2 T + \int_0^t E_u^N du\right)$$

$$=: A_T + B_T \int_0^t E_u^N du. \tag{6.21}$$

Gronwall's lemma (Lemma 6.30 below) applied to (6.21) now implies that $E_t^N \leq A_T e^{B_T t}$, thus

$$\mathbb{P}(\tau_N < t) \leq \frac{E_t^N}{N^2} \leq \frac{A_T e^{B_T T}}{N^2} \to 0$$

as $N \to \infty$.   □

Gronwall's lemma, just used, is a simple but very useful result which we call on again later.

Lemma 6.30 *Let f be some Lebesgue integrable real-valued function and suppose there exist some constants A and C such that*

$$f(t) \leq A + C \int_0^t f(u)\, du \tag{6.22}$$

for all $t \in [0, T]$. Then, for Lebesgue almost all $t \in [0, T]$,

$$f(t) \leq A e^{Ct}.$$

Proof: The function

$$h(t) := e^{-Ct} \int_0^t f(u) du, \qquad t \geq 0,$$

starts at $h(0) = 0$ and satisfies $h'(t) \leq A e^{-Ct}$ by (6.22). This implies that

$$\int_0^t f(u) du \leq \frac{A}{C}(e^{Ct} - 1).$$

Substitution back into (6.22) establishes the result.   □

Note that the SDE we have considered is a diffusion SDE but that time t could be one of the components so we already have results for the time-inhomogeneous case. The following theorem, one we shall appeal to later, explicitly identifies and separates out the time component.

Theorem 6.31 *Suppose the (Borel measurable) functions*

$$b : \mathbb{R}^n \times \mathbb{R}^+ \to \mathbb{R}^n,$$

$$\sigma : \mathbb{R}^n \times \mathbb{R}^+ \to \mathbb{R}^n \times \mathbb{R}^d$$

satisfy:

(i) *the local Lipschitz growth conditions, i.e. for each $N > 0$,*

$$\begin{aligned}
|b(x, t) - b(y, t)| &\leq K_N |x - y|, \\
|\sigma(x, t) - \sigma(y, t)| &\leq K_N |x - y|, \\
|b(x, s) - b(x, t)| &\leq K_N |t - s|, \\
|\sigma(x, s) - \sigma(x, t)| &\leq K_N |t - s|,
\end{aligned} \tag{6.23}$$

whenever $|x|, |y| \leq N$, $s \leq t \leq N$, *and some* $K_N > 0$;
(ii) the linear growth conditions

$$|b(x,t)| \leq K(1 + |x|),$$
$$|\sigma(x,t)| \leq K(1 + |x|), \tag{6.24}$$

for some $K > 0$.

Then strong existence and pathwise uniqueness hold for the SDE (σ, b).

6.6 STRONG MARKOV PROPERTY

We discussed the strong Markov property in Section 2.4, where we proved that Brownian motion is strong Markov. Here we show that the solution process of a locally Lipschitz SDE is also strong Markov, a property essentially inherited from the strong Markov property of the driving Brownian motion. Recall that the basic idea behind the strong Markov property is that a process X is strong Markov if, given any finite stopping time τ, the evolution of X beyond τ is dependent on the evolution prior to τ only through its value at τ. The formal definition is as follows.

Definition 6.32   Let $(\Omega, \{\mathcal{F}_t\}, \mathcal{F}, \mathbb{P})$ be a filtered probability space supporting an $\mathbb{R}^n$-valued process adapted to $\{\mathcal{F}_t\}$. We say that X is a strong Markov process if, given any a.s. finite $\{\mathcal{F}_t\}$ stopping time τ, any $\Gamma \in \mathcal{B}(\mathbb{R}^n)$, and any $t \geq 0$,

$$\mathbb{P}(X_{\tau+t} \in \Gamma | \mathcal{F}_\tau) = \mathbb{P}(X_{\tau+t} \in \Gamma | X_\tau) \quad \text{a.s.} \tag{6.25}$$

From (6.25) we can immediately conclude, using the tower property, that given any $t_1 < \ldots < t_k$, $\Gamma_i \in \mathcal{B}(\mathbb{R}^n), i = 1, \ldots, k$,

$$\mathbb{P}(X_{\tau+t_i} \in \Gamma_i, i = 1, \ldots, k | \mathcal{F}_\tau) = \mathbb{P}(X_{\tau+t_i} \in \Gamma_i, i = 1, \ldots, k | X_\tau) \quad \text{a.s.}$$

Further, as shown in Karatzas and Shreve (1991, p. 82),

$$\mathbb{P}(X_{\tau+\cdot} \in \Gamma | \mathcal{F}_\tau) = \mathbb{P}(X_{\tau+\cdot} \in \Gamma | X_\tau) \quad \text{a.s.}$$

for all $\Gamma \in \mathcal{B}((\mathbb{R}^n)^{[0,\infty)})$ (the Borel σ-algebra on the space of functions from $\mathbb{R}^+$ to $\mathbb{R}^n$. So (6.25) is a much stronger result than it may at first appear and is precisely the property that we described earlier. An equivalent formulation to (6.25), the one we shall appeal to when verifying the strong Markov property, is the following standard result (see, for example, Karatzas and Shreve (1991)).

Theorem 6.33 *The process X is strong Markov if and only if, for all a.s. finite $\{\mathcal{F}_t\}$ stopping times τ and all $t \geq 0$,*

$$\mathbb{E}[f(X_{\tau+t})|\mathcal{F}_\tau] = \mathbb{E}[f(X_{\tau+t})|X_\tau] \tag{6.26}$$

for all bounded continuous f.

As mentioned earlier, the strong Markov property of the solution process for a locally Lipschitz SDE (σ, b) is essentially inherited from the strong Markov property of the driving Brownian motion. Theorem 6.36, the main result of this section, establishes this fact. In order to prove this result we first need two lemmas which show that the solution process for a globally Lipschitz SDE is well behaved as a 'function' of the initial distribution.

Lemma 6.34 *Let X and Y be two solution processes, on some probability space $(\Omega, \{\mathcal{F}_t\}, \mathcal{F}, \mathbb{P})$, for the SDE (σ, b) and suppose that σ and b are globally Lipschitz (i.e. satisfy (6.13)). Then, for all $T > 0$ and all $t \in [0, T]$,*

$$\mathbb{E}[(X_t - Y_t)^{*2}] \leq D_T \mathbb{E}[(X_0 - Y_0)^2]$$

where

$$D_T = 2\exp(8K^2(4 + T)T).$$

Proof: Define

$$Z_t = X_t - Y_t,$$
$$\hat{Z}_t = (X_t - X_0) - (Y_t - Y_0),$$

and note that

$$Z_t^{*2} \leq 2|X_0 - Y_0|^2 + 2\hat{Z}_t^{*2}. \tag{6.27}$$

Since X and Y are solution processes it follows that $X = \mathcal{G}^{X_0}X$, $Y = \mathcal{G}^{Y_0}Y$, so applying (6.12) and the global Lipschitz conditions yields

$$\mathbb{E}[\hat{Z}_t^{*2}] \leq 2(4+t)\mathbb{E}\left[\int_0^t |b(X_u) - b(Y_u)|^2 + |\sigma(X_u) - \sigma(Y_u)|^2\, du\right]$$

$$\leq 4(4+t)K^2 \int_0^t \mathbb{E}[Z_u^{*2}]\, du. \tag{6.28}$$

Defining $h_t := \mathbb{E}[Z_t^{*2}]$, it follows from (6.27) and (6.28) that

$$h_t \leq 2\mathbb{E}[|X_0 - Y_0|^2] + 2\mathbb{E}[\hat{Z}_t^{*2}]$$

$$\leq 2\mathbb{E}[|X_0 - Y_0|^2] + 8(4+t)K^2 \int_0^t h_u\, du.$$

The result now follows from Gronwall's lemma (Lemma 6.30). $\qquad \square$

Lemma 6.35 *Let ξ be some $\mathcal{F}_0$-measurable random variable and suppose the (Borel measurable) functions σ and b are globally Lipschitz, i.e. satisfy (6.13). Now suppose that $(\Omega, \{\mathcal{F}_t\}, \mathcal{F}, \mathbb{P}, W, X^\xi)$ is a (strong) solution to the SDE (σ, b) with $X_0 = \xi$ a.s. Then, for all bounded continuous functions $f : \mathbb{R}^n \to \mathbb{R}$,*

$$\mathbb{E}[f(X_t^\xi)|\mathcal{F}_0] = v(\xi, t)$$

where, for all $x \in \mathbb{R}^n$, the function v (which is continuous in x) is given by

$$v(x, t) := \mathbb{E}[f(X_t^x)].$$

Proof: The solution process X^ξ solves

$$X_t^\xi = \xi + \int_0^t b(X_u)\, du + \int_0^t \sigma(X_u)\, dW_u. \tag{6.29}$$

Let $f : \mathbb{R}^n \to \mathbb{R}$ be some bounded continuous function. If $\xi = x$ a.s., for some $x \in \mathbb{R}^n$, it suffices to prove that

$$\mathbb{E}[f(X_t^x)|\mathcal{F}_0] = \mathbb{E}[f(X_t^x)] \quad \text{a.s.} \tag{6.30}$$

This is immediate since x is a constant and the Brownian motion in (6.29) is independent of $\mathcal{F}_0$.

If ξ can take only countably many values $\{\xi_j : j \geq 1\}$ the result follows from (6.30) by summation (note that ξ is $\mathcal{F}_0$-measurable),

$$\mathbb{E}[f(X_t^\xi)|\mathcal{F}_0] = \sum_{j=1}^\infty \mathbb{1}_{\{\xi=\xi_j\}} \mathbb{E}[f(X_t^{\xi_j})|\mathcal{F}_0]$$

$$= \sum_{j=1}^\infty \mathbb{1}_{\{\xi=\xi_j\}} v(\xi_j, t)$$

$$= v(\xi, t).$$

The general case now follows by approximation of ξ by a random variable taking at most countably many values. Given a general ξ, let $\{\xi_k\}$ be some sequence of random variables converging to ξ in $\mathcal{L}^2$ (for example, each ξ_k could be the random variable ξ rounded in each component to the nearest $1/k$). It follows from Lemma 6.34 that, for each $t \geq 0$, $X_t^{\xi_k} \to X_t^\xi$ in $\mathcal{L}^2$, and this in turn implies that $f(X_t^{\xi_k}) \to f(X_t^\xi)$ in $\mathcal{L}^2$. Consequently,

$$\mathbb{E}\left[\left(\mathbb{E}[f(X_t^{\xi_k})|\mathcal{F}_0] - \mathbb{E}[f(X_t^\xi)|\mathcal{F}_0]\right)^2\right] = \mathbb{E}\left[\left(\mathbb{E}[f(X_t^{\xi_k}) - f(X_t^\xi)|\mathcal{F}_0]\right)^2\right]$$

$$\leq \mathbb{E}\left[\left(f(X_t^{\xi_k}) - f(X_t^\xi)\right)^2\right] \to 0$$

as $k \to \infty$, so $\mathbb{E}[f(X_t^{\xi_k})|\mathcal{F}_0] \to \mathbb{E}[f(X_t^\xi)|\mathcal{F}_0]$ in $\mathcal{L}^2$. But $\mathcal{L}^2$ convergence implies a.s. convergence along a fast subsequence so, given any $\{\xi_k\}$ such that $\xi_k \to \xi$ in $\mathcal{L}^2$, there is some subsequence $\{\xi_{k'}\}$ along which

$$\mathbb{E}[f(X_t^{\xi_{k'}})|\mathcal{F}_0] \to \mathbb{E}[f(X_t^\xi)|\mathcal{F}_0] \quad \text{a.s.} \tag{6.31}$$

An immediate consequence of this last observation, taking ξ to be some arbitrary constant and $\{\xi_k\}$ to be an arbitrary sequence of constants converging to ξ, is that the function $v(x,t)$ is continuous in x. This continuity along with a second application of (6.31) now completes the proof, since given an arbitrary random variable ξ there is a fast subsequence $\{\xi_{k'}\}$ along which (6.31) holds and thus, for almost every ω,

$$\mathbb{E}[f(X_t^\xi)|\mathcal{F}_0](\omega) = \lim_{k' \to \infty} \mathbb{E}[f(X_t^{\xi_{k'}})|\mathcal{F}_0](\omega)$$
$$= \lim_{k' \to \infty} v(\xi_{k'}(\omega),t) = v(\xi(\omega),t).$$

$\square$

Armed with the regularity result established in Lemma 6.35, we are now able to establish the strong Markov property.

Theorem 6.36 *Suppose the SDE (σ,b) is globally Lipschitz and let $(\Omega, \{\mathcal{F}_t\}, \mathcal{F}, \mathbb{P}, W, X)$ be some solution. Then the solution process X is strong Markov, i.e. (6.26) holds for all bounded continuous f and all a.s. finite $\{\mathcal{F}_t\}$ stopping times τ.*

Proof: Given any $\{\mathcal{F}_t\}$ stopping time $\tau < \infty$ a.s., we have, for almost every ω and all $t \geq 0$,

$$X_{\tau+t}(\omega) = X_0(\omega) + \int_0^{\tau(\omega)+t} b(X_u(\omega))\,du + \int_0^{\tau(\omega)+t} \sigma(X_u(\omega))\,dW_u(\omega)$$
$$= X_{\tau(\omega)}(\omega) + \int_{\tau(\omega)}^{\tau(\omega)+t} b(X_u(\omega))\,du + \int_{\tau(\omega)}^{\tau(\omega)+t} \sigma(X_u(\omega))\,dW_u(\omega)$$
$$= X_\tau + \int_0^t b(X_{\tau+u})\,du + \int_0^t \sigma(X_{\tau+u})\,d\widetilde{W}_u \tag{6.32}$$

where $\widetilde{W}_t = W_{\tau+t} - W_\tau$ and in the last line we have suppressed ω in the notation. Note, by the strong Markov property of Brownian motion (Theorem 2.15), that $\widetilde{W}$ is an $\{\mathcal{F}_{\tau+t} : t \geq 0\}$ Brownian motion independent of $\mathcal{F}_\tau$. Thus (6.32) shows that the process $\widetilde{X}. := X_{\tau+}.$ is a strong solution of the SDE (σ,b) for the initial data $(\Omega, \{\tilde{\mathcal{F}}_t\}, \mathcal{F}, \mathbb{P}, \widetilde{W}, X_\tau)$ where $\tilde{\mathcal{F}}_t = \mathcal{F}_{\tau+t}$. It follows from Lemma 6.35 that

$$\mathbb{E}[f(X_{\tau+t})|\mathcal{F}_\tau] = \mathbb{E}[f(\widetilde{X}_t)|\tilde{\mathcal{F}}_0] = v(\widetilde{X}_0,t) = v(X_\tau,t). \tag{6.33}$$

Finally, observe that the right-hand side of (6.33) is $\sigma(X_\tau)$-measurable, hence so is the left. It follows that

$$\mathbb{E}[f(X_{\tau+t})|\mathcal{F}_\tau] = \mathbb{E}[f(X_{\tau+t})|X_\tau],$$

which establishes (6.26) as required. $\square$

Our final result in this section extends Theorem 6.36 to the case of locally Lipschitz coefficients subject to a linear growth constraint. This localization argument follows familiar lines.

Corollary 6.37   *Theorem 6.36 remains valid if the global Lipschitz requirement on the SDE is relaxed to the local Lipschitz conditions (6.18) combined with the linear growth condition (6.19).*

Proof: Let X be a solution process for the SDE (σ, b) and define $S_N = \inf\{t > 0 : |X_t| > N\}$. Denote by X^N the solution to the globally Lipschitz SDE (σ^N, b^N) defined in the proof of Corollary 6.28. By Theorem 6.36, X^N is strong Markov and thus, given any a.s. finite stopping time τ and bounded continuous f,

$$\mathbb{E}[f(X^N_{\tau+t})|\mathcal{F}_\tau] = v^N(X^N_\tau, t) \quad \text{a.s.}$$

where

$$v^N(x, t) := \mathbb{E}[f(X_t^{(N,x)})],$$

$X^{(N,x)}$ being the process with initial condition $X^N_0 = x$ a.s. Noting that $X(\omega)$ and $X^N(\omega)$ agree for $t \le S_N(\omega)$, and defining $v(x, t)$ as in Lemma 6.35, we find

$$\left|\mathbb{E}[f(X_{\tau+t})|\mathcal{F}_\tau] - v(X_\tau, t)\right| = \left|\mathbb{E}\left[(f(X_{\tau+t}) - f(X^N_{\tau+t}))|\mathcal{F}_\tau\right]\right.$$
$$\left. - \left(v(X_\tau, t) - v^N(X^N_\tau, t)\right)\right|$$
$$\le \left|\mathbb{E}\left[(f(X_{\tau+t}) - f(X^N_{\tau+t}))|\mathcal{F}_\tau\right]\right|$$
$$+ \left|v(X_\tau, t) - v^N(X_\tau, t)\right|$$
$$+ \left|v^N(X_\tau, t) - v^N(X^N_\tau, t)\right|$$
$$\le 2B\mathbb{E}[\mathbb{1}_{\{\tau+t>S_N\}}|\mathcal{F}_\tau] + \left|v(X_\tau, t) - v^N(X_\tau, t)\right|$$
$$+ 2B\mathbb{1}_{\{\tau>S_N\}}, \qquad (6.34)$$

where B is some constant bounding f. Letting $N \to \infty$, the third term on the right-hand side of (6.34) converges to zero a.s., as does the first (applying the dominated convergence theorem). Given any $x \in \mathbb{R}^n$,

$$|v(x, t) - v^N(x, t)| = \left|\mathbb{E}[f(X_t^x) - f(X_t^{(N,x)})]\right| \le 2B\mathbb{P}(S_N < t) \to 0$$

as $N \to \infty$. Hence the second term on the right-hand side of (6.34) also converges to zero a.s. and this establishes the result. □

6.7 MARTINGALE REPRESENTATION REVISITED

We established in Corollary 5.47 and Theorem 5.50 that uniqueness of a (local) martingale measure implied martingale representation. This in turn can be related to the uniqueness of the solution to an SDE which the underlying martingale satisfies. Theorem 5.49, the martingale representation theorem for Brownian motion, was a special case of this result.

Theorem 6.38 *Suppose that weak uniqueness holds for the SDE (σ, b). Then, given any solution $(\Omega, \{\mathcal{F}_t\}, \mathcal{F}, \mathbb{P}, W, X)$ of (σ, b), any $\{\mathcal{F}_t^X\}$ local martingale N can be represented in the form*

$$N_t = N_0 + \int_0^t \phi_u \cdot dX_u^{\mathrm{loc}, \mathbb{P}},$$

for some $\{\mathcal{F}_t^X\}$-predictable ϕ.

Proof: By Theorem 5.50, it suffices to prove that, if $\mathbb{Q} \sim \mathbb{P}$ and $X^{\mathrm{loc}, \mathbb{P}}$ is an $(\{\mathcal{F}_t^X\}, \mathbb{Q})$ local martingale, then $\mathbb{P} = \mathbb{Q}$. Note that

$$X_t^{\mathrm{loc}, \mathbb{P}} = X_t - X_0 - \int_0^t b(X_u) du$$

and thus, since X solves (σ, b), $X^{\mathrm{loc}, \mathbb{P}}$ satisfies the SDE

$$dX_t^{\mathrm{loc}, \mathbb{P}} = \sigma(X_t) dW_t.$$

By Girsanov's theorem (Corollary 5.26), since $X^{\mathrm{loc}, \mathbb{P}}$ is also an $(\{\mathcal{F}_t^X\}, \mathbb{Q})$ local martingale,

$$dX_t^{\mathrm{loc}, \mathbb{P}} = \sigma(X_t) d\widetilde{W}_t$$

and so

$$dX_t = b(X_t) dt + \sigma(X_t) d\widetilde{W}_t$$

for some $\mathbb{Q}$ Brownian motion $\widetilde{W}$. But (σ, b) admits a unique weak solution, and X satisfies the same SDE under both $\mathbb{P}$ and $\mathbb{Q}$, so X has the same law under $\mathbb{P}$ and $\mathbb{Q}$. This, in turn, implies that $\mathbb{P} = \mathbb{Q}$ on $\{(\mathcal{F}_t^X)^\circ\}$, and indeed on $\{\mathcal{F}_t^X\}$ since $\mathbb{P} \sim \mathbb{Q}$. □

7

Option Pricing in Continuous Time

At last we are able to develop the theory of option pricing in continuous time, and the probabilistic tools which we have developed in Chapters 2–6 are just what we need to do it. This theory, when developed carefully, is quite technical but the basic ideas at the centre of the whole theory are the same as for discrete time, replication and arbitrage. If you have not already done so you should now read Chapter 1 which contains all the important concepts and techniques which will be developed in this chapter, but in a much simpler and mathematically more straightforward setting. We will assume throughout this chapter that you are familiar with Chapter 1.

There is a conflict when writing about a technical theory. If you include all the technical conditions as and when they arise the development of the theory can be very dry and the main intuitive ideas can be missed. On the other hand, if the technicalities are omitted, or deferred until later, it can be difficult for the reader to know exactly what conditions apply and, for example, to decide whether any particular model is one to which the theory applies. If working through this chapter you feel you are missing some important intuition, Chapter 1 should help. If there are details which confuse you at first, read on and spend more time on them second time through.

This chapter is organized as follows. In Section 7.1 we define a continuous time economy. An economy consists of a set of assets and a set of admissible trading strategies. For technical reasons we will not be able to completely specify exactly which strategies are allowed until we have developed the theory a little further. We have therefore given a partial description of admissible strategies in Section 7.1.2 and leave until Section 7.3.3 the final definition. With the exception of this one detail, Section 7.1 defines the probabilistic set-up for our economy. The conditions we impose at this stage will be assumed in all results later in the chapter.

Section 7.2 contains most of the intuition behind the results of continuous time option pricing theory. We derive the standard partial differential

Financial Derivatives in Theory and Practice Revised Edition.   P. J. Hunt and J. E. Kennedy
© 2004 John Wiley & Sons, Ltd   ISBNs: 0-470-86358-7 (HB); 0-470-86359-5 (PB)

equations for the price of a European derivative and an expression for the value as an expectation. The development in this section is as free of technical details as we can manage without sacrificing accuracy. The price we pay is that we have to make various assumptions which we do not fully justify until Section 7.3. The restriction to European derivatives is for clarity and various generalizations are possible which we cover in Section 7.4.

In Section 7.3 we give a rigorous development of the theory. To do this we must introduce numeraires, martingale measures and admissible trading strategies. We develop here results which explain when an economy will be complete and hence when we will be able to price all the derivatives of interest to us. The treatment we provide is different to most others that we have encountered. We develop an $\mathcal{L}^1$ theory and highlight the importance of numeraires (which have recently become an increasingly important modelling tool). We also introduce a different (and more natural) notion of completeness than that usually considered. Finally, in Section 7.4, we consider several generalizations of the framework treated in Section 7.3. It turns out that the theory of Section 7.3 extends easily to derivatives in which an option holder is able to exercise controls (American options, for example), to assets which pay dividends, and to an economy over the infinite time horizon. The case where the number of underlying assets is infinite is discussed in Chapter 8.

7.1 ASSET PRICE PROCESSES AND TRADING STRATEGIES

Before we can start pricing derivatives we must first define the underlying economy in which we are working. An economy, which we will denote by $\mathcal{E}$, is defined by two components. The first, which we define in Section 7.1.1, is a model for the evolution of the prices of the assets in the economy. The second, the subject of Section 7.1.2, is the set of trading strategies that we will allow within the economy. Throughout we shall restrict attention to some finite time interval $[0, T]$. We discuss the case of an infinite horizon in Section 7.4.

7.1.1 A model for asset prices

Let $W = (W^{(1)}, W^{(2)}, \ldots, W^{(d)})$ be a standard d-dimensional Brownian motion on some filtered probability space $(\Omega, \{\mathcal{F}_t\}, \mathcal{F}, \mathbb{P})$ satisfying the usual conditions (recall that we need these to ensure that stochastic integrals are well defined). In addition, we suppose that $\mathcal{F}_0$ is the completion of the trivial σ-algebra, $\{\emptyset, \Omega\}$.

The economy $\mathcal{E}$ consists of n assets with price process given by $A = (A^{(1)}, A^{(2)}, \ldots, A^{(n)})$. The process A is assumed to be continuous and almost surely finite. We will insist that A_0 is measurable with respect $\mathcal{F}_0$, and thus some constant known at time zero, and that A satisfies an SDE of the form

$$dA_t = \mu(A_t, t)\, dt + \sigma(A_t, t)\, dW_t\,, \qquad (7.1)$$

for some Borel measurable μ and σ. Recall from Chapter 6 that it is implicit in (7.1) that for any solution process A to (7.1) the finiteness condition

$$\int_0^t \left\{ |\mu(A_u, u)| + |\sigma(A_u, u)|^2 \right\} du < \infty \qquad \text{a.s.}$$

holds.

Chapter 6 provided an extensive study of SDEs of the form (7.1). If we are using such an SDE to specify a model we need to ensure that the given SDE has a solution, at least a weak solution, and that some kind of uniqueness result holds so that there is no ambiguity about the asset price model under consideration. Throughout this chapter we shall keep the discussion general but usually in practice the SDE (σ, μ) will be taken to be locally Lipschitz and subject to a linear growth condition. Recall from Theorem 6.31 that this ensures that a strong solution exists which is pathwise unique and unique in law, and that the process (A_t, t) is strong Markov. All the processes we shall encounter satisfy these local Lipschitz and linear growth conditions, (6.23) and (6.24) (although, in practice, we will not always explicitly present an SDE satisfied by the asset price processes).

Example 7.1 This example is the model first considered by Black and Scholes (1973) in their original paper on the subject of option pricing. It is one that we shall develop throughout this chapter. Start with a probability space $(\Omega, \mathcal{F}, \mathbb{P})$ which satisfies the usual conditions and which supports a one-dimensional Brownian motion W. Define two assets in the economy, the stock and the bond, whose price processes S and D satisfy the SDE

$$dS_t = \mu S_t dt + \sigma S_t dW_t\,,$$

$$dD_t = r D_t dt\,.$$

It is easy to check that the conditions (6.23) and (6.24) are satisfied so SDE (7.1) has a pathwise unique strong solution which is (apply Itô's formula to check)

$$S_t = S_0 \exp\big((\mu - \tfrac{1}{2}\sigma^2)t + \sigma W_t\big),$$

$$D_t = D_0 \exp(rt)\,.$$

We suppose that S_0 is some given constant and that the bond has unit value at the terminal time T, which then yields

$$D_t = \exp\big(-r(T - t)\big)\,.$$

7.1.2 Self-financing trading strategies

In this section we discuss trading strategies for a continuous time economy and some of their basic properties, these being natural generalizations of those we met in Chapter 1 for a discrete time economy. We will later, in Section 7.3.3, impose further technical constraints on those strategies which we will allow, *admissible strategies*. Without these technical constraints most economies of practical interest would admit arbitrage.

The first step in defining a trading strategy is to decide what information will be available to make the trading decision at any time t. The flow of information available in the economy is represented by the asset filtration $\{\mathcal{F}_t^A\}$ which we define formally in Section 7.3.1. Intuitively, the information available in this filtration at any time t is the past realization of the price process.

A trading strategy will be specified by a stochastic process $\phi = (\phi^{(1)}, \ldots, \phi^{(n)})$ where $\phi_t^{(i)}$ is interpreted as the number of units of asset i held at time t, before time t trading. Consequently, the value at time t of the portfolio is given by $V_t = \phi_t \cdot A_t$. A decision to rebalance the portfolio at time t must be based on information available the instant before time t, thus we require ϕ to be *predictable* with respect to the filtration $\{\mathcal{F}_t^A\}$.

Associated with any trading strategy is its *capital gain*. This is just the change in value of the portfolio due to trading in the assets. Consider first a *simple trading strategy*, one which is piecewise constant and bounded, and thus of the form

$$\phi_t = \phi_\ell 1(\tau_\ell, \tau_{\ell+1}], \qquad \ell = 0, 1, \ldots, p,$$

where $\phi_\ell \in \mathbb{R}^n$ and $0 = \tau_0 < \ldots < \tau_{p+1}$ are $\{\mathcal{F}_t^A\}$ stopping times. For such a strategy we define the capital gain G_t^ϕ (which can be negative) in the obvious way:

$$G_t^\phi = \sum_{\ell=0}^p \phi_\ell \cdot \left(A_{t \wedge \tau_{\ell+1}} - A_{t \wedge \tau_\ell} \right). \tag{7.2}$$

Note that ϕ is predictable – decisions about holdings of assets are made the instant before subsequent price movements occur.

This class of simple strategies is too restrictive for modelling purposes as it does not enable us to replicate and thus price all derivatives of interest. We need to define the capital gain for other more general trading strategies and we *choose to do this* via the Itô integral.

Definition 7.2 *The capital gain G^ϕ associated with any $(\{\mathcal{F}_t^A\}$-predictable) trading strategy ϕ for the economy $\mathcal{E}$ is defined to be the Itô integral*

$$G_t^\phi = \int_0^t \phi_u \cdot dA_u.$$

Remark 7.3: Note that this definition of the capital gain is indeed a modelling assumption. For trading strategies where trading occurs infinitesimally one could in principle make an alternative definition which is consistent with (7.2), such as the Stratonovich integral. However this alternative definition would lack certain important properties which the capital gain process should possess (see equation (7.27) and Remark 7.5).

Remark 7.4: We saw how to define the Itô integral in Chapter 4 and discovered some of its properties. Recall the fact proved there, Theorem 4.15 and Remark 4.16, that if the integrand (trading strategy) ϕ is left-continuous then, for almost every ω,

$$\left(\int_0^t \phi_u \cdot dA_u \right)(\omega) = \lim_{n \to \infty} \sum_{k \geq 1} \phi_{t \wedge T_{k-1}^n}(\omega) \cdot \left[A_{t \wedge T_k^n}(\omega) - A_{t \wedge T_{k-1}^n}(\omega) \right],$$

where, for each n,

$$T_0^n \equiv 0, \ T_{k+1}^n = \inf\{t > T_k^n : |\phi_t - \phi_{T_k^n}| > 2^{-n}\}.$$

Remark 7.5: The Itô integral has a second important property which makes this choice the appropriate one for financial modelling: Itô integrals preserve local martingales (Remark 4.22). We shall see in Section 7.3.3 that it is precisely this property which enables us to construct an arbitrage-free model.

There is one further restriction that we impose on the trading strategies that we allow in the economy $\mathcal{E}$. We will insist that no money is either injected into or removed from a trading portfolio other than the initial amount at time zero. Consequently, the value of the portfolio at time t, $\phi_t \cdot A_t$, will be the initial amount invested plus the capital gain from trading. Such a strategy is called *self-financing*. In particular, note that we have assumed that the market is frictionless, meaning there is no cost associated with rebalancing a portfolio. This is an essential assumption which we continue to make throughout this book. Without it the ideas of replication and no arbitrage cannot be applied to price derivatives.

Definition 7.6 *A self-financing trading strategy ϕ for the economy $\mathcal{E}$ is a stochastic process $\phi = (\phi^{(1)}, \dots, \phi^{(n)})$ such that:*

(i) ϕ is $\{\mathcal{F}_t^A\}$-predictable;
(ii) ϕ has the self-financing property,

$$\phi_t \cdot A_t = \phi_0 \cdot A_0 + \int_0^t \phi_u \cdot dA_u.$$

7.2 PRICING EUROPEAN OPTIONS

A *European option* is a derivative which at some terminal time T has a value given by some known function of the asset prices at time T. Our aim in this section is to convey the essential ideas behind pricing and hedging such an option without getting bogged down in technicalities. In later sections we fill in the gaps and generalize the derivatives that we consider.

Let $\mathcal{E}$ be an economy which is arbitrage-free (we will define precisely what we mean by this in Section 7.3.3) and consider a European option having value $V_T = F(A_T)$ at time T, where V_T is some $\mathcal{F}_T^A$-measurable random variable. We assume that there exists some admissible self-financing trading strategy which replicates the option, and thus its price is well defined for all $t \leq T$. Our aim is to calculate both this price and the replicating strategy for the option.

We begin by assuming that the value at time t of this derivative is given by some as yet unknown function $V(A_t, t)$ which is $C^{2,1}(\mathbb{R}^n, [0, T))$. This assumption must be justified later, after we have deduced the form of V. By the assumption of no arbitrage the value of the option at time t must be the time-t value of the replicating strategy (Section 7.3.4 formalizes this argument) and so

$$V(A_t, t) = V(A_0, 0) + \sum_{i=1}^{n} \int_0^t \phi_u^{(i)} dA_u^{(i)}. \tag{7.3}$$

On the other hand, applying Itô's formula to V yields

$$V(A_t, t) = V(A_0, 0) + \sum_{i=1}^{n} \int_0^t \frac{\partial V(A_u, u)}{\partial A_u^{(i)}} dA_u^{(i)} + \int_0^t \mathcal{G}V(A_u, u) \, du \tag{7.4}$$

where

$$\mathcal{G}V(A_t, t) := \frac{\partial V(A_t, t)}{\partial t} + \frac{1}{2} \sum_{i,j} a_{ij}(A_t, t) \frac{\partial^2 V(A_t, t)}{\partial A_t^{(i)} \partial A_t^{(j)}}$$

and

$$a_{ij}(A_t, t) = \sum_{k=1}^{d} \sigma_{ik}(A_t, t)\sigma_{jk}(A_t, t).$$

Now consider equations (7.3) and (7.4) together. Both will hold a.s. if for all $t \in [0, T)$ with probability one we have

$$\phi^{(i)}(A_t, t) = \frac{\partial V(A_t, t)}{\partial A_t^{(i)}} \tag{7.5}$$

and

$$\mathcal{G}V(A_t, t) = 0. \tag{7.6}$$

Then in addition, if (7.5) holds, the self-financing property of V means that the following equality holds a.s. for all $t \in [0, T)$:

$$V(A_t, t) = \sum_{i=1}^{n} \frac{\partial V(A_t, t)}{\partial A_t^{(i)}} A_t^{(i)}. \tag{7.7}$$

7.2.1 Option value as a solution to a PDE

A $C^{2,1}(\mathbb{R}^n, [0, T))$ function V satisfying equations (7.5)–(7.7) is a candidate for the value of the European option. Suppose possible values of the process A lie in a region $U \subset \mathbb{R}^n$. In order for (7.6) and (7.7) to hold it is necessary that V satisfies the *partial differential equation* (PDE): for all $(x, t) \in U \times [0, T)$,

$$\mathcal{G}V(x, t) = 0, \tag{7.8}$$

$$\sum_{i=1}^{n} x^{(i)} \frac{\partial V(x, t)}{\partial x^{(i)}} = V(x, t). \tag{7.9}$$

The requirement that $V_T = F(A_T)$ gives the *boundary condition*

$$V(x, T) = F(x), \qquad x \in U.$$

Solving this PDE is one way of pricing the option. If we can find a solution V to the above then it will be $C^{2,1}(\mathbb{R}^n, [0, T))$ and indeed (7.5) then defines a self-financing trading strategy which replicates the option.

Usually there will at this stage be many solutions $V(x, t)$ to (7.8) and (7.9). Unless we eliminate all but at most one of these the economy admits arbitrage (consider the difference between two replicating strategies). As noted in Section 7.1.2, it is therefore necessary to restrict the strategies that we allow in order to ensure no arbitrage for the economy, and this requirement forms an extra restriction on the solutions to the above set of equations.

The requirement that the replicating strategy be *admissible*, as we define in Section 7.3.3, is sufficient to ensure no arbitrage. Consequently, there will be at most a unique solution $V(x, t)$ to (7.8)–(7.9) corresponding to an admissible strategy. Therefore, if we can show that one of our solutions to (7.8)–(7.9) corresponds to an admissible strategy then this is indeed the price of our option. Note that at this stage we do not know that any such solution will exist which is $C^{2,1}(\mathbb{R}^n, [0, T))$; we assumed only that the option was attainable.

The PDE approach to option pricing is commonly used but will not be the primary focus of this book. The reader is referred to Duffie (1996), Lamberton and Lapeyre (1996) and Willmot *et al.* (1993) for more on this approach. The equations satisfied by the option price, (7.8)–(7.9) above, are of linear second-order parabolic form and these have been extensively studied and are well understood. Mild regularity conditions ensure the existence and uniqueness of the solution for suitable boundary conditions. In practice they are often solved by numerical techniques.

Example 7.7 Consider the Black–Scholes economy introduced in Example 7.1 and an option which pays

$$F(S_T, D_T) = (S_T - K)_+$$

at time T. If we denote by $V(S_t, D_t, t)$ the value of this option at time t, the PDEs (7.8) and (7.9) for the value of the option become

$$\frac{\partial}{\partial t} V(S_t, D_t, t) + \tfrac{1}{2}\sigma^2 S_t^2 \frac{\partial^2}{\partial S_t^2} V(S_t, D_t, t) = 0 \tag{7.10}$$

$$S_t \frac{\partial}{\partial S_t} V(S_t, D_t, t) + D_t \frac{\partial}{\partial D_t} V(S_t, D_t, t) = V(S_t, D_t, t), \tag{7.11}$$

all other terms in $\mathcal{G}V$ being zero because the bond is a finite variation process. Equations (7.10) and (7.11) are equivalent to the celebrated Black–Scholes PDE and can be solved explicitly subject to the boundary condition

$$V(S_T, D_T, T) = (S_T - K)_+ \,.$$

Anyone already familiar with deriving the Black–Scholes PDE may at this point be a little confused because (7.10)–(7.11) are not in the standard form. The reason for this is that the value function V is here parameterized in terms of S_t, D_t and t. In the more usual development the value function is parameterized in terms only of S_t and t. Since D_t is a known function of time this can be done, but in doing it some of the underlying symmetry and structure of the problem is lost (although it is more straightforward to solve a PDE in two variables than in three). To recover the more usual Black–Scholes PDE, let $\widehat{V}(S_t, t) = V(S_t, D_t, t)$. Then

$$\begin{aligned}
\frac{\partial \widehat{V}}{\partial t} &= \frac{\partial V}{\partial t} + \frac{\partial V}{\partial D_t}\frac{\partial D_t}{\partial t} \\
&= \frac{\partial V}{\partial t} + rD_t \frac{\partial V}{\partial D_t} \,.
\end{aligned} \tag{7.12}$$

Further, since V is self-financing,

$$\begin{aligned}
\widehat{V}(S_t, t) = V(S_t, D_t, t) &= S_t \frac{\partial V}{\partial S_t} + D_t \frac{\partial V}{\partial D_t} \\
&= S_t \frac{\partial \widehat{V}}{\partial S_t} + D_t \frac{\partial V}{\partial D_t} \,.
\end{aligned} \tag{7.13}$$

Substituting (7.13) into (7.12) now gives

$$\frac{\partial V}{\partial t} = \frac{\partial \widehat{V}}{\partial t} - r\left(\widehat{V} - S_t \frac{\partial \widehat{V}}{\partial S_t}\right)$$

and so (7.10) becomes

$$\left(\tfrac{1}{2}\sigma^2 S_t^2 \frac{\partial^2}{\partial S_t^2} + rS_t \frac{\partial}{\partial S_t} + \frac{\partial}{\partial t}\right)\widehat{V}(S_t,t) = r\widehat{V}(S_t,t)$$

which is the Black–Scholes PDE in its more familiar form. There are many ways to solve a PDE such as this. We leave this until the next section, where we derive the solution using the martingale approach.

7.2.2 Option pricing via an equivalent martingale measure

Central to our probabilistic formulation of the discrete theory of Chapter 1 was the existence of a probability measure with respect to which the process A, suitably rebased, was a martingale. No arbitrage was shown to be equivalent to the existence of an equivalent martingale measure (EMM) and (under the assumption of no arbitrage) completeness held if and only if this measure was unique. Analogous results, modulo technicalities, hold for continuous time. We return to a discussion of the general picture later. For now we focus our attention on how to use the ideas of Chapter 1 to price a European option, assuming that the economy is arbitrage-free.

When we write down a model such as (7.1) we are describing the law of the asset price process A in some arbitrarily chosen units such as pounds or dollars. These are in themselves not assets in the economy – they are merely units. As such we are free to redefine these units in any way we like, including choosing units that may evolve randomly through time, as long as we introduce no real economic effects.

For reasons which will become clear later, we will restrict attention to units which are also *numeraires*. For now it is enough to think of these just as units; later, in Section 7.3.2, we will introduce numeraires formally and the reasons for this restriction will become clear. So suppose we change units to those given by the numeraire N, where N is a strictly positive continuous semimartingale adapted to $\{\mathcal{F}_t^A\}$. Let A^N be the price process A quoted in the new units, so $A_t^N = N_t^{-1} A_t$, which is also a continuous semimartingale.

Now consider a replicating strategy ϕ having the self-financing property,

$$V_t = \phi_t \cdot A_t = V_0 + \int_0^t \phi_u \cdot dA_u.$$

Since the numeraire represents only a change of units and since ϕ is self-financing, it is intuitively clear that

$$V_t^N := N_t^{-1} V_t = V_0^N + \int_0^t \phi_u \cdot dA_u^N \tag{7.14}$$

and the reader can check that this is indeed the case using Itô's formula (this is the unit invariance theorem which we present formally in Section 7.3.2). Immediately from (7.14), any self-financing strategy remains self-financing under a change of units.

The motivation for this change of units is, as was the case in Chapter 1, to enable us to change to an EMM corresponding to the numeraire N. That is, suppose there exists some measure $\mathbb{N} \sim \mathbb{P}$ such that A^N is an $\{\mathcal{F}_t^A\}$ martingale under $\mathbb{N}$. If such a measure exists, it follows from the representation (7.14) that V^N is an $\{\mathcal{F}_t^A\}$ local martingale under $\mathbb{N}$. If V^N is in fact a true martingale, and our definition of an admissible in Section 7.3.3 will ensure that it is, we can now use the martingale property to find the option value,

$$V_t^N = \mathbb{E}_{\mathbb{N}}\left[V_T^N | \mathcal{F}_t^A\right],$$

and thus

$$V_t = N_t \mathbb{E}_{\mathbb{N}}\left[\frac{F(A_T)}{N_T}\bigg| \mathcal{F}_t^A\right]. \tag{7.15}$$

In general, to calculate the option value we need to know the law of the process (A, N) under the measure $\mathbb{N}$, and Girsanov's theorem is the result which gives us this law. The problem is often simplified in practice by a judicious choice of numeraire which reduces the dimensionality of the problem and leaves us only needing to calculate the law of A^N.

A review of this section should convince you that we have relied neither on the Markovian form of A nor on any $C^{2,1}$ regularity conditions for the value function V. Consequently, this martingale approach is more general than the PDE approach and can be applied to other more general models and to other derivatives. In this book we will not extend the class of models we consider beyond those of Section 7.1 but we will later make considerable use of being able to price more general derivatives, in particular path-dependent and American products.

Example 7.8 Consider once more the Black–Scholes model of Example 7.1 and take D as numeraire. Then by Itô's formula, writing $S_t^{(D)} = D_t^{-1} S_t$, we have

$$dS_t^{(D)} = (\mu - r)S_t^{(D)} dt + \sigma S_t^{(D)} dW_t.$$

Suppose an EMM $\mathbb{Q}$ exists corresponding to this numeraire. We know from Girsanov's theorem (Theorem 5.20) that a change of measure only alters the drift of a continuous semimartingale. This, combined with the fact that $S^{(D)}$ is a martingale under $\mathbb{Q}$, implies that

$$dS_t^{(D)} = \sigma S_t^{(D)} d\widetilde{W}_t,$$

where $\widetilde{W}$ is a standard Brownian motion under $\mathbb{Q}$. The solution to this SDE is given by

$$S_t^{(D)} = S_0^{(D)} \exp(\sigma \widetilde{W}_t - \tfrac{1}{2}\sigma^2 t) \tag{7.16}$$

and so
$$S_t = S_0 \exp(\sigma \widetilde{W}_t + rt - \tfrac{1}{2}\sigma^2 t).$$

A direct calculation confirms that $S^{(D)}$ is indeed a true martingale, and the process $D^{(D)} \equiv 1$ is trivially also a martingale under $\mathbb{Q}$.

We now return to the valuation problem of Example 7.7. Note that, with respect to the measure $\mathbb{Q}$, the law of S no longer depends on μ and hence neither will the option value V. This should not surprise you. In Chapter 1 we saw that the role of the probability measure $\mathbb{P}$ was purely as a marker for those outcomes (sample paths) which are possible, and the exact probabilities are not important. The drift of a continuous semimartingale is dependent on which equivalent measure we are working in and so cannot be relevant. The volatility σ, on the other hand, is a sample path property and is invariant under a change of probability measure. Consequently, it can and does have an effect on the option value.

Writing $A = (S, D)$ clearly here $\mathcal{F}_t^A = \mathcal{F}_t^W$. From (7.15) and (7.16) we now have

$$\begin{aligned}
V_t &= D_t \mathbb{E}_{\mathbb{Q}}\big[(S_T - K)_+ D_T^{-1} | \mathcal{F}_t^W\big] \\
&= D_t \mathbb{E}_{\mathbb{Q}}\big[S_T^{(D)} \mathbb{1}_{\{S_T^{(D)} > K\}} | \mathcal{F}_t^W\big] - D_t K \mathbb{E}_{\mathbb{Q}}\big[\mathbb{1}_{\{S_T^{(D)} > K\}} | \mathcal{F}_t^W\big] \qquad (7.17) \\
&= S_t N(d_1) - K D_t N(d_2)
\end{aligned}$$

where
$$d_1 = \frac{\log(S_t / K D_t)}{\sigma \sqrt{T - t}} + \tfrac{1}{2}\sigma \sqrt{T - t},$$

$$d_2 = \frac{\log(S_t / K D_t)}{\sigma \sqrt{T - t}} - \tfrac{1}{2}\sigma \sqrt{T - t},$$

and (7.17) has been evaluated by explicit calculation.

7.3 CONTINUOUS TIME THEORY

We are now familiar with the basic underlying ideas of derivative pricing. First we must define the economy along with those trading strategies which are *admissible*; then we must prove that the economy is *arbitrage-free*. Finally, given any derivative, we must prove that there exists some admissible trading strategy which replicates it, and the value of this derivative is then the value of its replicating portfolio. Showing the existence of a suitable replicating strategy is usually done by proving a *completeness* result, i.e. proving that all derivatives in some set (which includes the one under consideration) can be replicated by some admissible trading strategy.

What we have done so far in this chapter is provide techniques to find the value and replicating strategy when it exists, which is the most important thing for a practitioner to know. Now we fill in the remaining parts of the theory.

We begin in Section 7.3.1 with a brief discussion of the information observable within the economy. Then, in Section 7.3.2, we provide precise definitions for numeraires and EMMs, and derive some of their properties. We already know that these are important tools in discrete time, but in continuous time they are even more central to the theory. The first instance of this added importance comes in Section 7.3.3 where we define admissible strategies and arbitrage. In continuous time we must eliminate certain troublesome strategies from the economy, otherwise virtually any model admits arbitrage and is thus rendered useless for the purposes of option pricing. With our careful choice of admissible strategies the arbitrage-free property follows quickly, so we now know that the replication arguments can be applied to the economy.

Section 7.3.4 is a brief summary of the argument that shows that the value of a derivative in an arbitrage-free economy is the value of any replicating portfolio. This section pulls together in one place several ideas that we have already met earlier in the chapter. The last part of the general theory needed is a discussion of completeness, and this is the topic of Section 7.3.5. It is only here that we can fully appreciate the reasons for our carefully chosen definition of a numeraire. As we shall see, completeness (in a precise sense to be defined) for an economy is equivalent to the existence of a unique EMM (for any given numeraire). When weak uniqueness holds for the SDE defining the asset price process, equation (7.1), completeness is then equivalent to the existence of a finite variation numeraire. The weak uniqueness property of the SDE can be checked using standard techniques such as those we met in Chapter 6. The existence or otherwise of a finite variation numeraire is usually easily verified in practice.

The presentation of the theory given here places much more emphasis on numeraires than is usual. This is justified by the role they play in completeness, but more importantly by the need in much recent work to be able to switch freely between different numeraires and know that we are allowed to do so.

7.3.1 Information within the economy

Part of the definition of a trading strategy is the amount of information available at any time t on which to base the strategy ϕ. Recall that we insisted that ϕ be $\{\mathcal{F}_t^A\}$-predictable. Here we formally define this filtration and two others that are relevant to the economy.

Definition 7.9 *For an economy $\mathcal{E}$ defined on the probability space $(\Omega, \{\mathcal{F}_t\}, \mathcal{F}, \mathbb{P})$, the natural filtration, $\{\mathcal{F}_t^E\}$, the asset filtration, $\{\mathcal{F}_t^A\}$, and the Brownian filtration, $\{\mathcal{F}_t^W\}$, are defined to be the $\mathbb{P}$-augmentations of the following:*

(i) $(\mathcal{F}_t^E)^\circ = \sigma(A_u^{(i)}/A_u^{(j)} : u \le t, 1 \le i,j \le n);$

(ii) $(\mathcal{F}_t^A)^\circ = \sigma(A_u^{(i)} : u \le t, 1 \le i \le n);$

(iii) $(\mathcal{F}_t^W)^\circ = \sigma(W_u^{(i)} : u \le t, 1 \le i \le d).$

It is important to appreciate the significance of each of these to derivative pricing. The Brownian filtration contains all the information about the Brownian motions which underlie the economy. This filtration is important only at the start of the modelling process when we specify the asset price processes (Section 7.1.1). The asset filtration contains all the information that is available to us if we watch the asset prices evolve through time. This is what can be observed in practice and so all trading strategies should be based on information contained in the asset filtration and no more. The third of the filtrations, the natural filtration, contains all the information in the economy about *relative* asset prices. This is arguably the most fundamental of the three since it contains only information intrinsic to the economy. The asset filtration, by contrast, will depend on the units initially chosen to set up the economy.

7.3.2 Units, numeraires and martingale measures

We met the idea of a unit in Section 7.2. We shall formally reintroduce units in this section along with the more refined notion of a *numeraire*. Numeraires play an important role in the theory of option pricing, but that role has not been fully appreciated and exploited until very recently.

Definition 7.10 *A unit U for the economy $\mathcal{E}$ defined on the probability space $(\Omega, \{\mathcal{F}_t\}, \mathcal{F}, \mathbb{P})$ is any a.s. positive continuous semimartingale adapted to the filtration $\{\mathcal{F}_t\}$.*

Remark 7.11: What is important in the definition of a unit is that it be defined on the probability space on which we set up the economy and that it be positive. That is, it must exist and we need to be able to divide through by it.

Example 7.12 Define an economy $\mathcal{E}$ on the probability space $(\Omega, \{\mathcal{F}_t\}, \mathcal{F}, \mathbb{P})$ which supports d-dimensional Brownian motion W, and suppose that the asset price process solves the SDE

$$dA_t = \mu(A_t, t)dt + \sigma(A_t, t)d\widehat{W}_t,$$

where $\widehat{W}$ is $(d-1)$-dimensional Brownian motion, the first $(d-1)$ components of W. We can now define the unit $U_t = \exp\left(W_t^{(d)}\right)$ which is independent of all the assets in the economy. We have been able to do this because the probability space we have used to set up the economy is richer than we need to specify the asset price process.

If we change the units in which we describe the prices of the assets in the economy this should have no economic effect – it is just a different way of presenting price information. This is the content of the following simple yet very important result.

Theorem 7.13   (Unit invariance) *For any unit U, a trading strategy is self-financing with respect to A if and only if it is self-financing with respect to A^U.*

Proof: A process U is a.s. positive and finite if and only if U^{-1} is also a.s. positive and finite, so it suffices to prove the implication in one direction. So suppose ϕ is self-financing with respect to A. Writing $V_t = \phi_t \cdot A_t$, we have

$$V_t = V_0 + \int_0^t \phi_u \cdot dA_u.$$

We must show that $V_t^U := \phi_t \cdot A_t^U$ satisfies

$$V_t^U = V_0^U + \int_0^t \phi_u \cdot dA_u^U.$$

This follows from an application of the integration by parts formula:

$$
\begin{aligned}
dV_t^U &= U_t^{-1}dV_t + V_t d(U_t^{-1}) + dV_t d(U_t^{-1}) \\
&= \phi_t \cdot [U_t^{-1}dA_t + A_t d(U_t^{-1}) + dA_t d(U_t^{-1})] \\
&= \phi_t \cdot dA_t^U.
\end{aligned}
$$

$\square$

So changing units makes no material difference to our economy. However, introducing a new unit may change the information in the economy (i.e. $\{\mathcal{F}_t^{A^U}\} \neq \{\mathcal{F}_t^A\}$) and this may complicate calculations. There is, however, a class of units which introduce no additional information over and above that already available through the asset price filtration $\{\mathcal{F}_t^A\}$, numeraires. Indeed, using a numeraire unit *may* reduce the amount of information.

Definition 7.14 *A numeraire N for the economy $\mathcal{E}$ is any a.s. strictly positive $\{\mathcal{F}_t^A\}$-adapted process of the form*

$$
\begin{aligned}
N_t &= N_0 + \int_0^t \alpha_u \cdot dA_u \\
&= \alpha_t \cdot A_t,
\end{aligned}
$$

where α is $\{\mathcal{F}_t^A\}$-predictable.

Remark 7.15: N_t is the value at t of a portfolio which starts with value N_0 and is traded according to the self-financing strategy α. Note, however, that we have not yet defined which strategies will be admissible within our economy and so there is no assumption that α be admissible.

Remark 7.16: N_t is adapted to $\mathcal{F}_t^A$ and so $\mathcal{F}_t^A = \mathcal{F}_t^{(A,N)}$.

We now prove a result which is, as was the case for the unit invariance theorem, simple but of great importance. It is central to deciding when an economy is complete. What this theorem shows is that if we can represent a random variable, working in units of some numeraire N, with a strategy that is not necessarily self-financing, then we can find some strategy which also replicates the random variable but is self-financing. That is, the theorem is the link between the martingale representation theorem of Chapter 6, which shows how to represent a local martingale as a stochastic integral, and completeness, which additionally requires that the replicating strategy be self-financing. It is the fact that the unit is a numeraire that makes this result hold and thus this result demonstrates the importance of numeraires to questions of completeness.

Theorem 7.17 *Let N be some numeraire process and, for $0 \le t \le T$, let*

$$Z_t = Z_0 + \int_0^t \phi_u \cdot dA_u^N,$$

for some $\{\mathcal{F}_t^A\}$-predictable ϕ. Then there exists some $\{\mathcal{F}_t^A\}$-predictable $\hat{\phi}$ such that

$$Z_t = Z_0 + \int_0^t \hat{\phi}_u \cdot dA_u^N$$
$$= \hat{\phi}_t \cdot A_t^N.$$

That is, $\hat{\phi}$ is self-financing.

Proof: Since N is a numeraire process, we can write

$$N_t = N_0 + \int_0^t \alpha_u \cdot dA_u,$$

for some self-financing $\alpha \neq 0$. In particular, it follows from the unit invariance theorem that

$$\alpha_t \cdot A_t^N = 1, \qquad \alpha_t \cdot dA_t^N = 0.$$

This immediately yields a suitable strategy,

$$\hat{\phi}_t = \phi_t + (Z_t - \phi_t \cdot A_t^N)\alpha_t.$$

$\square$

The next important concept is that of *martingale measures* for the economy $\mathcal{E}$. We saw in Section 7.2 that martingale measures are useful for calculation. In Chapter 1 we saw their role in deciding whether an economy admits arbitrage and is complete. We will in the sections that follow provide analogous results for a continuous time economy. The results are basically the same as those we have met already but with extra complexities which are present in the richer continuous time setting. In particular, we shall also use martingale measures as a tool to specify exactly which trading strategies are admissible (the detail we omitted in Section 7.1.2). We begin with the definition.

Definition 7.18   *The measure* $\mathbb{N}$ *defined on the probability space* $(\Omega, \{\mathcal{F}_t\}, \mathcal{F}, \mathbb{P})$ *is a martingale measure (or equivalent martingale measure) for the economy* $\mathcal{E}$ *if* $\mathbb{N} \sim \mathbb{P}$ *and there exists some numeraire* N *such that* A^N *is an* $\{\mathcal{F}_t^A\}$ *martingale under the measure* $\mathbb{N}$. *The pair* $(N, \mathbb{N})$ *is then referred to as a numeraire pair.*

Remark 7.19: This definition of an EMM should be compared with Definition 5.40 of Chapter 5. In Definition 7.18 above, the process A^N must be a martingale under $\mathbb{N}$ but not necessarily under $\mathbb{P}$. Note, however, that if $\mathbb{N}$ and $\mathbb{M}$ are two EMMs corresponding to the same numeraire, N, then they are EMMs for the martingale A^N in the sense of Definition 5.40 (note that $\mathcal{F}_0^A$ is trivial, which ensures that condition (ii) of the definition holds).

Recall from our discussion of change of measure in Chapter 6 that if $\mathbb{Q} \sim \mathbb{P}$ are two equivalent measures with respect to $\mathcal{F}_T$ then, for any $A \in \mathcal{F}_T$,

$$\mathbb{Q}(A) = 0 \text{ if and only if } \mathbb{P}(A) = 0\,.$$

That is, a change of measure does not alter the sets of paths which are possible, it just changes the probabilities of these sets (from one strictly positive value to another strictly positive value). This is an important observation for option pricing which, as we saw in Chapter 1, is concerned with replication on a sample path by sample path basis. We do not want a change of probability measure to alter the set of paths which the economy can exhibit.

The question of whether or not a numeraire pair exists for any given economy $\mathcal{E}$ can be tricky to decide. In practice this is not a problem. We usually specify a model for an economy already in some martingale measure $\mathbb{N}$. It is, however, important to understand how the evolution of asset prices will vary between different martingale measures and between a martingale measure and any other equivalent measure. Girsanov's theorem, which we met in Chapter 5, is the tool that allows us to do this.

Suppose $(N, \mathbb{N})$ is a numeraire pair for the economy $\mathcal{E}$, where $N_t = \alpha_t \cdot A_t$. Under the measure $\mathbb{P}$, A satisfies the SDE (7.1) and this can be rewritten in the form

$$dA^N = \mu^N(A_t, \alpha_t, t)dt + \sigma^N(A_t, \alpha_t, t)dW_t, \tag{7.18}$$

$$dN_t = \alpha_t \cdot \mu(A_t, t)dt + \alpha_t \cdot \sigma(A_t, t)dW_t, \tag{7.19}$$

where μ^N and σ^N can be derived from (7.1) using Itô's formula. By Girsanov's theorem, assuming the Radon–Nikodým derivative $\frac{d\mathbb{N}}{d\mathbb{P}}$ can be written in the form

$$\left.\frac{d\mathbb{N}}{d\mathbb{P}}\right|_{\mathcal{F}_t} = \exp\left(\int_0^t C_u \cdot dW_u - \tfrac{1}{2}\int_0^t |C_u|^2 du\right) \tag{7.20}$$

for some $\{\mathcal{F}_t\}$-predictable C, which is certainly true if the filtration $\{\mathcal{F}_t\}$ is the augmented natural filtration for some Brownian motion (Theorem 5.24), then

$$\widetilde{W}_t = W_t - \int_0^t C_u du$$

is an $\{\mathcal{F}_t\}$ Brownian motion under $\mathbb{N}$. Rewriting (7.18) and (7.19) in terms of $\widetilde{W}_t$, then

$$dA_t^N = \left(\mu^N(A_t, \alpha_t, t) + \sigma^N(A_t, \alpha_t, t)C_t\right)dt + \sigma^N(A_t, \alpha_t, t)d\widetilde{W}_t, \quad (7.21)$$

$$dN_t = \alpha_t \cdot \left(\mu(A_t, t) + \sigma(A_t, t)C_t\right)dt + \alpha_t \cdot \sigma(A_t, t)d\widetilde{W}_t. \quad (7.22)$$

Since $(N, \mathbb{N})$ is a numeraire pair, we can conclude from (7.21) that

$$dA_t^N = \sigma^N(A_t, \alpha_t, t)d\widetilde{W}_t, \quad (7.23)$$

otherwise A^N is not even a local martingale under $\mathbb{N}$. (We could have anticipated the form of (7.23) from the martingale property of A^N and the fact that quadratic variation is a sample path property, not a property of the probability measure.) We do not, however, know the SDE which N satisfies unless we know C.

In the case where there exists some finite variation numeraire $N_t^f = \alpha_t^f \cdot A_t$ we can find the SDE for N. To see this, note that since α^f is self-financing and N^f is finite variation,

$$\alpha_t^f \cdot dA_t^N = d\left(\frac{N_t^f}{N_t}\right) = \left(\frac{N_t^f}{N_t}\right)\left(\frac{dN_t^f}{N_t^f} + \frac{d[N]_t}{N_t^2} - \frac{dN_t}{N_t}\right). \quad (7.24)$$

The left-hand side of (7.24) is a local martingale under $\mathbb{N}$, and N^f, being a finite variation process, satisfies the same SDE under $\mathbb{P}$ and $\mathbb{N}$ (substitute $\sigma \equiv 0$ into (7.22)). It thus follows by rearranging (7.24) that

$$dN_t = \frac{N_t}{N_t^f}dN_t^f + \frac{d[N]_t}{N_t} + d(\text{local martingale}),$$

under $\mathbb{N}$ and thus, from (7.22),

$$dN_t = \frac{N_t}{N_t^f}dN_t^f + \frac{d[N]_t}{N_t} + \alpha_t \cdot \sigma(A_t, t)d\widetilde{W}_t.$$

This, together with (7.23), completely specifies the SDE for the assets under $\mathbb{N}$.

In practice we tend to specify a model by giving equations of the form (7.21) and (7.22). We assume the original measure $\mathbb{P}$ was such that we could change to an appropriate measure $\mathbb{N}$. If the reader is ever faced with the problem of proving, starting from some $\mathbb{P}$, that an EMM exists for a particular numeraire N, Girsanov's theorem points in the right direction. The Radon–Nikodým derivative will be of the form (7.20) for some C satisfying (from (7.21) and (7.23))

$$\mu^N(A_t, \alpha_t, t) + \sigma^N(A_t, \alpha_t, t)C_t = 0.$$

This is a necessary condition. In addition we must prove that (7.20) defines a true $\{\mathcal{F}_t\}$ martingale under $\mathbb{P}$, not just a local martingale, and that (7.23) defines a true $\{\mathcal{F}_t^A\}$ martingale under $\mathbb{N}$. Novikov's condition (Theorem 5.16) is one way to establish this in some cases. Example 7.20 shows things are straightforward for the Black–Scholes model.

Example 7.20 We return to the Black–Scholes example and take D as our numeraire as before. Define a measure $\mathbb{Q} \sim \mathbb{P}$ on $\mathcal{F}_T^W$ via

$$\left.\frac{d\mathbb{Q}}{d\mathbb{P}}\right|_{\mathcal{F}_t^W} = \rho_t$$

where

$$\rho_t = \exp(\phi W_t - \tfrac{1}{2}\phi^2 t),$$

$$\phi = \frac{r - \mu}{\sigma}.$$

Note that ρ is an $\{\mathcal{F}_t^W\}$ martingale under $\mathbb{P}$, $\rho_0 = 1$, so $d\mathbb{Q}/d\mathbb{P}$ is well defined.

Under the measure $\mathbb{Q}$, $D_t^{(D)} \equiv 1$ so is trivially a martingale. Furthermore, by Girsanov's theorem,

$$\widetilde{W}_t := W_t - \phi t$$

is an $\{\mathcal{F}_t^W\}$ Brownian motion under $\mathbb{Q}$, and we have

$$dS_t^D = \sigma S_t^D d\widetilde{W}_t.$$

The solution to this SDE (see equation (7.16)) is an $\{\mathcal{F}_t^W\}$ martingale. Thus $\mathbb{Q}$ is an EMM for numeraire D.

7.3.3 Arbitrage and admissible strategies

A prerequisite for being able to price derivatives using replication arguments is that the economy does not allow arbitrage. We saw this in Chapter 1 and will see it again in Section 7.3.4. An *arbitrage* is a self-financing trading strategy which generates a profit with positive probability but which cannot generate a loss. This is summarized in the following definition.

Definition 7.21 *An arbitrage for the economy $\mathcal{E}$ is an admissible trading strategy ϕ such that one of the following conditions holds:*

(i) $\phi_0 \cdot A_0 < 0$ and $\phi_T \cdot A_T \geq 0$ a.s.;

(ii) $\phi_0 \cdot A_0 \leq 0$ and $\phi_T \cdot A_T \geq 0$ a.s. with $\phi_T \cdot A_T > 0$ with strictly positive probability.

If there is no such ϕ then the economy is said to be arbitrage-free.

The first step in derivative pricing is to ensure that the model for the economy is arbitrage-free. This imposes conditions *both* on the law for the evolution of the asset prices *and* the set of trading strategies which are allowed. The following two examples illustrate each of these points.

Example 7.22 Consider an economy $\mathcal{E}$ consisting of two assets satisfying the SDE

$$dA_t^{(i)} = \mu^{(i)} A_t^{(i)} dt + \sigma A_t^{(i)} dW_t, \qquad i = 1, 2,$$

where $\mu^{(1)} > \mu^{(2)}$ and $A_0^{(1)} = A_0^{(2)}$. Let ϕ be the 'buy and hold' strategy $\phi_t^{(1)} = -\phi_t^{(2)} = 1, 0 \leq t \leq T$. At time zero the portfolio value is $A_0^{(1)} - A_0^{(2)} = 0$, whereas at any subsequent time t,

$$\begin{aligned} V_t &= \phi_t^{(1)} A_t^{(1)} + \phi_t^{(2)} A_t^{(2)} \\ &= (1 - e^{(\mu^{(2)} - \mu^{(1)})t}) A_t^{(1)} \\ &> 0 \qquad \text{a.s.} \end{aligned}$$

In particular $V_T > 0$ a.s., hence ϕ is an arbitrage.

Example 7.23 The following example of Karatzas (1993) is taken from Duffie (1996). Consider the special case of the Black–Scholes economy when $S_0 = 1, r = \mu = 0$ and $\sigma = 1$. Then

$$dS_t = S_t dW_t,$$

$$D_t = 1,$$

where W is a standard one-dimensional Brownian motion. We will construct a self-financing trading strategy $\hat{\phi}$ which is an arbitrage for this economy. First define a trading strategy $\phi_t = (\phi_t^{(S)}, \phi_t^{(D)})$, $0 < t < T$, via

$$\phi_t^{(S)} = \frac{1}{S_t \sqrt{T - t}}, \qquad 0 < t < T,$$

and choose $\phi^{(D)}$ to make ϕ self-financing,

$$\phi_t^{(D)} = -\phi_t^{(S)} S_t + \int_0^t \phi_u^{(S)} \, dS_u.$$

The initial portfolio value, $\phi_0^{(S)} S_0 + \phi_0^{(D)} D_0$, is zero and the gain process for this strategy is given by

$$G_t^\phi = \int_0^t \phi_u^{(S)} dS_u = \int_0^t \frac{1}{\sqrt{T-u}} dW_u. \qquad (7.25)$$

Now define an $\{\mathcal{F}_t^A\}$ stopping time τ by

$$\tau := \inf\{t > 0 : G_t^\phi = \alpha\}, \qquad (7.26)$$

the first time the gain hits some positive level α. By Exercise 7.24, $0 < \tau < T$ a.s. and thus the gain will at some point on $(0, T)$ hit any positive level α. We can use this observation to modify ϕ to construct an arbitrage as follows. Define the self-financing trading strategy $\hat\phi$ via

$$\hat\phi_t^{(S)} = \begin{cases} \phi_t^{(S)}, & t \leq \tau \\ 0, & \text{otherwise}, \end{cases}$$

$$\hat\phi_t^{(D)} = \begin{cases} \phi_t^{(D)}, & t \leq \tau \\ \alpha, & \text{otherwise}. \end{cases}$$

This strategy starts with zero wealth and guarantees a final portfolio value at T of $\alpha > 0$, hence it is an arbitrage.

Exercise 7.24: Show that the process $\widehat{W}$ defined by

$$\widehat{W}_t := G_{T(1-e^{-t})}^\phi$$

is a Brownian motion, where G^ϕ is defined by (7.25). Hence show that the stopping time τ defined in (7.26) is strictly less that T with probability one.

In Example 7.22 the arbitrage was a simple strategy, and we certainly want to include all simple strategies within any model. The problem here is with the model for the evolution of the asset prices. In Chapter 1 we saw that the existence of a numeraire pair was sufficient (and necessary) to ensure that the asset price process is not one which introduces arbitrage. The model in Example 7.22 does not admit a numeraire pair.

The strategy of Example 7.23 is not restricted to the asset price model given there. The strategy is referred to as a 'doubling strategy' because the holding of one asset keeps on doubling up (indeed in the classical example of such a strategy the holding is precisely doubled at each of a set of stopping times). Such a doubling strategy is not particular to this economy. Consequently we can see that, unlike in the discrete time setting, it is not sufficient to impose conditions just on the asset price process but we must also impose further conditions other than $\{\mathcal{F}_t^A\}$-predictability and the self-financing property on the trading strategies that we allow. Of course, we must also ensure that we do not remove too many strategies.

There is no unique way to eliminate these troublesome trading strategies. The restrictions that we choose have the advantage of being 'numeraire-friendly'. We are able to change between different numeraires and EMMs for modelling and calculation purposes without worrying about the implications for trading strategies. We now formally introduce our definition of *admissible* trading strategies.

Definition 7.25 *If the economy $\mathcal{E}$ does not admit any numeraire pair the set of admissible strategies is taken to be the empty set. Otherwise, an $\{\mathcal{F}_t^A\}$-predictable process ϕ is admissible for the economy $\mathcal{E}$ if it is self-financing,*

$$\phi_t \cdot A_t = \phi_0 \cdot A_0 + \int_0^t \phi_u \cdot dA_u \,,$$

and if, for all numeraire pairs $(N, \mathbb{N})$, the numeraire-rebased gain process

$$\frac{G_t^\phi}{N_t} = \int_0^t \phi_u \cdot dA_u^N \tag{7.27}$$

is an $\{\mathcal{F}_t^A\}$ martingale under $\mathbb{N}$.

Definition 7.26 *If ϕ is admissible then $V_t = \phi_t \cdot A_t$ is said to be a price process.*

Remark 7.27: The martingale requirement that we have imposed on trading strategies is sufficient to remove the doubling strategy arbitrage, as Theorem 7.32 demonstrates. There are two different ways commonly employed to remove such strategies. The first is to impose some kind of integrability condition on the set of trading strategies, and our conditions fall into this category. Other integrability conditions are often more restrictive, with our choice being precisely enough to ensure that the standard proof of no arbitrage (Theorem 7.32) can be applied. The second method is to impose some absolute lower bound on the value of the portfolio, which clearly removes doubling (this makes the gain process in (7.27) a supermartingale). However, for some economies this approach also removes some simple strategies.

Remark 7.28: One reason why we chose, in Section 7.1.2, to model the gain process via an Itô integral was that it preserves the local martingale property. The additional constraint on the gain G_t^ϕ of an admissible strategy ensures that trading in assets which have zero expected change in price over any time interval results in zero expected gain. This behaviour is intuitively what one would expect. If one placed no restrictions on ϕ this would not be the case in general.

Remark 7.29: The set of admissible strategies posseses several important properties. First note that it is linear: if ϕ_1 and ϕ_2 are admissible then so is $\alpha\phi_1 + \beta\phi_2$ for $\alpha, \beta \in \mathbb{R}$. This is important in the replication argument to price a derivative (see Section 7.3.4) in which we need to consider the difference between two strategies. This is why we have chosen the intersection over numeraire pairs rather than the union in our definition of 'admissible'.

A second property of our definition is that it is independent of the particular measure $\mathbb{P}$ in which the economy is initially specified. Recall that option pricing is about replication not probability. The role of the probability measure is as a marker for those sample paths which are allowed and the precise probabilities are not important. It is mathematically more pleasing for the strategies to also be independent of the measure $\mathbb{P}$ and in practice this means we avoid tricky questions on which strategies are admissible when we change numeraire, something we do often in applications. This is also important when we come to consider completeness in Section 7.3.5.

Remark 7.30: We shall see in Section 7.3.5 that, when the economy is complete, the martingale condition (7.27) holds for one numeraire pair $(N, \mathbb{N})$ if and only if it holds for all numeraire pairs. Note that the martingale property of an admissible strategy ensures in particular that, for all numeraire pairs $(N, \mathbb{N}), V_T^N = \phi_T \cdot A_T^N$ satisfies

$$\mathbb{E}_{\mathbb{N}}[|V_T^N|] < \infty. \qquad (7.28)$$

Completeness questions involve deciding which $\mathcal{F}_T^A$-measurable random variables V_T satisfying (7.28) can be replicated by an admissible trading strategy.

Remark 7.31: Note that if there is no numeraire pair then the economy we have defined is particularly uninteresting because there are no admissible trading strategies. A natural concern at this point is that we may have defined an otherwise perfectly reasonable model for an economy and then eliminated all the trading strategies – and we would at the very least like to have simple strategies included, those which could be implemented in practice. There are two things which alleviate this concern. The first is that when modelling in practice we usually specify the initial model in a martingale measure, so we know one exists and we have a set of strategies that at least includes all the simple strategies. The second reason is the close relationship between the existence of a martingale measure and absence of arbitrage (which we proved in the discrete setting of Chapter 1). This topic is much more technical in the continuous time setting than in discrete time and we omit any further discussion of this point. Roughly speaking, the continuous time result states that if there is no *approximate arbitrage* (a technical term we shall not define) then there exists a numeraire pair. Therefore we have only ruled out economies that admit arbitrage. Delbaen and Schachermayer (1999) and references therein contain a fuller discussion of this point. With our definition we do not need to worry about such matters since the following theorem guarantees that our economy does not admit an arbitrage strategy.

Theorem 7.32 *The economy $\mathcal{E}$ is arbitrage-free.*

Proof: Suppose ϕ is an arbitrage for the economy $\mathcal{E}$ and let

$$V_t = \phi_0 \cdot A_0 + \int_0^t \phi_u \cdot dA_u.$$

Since ϕ is admissible there exists some numeraire pair $(N, \mathbb{N})$ such that V^N is a martingale under $\mathbb{N}$. In particular,

$$\mathbb{E}_{\mathbb{N}}[V_T^N] = V_0^N.$$

It is straightforward to show that this contradicts ϕ being an arbitrage. $\square$

7.3.4 Derivative pricing in an arbitrage-free economy

Here we group together the ideas of option pricing via replication which we have already encountered throughout the text. We summarize this in the following theorem.

Theorem 7.33 *Let $\mathcal{E}$ be an arbitrage-free economy and let V_T be some $\mathcal{F}_T^A$-measurable random variable. Consider a derivative (contingent claim) which has value V_T at the terminal time T. If V_T is attainable, i.e. if there exists some admissible ϕ such that*

$$V_T = \phi_0 \cdot A_0 + \int_0^T \phi_u \cdot dA_u \qquad a.s.,$$

then the time-t value of this derivative is given by $V_t = \phi_t \cdot A_t$.

Proof: The value of the derivative at time t is just the time-t wealth needed to enable us to guarantee, for each $\omega \in \Omega$ (actually, for all ω in some set of probability one), an amount $V_T(\omega)$ at the terminal time. The trading strategy ϕ generates a portfolio which has value at t given by $\phi_t \cdot A_t$ and, in particular, value $V_T = \phi_T \cdot A_T$ at the terminal time. Clearly then $\phi_t \cdot A_t$ is a candidate for the time-t price of the derivative. To show that this is indeed the correct price we must show that if ψ is any other admissible strategy which replicates the option at T then, with probability one,

$$\phi_t \cdot A_t = \psi_t \cdot A_t \quad \text{for all } t.$$

Suppose ψ is some other strategy which replicates V_T and let α be some admissible trading strategy corresponding to a numeraire N. (See the proof of Theorem 7.48 (i) for the construction of one such strategy and numeraire.) For fixed $t^* \geq 0$, consider the admissible trading stategy φ defined as follows:

$$\varphi_t = \begin{cases} (\psi_t - \phi_t) + 2\dfrac{(\phi_{t^*} - \psi_{t^*}) \cdot A_{t^*}}{N_{t^*}}\alpha_t & \text{if } t > t^* \text{ and } (\phi_{t^*} - \psi_{t^*}) \cdot A_{t^*} > 0 \\[2mm] (\phi_t - \psi_t) & \text{otherwise.} \end{cases}$$

Since $\mathcal{F}_0$ is trivial we may assume, without loss of generality, that $\varphi_0 \cdot A_0 = c \leq 0$ a.s., for some constant c. Furthermore,

$$\varphi_T \cdot A_T = 2\frac{(\phi_{t^*} - \psi_{t^*}) \cdot A_{t^*}}{N_{t^*}} N_T \mathbb{1}_{\{(\phi_{t^*} - \psi_{t^*}) \cdot A_{t^*} > 0\}}.$$

Thus $\varphi_T \cdot A_T \geq 0$ a.s., and so the absence of arbitrage implies that $\varphi_0 \cdot A_0 = \varphi_T \cdot A_T = 0$ a.s. But this in turn implies that $\psi_{t^*} \cdot A_{t^*} \geq \phi_{t^*} \cdot A_{t^*}$ a.s. A similar argument reversing the roles of ϕ and ψ shows that $\phi_{t^*} \cdot A_{t^*} \geq \psi_{t^*} \cdot A_{t^*}$ a.s., thus $\phi_{t^*} \cdot A_{t^*} = \psi_{t^*} \cdot A_{t^*}$ a.s., as required. $\qquad\square$

Corollary 7.34 *Let $\mathcal{E}$ be an economy with a numeraire pair $(N, \mathbb{N})$ and let V_T be some attainable contingent claim. Then V_t admits the representation*

$$V_t = N_t \mathbb{E}_\mathbb{N}[V_T^N | \mathcal{F}_t^A].$$

Proof: The existence of an EMM along with our definition of an admissible trading strategy ensures that the economy is arbitrage-free. The result is now immediate from the martingale property of an admissible strategy (Definition 7.25). $\qquad\square$

7.3.5 Completeness

In Chapter 1 we learnt that the concept of completeness is tied in with what we mean by an attainable claim. The following definition is broad enough to cover the cases of interest in our current context. Let S be a collection of random variables.

Definition 7.35 *The economy $\mathcal{E}$ is said to be S-complete if for every random variable $X \in S$ there exists some admissible trading strategy ϕ such that*

$$X = \phi_0 \cdot A_0 + \int_0^T \phi_u \cdot dA_u.$$

Typically the set S is specified as a collection of random variables measurable with respect to one of $\mathcal{F}_T^A$ or $\mathcal{F}_T^W$ and satisfying some integrability condition. The choice of S in the following definition is a natural one, and for this special choice we will simply refer to the economy $\mathcal{E}$ as being complete.

Definition 7.36 *Let $\mathcal{E}$ be an economy which admits some numeraire pair $(N, \mathbb{N})$. The economy $\mathcal{E}$ is said to be complete if for every $\mathcal{F}_T^A$-measurable random variable X satisfying*

$$\mathbb{E}_\mathbb{N}\left[\left|\frac{X}{N_T}\right|\right] < \infty$$

there exists some admissible trading strategy ϕ such that

$$X = \phi_0 \cdot A_0 + \int_0^T \phi_u \cdot dA_u.$$

Remark 7.37: When we wish to distinguish this from any other S-completeness we will refer to it as $\mathcal{F}_T^A$-completeness.

Remark 7.38: This definition of completeness is by no means a standard one. It is more usual to work with the augmented Brownian filtration $\{\mathcal{F}_t^W\}$ and then take S to be the set of $\mathcal{F}_T^W$-measurable random variables having finite variance under $\mathbb{P}$. Admissible strategies are taken to be $\{\mathcal{F}_t^W\}$-predictable and to satisfy suitable integrability conditions (usually expressed in the original measure $\mathbb{P}$) which ensure no arbitrage. For details of two possible choices of admissible strategies the reader is referred to Duffie (1996).

It is implicit, but not immediately obvious, that Definition 7.36 is independent of the numeraire pair $(N, \mathbb{N})$. This we shall shortly prove in Corollary 7.40. Corollary 7.40 also shows that admissibility of a strategy in a complete economy, as defined in Definition 7.25, also reduces to a requirement for a single, arbitrary numeraire pair – if (7.27) defines a martingale for one numeraire pair then, in a complete economy, it defines a martingale for any numeraire pair. These results both follow immediately from the stronger result that the Radon–Nikodým derivative connecting two EMMs in a complete economy is just the ratio of the corresponding numeraires.

Theorem 7.39 *Let $\mathcal{E}$ be some economy and suppose that the following partial completeness result holds: for some numeraire pair $(N, \mathbb{N})$, given any random variable X for which X/N_T is $\mathcal{F}_T^A$-measurable and*

$$\mathbb{E}_{\mathbb{N}}\left[\left|\frac{X}{N_T}\right|\right] < \infty,$$

there exists some self-financing $\{\mathcal{F}_t^A\}$-predictable ϕ such that

$$X = \phi_0 \cdot A_0 + \int_0^T \phi_u \cdot dA_u,$$

and

$$V_t^N := \phi_0 \cdot A_0^N + \int_0^t \phi_u \cdot dA_u^N$$

is a $(\{\mathcal{F}_t^A\}, \mathbb{N})$ martingale (recall that full admissibility requires that V^N be a $(\{\mathcal{F}_t^A\}, \mathbb{N})$ martingale for all numeraire pairs $(N, \mathbb{N})$). Then if $(M, \mathbb{M})$ is any other numeraire pair, the measures $\mathbb{N}$ and $\mathbb{M}$ are related via

$$\left.\frac{d\mathbb{M}}{d\mathbb{N}}\right|_{\mathcal{F}_T^A} = \frac{M_T/M_0}{N_T/N_0} \quad \mathbb{P}\text{-a.s.} \tag{7.29}$$

As a consequence, M^N is an $(\{\mathcal{F}_t^A\}, \mathbb{N})$ martingale.

Conversely, if $(N, \mathbb{N})$ *is a numeraire pair and if* M *a numeraire such that* M^N *is an* $(\{\mathcal{F}_t^A\}, \mathbb{N})$ *martingale, then (7.29) defines a martingale measure* $\mathbb{M}$ *on the* σ-*algebra* $\mathcal{F}_T^A$ *with associated numeraire* M.

Proof: We prove (7.29) by showing, equivalently, that

$$\frac{d\mathbb{N}}{d\mathbb{M}}\bigg|_{\mathcal{F}_T^A} = \frac{N_T/N_0}{M_T/M_0} \quad \mathbb{P}\text{-a.s.},$$

and to prove this we show that, for any $S \in \mathcal{F}_T^A$,

$$\mathbb{E}_{\mathbb{M}}\left[\mathbb{1}_S \frac{N_T/N_0}{M_T/M_0}\right] = \mathbb{E}_{\mathbb{M}}\left[\mathbb{1}_S \frac{d\mathbb{N}}{d\mathbb{M}}\right].$$

Define a sequence of $\{\mathcal{F}_t^A\}$ stopping times T_n via

$$T_n = \inf\left\{t > 0 : N_t^M > n\right\}$$

and let

$$V^n = \mathbb{1}_S \mathbb{1}_{\{T < T_n\}}.$$

By the partial completeness hypothesis (applied to $X = V^n N_T$), combined with the unit invariance theorem, we can write

$$V^n = \phi_0^n \cdot A_0^N + \int_0^T \phi_u^n \cdot dA_u^N$$

for some self-financing $\{\mathcal{F}_t^A\}$-predictable ϕ^n, where

$$Y_t^n := \phi_0^n \cdot A_0^N + \int_0^t \phi_u^n \cdot dA_u^N$$

is a $(\{\mathcal{F}_t^A\}, \mathbb{N})$ martingale.

A further application of the unit invariance theorem yields

$$\hat{Y}_t^n = \hat{Y}_0^n + \int_0^t \phi_u^n \cdot dA_u^M,$$

where $\hat{Y}_t^n = Y_t^n N_t^M$. This representation shows $\hat{Y}^n$ to be an $\{\mathcal{F}_t^A\}$ local martingale under $\mathbb{M}$ which, since $|\hat{Y}_t^n| = |Y_t^n| \|N_t^M\| \leq n$, is also an $\{\mathcal{F}_t^A\}$

martingale. We can conclude (remembering that $\mathcal{F}_0^A$ is trivial) that

$$\mathbb{E}_M\left[Y_T^n \frac{d\mathbb{N}}{d\mathbb{M}}\right] = \mathbb{E}_N[Y_T^n] = Y_0^n = \frac{M_0}{N_0}\hat{Y}_0^n$$

$$= \frac{M_0}{N_0}\mathbb{E}_M[\hat{Y}_T^n] = \frac{M_0}{N_0}\mathbb{E}_M\left[Y_T^n \frac{N_T}{M_T}\right]. \tag{7.31}$$

Now take the limit as $n \to \infty$. Since $Y_T^n \le 1$ and both $\frac{d\mathbb{N}}{d\mathbb{M}}$ and N_T^M have finite expectation under $\mathbb{M}$ (the latter being a positive local martingale integrable at zero and hence a supermartingale), we can apply the dominated convergence theorem to both sides of (7.31) to obtain

$$\mathbb{E}_M\left[1_S \frac{d\mathbb{N}}{d\mathbb{M}}\right] = \mathbb{E}_M\left[1_S \frac{N_T/N_0}{M_T/M_0}\right],$$

which establishes (7.30).

An immediate consequence of (7.29) is that $\mathbb{E}_N[(M_T/M_0)/(N_T/N_0)] = 1$ which, combined with the fact that M^N is a positive $(\{\mathcal{F}_t^A\}, \mathbb{N})$ local martingale (hence supermartingale), implies that M^N is a true $(\{\mathcal{F}_t^A\}, \mathbb{N})$ martingale.

The converse part of the theorem is immediate.   □

This yields the following important corollary.

Corollary 7.40 *Suppose the hypotheses of Theorem 7.39 hold. Then the following are true.*

(i) If L is a $(\{\mathcal{F}_t^A\}, \mathbb{M})$ martingale, then LM^N is a $(\{\mathcal{F}_t^A\}, \mathbb{N})$ martingale. In particular, if (7.27) in the definition of an admissible trading strategy holds for one numeraire pair then it holds for all numeraire pairs.

(ii) If $(N, \mathbb{N})$ and $(M, \mathbb{M})$ are two numeraire pairs then

$$\mathbb{E}_M\left[\left|\frac{X}{M_T}\right|\right] = \mathbb{E}_N\left[\left|\frac{X}{N_T}\right|\right] < \infty,$$

so Definition 7.36 is independent of the numeraire pair $(N, \mathbb{N})$ chosen in that definition.

Theorem 7.39 is half of the continuous time equivalent to Theorem 1.21 for the discrete economy of Chapter 1. The full result, which is mathematically appealing but not the primary result to which we appeal in practice, is as follows.

Theorem 7.41 *Suppose the economy $\mathcal{E}$ admits some numeraire pair. Then $\mathcal{E}$ is complete if and only if, given any two numeraire pairs $(N, \mathbb{N}_1)$ and $(N, \mathbb{N}_2)$ with common numeraire N, $\mathbb{N}_1$ and $\mathbb{N}_2$ agree on $\mathcal{F}_T^A$.*

Proof: If the economy $\mathcal{E}$ is complete then it follows from Theorem 7.39 that $\frac{dN_1}{dN_2}\big|_{\mathcal{F}_T^A} = 1$ a.s., so $N_1 = N_2$ on $\mathcal{F}_T^A$. Conversely, if N_1 is the unique measure on $\mathcal{F}_T^A$ such that A^N is an $\{\mathcal{F}_t^A\}$ martingale, it follows from the martingale representation theorem (Corollary 5.47) that, given any $(\{\mathcal{F}_t^A\}, N_1)$ local martingale M, there exists some $\{\mathcal{F}_t^A\}$-predictable ϕ such that

$$M_t = M_0 + \int_0^t \phi_u \cdot dA_u^N. \tag{7.32}$$

By Theorem 7.17, ϕ can be assumed to be self-financing. Let X be an $\mathcal{F}_T^A$-measurable random variable satisfying $\mathbb{E}_{N_1}[|X/N_T|] < \infty$. Setting $M_t = \mathbb{E}_{N_1}[X/N_T|\mathcal{F}_t^A]$, a true martingale, (7.32) shows that the contingent claim X can be replicated by a self-financing trading strategy. It remains to check that ϕ is admissible. But the gain process in (7.32) is a martingale under N_1 since M is a martingale, and so it follows from Corollary 7.40 that ϕ is admissible. $\qquad\square$

Whereas Theorem 7.41 gives a precise characterization of when a model is complete, it is not easy to check in practice. The question of whether there exists a unique EMM for the process A^N on the filtration $\{\mathcal{F}_t^A\}$ is non-standard, the more usual question being whether the EMM is unique when working on the filtration $\{\mathcal{F}_t^{A^N}\}$, which may be less refined. Indeed, some of the essential intuition and structure is masked in the apparently simple statement of Theorem 7.41.

Theorem 7.43 below addresses the issues above. It gives criteria for completeness which are much more easily checked in practice and which are stated in terms of the process A rather than A^N. What becomes apparent also is the extra requirement, the existence of a finite variation numeraire, which is a consequence of the fact that any replicating strategy must be self-financing.

We shall need the following lemma in the proof of Theorem 7.43. It is similar in character to Theorem 7.17 and shows how the self-financing property of a strategy leads to an alternative form of representation result for local martingales.

Lemma 7.42 *Let $\mathcal{E}$ be a complete economy and (N, N) any associated numeraire pair. Then for any $(\{\mathcal{F}_t^A\}, N)$ local martingale M there exists some self-financing $\{\mathcal{F}_t^A\}$-predictable ϕ such that*

$$M_t = M_0 + \int_0^t \phi_u \cdot dA_u^N.$$

Proof: Let $\{T_n\}$ be a reducing sequence for the local martingale M and define $M^n := M^{T_n}$. Since M^n is a true martingale under N, $\mathbb{E}_N\left[\left|\frac{M_T^n N_T}{N_T}\right|\right] = \mathbb{E}_N[|M_T^n|] < \infty$ so, by completeness,

$$M_T^n N_T = \phi_0^n \cdot A_0 + \int_0^T \phi_u^n \cdot dA_u$$

for some admissible ϕ^n, and by the unit invariance theorem,

$$M_T^n = \phi_0^n \cdot A_0^N + \int_0^T \phi_u^n \cdot dA_u^N. \tag{7.33}$$

The integral on the right-hand side of (7.33), with the upper limit being replaced by the variable t, is a martingale (since ϕ^n is admissible), so taking expectations on both sides of (7.33), conditional on $\mathcal{F}_t^A$, we can conclude that

$$M_t^n = M_0 + \int_0^t \phi_u^n \cdot dA_u^N.$$

Defining the strategy ϕ via

$$\phi_u(\omega) = \phi_u^{n^*}(\omega),$$

where

$$n^*(\omega, u) = \min\{n : u \leq T_n(\omega)\},$$

yields

$$M_t^n = M_0 + \int_0^{t \wedge T_n} \phi_u \cdot dA_u^N.$$

Letting $n \uparrow \infty$ now yields the result. $\qquad\square$

We now come to the result on completeness to which we most often appeal in practice.

Theorem 7.43 *Let $\mathcal{E}$ be an economy admitting some numeraire pair and suppose that weak uniqueness holds for the SDE (7.1) (i.e. any two solution processes must have the same law). Then the following are equivalent.*

(i) $\mathcal{E}$ is complete.
(ii) There exists some $\{\mathcal{F}_t^A\}$-predictable ϕ satisfying:
    (a) $\phi_t \cdot \sigma(A_t, t) = 0$ for all t a.s.;
    (b) $\phi_t \cdot A_t > 0$ for all t a.s.;
    (c) $-\infty < \int_0^t \frac{\phi_u}{\phi_u \cdot A_u} \cdot dA_u < \infty$ for all t a.s.
(iii) There exists some finite variation numeraire N_t^f.

Remark 7.44: When one of the assets in an economy is of finite variation and positive it is common practice to take this asset as the numeraire. In these circumstances the converse part of Theorem 7.39 ensures that if an EMM exists for some numeraire then one will exist for the finite variation numeraire. However, this is not true in general when there is no finite variation asset. Part (iii) of the above theorem does not guarantee the existence of an EMM corresponding to the finite variation numeraire N^f. In particular, it is not necessarily the case that N^f is a price process (i.e. that it can be replicated by an admissible strategy). Example 8.11 provides such an example.

Remark 7.45: When working with the Brownian filtration $\{\mathcal{F}_t^W\}$, establishing that the economy is $\mathcal{F}_T^W$-complete in the sense of Remark 7.38 reduces to checking that $\mathrm{rank}(\sigma_t) = d$ (d being the dimension of the Brownian motion W) for all t a.s., *under the assumption that there is a finite variation asset in the economy.* If σ does have full rank d then $\{\mathcal{F}_t^A\} = \{\mathcal{F}_t^W\}$ and so the economy will also be complete in the sense discussed above. Clearly we can have an economy which is complete working with the $\{\mathcal{F}_t^A\}$ filtration but not with respect to the Brownian filtration $\{\mathcal{F}_t^W\}$. But it is the $\{\mathcal{F}_t^A\}$ filtration that makes sense from a modelling perspective. Many models specified by an SDE for which weak uniqueness holds admit a finite variation numeraire, and the treatment here shows that these economies are complete in a sense which is all that matters in practice.

Proof:
(i)$\Rightarrow$(ii): Let $(N, \mathbb{N})$ be some numeraire pair for the economy $\mathcal{E}$, and $N_t = \alpha_t \cdot A_t$ for some $\{\mathcal{F}_t^A\}$-predictable process α. Then, by the uniqueness of the Doob–Meyer decomposition, we can write the local martingale part under $\mathbb{N}$ of the numeraire N as

$$N_t^{\mathrm{loc},\mathbb{N}} = \int_0^t \alpha_u \cdot dA_u^{\mathrm{loc},\mathbb{N}}.$$

Since $\mathcal{E}$ is complete it follows from Lemma 7.42 that we can find some $\{\mathcal{F}_t^A\}$-predictable self-financing ψ_t such that

$$N_t^{\mathrm{loc},\mathbb{N}} = \int_0^t \psi_u \cdot dA_u^N = \psi_t \cdot A_t^N.$$

We now show that

$$\phi_t = \alpha_t - \psi_t^N + (\psi_t^N \cdot A_t^N)\alpha_t \tag{7.34}$$

has the required properties in part (ii) of the theorem.

Predictability is immediate. To check (a), it is clearly sufficient to prove

$$\phi_t \cdot dA_t^{\mathrm{loc},\mathbb{N}} = 0.$$

Noting that

$$dA_t = d(N_t A_t^N) = N_t dA_t^N + A_t^N dN_t + dN_t dA_t^N$$

and appealing to the Doob–Meyer decomposition yields

$$A_t^N \alpha_t \cdot dA_t^{\mathrm{loc},\mathbb{N}} = dA_t^{\mathrm{loc},\mathbb{N}} - N_t dA_t^N.$$

With ϕ as in (7.34) we have

$$
\begin{aligned}
\phi_t \cdot dA_t^{\mathrm{loc},\mathbb{N}} &= \alpha_t \cdot dA_t^{\mathrm{loc},\mathbb{N}} - \psi_t^N \cdot dA_t^{\mathrm{loc},\mathbb{N}} + (\psi_t^N \cdot A_t^N)\alpha_t \cdot dA_t^{\mathrm{loc},\mathbb{N}} \\
&= \alpha_t \cdot dA_t^{\mathrm{loc},\mathbb{N}} - \psi_t^N \cdot dA_t^{\mathrm{loc},\mathbb{N}} + \psi_t^N \cdot (dA_t^{\mathrm{loc},\mathbb{N}} - N_t dA_t^N) \\
&= \alpha_t \cdot dA_t^{\mathrm{loc},\mathbb{N}} - \psi_t \cdot dA_t^N \\
&= 0
\end{aligned}
$$

as required. Finally, note that α is self-financing, and thus

$$
\phi_t \cdot A_t = \alpha_t \cdot A_t - \psi_t^N \cdot A_t + (\psi_t^N \cdot A_t^N)\alpha_t \cdot A_t = N_t \,.
$$

Property (b) follows from the positivity of N_t.

To establish property (c), the reader can check using the self-financing property of ψ and α that

$$
\begin{aligned}
\frac{\phi_t}{\phi_t \cdot A_t} \cdot dA_t &= \frac{\phi_t}{N_t} \cdot dA_t \\
&= \frac{dN_t}{N_t} - \frac{(dN_t)^2}{N_t^2} - \frac{dN_t^{\mathrm{loc},\mathbb{N}}}{N_t} \\
&= d(\log N_t) - d\big((\log N_t)^{\mathrm{loc},\mathbb{N}}\big) \,.
\end{aligned}
$$

Property (c) is now immediate from the finiteness and positivity of N_t.

(ii)$\Rightarrow$(iii): Given ϕ_t satisfying (a),(b) and (c) of (ii), define

$$
\phi_t^* = \frac{\phi_t}{\phi_t \cdot A_t}
$$

and

$$
N_t^f = \exp\left(\int_0^t \phi_u^* \cdot dA_u \right). \tag{7.35}
$$

By Property (a), N^f is of finite variation and is a.s. positive for all t. It remains to check N^f has the form of a numeraire. But trivially

$$
N_t^f = (N_t^f \phi_t^*) \cdot A_t \,,
$$

and, since $\phi_t \cdot dA_t^{\mathrm{loc},\mathbb{N}} = 0$, we have, by applying Itô's formula to (7.35),

$$
dN_t^f = N_t^f \phi_t^* \cdot dA_t \,.
$$

We are done.

(iii)$\Rightarrow$(i): Let $(N, \mathbb{N})$ be some numeraire pair for the economy $\mathcal{E}$ and let X be any $\mathcal{F}_T^A$-measurable random variable satisfying

$$\mathbb{E}_\mathbb{N}\left[\left|\frac{X}{N_T}\right|\right] < \infty.$$

Consider the $\{\mathcal{F}_t^A\}$ martingale under $\mathbb{N}$,

$$M_t = \mathbb{E}_\mathbb{N}\left[\frac{X}{N_T}\bigg|\mathcal{F}_t^A\right].$$

By Theorem 6.38 there exists some $\{\mathcal{F}_t^A\}$-predictable η such that

$$M_t = \mathbb{E}_\mathbb{N}\left[\frac{X}{N_T}\right] + \int_0^t \eta_u \cdot dA_u^{\mathrm{loc},\mathbb{N}}.$$

In particular,

$$\frac{X}{N_T} = \mathbb{E}_\mathbb{N}\left[\frac{X}{N_T}\right] + \int_0^T \eta_u \cdot dA_u^{\mathrm{loc},\mathbb{N}}. \tag{7.36}$$

We now seek an expression for the local martingale part of A under $\mathbb{N}$. Recall that

$$dA_t = d(N_t A_t^N) = A_t^N dN_t + N_t dA_t^N + dN_t dA_t^N. \tag{7.37}$$

The second term in (7.37) is a local martingale under $\mathbb{N}$ and the last term is of finite variation, so we can deduce that

$$dA_t^{\mathrm{loc},\mathbb{N}} = (A_t^N dN_t)^{\mathrm{loc},\mathbb{N}} + N_t dA_t^N. \tag{7.38}$$

To find the local martingale part of $A^N dN$, write

$$Z_t = \frac{N_t^f}{N_t}$$

and observe that

$$A_t^N dN_t = A_t^{N^f} Z_t dN_t.$$

By Itô's formula

$$dN_t^f = Z_t dN_t + N_t dZ_t + dZ_t dN_t,$$

and so

$$(A_t^N dN_t)^{\mathrm{loc},\mathbb{N}} = -A_t^{N^f} N_t dZ_t.$$

Subsituting into (7.38) now gives

$$dA_t^{\mathrm{loc},\mathbb{N}} = N_t dA_t^N - A_t^{N^f} N_t dZ_t. \tag{7.39}$$

Since N^f is a numeraire, unit invariance implies that we can find an $\{\mathcal{F}_t^A\}$-predictable α^f such that

$$dZ_t = \alpha_t^f \cdot dA_t^N$$

and thus (7.39) becomes

$$dA_t^{\text{loc},\mathbb{N}} = N_t dA_t^N - A_t^{N^f} N_t \alpha_t^f \cdot dA_t^N.$$

Substituting into (7.36), we have

$$\frac{X}{N_T} = \mathbb{E}_\mathbb{N}\left[\frac{X}{N_T}\right] + \int_0^T N_u[\eta_u - (\eta_u \cdot A_u^{N^f})\alpha_u^f] \cdot dA_u^N.$$

Thus by Theorem 7.17, the martingale property of M and Corollary 7.40, we can find an admissible strategy replicating X and so we have shown that $\mathcal{E}$ is complete. $\square$

Our results from Chapter 6 for locally Lipschitz SDEs provide sufficient conditions for weak uniqueness to hold for an SDE and thus we have the following very useful corollary.

Corollary 7.46 *If the SDE (7.1) satisfies the local Lipschitz conditions (6.23) and the linear growth restriction (6.24), then the economy $\mathcal{E}$ is complete if and only if there exists some finite variation numeraire.*

Proof: Theorems 6.12 and 6.31 show that weak uniqueness holds for (7.1), so the result follows from Theorem 7.43. $\square$

7.3.6 Pricing kernels

In Chapter 1 we introduced the idea of a pricing kernel and only later did we meet numeraires. We saw there the close relationship between pricing kernels and numeraire pairs (although we did not use this term at that stage). We have, in this chapter, chosen to develop the theory using numeraires in preference to pricing kernels because of the intuitively appealing interpretation of a numeraire as a unit, because of results such as Theorem 7.43, and because it is the most common way models and pricing results are presented in the finance literature. However, the theory can be developed and presented in terms of pricing kernels, and Chapter 8 is an occasion when using pricing kernels is sometimes more convenient. Hence we establish the link between the two approaches.

To begin we must define what we mean by a pricing kernel.

Definition 7.47 *Let $\mathcal{E}$ be an economy defined on the probability space $(\Omega, \{\mathcal{F}_t\}, \mathcal{F}, \mathbb{P})$. We say that the strictly positive, $(\{\mathcal{F}_t^A\}, \mathbb{P})$ continuous semimartingale Z is a pricing kernel for the economy $\mathcal{E}$ if the process ZA is an $(\{\mathcal{F}_t^A\}, \mathbb{P})$ martingale.*

The following theorem demonstrates the equivalence of, on the one hand, a theory developed using pricing kernel terminology and ideas and, on the other, one developed using numeraires.

Theorem 7.48 *Let $\mathcal{E}$ be an economy defined on the probability space $(\Omega, \{\mathcal{F}_t\}, \mathcal{F}, \mathbb{P})$. Then:*

(i) *$\mathcal{E}$ admits a numeraire pair if and only if there exists a pricing kernel;*

(ii) *for a self-financing trading strategy ϕ, $\int_0 \phi_u \cdot dA_u^N$ is an $(\{\mathcal{F}_t^A\}, \mathbb{N})$ martingale for all numeraire pairs $(N, \mathbb{N})$ if and only if $\int_0 \phi_u \cdot d(ZA)_u$ is an $(\{\mathcal{F}_t^A\}, \mathbb{P})$ martingale for all pricing kernels Z;*

(iii) *any two numeraire pairs, $(N, \mathbb{N}_1)$ and $(N, \mathbb{N}_2)$, with common numeraire N agree on $\mathcal{F}_T^A$ if and only if any two pricing kernels, $Z^{(1)}$ and $Z^{(2)}$, agree up to $\mathcal{F}_T^A$ (i.e. $\mathbb{E}_{\mathbb{P}}[Z_T^{(1)} - Z_T^{(2)} | \mathcal{F}_T^A] = 0$ a.s.).*

Remark 7.49: What Theorem 7.48 shows is that all the results regarding arbitrage, admissible strategies and completeness carry over immediately to a corresponding result for pricing kernels (see Theorems 7.32 and 7.41 for the relevant results stated in numeraire form).

Proof:
Proof of (i): Suppose $(N, \mathbb{N})$ is a numeraire pair for $\mathcal{E}$. Define the a.s. positive, $\{\mathcal{F}_t^A\}$-adapted process $Z = (N\rho)^{-1}$ where

$$\rho_t := \frac{d\mathbb{P}}{d\mathbb{N}}\bigg|_{\mathcal{F}_t^A} = \mathbb{E}_{\mathbb{N}}\left[\frac{d\mathbb{P}}{d\mathbb{N}} \bigg| \mathcal{F}_t^A\right].$$

Then

$$
\begin{aligned}
\mathbb{E}_{\mathbb{P}}\left[(ZA)_T | \mathcal{F}_t^A\right] &= \frac{\mathbb{E}_{\mathbb{N}}\left[(ZA)_T \frac{d\mathbb{P}}{d\mathbb{N}} | \mathcal{F}_t^A\right]}{\mathbb{E}_{\mathbb{N}}\left[\frac{d\mathbb{P}}{d\mathbb{N}} | \mathcal{F}_t^A\right]} \\
&= \frac{\mathbb{E}_{\mathbb{N}}\mathbb{E}_{\mathbb{N}}\left[(ZA)_T \frac{d\mathbb{P}}{d\mathbb{N}} | \mathcal{F}_T^A | \mathcal{F}_t^A\right]}{\rho_t} \\
&= \frac{\mathbb{E}_{\mathbb{N}}[A_T^N | \mathcal{F}_t^A]}{\rho_t} \\
&= \frac{A_t^N}{\rho_t} = (ZA)_t .
\end{aligned}
$$

A similar argument shows that $\mathbb{E}_{\mathbb{P}}[|(ZA)_t|] < \infty$ and ZA is clearly $\{\mathcal{F}_t^A\}$-adapted. Thus ZA is an $(\{\mathcal{F}_t^A\}, \mathbb{P})$ martingale and Z is a pricing kernel.

Conversely, suppose Z is a pricing kernel for $\mathcal{E}$. Define a numeraire N as follows. Let

$$i_t := \left\{ \min i \in \{1, \dots, n\} : |A_t^{(i)}| \geq |A_t^{(j)}|, \text{ for all } 1 \leq j \leq n \right\},$$

the index of the asset which has the price at t with maximum absolute value. Define the sequence of $\{\mathcal{F}_t^A\}$ stopping times τ_m and the index I_t via

$$\tau_0 = 0, \qquad \tau_m = \inf\{t > \tau_{m-1} : |A_t^{(I_{\tau_{m-1}})}| < \tfrac{1}{2}|A_t^{i_t}|\},$$

$$I_t = i_{\tau_{k(t)}} \text{ where } k(t) \text{ is such that } \tau_{k(t)} \le t < \tau_{k(t)+1}.$$

That is, we construct the index process I and the sequence τ_m inductively as follows. Start with I_0 being the index of the asset with price of maximum absolute value and $\tau_0 = 0$. Wait until this asset price is less than half the value (in absolute terms) of the new maximum price, then switch the index I to the current maximum priced asset. The time of this switch is τ_1. Continue in this way. Now take the numeraire N to be

$$N_t := \left| A_t^{(I_t)} \prod_{j=1}^{k(t)} \frac{A_{\tau_j}^{(I_{\tau_{j-1}})}}{A_{\tau_j}^{(I_{\tau_j})}} \right| = \left(\frac{1}{2}\right)^{k(t)} |A_t^{(I_t)}|,$$

the portfolio produced by starting with a unit amount of the most valuable asset (in absolute terms) and switching to the most valuable asset (again in absolute terms) at each $\{\mathcal{F}_t^A\}$ stopping time τ_m.

The process ZA is an $(\{\mathcal{F}_t^A\}, \mathbb{P})$ martingale so ZN is an $(\{\mathcal{F}_t^A\}, \mathbb{P})$ local martingale. The sequence $\{\tau_m\}$ is a localizing sequence for NZ. Furthermore, for all $m > 0, (ZN)_t^{\tau_m} \le \sum_{i=1}^n |Z_t A_t^{(i)}|$ and $\sum_{i=1}^n \mathbb{E}_\mathbb{P}[|Z_t A_t^{(i)}|] < \infty$ so the dominated convergence theorem (applied to $Z_t N_t^{\tau_m}, m \ge 1$) ensures that ZN is a true $(\{\mathcal{F}_t^A\}, \mathbb{P})$ martingale. Defining the measure $\mathbb{N}$ via

$$\frac{d\mathbb{N}}{d\mathbb{P}} := \frac{(NZ)_T}{(NZ)_0},$$

it follows easily that $(N, \mathbb{N})$ is a numeraire pair.

Proofs of (ii) and (iii): The proofs of (ii) and (iii) are relatively straightforward using the construction in the proof of (i). We prove (iii) in one direction for illustration. Suppose $\mathbb{N}_1 = \mathbb{N}_2$ on $\mathcal{F}_t^A$ whenever $(N, \mathbb{N}_1)$ and $(N, \mathbb{N}_2)$ are two numeraire pairs with common numeraire N, and let $Z^{(1)}$ and $Z^{(2)}$ be two pricing kernels. Define the measures $\mathbb{N}_1$ and $\mathbb{N}_2$ by

$$\frac{d\mathbb{N}_i}{d\mathbb{P}} := N_T Z_T^{(i)}, \qquad i = 1, 2, \tag{7.40}$$

where the numeraire N is as constructed in the proof of (i). As shown there, $(N, \mathbb{N}_1)$ and $(N, \mathbb{N}_2)$ are both numeraire pairs, thus $\mathbb{N}_1 = \mathbb{N}_2$ on $\mathcal{F}_T^A$ by hypothesis. It follows from (7.40) that $Z^{(1)} = Z^{(2)}$ up to $\mathcal{F}_T^A$, as required. $\square$

The equivalence of the pricing kernel and numeraire approaches also leads to a pricing kernel valuation formula. We shall appeal to this in Chapter 8. The proof follows immediately from Theorem 7.48 and Corollary 7.34, the martingale valuation formula using numeraires.

Corollary 7.50 *Let $\mathcal{E}$ be an economy with a pricing kernel Z and let V_T be some attainable contingent claim. Then V_t admits the representation*

$$V_t = Z_t^{-1}\mathbb{E}_{\mathbb{P}}\left[V_T Z_T | \mathcal{F}_t^A\right].$$

7.4 EXTENSIONS

The theory developed in Section 7.3 is restrictive in several senses. In particular,

(i)   derivatives were restricted to making a single payment, V_T, at the deterministic terminal time T;

(ii) the payment amount V_T is an $\mathcal{F}_T^A$-measurable random variable and no control can be applied by the purchaser of the option;

(iii) the economy comprises only a finite number of assets, each of which is a continuous process which pays no dividends;

(iv) the economy and trading strategies are defined only over a finite time interval $[0, T]$.

Each of these restrictions can be removed in a more general theory. For the applications treated in this book we will need some of these generalizations but not others. Below we discuss each in turn and show how to carry out the extension or point towards other references which do so.

7.4.1 General payout schedules

The obvious extension of the payout schedule treated in Section 7.3 (a single payment at T) is to replace it by a payout process $\{H_t, 0 < t \leq T\}$, where H is some $\{\mathcal{F}_t^A\}$-adapted, càdlàg semimartingale, H_t being the total payments made up to and including time t. To treat a payoff as general as this we would need to have defined the stochastic integral for a general semimartingale integrator. This is done, for example, in Chung and Williams (1990), Elliott (1982) and Protter (1990).

We shall not need this generality. For our purposes it will be sufficient to consider processes H of the form

$$H_t = C_t + \sum_{i=1}^{m} V^{(i)} \mathbb{1}_{\{t \geq \tau_i\}}$$

where C is some continuous $\{\mathcal{F}_t^A\}$-adapted process, each τ_i is an $\{\mathcal{F}_t^A\}$ stopping time and each $V^{(i)}$ is $\mathcal{F}_{\tau_i}^A$-measurable. Because option valuation is linear, to treat this type of payoff it is sufficient to separately value a continuous payment stream C and a single $\mathcal{F}_\tau^A$-measurable payment V made at some $\{\mathcal{F}_t^A\}$ stopping time τ.

The need to treat a continuous payment stream arises from futures contracts. This is our only application of this and so we defer a discussion until then, Chapter 20. This leaves the problem of valuing a single $\mathcal{F}_\tau^A$-measurable payment V made at the $\{\mathcal{F}_t^A\}$ stopping time $\tau \leq T$. Two important examples of products of this type are *barrier rebates*, payments of known amount paid at the moment an asset price hits some barrier level, and *American options*, which we discuss in more detail in Section 7.4.2 below.

Let $(N, \mathbb{N})$ be some numeraire pair for the economy $\mathcal{E}$. Given any (V, τ) with

$$\mathbb{E}_{\mathbb{N}}\left[\left|\frac{V}{N_\tau}\right|\right] < \infty,$$

we can define

$$V_T := V + \int_0^T \left(\frac{V}{N_\tau}\right) 1_u(\tau, T] dN_u = V \frac{N_T}{N_\tau}. \tag{7.41}$$

Note that the integrand in (7.41) is $\{\mathcal{F}_t^A\}$-predictable (being left-continuous with right limits and adapted) so the integral in (7.41) is a standard stochastic integral. The payoff V_T is precisely that which results from investing the payoff V, made at τ, in the numeraire strategy N.

In Section 7.3 we treated payoffs V_T for which

$$\mathbb{E}_{\mathbb{M}}\left[\left|\frac{V_T}{M_T}\right|\right] < \infty,$$

for all numeraire pairs $(M, \mathbb{M})$. Note that with V and V_T defined as above, for any $\{\mathcal{F}_t^A\}$ stopping time τ and any numeraire pair $(M, \mathbb{M})$,

$$\mathbb{E}_{\mathbb{M}}\left[\left|\frac{V_T}{M_T}\right|\right] = \mathbb{E}_{\mathbb{M}}\mathbb{E}_{\mathbb{M}}\left[\left|\frac{V}{M_\tau}\frac{M_\tau}{N_\tau}\frac{N_T}{M_T}\right| \, \Big| \, \mathcal{F}_\tau^A\right] \leq \mathbb{E}_{\mathbb{M}}\left[\left|\frac{V}{M_\tau}\right|\right], \tag{7.42}$$

(the inequality following by applying (the conditional form of) Fatou's lemma to the positive local martingale M/N). Thus if $\mathbb{E}_{\mathbb{M}}[|\frac{V}{M_\tau}|] < \infty$ for all numeraire pairs we can price this option using the theory of Section 7.3. If the economy is complete then the inequality in (7.42) is an equality and (V, τ) can be replicated if and only if (V_T, T) can be replicated.

We conclude from the above discussions that, by immediately investing any payment in the numeraire N, the valuation of (V, τ) is identical to the valuation of (V_T, T), and this latter problem was the topic of Section 7.3. The martingale valuation formula becomes, for $t \leq \tau$,

$$V_t = N_t \mathbb{E}_{\mathbb{N}}\left[\frac{V}{N_\tau} \, \Big| \, \mathcal{F}_t^A\right].$$

7.4.2 Controlled derivative payouts

A particularly important problem is the valuation of derivatives whose payout depends on actions taken by the buyer of the option (the more general case where both buyer and seller can apply controls can also be handled, but we shall not need these results). An American put option, for example, allows the option holder to receive the amount $(K - S_\tau)_+$ at time τ, where K is some constant, S is the price of a stock and the stopping time τ is chosen by the holder of the option.

More generally, the control problem can be formulated as follows. Let U be the set of controls (for the American option this would be the set of all $\{\mathcal{F}_t^A\}$ stopping times). By the techniques of Section 7.4.1 we can assume a single payment V_T is made at time T. For each $u \in U$ let $V_T(u)$ denote the payment made at T if the option holder follows the strategy u. The set of controls must be such that, for each $u \in U$, $V_T(u)$ is an attainable claim and this will impose constraints on the set U. For example, in the case of an American option the time τ must be an $\{\mathcal{F}_t^A\}$ stopping time and not a more general $\{\mathcal{F}_t^W\}$ stopping time. Let $V_t(u)$ be the value at t of the payout $V_T(u)$ and let

$$V_0^* := \sup_{u \in U} V_0(u).$$

Suppose, for some $u^* \in U$, $V_0(u^*) = V_0^*$. Then (subject to regularity conditions on the set U as described below), we claim, V_T is attainable, has initial value $V_0 = V_0^*$, and is replicated by the strategy $\phi(u^*)$ which replicates $V_T(u^*)$. To see this, suppose that we buy the option for $V < V_0^*$. We choose to follow the strategy u^* and simultaneously, at a cost of $V_0(u^*)$, replicate the payoff $V_T(u^*)$ dynamically. Doing so, we have a time zero profit of $V_0(u^*) - V$ and no other cashflows – an arbitrage. Thus $V_0 \geq V(u^*) = V_0^*$.

To see that $V_0 \leq V(u^*)$ is a little more involved. Suppose we sell the option for $V > V_0^*$. We must assume that the set of strategies, and the information available to us about the strategy which the option holder is following, together are sufficient for us to be able to follow a strategy which performs no worse than that of the option holder. This will not be the case in general. One example where it is the case is if the holder must declare the strategy u at time zero, in which case we can replicate the payout and receive an arbitrage profit of $V - V(u) \geq V - V_0^*$ at time zero. A second more realistic and more important example is the American option. In this case, if τ is an arbitrary $\{\mathcal{F}_t^A\}$ stopping time and if τ^* is the optimal stopping time,

$$\mathbb{P}\big(V_t(\tau^*) \geq V_t(\tau) \text{ for all } t\big) = 1,$$

so we can make a profit of $V - V(\tau^*)$ at a time zero and an additional profit of $V_\tau(\tau^*) - V_\tau(\tau)$ at the holder's exercise time τ. Once again this, would be an arbitrage and thus we can conclude that $V_0 \geq V_0^*$.

There are two issues not addressed in the discussions above. The first is that the supremum over $u \in U$ may not be attainable. In this case the payoff is not attainable and cannot be valued. The second, one of considerable theoretical and practical importance, is how we find the optimal strategy u^* for any particular problem. This topic, optimal control theory, is not one we shall tackle, but Karatzas and Shreve (1998) and Oksendal (1989) are suitable places to look.

7.4.3 More general asset price processes

There are several ways to generalize the model for the assets in the economy. We have assumed that all price processes are continuous semimartingales driven by a finite number of Brownian motions. This can be generalized to the case of a general semimartingale. Continuous models driven by more than a finite number of Brownian motions have been considered by, amongst others, Kennedy (1994) and Jacka et al. (1998). Models which allow jumps are particularly important when modelling assets which are subject to default risk and thus which can suddenly devalue. References covering this more general situation include Jarrow and Turnbull (1995), Jarrow and Madan (1995) and Duffie and Singleton (1997).

A second restriction we have imposed on the assets in an economy is that they do not pay dividends. This is a significant omission as most stocks pay dividends. However, these often fall under the results of Section 7.3 as we now show.

Suppose A is the price process, with jumps, of the dividend-paying assets of an economy. Suppose the dividends are paid at times $T_1, \ldots, T_m$, and the dividend amounts at T_i form an $\mathcal{F}_{T_i}^A$-measurable random vector G_i. Let

$$\hat{A}_t := A_t + \sum_{i=1}^{m} G_i \mathbb{1}_{\{t \geq T_i\}} \frac{N_t}{N_{T_i}}$$

where N is some numeraire. We assume the process $\hat{A}$ is continuous. We can interpret $\hat{A}_t^{(i)}$ as the value at t of buying and holding the asset $A^{(i)}$ at time zero and reinvesting all dividends in the numeraire N. If we treat $\hat{A}$ as the 'true' asset price process of the economy then, noting that $\mathcal{F}_t^A = \mathcal{F}_t^{\hat{A}}$, all the results of Section 7.3 apply if we consider the process $\hat{A}$ in preference to A.

This analysis does not treat the case when $\hat{A}$ is not continuous (so the asset prices are discontinuous even after correcting for dividend payments) or when the dividends, unrealistically, are a continuous cashflow stream. The former case is beyond the scope of this book. We do not discuss the latter, but a formal treatment can be developed easily from the ideas of Section 7.4.1 and Chapter 20.

The final extension we shall consider is to the case of a continuum of assets rather than just finitely many. This is relevant to the modelling of interest rate products and is the subject of Chapter 8.

7.4.4 Infinite trading horizon

There are some situations in which we would like the economy to be defined over an infinite horizon. We shall meet one such situation in the next chapter where we discuss term structure models. Doing this introduces a few extra problems, as the following example illustrates.

Example 7.51 Return to the basic Black–Scholes economy introduced in Example 7.1 and suppose that $\mu \geq r$. Define the stopping time τ to be

$$\tau := \inf\{t > 0 : S_t \exp(-rt) > 2S_0\}.$$

It follows from Theorem 2.8 that $\tau < \infty$ a.s. and thus the strategy of buying one unit of stock, S, and selling one unit of bond, D, at time zero yields a guaranteed profit of S_0 at τ with no risk. Similarly, if $\mu < r$ the strategy of buying one unit of bond and selling one unit of stock at time zero and holding this portfolio until

$$\eta := \inf\{t > 0 : S_t \exp(-rt) < S_0/2\}$$

guarantees a riskless profit of $S_0/2$ at $\eta < \infty$.

A strategy such as the one above is an arbitrage and as such presents problems when we come to pricing. There are two obvious ways to deal with this problem. The first is to eliminate the strategy completely from the economy by modifying our admissibility definition (Definition 7.25) and that of a numeraire pair to insist that all gain processes are *uniformly integrable* martingales (recall that if M is a UI martingale then it follows from the optional sampling theorem that $\mathbb{E}[M_\tau] = M_0$ for all stopping times τ). If we adopt this approach the theory of this chapter is altered only by replacing the finite time interval $[0, T]$ by the (closed) time interval $[0, \infty]$.

The problem with adopting the approach above is that it would rule out many otherwise acceptable models, such as the Black–Scholes model, and may be too restrictive for many applications. We shall adopt an alternative approach, one which effectively reduces the general modelling and valuation problem to the case we have already considered. We allow the model, the economy and trading strategies to be defined over the time horizon $[0, \infty)$, thus incorporating the yield curve models introduced in Chapter 8, but will only value derivatives for which the payout is made on or before some predetermined time T, a time that will depend on the derivative in question. This is sufficiently general to treat all derivatives of practical importance.

If we adopt this latter approach we are effectively reduced to the framework of Section 7.3. One could (we shall not) introduce the terminology *T-admissible* (admissible when the strategy and economy are considered only on

$[0, T])$, *T-complete* (complete when the strategy and economy are considered only on $[0, T])$ and *locally complete* (*T*-complete for every T). An economy such as this would admit an arbitrage in the sense of Example 7.51 but this would not violate any of the replication arguments used to price a derivative which must settle on or before $T < \infty$.

8

Dynamic Term Structure Models

8.1 INTRODUCTION

The theory of Chapter 7 applies to models for which there are only finitely
many assets. The main applications treated in this book are pricing problems
for interest rate derivatives, in which case the underlying assets are pure
discount bonds. One can define a pure discount bond for each and every
maturity date, a continuum of assets, and in this context the theory of
Chapter 7 may be inadequate. In actual fact, for the vast majority of
applications it is only necessary to model a (small) finite number of these
bonds and in this sense the finite-asset theory is usually sufficient even for
interest rate products. However, there are occasions, the valuation of futures
contracts being a case in point, when it is convenient to model an infinite
number of bonds. Furthermore, there is considerable economic and theoretical
understanding to be gained from studying models of the whole term structure
(of interest rates), and it is important to know even for models involving
finitely many bonds that there is a (realistic) whole term structure model
consistent with the model. This is the topic of this chapter.

Section 8.2 briefly describes how, by restricting the class of admissible
trading strategies, a study of derivative valuation for an economy comprising a
continuum of bonds can effectively be reduced to the finite-asset case studied
in Chapter 7. Then, in Section 8.3, term structure models are studied in
detail, at the abstract level. Several ways to specify a term structure model
are presented, including the celebrated Heath-Jarrow-Morton framework, and
these are related to each other. Very little is said about particular models,
this discussion being deferred until Part II.

8.2 AN ECONOMY OF PURE DISCOUNT BONDS

As was the case with finitely many assets, to specify an economy of pure
discount bonds we must define both the joint law of all the assets within

Financial Derivatives in Theory and Practice Revised Edition. P. J. Hunt and J. E. Kennedy
© 2004 John Wiley & Sons, Ltd   ISBNs: 0-470-86358-7 (HB); 0-470-86359-5 (PB)

the economy and the set of admissible trading strategies. One could do this in considerable generality. For example, Kennedy (1994) considers a general class of Gaussian term structure models which includes models generated by a Brownian sheet, and Jacka *et al.* (1998) also consider continuous models more general than those generated by only finitely many Brownian motions. Neither of these, however, discusses admissible strategies and questions of completeness. The latter issues are considered by Björk *et al.* (1997a; 1997b) who develop term structure models driven by a finite number of Brownian motions and a marked point process. A trading strategy is then a predictable process L with L_t being a signed finite Borel measure on $[t, \infty)$, $L_t(A)$ representing the holding at t of bonds with maturities in the set $A \in \mathcal{B}([t, \infty))$. This approach is mathematically very appealing, but we shall not have need of such generality.

In keeping with the continuous theory of Chapter 7 we shall develop term structure models under the following assumption (which, through the filtration restriction, is stronger than the conditions imposed in Chapter 7):

the underlying 'real-world' probability space $(\Omega, \{\mathcal{F}_t\}, \mathcal{F}, \mathbb{P})$ *satisfies the usual conditions, supports a d-dimensional* $(\{\mathcal{F}_t\}, \mathbb{P})$ *Brownian motion* W, *and the filtration* $\{\mathcal{F}_t\}$ *is the (augmented) natural filtration generated by* W, $\{\mathcal{F}_t\} = \{\mathcal{F}_t^W\}$.

Now denote by D_{tT} the value at t of a pure discount bond with maturity T, an asset which pays a unit amount on its maturity date. We shall, in Section 8.3, describe in detail several ways to model the complete term structure of pure discount bonds, $\{D_{tT} : 0 \le t \le T < \infty\}$. If, in keeping with the notation of Chapter 7, we denote by $\{\mathcal{F}_t^A\}$ the filtration generated by the asset prices of the economy,

$$\mathcal{F}_t^A := \sigma(D_{uT} : 0 \le u \le t, u \le T < \infty),$$

then every model we shall develop has the following four properties:

Definition 8.1 *A term structure (or (TS)) model is a model of the dynamics of the discount curve,* $\{D_{.T} : 0 \le T < \infty\}$, *with the following properties:*
(TS1) *For all* $T \ge 0$, *the process* $\{D_{tT}, t \le T\}$, *is a continuous semimartingale with respect to the filtration* $\{\mathcal{F}_t\}$.
(TS2) $D_{tT} \ge 0$ *for all* $t \le T$.
(TS3) $D_{TT} = 1$ *for all* T.
(TS4) *The model admits a pricing kernel: there exists some strictly positive continuous semimartingale* Z *such that* $Z_t D_{tT}$ *is an* $(\{\mathcal{F}_t^A\}, \mathbb{P})$ *martingale for all* $T \ge 0$.

The focus of this book is on models in which asset prices are continuous and which we can integrate against (so that the gain from trading is well defined), thus (TS1) is a natural assumption for us to make (recall Theorem

4.27). Conditions (TS2) and (TS3) are obviously required. The final condition, (TS4), requires a little more comment. We saw for the discrete economy of Chapter 1 that, in that context, the existence of a pricing kernel is equivalent to the absence of arbitrage (Theorem 1.7). We saw in Chapter 7 that, as long as we eliminate troublesome 'doubling strategies', the existence of a numeraire pair (which, over a finite time horizon, is equivalent to the existence of a pricing kernel) implies no arbitrage (Theorem 7.32), and we commented that the absence of an 'approximate arbitrage' implies the existence of a numeraire pair (Remark 7.31). Condition (TS4) is thus a natural condition to impose and will guarantee the absence of arbitrage (assuming, as before, that doubling strategies are removed). However, there is currently no proof that no approximate arbitrage implies the existence of a numeraire pair under which *all* bonds are simultaneously martingales.

Given a model for the asset price evolution satisfying (TS1)–(TS4), we can proceed to define an admissible trading strategy for the economy which mirrors the one used in Chapter 7. Note the significant restriction to the generality of an admissible trading strategy imposed by condition (i) of Definition 8.2 below. Note also that, whereas in Chapter 7 the definition of an admissible strategy was stated in terms of numeraire pairs, here it is presented in terms of pricing kernels. We showed in Section 7.3.6 that the two are equivalent over a finite horizon. Here we prefer to use pricing kernels rather than numeraire pairs because of the technical issues which arise when considering a model over the infinite time horizon $[0, \infty)$ (see Theorem 8.9).

Definition 8.2 *An admissible trading strategy for the term structure economy $\mathcal{E}$ is an $\{\mathcal{F}_t^A\}$-predictable vector process $\phi = (\phi^{(1)}, \phi^{(2)}, \dots)$ and a sequence of times (not necessarily increasing) $(T_1(\phi), T_2(\phi), \dots)$ such that:*

(i)   *for each $T > 0$, there exists some $M(\phi, T)$ such that $\phi_t^{(i)} = 0$ for all $i > M(\phi, T)$, $t \le T$;*

(ii) *the strategy is self-financing,*

$$V_t = \sum_{i=1}^{\infty} \phi_t^{(i)} D_{tT_i(\phi)} = \sum_{i=1}^{M(\phi,t)} \phi_t^{(i)} D_{tT_i(\phi)}$$

$$= \sum_{i=1}^{M(\phi,t)} \phi_0^{(i)} D_{0T_i(\phi)} + \int_0^t \sum_{i=1}^{M(\phi,t)} \phi_{uT_i(\phi)}^{(i)} dD_{uT_i(\phi)},$$

where V_t is value at t of the portfolio corresponding to the strategy ϕ;

(iii) *for all pricing kernels Z, the kernel-rescaled gain process*

$$(ZG^\phi)_t = \int_0^t \sum_{i=1}^{M(\phi,t)} \phi_u^{(i)} d(Z.D_{.T_i(\phi)})_u$$

is an $(\{\mathcal{F}_t^A\}, \mathbb{P})$ martingale.

Remark 8.3: We have imposed the condition on the trading strategy ϕ that it involves only a predetermined and finite set of bonds up to any time T. This reflects the way these models are used in practice – for most derivatives there is a known finite set of bonds which is required to price and hedge the product. In this context the existence of a continuum of bonds is more of a consistency condition on the model than an essential requirement.

Remark 8.4: Under the terms of an interest rate futures contract, discussed in Chapter 20, cashflows occur throughout the life of the contract and with at most one cashflow each day. As such, a trading strategy need only involve finitely many bonds (one per day until the futures expiry date). However, from a modelling perspective it is convenient to treat a futures contract as if resettlement occurred continuously, just as we allow continuous rebalancing in the definition of a trading strategy, something which is impossible in practice. To replicate a (theoretical) futures contract such as this would, in a general term structure model, require trading in a continuum of bonds, and such a strategy is not within the scope of the development here. However, the degeneracy of the models we consider, in which a continuum of bonds is driven by finitely many Brownian motions, will allow us to overcome this problem.

With the above definition of an admissible strategy the martingale property of the gain process precludes any arbitrage within the economy, exactly as it did when the economy only had finitely many assets. Furthermore, the martingale valuation formula for an attainable claim is also valid (Corollaries 7.34 and 7.50). The remaining question is: under what conditions is the economy complete? The following result is adequate for all the applications we have encountered.

Theorem 8.5 *Let $\mathcal{E}$ be a term structure economy defined on the probability space $(\Omega, \{\mathcal{F}_t\}, \mathcal{F}, \mathbb{P})$, with asset filtration $\{\mathcal{F}_t^A\}$ and admitting a pricing kernel Z. Suppose that, for each $T > 0$, there exists a finite set of times $(T_1, \ldots, T_{n(T)})$ such that, for all $t \leq T$,*

$$\mathcal{F}_t^A = \sigma(D_{tT_1}, \ldots, D_{tT_{n(T)}}). \tag{8.1}$$

Then the economy is $\mathcal{F}_T^A$-complete if any two pricing kernels Z_1 and Z_2 for the finite economy comprising the bonds $(D_{tT_1}, \ldots, D_{tT_{n(T)}})$ agree on $\mathcal{F}_T^A$ (or equivalently, if, whenever $(N, \mathbb{N}_1)$ and $(N, \mathbb{N}_2)$ are two numeraire pairs with common numeraire N, $\mathbb{N}_1$ and $\mathbb{N}_2$ agree on $\mathcal{F}_T^A$).

Suppose, further, that for each $T > 0$ the process $D_t := (D_{tT_1}, \ldots, D_{tT_{n(T)}})$ satisfies an SDE of the form

$$dD_t = \mu(D_t, t)\, dt + \sigma(D_t, t)\, dW_t \tag{8.2}$$

for some Borel measurable μ and σ, and that weak uniqueness holds for the SDE (8.2). Then the economy is complete if either of the two following conditions holds:

(i) *There exists some $\{\mathcal{F}_t^A\}$-predictable ϕ satisfying:*

    (a) $\phi_t \cdot \sigma(D_t, t) = 0$ *for all t a.s.;*

    (b) $\phi_t \cdot D_t > 0$ *for all t a.s.;*

    (c) $-\infty < \int_0^t \dfrac{\phi_u}{\phi_u \cdot D_u} \cdot dD_u < \infty$ *for all t a.s.*

(ii) *There exists some finite variation numeraire N_t^f.*

Proof: Consider the finite economy comprising the assets $(D_{tT_1}, \ldots, D_{tT_{n(T)}})$. That this economy is $\mathcal{F}_T^A$-complete follows from the filtration restriction (8.1) and Theorems 7.41 and 7.43 (see Theorem 7.48 and Remark 7.49 for the pricing kernel statement of these numeraire-pair results). This immediately implies that the term structure economy is $\mathcal{F}_T^A$-complete since replication can be performed with this finite set of assets. (Note that Theorem 7.39 ensures that the larger set of numeraires in the term structure economy, which could potentially have made a strategy which is admissible for the finite economy inadmissible for the term structure economy, presents no problem.) $\qquad\square$

Remark 8.6: If the conditions of Theorem 8.5 are satisfied the model is complete and replication over the interval $[0,\ T]$ can be performed by a given finite set of bonds. In particular, a unit payoff at any time $S < T$ can be replicated. Thus the zero coupon bond of maturity S (which is not in the set $(T_1, T_2, \ldots, T_{n(T)})$) is redundant. Clearly this is not the case in practice and care must be taken when using a (necessarily simple) model to ensure that an inappropriate replicating portfolio is not used (for example, replicating a one-week bond with a combination of bonds with maturities all over ten years!).

8.3 MODELLING THE TERM STRUCTURE

The remainder of this chapter is devoted to discussing eight closely related ways to specify a model for the term structure satisfying (TS1)-(TS4), and it is based primarily on the papers of Baxter (1997) and Jin and Glasserman (2001). Each of these techniques has its place in the development of the theory or as a tool for applications.

In this introduction we will briefly introduce each technique and summarize the relationships between them. In the remaining sections we shall go into more detail on each and establish these connections. We will, in passing, mention some of the most important example models but will not study them in detail until Part II of the book when they will be introduced in the context of an appropriate application. The eight techniques are as follows.

(PDB) *Pure discount bond*: The most direct way to specify a term structure model is by explicitly stating the law of the pure discount bonds or an SDE that they satisfy. It is also common to give an SDE for $\{D_{tT}/N_t,$

$0 \leq t \leq T < \infty\}$, where N is some numeraire process. In the latter situation the numeraire is often a given bond, and the model specification is usually incomplete in that the law of the numeraire bond is not given and the economy is only modelled until some fixed time T because this is all that is needed for the application to hand.

Clearly all models satisfying (TS1)–(TS4) can be specified in this way. Any model which satisfies (TS1)–(TS4) which is specified this way is said to be (PDB).

(PK) *Pricing kernel:* Take *any* strictly positive continuous semimartingale $Z \in \mathcal{L}^1(\Omega, \{\mathcal{F}_t\}, \mathcal{F}, \mathbb{P})$ and define

$$D_{tT} = Z_t^{-1} \mathbb{E}_{\mathbb{P}}[Z_T | \mathcal{F}_t]. \tag{8.3}$$

The process Z may be specified either explicitly or implicitly, depending on the application.

One might think of generalizing this class by replacing the 'real-world' measure $\mathbb{P}$ in (8.3) by some other locally equivalent measure $\mathbb{Z}$, something we shall do frequently in the models below. For a (PK) model, however, this offers no generalization since, if $(\hat{Z}, \mathbb{Z})$ is such that

$$D_{tT} = \hat{Z}_t^{-1} \mathbb{E}_{\mathbb{Z}}[\hat{Z}_T | \mathcal{F}_t]$$

for all $t \leq T$, then

$$D_{tT} = Z_t^{-1} \mathbb{E}_{\mathbb{P}}[Z_T | \mathcal{F}_t],$$

where

$$Z_t := \hat{Z}_t \frac{d\mathbb{Z}}{d\mathbb{P}}\Big|_{\mathcal{F}_t}.$$

(N) *Numeraire:* Define the law of some asset or, more generally, some numeraire, N, in a martingale measure $\mathbb{N}$ which is locally equivalent to $\mathbb{P}$, then extend to a definition for the whole term structure via

$$D_{tT} = N_t \mathbb{E}_{\mathbb{N}}[N_T^{-1} | \mathcal{F}_t].$$

Note that under this approach, which is similar to (PK), there will be additional restrictions on N. For example, if we claim that N is the value of a bond maturing at T then $N_T = 1$.

(FVK) *Finite variation kernel:* This is the subclass of (PK) for which the (continuous) pricing kernel is also taken to be a finite variation process. We will denote such a kernel by ζ rather than Z to distinguish it, and so (8.3) becomes

$$D_{tT} = \zeta_t^{-1} \mathbb{E}_{\mathbb{Z}}[\zeta_T | \mathcal{F}_t].$$

Observe that the real-world measure $\mathbb{P}$ in (8.3) is here replaced by some arbitrary $\mathbb{Z}$ which is locally equivalent to $\mathbb{P}$. It was not necessary, for a general (PK) model, to allow this extra freedom in model specification because a change of measure from an arbitrary $\mathbb{Z}$ to $\mathbb{P}$ left the form (8.3) unaltered, albeit with a different kernel. By contrast, the finite variation property of the kernel is dependent on the measure in question.

(acFVK) *Absolutely continuous (FVK):* This is the subset of (FVK) models for which the finite variation kernel ζ is absolutely continuous with respect to Lebesgue measure. In practice all models encountered which are of type (FVK) are also of type (acFVK). However, examples do exist which are (FVK) but not (acFVK), as we shall show later.

(SR) *Short rate:* Let r be some adapted process, not necessarily continuous, such that the process $\int_0^\cdot r_u \, du$ exists. Let $\mathbb{Q}$ be some measure locally equivalent to $\mathbb{P}$ and suppose that

$$\mathbb{E}_{\mathbb{Q}}\left[\exp\left(-\int_0^t r_u \, du\right)\right] < \infty,$$

for all $t \geq 0$. Now define the term structure via

$$D_{tT} = \mathbb{E}_{\mathbb{Q}}\left[\exp\left(-\int_t^T r_u \, du\right)\Big|\mathcal{F}_t\right].$$

The measure $\mathbb{Q}$ is referred to as the *risk-neutral* measure. Clearly such a model is also an (acFVK) model. We say that this model is a short-rate model, (SR), if the process r has the additional interpretation as the short-term interest rate, or short rate,

$$r_t = -\lim_{h\downarrow 0} \frac{1}{h} \log(D_{t,t+h}).$$

(HJM) *Heath–Jarrow–Morton:* A requirement of any reasonable term structure model is that the discount curve be almost everywhere differentiable in the maturity parameter T. If it is also absolutely continuous with respect to Lebesgue measure, another reasonable requirement, then we have the representation

$$D_{tT} = \exp\left(-\int_t^T f_{tS} \, dS\right),$$

where the f_{tS} are the *instantaneous forward rates*. The (HJM) approach is to model these instantaneous forward rates. By definition every (HJM) model is an (SR) model.

An (HJM) model is usually specified via an SDE which the instantaneous forwards satisfy. For the resulting model to satisfy (TS1)–(TS4), in particular (TS4), additional regularity conditions need to be imposed on this SDE. These are discussed in detail in Section 8.3.7.

(FH) *Flesaker–Hughston*: The Flesaker–Hughston approach is specifically intended to produce term structure models for which all interest rates are positive (meaning D_{tT} is non-increasing in T for all $t \geq 0$ a.s.). Let $\mathbb{Z}$ be locally equivalent to $\mathbb{P}$ and let $\{M_{\cdot S}, 0 \leq S < \infty\}$, be a family of (jointly measurable) positive $(\{\mathcal{F}_t\}, \mathbb{Z})$ martingales (automatically continuous given the Brownian restrictions on the original set-up). Now define the term structure via

$$D_{tT} = \frac{\int_T^\infty M_{tS}\, dS}{\int_t^\infty M_{tS}\, dS}.$$

Note that D_{tT} is decreasing in T so interest rates are always positive, and that $\lim_{T \to \infty} D_{tT} = 0$.

Each of the procedures above will generate a model satisfying (TS1)–(TS4). The different approaches do not yield exactly the same set of models, but they are very closely related. Using a superscript $+$ to denote a model with positive interest rates (i.e. D_{tT} is decreasing in T for all t a.s.) and a superscript 0 to denote a model for which the discount curve decreases to zero as the maturity parameter $T \to \infty$, a summary of these relationships is as follows:

(TS)   = (PDB)   = (PK)   =* (N)   ⊃   (FVK)   ⊃ (acFVK)   ⊃ (SR)   ⊃ (HJM)

(TS)$^+$ = (PDB)$^+$ = (PK)$^+$ =* (N)$^+$ ⊃   (FVK)$^+$ ⊃ (acFVK)$^+$ = (SR)$^+$ = (HJM)$^+$

(TS)$^{+0}$ = (PDB)$^{+0}$ = (PK)$^{+0}$ =* (N)$^{+0}$ ⊃ $\left\{\begin{matrix}(FVK)^{+0} \\ (FH)\end{matrix}\right\}$ ⊃ (acFVK)$^{+0}$ = (SR)$^{+0}$ = (HJM)$^{+0}$

The starred equalities, (PK) =* (N), and consequential identities for the restriction to positive rates, do not strictly hold in general if the economy is considered over the infinite horizon on a probability space which satisfies the usual conditions. If the model is set up with the filtration being the (uncompleted) natural filtration for a d-dimensional Brownian motion, or if the horizon is finite, this is an equality. We discuss this technical point further below.

If interest rates are constrained to be positive, then (acFVK), (SR) and (HJM) coincide. In addition, (FH) contains all (acFVK) models for which interest rates are positive and $\lim_{T \to \infty} D_{tT} = 0$ a.s. There are examples of models which are (FH) but not (FVK)$^{+0}$ and vice versa, so no more can be said on this matter. However, in Section 8.3.8 we give an alternative representation of an (FH) model, one which readily generalizes to a class we call (eFH), *extended Flesaker–Hughston*. This broader class is also contained strictly within (N)$^{+0}$ but strictly contains all (FVK)$^{+0}$ models.

In the remainder of this chapter we will prove all these facts and provide examples to show where inclusions are strict.

For completeness we note here that, for each model class (M), $(M)^{+0} \subset (M)^+ \subset (M)$. The inclusion is immediate. That this inclusion is strict is also obvious. Two example discount curve models which establish this, chosen because they are easy to explain rather than because they are realistic, are as follows. Each is (HJM), the smallest class we consider, and thus establishes the strict inclusion for all classes. The first, a model which is (HJM) but not $(HJM)^+$, is the non-stochastic model defined by the initial discount curve $D_{0T} = 1 - \frac{1}{2}e^{-T}\sin(T)$, a model in which interest rates oscillate between positive and negative. The second, a model which is $(HJM)^+$ but not $(HJM)^{+0}$ is the non-stochastic model defined by $D_{0T} = \exp\left(-\frac{T}{1+T}\right)$. In each case the curve at any future time is given by $D_{tT} = D_{0T}/D_{0t}$.

On an historical note, one of the major breakthroughs in term structure models was the work of Heath *et al.* (1992). However, as the above discussion makes clear, (HJM) is the smallest class of model that we consider. It turns out that the manner in which an (HJM) model is specified often makes it difficult to work with. Furthermore, the technical conditions which are usually imposed on (HJM) models (to ensure that they are (HJM)) make it hard to prove that any given model is of (HJM) type. By contrast, proving that a model is (PK) is usually straightforward. Furthermore, the (HJM) framework (via its technical regularity conditions) excludes some models of the term structure which are otherwise perfectly acceptable and endowed with all the properties that one would like for a term structure model. Example 8.11 below illustrates this point.

8.3.1 Pure discount bond models

We need say very little about this approach, which is merely a way to specify a model satisfying (TS1)–(TS4), i.e. directly in terms of the bond prices. Using this approach it is not obvious what restrictions must be imposed on the directly modelled asset prices to ensure that (TS1)–(TS4) are satisfied. The approach is, however, one of the most natural to use for modelling since bond prices and their dynamics are directly observable. Some of the most important examples defined in this way are the recent 'market models' developed by Miltersen *et al.* (1997), Brace *et al.* (1997) and Jamshidian (1997). These models define the (numeraire-rebased) dynamics of a finite set of bonds. The existence of a model for the whole term structure consistent with these dynamics then needs to be established. We shall discuss this further in Chapter 18.

8.3.2 Pricing kernel approach

Let $Z \in \mathcal{L}^1(\Omega, \{\mathcal{F}_t\}, \mathcal{F}, \mathbb{P})$ be a strictly positive continuous semimartingale.

Define

$$D_{tT} = Z_t^{-1}\mathbb{E}_{\mathbb{P}}[Z_T|\mathcal{F}_t].\qquad(8.4)$$

This is a candidate term structure model and it clearly satisfies (TS2)–(TS4). The expectation in (8.4) is an $(\{\mathcal{F}_t\},\mathbb{P})$ martingale, hence is continuous, by Theorem 5.49, and Z^{-1} is a continuous semimartingale, thus D_{tT} is a continuous semimartingale for all $T \geq 0$. This establishes (TS1) and thus we conclude that (PK) $\subseteq$ (TS).

That (TS) $\subseteq$ (PK) is immediate since we can take the pricing kernel in (TS4) to be the process Z in (8.4). We have thus established that (TS) = (PK).

Remark 8.7: Had we allowed the process Z to be more general than a continuous semimartingale, the process D_{tT} would not have been a continuous semimartingale. However, the process $Z_t D_{tT}$, which has the interpretation of the value at t of a bond paying unity at T but stated in the units $U = Z^{-1}$, is a continuous martingale (continuous because we are working on a Brownian filtration). Thus even in this more general case there is a unit in which the bond price process is a continuous semimartingale under $\mathbb{P}$.

Remark 8.8: We met pricing kernels in Chapter 1 and Section 7.3.6. Recall that the reason, obvious in earlier discussions but less so in this context, for calling Z a pricing kernel arises from the derivative valuation formula $V_t = Z_t^{-1}\mathbb{E}_{\mathbb{P}}[V_T Z_T|\mathcal{F}_t]$. That is (recalling that taking an expectation is the same as performing an integration), Z takes on the role of a classical kernel.

8.3.3 Numeraire models

An alternative to (PK) is to use a numeraire pair $(N,\mathbb{N})$ to define a term structure model,

$$D_{tT} = N_t\mathbb{E}_{\mathbb{N}}[N_T^{-1}|\mathcal{F}_t].\qquad(8.5)$$

All numeraire models are (TS) models (i.e. satisfy conditions (TS1)–(TS4)), or equivalently are of class (PK). To see that (N) $\subseteq$ (PK), define

$$Z_t = N_t^{-1}\frac{d\mathbb{N}}{d\mathbb{P}}\Big|_{\mathcal{F}_t}.\qquad(8.6)$$

Substitution of (8.6) into (8.5) yields (8.4).

We are also able to establish a (partial) converse to this result and show (modulo technicalities at $t = \infty$) that (PK) $\subseteq$ (N) and thus (PK) = (N). More precisely, we have the following theorem.

Theorem 8.9 *Let Z be some pricing kernel defined on the probability space $(\Omega,\{\mathcal{F}_t\},\mathcal{F},\mathbb{P})$. As previously, suppose that $\{\mathcal{F}_t\} = \{\mathcal{F}_t^W\}$, the augmented filtration generated by a d-dimensional Brownian motion W. Consider now a*

(PK) model defined by equation (8.4). Define the unit-rolling numeraire N^u via

$$N_t^u = \frac{D_{t\lfloor t+1 \rfloor}}{\prod_{i=1}^{\lfloor t \rfloor} D_{i-1,i}}$$

($\lfloor x \rfloor$ being the largest integer no greater than x), the portfolio constructed by repeatedly buying the bond expiring one time unit in the future. Then we have the following:

(i) Given any $T^* > 0$, we have that

$$D_{tT} = N_t^u \mathbb{E}_{\mathbb{N}^{T^*}} \left[(N_T^u)^{-1} | \mathcal{F}_t \right],$$

for all $t \le T \le T^*$, where the measure $\mathbb{N}^{T^*} \sim \mathbb{P}$ is defined by

$$\left. \frac{d\mathbb{N}^{T^*}}{d\mathbb{P}} \right|_{\mathcal{F}_t} := (ZN^u)_t^{T^*}, \tag{8.7}$$

$(ZN^u)^{T^*}$ being the martingale (ZN^u) stopped at T^*. That is, the model is of class (N) over the time interval $[0, T^*]$.

(ii) If the filtration $\{\mathcal{F}_t\}$ is taken instead to be the uncompleted filtration $\{(\mathcal{F}_t^W)^\circ\}$ then, for all $T \in [0, \infty)$,

$$D_{tT} = N_t^u \mathbb{E}_{\mathbb{N}} \left[(N_T^u)^{-1} | \mathcal{F}_t \right],$$

where the measure $\mathbb{N}$, locally equivalent to $\mathbb{P}$ with respect to $(\mathcal{F}_\infty^W)^\circ$, is defined via

$$\left. \frac{d\mathbb{N}}{d\mathbb{P}} \right|_{\mathcal{F}_t} := (ZN^u)_t,$$

for all $t \ge 0$.

Remark 8.10: The reason for this somewhat involved statement of the converse lies in the results and discussion of Section 5.1.2 and, in particular, Theorems 5.14 and 5.15. The first part of this converse result is sufficient for all practical purposes since in practice we only ever need to consider an economy over a finite horizon.

Proof: The proof of (i) is immediate by carrying out the change of measure prescribed by (8.7). The proof of (ii) follows from Theorem 5.15. □

The disadvantage of the numeraire approach compared with (PK) is that, in claiming N to be a numeraire, we must ensure consistency conditions hold within the model. For example, if N_t, $t \le T$, is the value of the pure discount bond maturing at T then $N_T = 1$. On the other hand, modelling within a martingale measure is more direct and intuitive. Furthermore, applications often dictate which measure we should work in and it usually corresponds to a particular numeraire. All the models we shall cover are developed within a martingale measure $\mathbb{N}$ rather than the 'real-world' measure $\mathbb{P}$.

For our purposes the most important examples of models specified via a numeraire are Markov-functional models. We shall meet these formally in Chapter 19 but, in fact, most models used in practice are Markov-functional. In the examples introduced in Chapter 19 the law of the numeraire is only calculated implicitly and no SDE is ever specified which is satisfied by the numeraire.

8.3.4 Finite variation kernel models

An (FVK) model is one for which the prices of all pure discount bonds can be expressed in the form

$$D_{tT} = \zeta_t^{-1} \mathbb{E}_{\mathbb{Z}}[\zeta_T | \mathcal{F}_t] \tag{8.8}$$

for some adapted and continuous finite variation process ζ and some measure $\mathbb{Z}$ locally equivalent to $\mathbb{P}$.

Historically, most models that have been studied are (FVK) models, indeed short-rate models (although they are increasingly not parameterized in terms of the short rate). Does this restriction eliminate any realistic models of the term structure, or can all models be cast in this form? The answer to this question is provided by the following example, a model which possesses all the properties one might realistically demand of an interest rate model but one which is not (FVK).

Example 8.11 Consider the (PK) model generated by taking the pricing kernel Z to be the solution, with $Z_0 = 1$ a.s., to

$$dZ_t = Z_t^2 \, dW_t \,, \tag{8.9}$$

where W is a one-dimensional Brownian motion. The process Z is the reciprocal of the Bessel(3) process and various of its properties are developed in Exercise 8.14 below. In particular, it is well known as an example of a process which is a uniformly integrable local martingale but which is not a true martingale. This kernel, being a.s. positive, is also a (strict) supermartingale and, furthermore, $\mathbb{E}_{\mathbb{P}}[Z_T | \mathcal{F}_t] \downarrow\downarrow 0$ as $T \to \infty$. Consequently, (8.9) can be used to define a term structure model in which rates are a.s. strictly positive for all t and the discount curve decays to zero as $T \to \infty$, one of the nicest (single-factor) interest rate models one could imagine. Indeed, applying a time-change technique which applies equally well to many other pricing kernel models, given any continuous initial discount curve $\{D_{0T}, T \geq 0\}$, the model can be adapted to fit this initial term structure. To do this, let

$$E(T) := \mathbb{E}_{\mathbb{P}}[Z_T] \,,$$

a decreasing function of T, and define the process $\hat{Z}$ by

$$\hat{Z}_t := Z\big(E^{-1}(D_{0t})\big).$$

Then $\hat{Z}$ is a pricing kernel on the the probability space $(\Omega, \{\tilde{\mathcal{F}}_t\}, \mathcal{F}, \mathbb{P})$ where $\tilde{\mathcal{F}}_t := \mathcal{F}_{E^{-1}(D_{0t})}$, and the model is consistent with the initial discount curve $\{D_{0T}, T \geq 0\}$.

We show that this model is not (FVK). To do so, we first show that the model is $\mathcal{F}_T^A$-complete for any $T > 0$. To see this, work on the time interval $t \in [0, T]$ and take $N_t = D_{tT}$ as numeraire. Under the measure $\mathbb{N}$ defined by

$$\left.\frac{d\mathbb{N}}{d\mathbb{P}}\right|_{\mathcal{F}_t} = Z_t D_{tT},$$

all N-rebased assets are $\{\mathcal{F}_t\}$ martingales. Consider now the process D_{tS}^N for some $S > T$. This is an $(\{\mathcal{F}_t\}, \mathbb{N})$ martingale, hence, by the martingale representation theorem for Brownian motion (Theorem 5.49),

$$D_{tS}^N = D_{0S}^N + \int_0^t \phi_u \, d\widetilde{W}_u \qquad (8.10)$$

for some $\{\mathcal{F}_t\}$-predictable process ϕ, where the $(\{\mathcal{F}_t\}, \mathbb{N})$ Brownian motion $\widetilde{W}$ is defined by

$$\widetilde{W}_t = W_t - \int_0^t C_u \, du$$

for some $\{\mathcal{F}_t\}$-predictable process C (defined precisely by Theorem 5.24 with $\mathbb{Q} = \mathbb{N}$). Given any $\mathcal{F}_T^A$-measurable random variable V_T with $\mathbb{E}_{\mathbb{N}}[\|V_T/N_T\|] < \infty$ (note, in fact, that $N_T = 1$), the Brownian martingale representation theorem can be applied again to conclude that

$$\frac{V_T}{N_T} = \mathbb{E}_{\mathbb{N}}\left[\frac{V_T}{N_T}\right] + \int_0^T \psi_u \, d\widetilde{W}_u$$

for some $\{\mathcal{F}_t\}$-predictable process ψ. Combining this with (8.10) yields

$$\frac{V_T}{N_T} = \mathbb{E}_{\mathbb{N}}\left[\frac{V_T}{N_T}\right] + \int_0^T \psi_u \phi_u^{-1} \, dD_{uS}^N.$$

It now follows from Theorem 7.17 that this strategy can be made to be self-financing (if we also trade in the numeraire asset D_{tT}), and from Corollary 7.40 that the strategy is admissible, hence the economy is complete.

Theorems 7.41 and 7.48 showed that if an economy is complete then any two pricing kernels agree up to $\mathcal{F}_T^A$, which in this case is all of $\mathcal{F}_T$. This contradicts the existence of a finite variation pricing kernel. For suppose there does exist some measure $\mathbb{Z}$, locally equivalent to $\mathbb{P}$, and a finite variation process ζ which is a pricing kernel under the measure $\mathbb{Z}$. Defining the process ρ by

$$\rho_t := \left.\frac{d\mathbb{Z}}{d\mathbb{P}}\right|_{\mathcal{F}_t},$$

it follows that $\zeta\rho$ is a pricing kernel under $\mathbb{P}$. But the economy is complete and as such admits a unique pricing kernel. Since Z is also a pricing kernel under $\mathbb{P}$ we can conclude that $Z = \zeta\rho$. Write $Z = Z_0\mathcal{E}(X)$ and $\rho = \rho_0\mathcal{E}(\hat{X})$ for local martingales X and $\hat{X}$ (as we did in Lemma 5.17). Taking logs of Z and $\rho\zeta$, and appealing to the uniqueness of the Doob–Meyer decomposition, it follows that $X = \hat{X}$ and ζ is the constant 1. Thus $Z = \rho$ which is a contradiction since ρ is a martingale and yet Z is not.

Remark 8.12: The fact that the discount curve for Example 8.11 is strictly decreasing and converges to zero as the maturity tends to infinity means that it also proves that the inclusions $(\text{FVK})^{+0} \subset (\text{PK})^{+0}$ and $(\text{FVK})^{+} \subset (\text{PK})^{+}$ are strict.

Remark 8.13: Note, in particular, that this is an example of a model where there is no arbitrage, where the model is complete but where there is no numeraire pair $(N^f, \mathbb{N})$ for which the numeraire is finite variation. However, it is easy to see that there are many finite variation numeraires. See Remark 7.44 for more on the significance of this.

Exercise 8.14: Let $(W^{(1)}, W^{(2)}, W^{(3)})$ be a three-dimensional Brownian motion, but starting at the point $(W_0^{(1)}, W_0^{(2)}, W_0^{(3)}) = (1, 0, 0)$. Define

$$R_t := \sqrt{(W_t^{(1)})^2 + (W_t^{(2)})^2 + (W_t^{(3)})^2}.$$

Show that R, the Bessel(3) process, satisfies the SDE

$$dR_t = -dW_t + R_t^{-1}dt,$$

where the Brownian motion W is defined by

$$W_t := \int_0^t \sum_{i=1}^3 \frac{W_u^{(i)}}{R_u}\, dW_u^{(i)}.$$

Hence show that the kernel Z of Example 8.11 is given by $Z = R^{-1}$, and derive an expression for

$$f(Z_t, T - t) := \mathbb{E}_{\mathbb{P}}[Z_T | \mathcal{F}_t].$$

Establish the following facts which have been used above (the first of which shows that Z is not a martingale):

(i) $f(Z, T - t)$ is a.s. continuous and strictly decreasing in T, with $\lim_{T \to \infty} f(Z, T - t) = 0$;

(ii) $f(Z, T - t)$ is twice differentiable in the variable Z;

(iii) for all $t < T < S$, the diffusion term in the SDE for (D_{tS}/D_{tT}) is a.s. non-zero.

8.3.5 Absolutely continuous (FVK) models

Recall that a finite variation process is one which is finite, adapted and with sample paths which are finite variation functions. It is a standard result from analysis that any finite variation function can be written as a sum of two components, one which is absolutely continuous (with respect to Lebesgue measure) and one which is singular with respect to Lebesgue measure. The most common and important examples of (FVK) models are short-rate models, discussed shortly, for which the kernel is absolutely continuous. This leaves the question whether or not there are, in theory at least, (FVK) models in which the kernel is not absolutely continuous. The answer is yes, as the following (non-stochastic) example shows.

Example 8.15 Let F be the standard Cantor distribution function on $[0, 1]$, as defined, for example, in Chung (1974). This is a function $F : [0, 1] \to [0, 1]$ which is everywhere continuous and non-decreasing, almost everywhere differentiable with $F' = 0$, but singular with respect to Lebesgue measure. Now define the finite variation function (which we take as a deterministic pricing kernel process) ζ by

$$\zeta(t) = e^{-rt}\left(1 - F\left(\frac{t}{1+t}\right)\right)$$

for some $r > 0$. Clearly ζ is strictly decreasing and positive, thus corresponds to a pricing kernel, and the resulting model has all interest rates strictly positive. However, ζ is singular with respect to Lebesgue measure so the model is not (acFVK).

8.3.6 Short-rate models

A short-rate model is one for which the prices of all pure discount bonds can be expressed in the form

$$D_{tT} = \mathbb{E}_{\mathbb{Q}}\left[\exp\left(-\int_t^T r_u\, du\right)\Big|\mathcal{F}_t\right], \tag{8.11}$$

for some measure (the risk-neutral measure) $\mathbb{Q}$ locally equivalent to $\mathbb{P}$ and some adapted process r. In addition, it is a requirement that r has the interpretation as the instantaneous spot interest rate,

$$r_t = -\lim_{h \to 0} \frac{1}{h} \log D_{t,t+h}. \tag{8.12}$$

It is implicit in this definition that, with probability one, $\int_0^t r_u du$ and the expectation in (8.11) are finite. Note that any (acFVK) model can be written in the form (8.11) but does not necessarily satisfy (8.12).

Short-rate models are very common and have been very popular. This popularity is decreasing somewhat amongst (front-office) practitioners with the advent of 'market models', but they are still a very valuable tool and most banks have various examples implemented in their trading systems. Some of the best-known examples are the Vasicek–Hull–White model and the Black-Karasinski model, both of which we will describe in Chapter 17. (Actually most if not all models used in practice, even market models, are (SR) but not all models are parameterized as such.) Both of these examples are driven by a one-dimensional Brownian motion, but this need not be the case in general.

It is immediate, setting

$$\zeta_t = \exp\left(-\int_0^t r_u du\right), \tag{8.13}$$

that (SR) $\subseteq$ (acFVK). The question remains as to whether there are any (acFVK) models which are not (SR), i.e. for which the process r in (8.13) does not satisfy (8.12). Sufficient conditions for an (acFVK) model to be (SR) (and (HJM)) will be given in Theorem 8.18 below, conditions which mean that any 'reasonable' (acFVK) model must indeed be (SR). However, models do exist which are (acFVK) but not (SR) as the following example demonstrates.

Example 8.16 The example we are about to present is somewhat pathological and unrealistic. Theorem 8.18 below and the remark following it show this has to be the case. This contrasts with Example 8.11 which presented a reasonable and not unrealistic model that was (PK) but not (FVK).

Work on a filtered probability space $(\Omega, \{\mathcal{F}_t\}, \mathcal{F}, \mathbb{P})$ supporting a univariate Brownian motion W. Let $\{\alpha_n, n \geq 1\}$ be a set of probabilities and define the functions $R_n(t)$ via

$$R_n(t) = \begin{cases} \alpha_n^{-1}, & \text{if} \quad 0 \leq t < 2^{-n}, \\ -\alpha_n^{-1}, & \text{if} \quad 2^{-n} \leq t < 2^{-(n-1)}, \\ 1, & \text{if} \quad 2^{-(n-1)} \leq t \in [q2^{-n}, (q+1)2^{-n}), \quad q \text{ even}, \\ -1, & \text{if} \quad 2^{-(n-1)} \leq t \in [q2^{-n}, (q+1)2^{-n}), \quad q \text{ odd}. \end{cases}$$

Now define the stopping time τ_n via

$$\tau_n = \inf\{t = q2^{-(n-1)}, \text{ some integer } q : |W_{2q2^{-n}} - W_{(2q-1)2^{-n}}| > \ell_n\}$$

where ℓ_n is chosen such that $\mathbb{P}(\tau_n = 2^{-(n-1)}) = \alpha_n$. Note that $2^{n-1}\tau_n$ is a geometrically distributed random variable. Now let

$$\zeta_n(t) = \begin{cases} 0, & t < \tau_n, \\ \displaystyle\int_{\tau_n}^t R_n(u - \tau_n)\,du, & t \geq \tau_n, \end{cases}$$

and note that, for each n, ζ_n is an absolutely continuous finite variation process and $\zeta_n(t) \geq 0$ for all t. Furthermore, letting $\lfloor x \rfloor$ be the largest integer no larger than x,

$$\mathbb{E}_\mathbb{P}\left[\zeta_n(T)|\mathcal{F}_t\right] = \zeta_n(T)\mathbb{1}_{\{t \geq \tau_n\}} + 2^{-n}E\left(2^n T - 2\lfloor 2^{(n-1)}t \rfloor, \beta_n(t)/\alpha_n\right)\mathbb{1}_{\{t < \tau_n\}}$$

where $E(t, x) = \int_0^t e(u, x)du$,

$$e(t, x) = \begin{cases} 0, & t < 2, \\ x, & 2 \leq t < 3, \\ -x, & 3 \leq t < 4, \\ 1, & 4 \leq t \in [q, q+1), & q \text{ even}, \\ -1, & 4 \leq t \in [q, q+1), & q \text{ odd}, \end{cases}$$

and

$$\beta_n(t) = \mathbb{P}\left(\tau_n = \frac{\lfloor 2^{(n-1)}t \rfloor + 1}{2^{(n-1)}}\Big|\mathcal{F}_t\right).$$

Now defining

$$\zeta_t = 1 + \sum_{n=1}^\infty \zeta_n(t),$$

it follows from the monotone convergence theorem that $\mathbb{E}_\mathbb{P}[\zeta(T)] < 2$ and thus $\zeta \in \mathcal{L}^1$ and may be used as a pricing kernel (it is also strictly positive). Furthermore, the monotone convergence theorem also implies that

$$\mathbb{E}_\mathbb{P}\left[\sum_{n=1}^\infty \mathbb{1}_{\{\tau_n \leq t\}}\right] = \sum_{n=1}^\infty \mathbb{P}(\tau_n \leq t)$$

$$= \sum_{n=1}^\infty \mathbb{P}\left(\tau_n \leq \frac{\lfloor 2^{(n-1)}t \rfloor}{2^{(n-1)}}\right)$$

$$\leq \sum_{n=1}^\infty \left(1 - (1 - \alpha_n)^{2^{(n-1)}t}\right)$$

$$\leq t \sum_{n=1}^\infty 2^{(n-1)}\alpha_n.$$

Choosing $\{\alpha_n\}$ such that this sum is finite ensures that only finitely many τ_n will have occurred by any time t, and thus ζ is absolutely continuous and of finite variation a.s. But it is not difficult to verify (and this is left as an exercise) that with this pricing kernel D_{tT} is a.s. nowhere differentiable, for any t and T. In particular, it is not differentiable at $t = T$, thus the short rate does not exist.

8.3.7 Heath–Jarrow–Morton models

All term structure models of practical relevance have the property that the discount curve is almost everywhere differentiable in the maturity parameter T, a property that automatically holds for any model in which interest rates are positive. Furthermore, if the discount curve is absolutely continuous this then yields

$$D_{tT} = \exp\Big(-\int_t^T f_{tS}dS\Big),$$

where

$$f_{tT} = -\frac{\partial}{\partial T}\log(D_{tT}).$$

The idea of Heath *et al.* (1992) was to exploit this fact and to formulate a model by specifying the processes $\{f_{\cdot T} : 0 \le T < \infty\}$. This they did by giving SDEs satisfied by the forward rate processes

$$df_{tT} = \mu_{tT}dt + \sigma_{tT} \cdot dW_t, \tag{8.14}$$

for suitable μ and σ.

For this approach to be useful it is necessary to have a way to move from a specification of the instantaneous forwards to the discount bond price processes, i.e. to derive an SDE for $\int_t^T f_{tS}dS$ from the SDEs for the forwards f_{tT}. To do this systematically requires an application of a stochastic Fubini theorem, a result which allows the interchange of the order of integration when one integral is a Lebesgue integral and the other is a stochastic integral. For this result to be applicable, regularity conditions need to be imposed on the SDEs (8.14). If these regularity conditions do hold then the next step in the development of an (HJM) model is to determine conditions on the SDEs (8.14) to ensure that the model satisfies (TS1)–(TS4).

In their original paper, Heath *et al.* (1992) present a general SDE for the forwards, such as (8.14), and sufficient conditions for the stochastic Fubini result to apply. They then show that (TS1)–(TS4) hold, that the resulting model is (SR), and derive the corresponding SDEs for the pure discount bond prices. One could, generally, define an (HJM) model to be one for which the stochastic Fubini result and (TS1)–(TS4) hold. This is precisely the viewpoint we take, defining an (HJM) model to be one for which the *conclusions* of Theorem 8.21 are valid.

Definition 8.17 *An (HJM) is an (SR) model for which, in the risk-neutral measure* $\mathbb{Z}$, *the instantaneous forward rates obey SDEs of the form*

$$df_{tT} = \left(\frac{\partial \Sigma_{tT}}{\partial T} \cdot \Sigma_{tT}\right) dt - \left(\frac{\partial \Sigma_{tT}}{\partial T}\right) \cdot dW_t,$$

where each $\Sigma_{.T}$ *is* $\{\mathcal{F}_t\}$*-predictable, and for which*

$$dD_{tT} = D_{tT}\left(r_t dt + \Sigma_{tT} \cdot dW_t\right).$$

In this presentation of the (HJM) framework we will approach things from a slightly different perspective than Heath *et al.* (1992), following closely the approach of Baxter (1997). We start with the class (SR) of short-rate models and provide a sufficient condition on the corresponding pricing kernel ζ for the stochastic Fubini result to apply and thus for the model to be (HJM). This sufficient condition is not necessary but it is such that any model for which it fails to hold is highly unrealistic and thus of no practical relevance. This contrasts with the fact that there do exist realistic (PK) models, such as Example 8.11, which are not (FVK). We then derive the corresponding SDE for the instantaneous forwards and the interrelationships implied by the fact that (TS1)–(TS4) hold.

We begin with a statement of the sufficient condition, inequality (8.15), and the consequential relationship between the forwards and the finite variation kernel. Note in the statement of this result that we only require the model to be (acFVK). The (SR) property is then a consequence of the regularity condition (8.15).

Theorem 8.18 *Let* ζ *be a finite variation pricing kernel associated with an (acFVK) model (not assumed to be (SR)) with the minimal cononical decomposition* $\zeta = \zeta_0 + \zeta^+ - \zeta^-$ *(i.e.* ζ^+ *and* ζ^- *are the minimal non-decreasing functions, null at zero, for which this holds). Then a sufficient condition for the existence of the instantaneous forwards,* f, *is*

$$\mathbb{E}_{\mathbb{Z}}[\zeta_T^+] + \mathbb{E}_{\mathbb{Z}}[\zeta_T^-] < \infty, \tag{8.15}$$

for all $T > 0$, *in which case*

$$f_{tT} = -\frac{\mathbb{E}_{\mathbb{Z}}\left[\frac{\partial \zeta_T}{\partial T}\middle|\mathcal{F}_t\right]}{\mathbb{E}_{\mathbb{Z}}\left[\zeta_T\middle|\mathcal{F}_t\right]}. \tag{8.16}$$

In particular, writing $\zeta_t = \exp(-\int_0^t r_u\, du)$, $f_{tt} = r_t$ and the model is (SR).

Remark 8.19: The regularity condition (8.15) can easily be seen to be equivalent to the condition given in Baxter (1997), namely

$$\mathbb{E}_{\mathbb{Z}}\left[\int_0^T |r_u| \exp\left(\int_0^u -r_v\, dv\right) du\right] < \infty.$$

Proof: Condition (8.15) and the fact that ζ is absolutely continuous ensure that we can apply (the conditional form of) Fubini's theorem to (8.8) to obtain

$$D_{tT} = \zeta_t^{-1} \mathbb{E}_{\mathbb{Z}}[\zeta_T | \mathcal{F}_t]$$

$$= \zeta_t^{-1}\left(\mathbb{E}_{\mathbb{Z}}[\zeta_0 | \mathcal{F}_t] + \int_0^T \mathbb{E}_{\mathbb{Z}}\left[\frac{\partial \zeta_S}{\partial S}\Big| \mathcal{F}_t\right] dS\right).$$

It follows that

$$\frac{\partial D_{tT}}{\partial T} = \zeta_t^{-1} \mathbb{E}_{\mathbb{Z}}\left[\frac{\partial \zeta_T}{\partial T}\Big| \mathcal{F}_t\right],$$

$$f_{tT} := -\frac{\partial \log(D_{tT})}{\partial T} = -\frac{\mathbb{E}_{\mathbb{Z}}\left[\frac{\partial \zeta_T}{\partial T}\big| \mathcal{F}_t\right]}{\mathbb{E}_{\mathbb{Z}}\left[\zeta_T | \mathcal{F}_t\right]}.$$

The last statement of the theorem is immediate. ☐

Remark 8.20: It is implicit from the fact that ζ is a pricing kernel that ζ itself is in $\mathcal{L}^1(\Omega, \{\mathcal{F}_t\}, \mathcal{F}, \mathbb{Z})$. As a result, the left-hand side of (8.15) is infinite if and only if both expectations are infinite. But $\zeta(\omega) \geq 0$ for almost all ω and thus any model for which (8.15) does not hold must have a wildly fluctuating short rate (if $\zeta_t^-(\omega) > K$ then $\zeta_t^+(\omega) > K - \zeta_0(\omega)$) and it is not possible that some paths have highly positive short rates and others have very negative short rates; if large negative rates occur then large positive rates must also occur on the same path. Consequently, any (acFVK) model that is not (SR), or any (SR) model that is not (HJM), must be highly unrealistic and of no practical relevance. Example 8.16 was one such example which separated the classes (acFVK) and (SR). Example 8.24 at the end of this section is (SR) but not (HJM).

It remains for us to connect the SDEs that are satisfied by the bond prices D_{tT} and the instantaneous forwards f_{tT} under the regularity condition (8.15). This is the content of the next two theorems. For the first we work in the 'risk-neutral' measure $\mathbb{Z}$ corresponding to the kernel ζ; the general case then follows from Girsanov's theorem.

Theorem 8.21 *Given a general (acFVK) model with pricing kernel $\zeta_t = \exp(-\int_0^t r_u du)$, there exists, for each $T > 0$, some d-dimensional $\{\mathcal{F}_t\}$-predictable $\Sigma_{.T}$ such that*

(i) $\mathbb{E}_\mathbb{Z}[\zeta_T|\mathcal{F}_t] = \mathbb{E}_\mathbb{Z}[\zeta_T|\mathcal{F}_0] + \int_0^t \zeta_u D_{uT} \Sigma_{uT} \cdot dW_u,$

(ii) $dD_{tT} = D_{tT}(r_t dt + \Sigma_{tT} \cdot dW_t),$

where W is a d-dimensional $(\{\mathcal{F}_t\}, \mathbb{Z})$ Brownian motion.

If, in addition, (8.15) holds for each $T > 0$, then

(iii) $\mathbb{E}_\mathbb{Z}\left[\dfrac{\partial\zeta_T}{\partial T}\Big|\mathcal{F}_t\right] = \mathbb{E}_\mathbb{Z}\left[\dfrac{\partial\zeta_T}{\partial T}\Big|\mathcal{F}_0\right] - \int_0^t \zeta_u D_{uT}(\sigma_{uT} + f_{uT}\Sigma_{uT}) \cdot dW_u,$

(iv) $df_{tT} = -(\sigma_{tT} \cdot \Sigma_{tT})\, dt + \sigma_{tT} \cdot dW_t,$

where the processes σ and Σ are related by

$$\sigma_{tT} = -\frac{\partial \Sigma_{tT}}{\partial T},$$

and the model is (HJM).

Proof: The existence of Σ_{tT} in (i) for each T is just the Brownian martingale representation theorem (Theorem 5.49) applied to the martingale $M_t = \mathbb{E}_\mathbb{Z}[\zeta_T|\mathcal{F}_t]$. Equation (ii) now follows from (i) by applying Itô's formula to (8.8).

Turning to the second part of the theorem, the regularity condition (8.15) establishes the existence of the process $\frac{\partial\zeta_T}{\partial T}$, and it then follows by the Brownian martingale representation theorem again that, for each T, there exists some predictable process $\Lambda_{.T}$ such that

$$\mathbb{E}_\mathbb{Z}\left[\frac{\partial\zeta_T}{\partial T}\Big|\mathcal{F}_t\right] = \mathbb{E}_\mathbb{Z}\left[\frac{\partial\zeta_T}{\partial T}\Big|\mathcal{F}_0\right] + \int_0^t \Lambda_{uT} \cdot dW_u. \tag{8.17}$$

We must prove that Λ has the representation in (iii). We will only sketch the proof of this part of the theorem. The complete proof requires several technical lemmas which we will not include. The interested reader is referred to Baxter (1997) for details of this part of the argument. The key to the proof is to observe that (i), (8.17) and an application of Fubini's theorem yield two alternative representations for $\mathbb{E}_\mathbb{Z}[\zeta_T - \zeta_0|\mathcal{F}_t]$, namely

$$\mathbb{E}_\mathbb{Z}[\zeta_T - \zeta_0|\mathcal{F}_t] = \mathbb{E}_\mathbb{Z}[\zeta_T - \zeta_0|\mathcal{F}_0] + \int_0^t \zeta_u D_{uT} \Sigma_{uT} \cdot dW_u,$$

$$\mathbb{E}_\mathbb{Z}[\zeta_T - \zeta_0|\mathcal{F}_t] = \mathbb{E}_\mathbb{Z}\left[\int_0^T \frac{\partial\zeta_S}{\partial S}\, dS\Big|\mathcal{F}_t\right]$$

$$= \int_0^T \mathbb{E}_\mathbb{Z}\left[\frac{\partial\zeta_S}{\partial S}\Big|\mathcal{F}_t\right] dS$$

$$= \int_0^T \left(\mathbb{E}_\mathbb{Z}\left[\frac{\partial\zeta_S}{\partial S}\Big|\mathcal{F}_0\right] + \int_0^t \Lambda_{uS} \cdot dW_u\right) dS$$

$$= \mathbb{E}_\mathbb{Z}[\zeta_T - \zeta_0|\mathcal{F}_0] + \int_0^T \left(\int_0^t \Lambda_{uS} \cdot dW_u\right) dS.$$

Equating these two expressions yields

$$\int_0^T \left(\int_0^t \Lambda_{uS} \cdot dW_u \right) dS = \int_0^t \zeta_u D_{uT} \Sigma_{uT} \cdot dW_u \,. \tag{8.18}$$

A fact we shall not prove, which follows from condition (8.15), is that a stochastic Fubini theorem can be applied which allows us to interchange the order of the two integrations on the left-hand side of (8.18). Doing so yields

$$\int_0^t \left(\int_0^T \Lambda_{uS} \, dS \right) \cdot dW_u = \int_0^t \zeta_u D_{uT} \Sigma_{uT} \cdot dW_u \,,$$

$$\int_0^T \Lambda_{tS} \, dS = \zeta_t D_{tT} \Sigma_{tT} \,. \tag{8.19}$$

Differentiating both sides of (8.19) yields the representation in (iii). The final identity (iv) is now just an application of Itô's formula to (8.16). $\qquad\square$

Remark 8.22: It is implicit in the last statement that we can indeed integrate σ_{tT} over the maturity parameter T (and we made similar assumptions for the instantaneous forwards f_{tT}). For this to be possible σ must be measurable in this parameter. It is indeed possible to establish this fact, that there exists a version of σ which is *jointly measurable* as a map

$$\sigma : \Omega \times [0, T^*] \times [0, T^*] \to \mathbb{R}$$
$$(\omega, t, T) \mapsto \sigma_{tT}(\omega) \,,$$

for each $T^* > 0$. That is, σ (and f) is *jointly progressively measurable* (in t and T). The details can be found in Baxter (1997).

The relationship in part (iv) of Theorem 8.21 between the drift and diffusion terms of the instantaneous forward rates in the measure $\mathbb{Z}$ is a consequence of insisting, via (TS4), that the model is arbitrage-free. It is now a simple matter to see the form these equations must take in any other measure which is locally equivalent to $\mathbb{Z}$, in particular the real-world measure $\mathbb{P}$. The result is an immediate consequence of Theorem 8.21 and Girsanov's theorem (Theorem 5.24).

Theorem 8.23 *Under any measure $\mathbb{P}$ which is locally equivalent to $\mathbb{Z}$, the model of Theorem 8.21 obeys*

$$(ii') \ dD_{tT} = D_{tT}\Big((r_t + \Sigma_{tT} \cdot C_t)dt + \Sigma_{tT} \cdot d\widetilde{W}_t \Big).$$

If (8.15) also holds then (iv) becomes

$$(iv') \ df_{tT} = \left(\frac{\partial \Sigma_{tT}}{\partial T} \cdot (\Sigma_{tT} - C_t)\right) dt - \left(\frac{\partial \Sigma_{tT}}{\partial T}\right) \cdot d\widetilde{W}_t.$$

In the above $\widetilde{W}$ is a Brownian motion under $\mathbb{P}$ and is given by

$$\widetilde{W}_t = W_t - \int_0^t C_u \, du,$$

where C is the unique $\{\mathcal{F}_t\}$-predictable process such that

$$\frac{d\mathbb{P}}{d\mathbb{Z}}\Big|_{\mathcal{F}_t} = \exp\left(\int_0^t C_u \cdot dW_u - \tfrac{1}{2} \int_0^t |C_u|^2 \, du\right).$$

We conclude our discussion of (HJM) models with an example of a model that is (SR) but not (HJM), by necessity a model which fails to satisfy (8.15) and which is thus unrealistic.

Example 8.24 Work on a filtered probability space $(\Omega, \{\mathcal{F}_t\}, \mathcal{F}, \mathbb{P})$ supporting a univariate Brownian motion W. For each $n \geq 1$ define the function R_n via

$$R_n(t) = \begin{cases} 0, & \text{if } t < 1, \\ 2^n, & \text{if } t - 1 \in [q2^{-n}, (q+1)2^{-n}), \quad q \text{ odd}, \\ -2^n, & \text{if } t - 1 \in [q2^{-n}, (q+1)2^{-n}), \quad q \text{ even}. \end{cases}$$

Now define a random variable $\hat{n}$ by $\hat{n} = j$ if $|W_1| \in [L_j, L_{j+1})$, where the sequence L_j is defined so that $L_1 = 0$ and $2(N(L_{j+1}) - N(L_j)) = 2^{-j}$. Fix $r \geq 0$, $\alpha < 1$ and define the finite variation kernel ζ via

$$\zeta_t = \exp(-rt)\left(1 + \alpha \int_1^{t \vee 1} R_{\hat{n}}(u) \, du\right)$$

and, as usual,

$$D_{tT} = \zeta_t^{-1} \mathbb{E}_{\mathbb{Z}}[\zeta_T | \mathcal{F}_t]$$

(note that $0 \leq \zeta_t \leq \exp(-rt)$ so ζ is certainly in $\mathcal{L}^1$).

This is clearly an (acFVK) model. Furthermore, for all $T \geq 0$, there exists some $\varepsilon(T) > 0$ such that ζ_t is $\mathcal{F}_T$-measurable for all $t \in [T, T + \varepsilon(T)]$. Hence, writing $\zeta_t = \exp(-\int_0^t r_u du)$, the short rate exists and is exactly r_t, so the model is (SR). However, for $t > 1$ neither ζ_t^+ nor ζ_t^- is integrable and (exercise for the reader) D_{tT} is nowhere differentiable for $t < 1 \leq T$. Figure 8.1 shows the initial discount curve in the case $r = 10\%, \alpha = 25\%$.

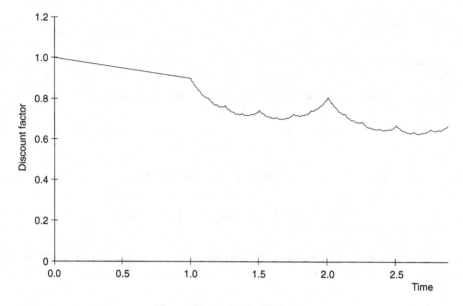

Figure 8.1 Non-differentiable (SR) discount curve

8.3.8 Flesaker–Hughston models

In 1996 Flesaker and Hughston proposed a general class of models in which interest rates are guaranteed to be positive. In their framework, the term structure is of the form

$$D_{tT} = \frac{\int_T^\infty M_{tS}\,dS}{\int_t^\infty M_{tS}\,dS}\,, \tag{8.20}$$

where $\{M_{.S} : 0 \le S < \infty\}$ is a family of positive $\{\mathcal{F}_t\}$ martingales under some measure $\mathbb{Z}$ locally equivalent to $\mathbb{P}$. Implicit in this definition is the fact that the family M is jointly measurable and that both integrals in (8.20) are finite a.s.

The best-known example of a model developed in this framework is the *rational log-normal model* defined by

$$D_{tT} = \frac{A_T + B_T M_t}{A_t + B_t M_t}\,.$$

Here the process M is a log-normal martingale with initial condition $M_0 = 1$ and the deterministic functions A and B are both positive, absolutely continuous, and decrease to zero at infinity. For consistency with the initial term structure the constraint

$$D_{0T} = \frac{A_T + B_T}{A_0 + B_0}$$

also applies. This still leaves freedom to tailor the model for any particular application.

Two noteworthy features of this model are as follows. First, it produces closed-form prices for both caps and swaptions. We will not derive these results, but the interested reader is referred to Flesaker and Hughston (1996). Second, given any t and T, D_{tT} is monotone in M_t and is bounded by A_T/A_t and B_T/B_t ($M_t = 0$ and $M_t = \infty$). Consequently, interest rates are also bounded both above and below.

To understand how (FH) relates to other ways of specifying the term structure we give an alternative representation, first presented in 1997 by, amongst others, Rutkowski (1997).

Theorem 8.25 *Given a jointly measurable family of positive* $(\{\mathcal{F}_t\}, \mathbb{Z})$ *martingales* $\{M_{\cdot S} : 0 \leq S < \infty\}$, *let*

$$Z_t = \int_t^\infty M_{tS} dS, \qquad (8.21)$$

$$A_t = \int_0^t M_{SS} dS.$$

Note that A is an absolutely continuous, increasing, finite variation process with $A_0 = 0$. If $Z_0 < \infty$ then both $Z \in \mathcal{L}^1(\Omega, \{\mathcal{F}_t\}, \mathcal{F}, \mathbb{Z})$ and $A_\infty \in \mathcal{L}^1(\Omega, \mathcal{F}, \mathbb{Z})$. Thus Z can be used as a pricing kernel and the resulting model has the form (8.20), and is thus one for which all interest rates are positive and, for all $t \geq 0$,

$$\lim_{T \to \infty} D_{tT} = 0, \quad a.s. \qquad (8.22)$$

Furthermore, in this case Z has the representation

$$Z_t = \mathbb{E}_{\mathbb{Z}}[A_\infty | \mathcal{F}_t] - A_t. \qquad (8.23)$$

Conversely, given any absolutely continuous, increasing A with $A_0 = 0$ and $\mathbb{E}_{\mathbb{Z}}[A_\infty] < \infty$, (8.23) defines a positive supermartingale $Z \in \mathcal{L}^1(\Omega, \{\mathcal{F}_t\}, \mathcal{F}, \mathbb{Z})$ and (8.4) then defines a model in which interest rates are positive and for which (8.22) holds. Furthermore, Z has the representation given in (8.21) for a (jointly measurable) family of positive martingales $\{M_{\cdot S} : 0 \leq S < \infty\}$.

Remark 8.26: What this theorem shows is that (FH) models are precisely the class of (PK) models with kernels of the form (8.23) for some increasing absolutely continuous A. If the continuity requirement is removed, (8.23) does in fact define a general *potential of class (D)*, i.e. a right-continuous, positive supermartingale Z with the property that $Z_\infty = 0$ (a potential) and for which the family $\{Z_\tau : \tau$ an $\{\mathcal{F}_t\}$stopping time$\}$ is uniformly integrable (of class (D)). Note that the kernel Z of Example 8.11 is also a potential, but the family $\{Z_\tau\}$ is not in that case uniformly integrable.

Proof: Given the family of positive martingales $\{M_{.S}\}$, it is clear that both Z and A exist, albeit possibly taking the value $+\infty$. However, if $Z_0 < \infty$ then it follows immediately from Fubini's theorem that $Z \in \mathcal{L}^1(\Omega, \{\mathcal{F}_t\}, \mathcal{F}, \mathbb{Z})$ and $A_\infty \in \mathcal{L}^1(\Omega, \mathcal{F}, \mathbb{Z})$, since

$$Z_0 = \int_0^\infty M_{0S}\, dS = \int_0^\infty \mathbb{E}_{\mathbb{Z}}[M_{SS}]dS = \mathbb{E}_{\mathbb{Z}}\left[\int_0^\infty M_{SS}\, dS\right] = \mathbb{E}_{\mathbb{Z}}[A_\infty],$$

and

$$Z_0 > \int_T^\infty M_{0S}\, dS = \int_T^\infty \mathbb{E}_{\mathbb{Z}}[M_{TS}]dS = \mathbb{E}_{\mathbb{Z}}\left[\int_T^\infty M_{TS}\, dS\right] = \mathbb{E}_{\mathbb{Z}}[Z_T],$$

the third relation in each case being the application of Fubini. Hence $Z \in \mathcal{L}^1(\Omega, \{\mathcal{F}_t\}, \mathcal{F}, \mathbb{Z})$ and, since it is also a strictly positive continuous semimartingale, can be used as a pricing kernel. A further application of Fubini, this time the conditional version, now yields

$$\mathbb{E}_{\mathbb{Z}}\left[Z_T|\mathcal{F}_t\right] = \mathbb{E}_{\mathbb{Z}}\left[\int_T^\infty M_{TS}\, dS|\mathcal{F}_t\right] = \int_T^\infty \mathbb{E}_{\mathbb{Z}}\left[M_{TS}|\mathcal{F}_t\right]dS = \int_T^\infty M_{tS}\, dS$$

for all $0 \le t \le T < \infty$. Therefore, substituting into (8.4),

$$D_{tT} = Z_t^{-1}\mathbb{E}_{\mathbb{Z}}[Z_T|\mathcal{F}_t] = \frac{\int_T^\infty M_{tS}\, dS}{\int_t^\infty M_{tS}\, dS},$$

as required.

The representation of Z in terms of A follows, once more, from conditional Fubini,

$$Z_t := \int_t^\infty M_{tS}\, dS = \int_t^\infty \mathbb{E}_{\mathbb{Z}}[M_{SS}|\mathcal{F}_t]\, dS$$

$$= \mathbb{E}_{\mathbb{Z}}\left[\int_t^\infty M_{SS}\, dS|\mathcal{F}_t\right] = \mathbb{E}_{\mathbb{Z}}\left[A_\infty - A_t|\mathcal{F}_t\right].$$

This completes the first part of the proof.

The converse result is immediate with the exception of the last part. But the fact that A is absolutely continuous, increasing and null at zero ensures that

$$A_t = \int_0^t a_u\, du$$

for some positive process a. Defining $M_{tS} = \mathbb{E}_{\mathbb{Z}}[a_S|\mathcal{F}_t]$, the fact that $A_\infty \in \mathcal{L}^1(\Omega, \mathcal{F}, \mathbb{Z})$ allows us to apply Fubini once more to obtain the representation (8.21). (That the family $\{M_{.T} : T \ge 0\}$ is jointly measurable is established in Baxter (1997).) $\square$

The representation (8.23) suggests an obvious generalization of the (FH) class of models.

Definition 8.27 *A term structure model is said to be extended Flesaker–Hughston, (eFH), if it is a (PK) model with pricing kernel of the form (8.23) for some adapted, increasing and continuous process A with $A_0 = 0$ and $A_\infty \in \mathcal{L}^1(\Omega, \mathcal{F}, \mathbb{Z})$.*

Remark 8.28: The only difference from a standard (FH) model is the relaxation of the absolute continuity requirement on A to one of continuity.

It is immediate from the defining equation (8.20) that the discount curve for an (FH) model is continuous in the maturity parameter T. The continuity property also carries over to the class (eFH). This is not true for a general (PK) model, even when interest rates are positive.

Corollary 8.29 *For all $t \geq 0$, the discount curve for an (eFH) model is a.s. continuous in the maturity parameter T.*

Proof: It follows from (8.23) and (8.4) that, for an (eFH) model,

$$D_{tT} = \frac{\mathbb{E}_{\mathbb{Z}}[A_\infty | \mathcal{F}_t] - \mathbb{E}_{\mathbb{Z}}[A_T | \mathcal{F}_t]}{\mathbb{E}_{\mathbb{Z}}[A_\infty | \mathcal{F}_t] - A_t}. \tag{8.24}$$

The continuity of A ensures that $\lim_{S \to T} A_S = A_T$ a.s., for all $T \geq 0$. Since $0 \leq A_S \leq A_\infty \in \mathcal{L}^1(\Omega, \mathcal{F}, \mathbb{Z})$, the dominated convergence theorem can be applied to the second expectation in the numerator of (8.24) and yields

$$\lim_{S \to T} D_{tS} = D_{tT},$$

as required. $\square$

We conclude our discussions on the theory of term structure models by proving the relationship between (FH), (eFH) and the other model classes already considered. First we show that (FH) contains all (acFVK)$^{+0}$ models.

Theorem 8.30 *The class (FH) of Flesaker–Hughston models contains all (acFVK) models for which interest rates are positive and $\lim_{T \to \infty} D_{tT} = 0$ for all $t \geq 0$ a.s. The class (eFH) contains all of (FVK)$^{+0}$.*

Proof: Suppose ζ is the pricing kernel for an (acFVK)$^{+0}$ model. Then, for all $t \leq T$,

$$\zeta_t \geq \mathbb{E}_{\mathbb{Z}}[\zeta_T | \mathcal{F}_t] \downarrow 0, \tag{8.25}$$

as $T \to \infty$. Thus ζ is a supermartingale and, since ζ is also continuous and of finite variation, it follows from the Doob–Meyer theorem (Corollary 3.94) that ζ is actually decreasing. Furthermore, since $\zeta > 0$ a.s., it follows from (8.25) that $\zeta_\infty = 0$.

Define $A = \zeta_0 - \zeta$. Clearly A is increasing, absolutely continuous, and $A_0 = 0$. Furthermore, $A_\infty = \zeta_0 \in \mathcal{L}^1(\Omega, \mathcal{F}, \mathbb{Z})$ so A can be used to define an (FH) pricing kernel Z, given by

$$Z_t := \mathbb{E}_{\mathbb{Z}}[A_\infty - A_t | \mathcal{F}_t] = \zeta_t.$$

Hence the model is (FH) as required.

When the model is only $(FVK)^{+0}$ the argument above carries through with the exception of the fact that A is now not necessarily absolutely continuous. The resulting model is thus (eFH) but not necessarily (FH).   □

It is clear that (FH) $\subset$ (eFH) with the inclusion being strict. We have just shown that $(FVK)^{+0} \subseteq$ (eFH) and $(acFVK)^{+0} \subseteq$ (FH). It is immediate that (eFH) $\subseteq (PK)^{+0}$. It remains to prove that these last three inclusions are strict. The next two examples do precisely that.

Example 8.31 For an example of a model that is $(PK)^{+0}$ but not (eFH) we return to Example 8.11. Recall that the underlying kernel Z for this model is defined by $Z_0 = 1$, a.s., and

$$dZ_t = Z_t^2 dW_t$$

where W is a one-dimensional Brownian motion on some probability space $(\Omega, \{\mathcal{F}_t\}, \mathcal{F}, \mathbb{P})$. We showed in Example 8.11 that the model is not (FVK). That it is not (eFH) follows by a similar argument.

Suppose there exists some measure $\mathbb{Z}$, locally equivalent to $\mathbb{P}$, and some increasing process A such that

$$\hat{Z}_t := \mathbb{E}_{\mathbb{Z}}[A_\infty | \mathcal{F}_t] - A_t \qquad (8.26)$$

is a pricing kernel under $\mathbb{Z}$. Since Z is a pricing kernel under $\mathbb{P}$ it follows immediately that ρZ is a pricing kernel under $\mathbb{Z}$, where

$$\rho_t := \left.\frac{d\mathbb{P}}{d\mathbb{Z}}\right|_{\mathcal{F}_t}.$$

As we saw in Example 8.11, the model is complete and hence the pricing kernel is unique, thus $\rho Z = \hat{Z}$. But Z is an $(\{\mathcal{F}_t\}, \mathbb{P})$ local martingale, so ρZ is an $(\{\mathcal{F}_t\}, \mathbb{Z})$ local martingale by Lemma 5.19. On the other hand, $\hat{Z}$ is defined in (8.26) as the sum of a martingale plus a finite variation term. Hence the finite variation term is constant which means the process A is constant and $\hat{Z} \equiv 0$, a contradiction. Thus the model is not (eFH).

The key to the example above was the fact that the pricing kernel was a local martingale but not a true martingale. A similar idea is used to generate a model that is (FH) but not $(FVK)^{+0}$. This thus demonstrates the strict containment in the relations $(FVK)^{+0} \subset$ (eFH) and $(acFVK)^{+0} \subset$ (eFH). Note that the Cantor example of Section 8.15 is $(FVK)^{+0}$ but not (FH) (all (FH) models have an absolutely continuous discount curve), so this example also shows that there are no additional containment relationships between the classes (FH) and $(FVK)^{+0}$.

Example 8.32 Let W be a Brownian motion on the probability space $(\Omega, \{\mathcal{F}_t\}, \mathcal{F}, \mathbb{P})$ and define $\{\hat{L}_t : 0 \leq t < 1\}$ by

$$\hat{L}_t = \int_0^t \frac{dW_u}{1 - u}.$$

The process $\hat{L}$ is a continuous local martingale, null at zero, and $[\hat{L}]_t = \left(\frac{t}{1-t}\right)$, so Lévy's theorem implies that $L^*(t) := L\left(\frac{t}{1+t}\right)$ is a Brownian motion (relative to the filtration $\tilde{\mathcal{F}}_t := \mathcal{F}_{\left(\frac{t}{1+t}\right)}$).

Now define the stopping time $\tau = \inf\{t > 0 : \hat{L}_t < -1\}$ and let $L = 2 + \hat{L}^\tau$. The fact that $L^*(t)$ is Brownian motion ensures that $\tau < 1$ a.s., and so the process L can be defined over $[0, \infty)$ (there will be no problem with the explosion of $\hat{L}$ at $t = 1$). Note that $\hat{L}$ is a martingale on any interval $[0, T]$ when $T < 1$, and thus L is a martingale on $[0, T]$. Combining these observations,

$$\mathbb{E}[L_T | \mathcal{F}_t] = L_t \mathbb{1}_{\{T<1\}} + \mathbb{1}_{\{T \geq 1\}}.$$

Now, for some fixed $r > 0$, define the continuous and increasing process G by

$$G_t = \frac{1}{1-t} \mathbb{1}_{\{t \leq \tau\}} + \frac{e^{r(t-\tau)}}{1 - \tau} \mathbb{1}_{\{t > \tau\}},$$

and observe L has the representation

$$L_t := 2 + \hat{L}_t^\tau = 2 + \int_0^t G_u dW_u^\tau$$

(note that the behaviour of G on $\{t > \tau\}$ is not material to this representation). We now use L and G to construct a model that is (FH) but not (FVK), letting

$$Z_t := \frac{L_t}{G_t}. \tag{8.27}$$

Applying Itô's formula to (8.27),

$$dZ_t = G_t^{-1} dL_t + L_t dG_t^{-1}$$
$$= dW_t^\tau - dA_t,$$

where the absolutely continuous process A is defined by

$$A_t := -\int_0^t L_u dG_u^{-1},$$

and thus

$$Z_t = Z_0 + W_t^\tau - A_t. \tag{8.28}$$

The fact that $Z_\infty = 0$ and $\tau < 1$, a.s., now yields, from (8.28),

$$A_\infty = Z_0 + W_\infty^\tau,$$

and, taking expectations,

$$\begin{aligned}
\mathbb{E}_\mathbb{P}[A_\infty|\mathcal{F}_t] &= Z_0 + \mathbb{E}_\mathbb{P}[W_\infty^\tau|\mathcal{F}_t] \\
&= Z_0 + W_t^\tau \\
&= Z_t + A_t\,.
\end{aligned}$$

Thus Z has the form of an (FH) pricing kernel under the measure $\mathbb{P}$. We show that this kernel does not generate an (FVK) model, essentially because L is a local martingale but not a true martingale.

Note that for this model the asset filtration $\{\mathcal{F}_t^A\}$ satisfies $\{\mathcal{F}_t^A\} = \{\mathcal{F}_t^{W^\tau}\} = \{\mathcal{F}_{t\wedge\tau}\}$. A similar argument to that used in Example 8.11 can be employed to show that this model is $\mathcal{F}_T^A$-complete for all $T > 0$ (filling in the details of this argument is left as an exercise). Now suppose there exists a strictly positive, continuous, finite variation process ζ and a measure $\mathbb{Z}$ locally equivalent to $\mathbb{P}$ such that

$$D_{tT} = \zeta_t^{-1}\mathbb{E}_\mathbb{Z}[\zeta_T|\mathcal{F}_t]\,. \tag{8.29}$$

Changing measure in (8.29) from $\mathbb{Z}$ to $\mathbb{P}$ yields

$$D_{tT} = (\zeta\rho)_t^{-1}\mathbb{E}_\mathbb{P}[(\zeta\rho)_T|\mathcal{F}_t]\,,$$

where

$$\rho_t := \left.\frac{d\mathbb{Z}}{d\mathbb{P}}\right|_{\mathcal{F}_t},$$

so $\zeta\rho$ is also a pricing kernel for the model under the measure $\mathbb{P}$. But the completeness of the model implies that the pricing kernel is unique, thus

$$\zeta\rho = Z = \frac{L}{G}$$

on $\{\mathcal{F}_{t\wedge\tau}\}$. It now follows from the uniqueness of the Doob–Meyer decomposition that $\rho = L$ and $\zeta = G^{-1}$ up to $\{\mathcal{F}_{t\wedge\tau}\}$ (the argument here is similar to that used in Example 8.11). But ρ^τ is an $(\{\mathcal{F}_{t\wedge\tau}\}, \mathbb{P})$ martingale and L^τ is not, a contradiction.

Part II
Practice

9

Modelling in Practice

9.1 INTRODUCTION

Part I was devoted to the theory of derivative pricing. The development given was very much a theoretical mathematical approach and showed how to move from a model for an economy to the price of a derivative within that economy.

When it comes to modelling derivatives in practice there is much more to do and knowing (to some degree) the mathematical theory is only the first step. If we are faced with the problem of pricing a particular derivative it would be unusual to have a model for the economy already given to us. We must decide on what that model should be and this will depend on many things, including the particular derivative in question and the need to get numbers out of this model in a reasonable time. How to get numbers out efficiently is another topic in itself, something we shall cover only in one particular case (see Section 17.4).

The problem of how to select and develop a model of (relevant parts of) the real-world economy is something we shall be considering throughout this part of the book. In this chapter we aim to give a few pointers as to how we approach this problem and a few warnings of the pitfalls to avoid. There is no recipe for doing this and what works well in one situation may fail dismally in another. All you can do is keep your wits about you, work hard to understand the essence of any given product and try to focus on the really important issues.

9.2 THE REAL WORLD IS NOT A MARTINGALE MEASURE

The mathematical theory of derivative pricing is clear. One starts with a model of the asset price processes in an economy, usually specified via some SDE in a real-world probability measure $\mathbb{P}$. Then one chooses a numeraire N and changes probability measure, from the original measure $\mathbb{P}$ to an equivalent

---

Financial Derivatives in Theory and Practice Revised Edition. P. J. Hunt and J. E. Kennedy
© 2004 John Wiley & Sons, Ltd ISBNs: 0-470-86358-7 (HB); 0-470-86359-5 (PB)

martingale measure $\mathbb{N}$ under which all N-rebased assets are martingales. Having done this, the value of any derivative can be calculated by taking expectations in this measure $\mathbb{N}$.

For most products encountered in practice the value is determined by the joint distribution of a finite number of asset prices on a finite number of dates *in a martingale measure*. On the other hand, a model for the asset price processes can only be formulated based on information and intuition available *in the real world* – we do not live in a martingale measure. So the question arises as to how we should use real-world information to formulate a model which, when we have changed to a martingale measure, will give an asset price distribution which in turn yields a reasonable price for the derivative under consideration. To answer this question we need to consider what is common to both the real-world measure and any EMM. This includes the following:

(i)   the quadratic variation of any continuous process;
(ii) the prices of traded instruments;
(iii) null sets.

9.2.1 Modelling via infinitesimals

Each of the above quantities (i)–(iii) gives useful information about the martingale measure. In theory, for an economy containing a finite variation asset, it is enough to know the quadratic covariation of all the asset price processes and the drift of the finite variation asset in the real-world measure. If we do, and if we assume the market is arbitrage-free, then these completely determine all the joint distributions in a martingale measure. To see why this is, consider the SDE which determines the evolutions of the asset price processes. If we change to an EMM the drifts will change to make numeraire-rebased assets into martingales. This fact and the quadratic covariations uniquely determine the SDE for numeraire-rebased assets in the EMM. The SDE satisfied by the finite variation asset is unchanged from the real world. These new SDEs, assuming they have a unique solution, completely determine the law of all the asset price processes in the EMM, and hence all joint distributions and all derivative prices.

Whereas the logic of the previous paragraph is valid and infinitesimals determine everything in theory, in practice we do not attempt to determine price processes via their infinitesimals. There are a number of reasons for this.

Even when practitioners use time series to estimate the infinitesimals, the first step in doing so is to postulate a simple functional form for the diffusion term of a process. Then various parameters are estimated from the data. An obvious example of this is using a log-normal model for a stock price process. The price A is assumed to follow an SDE of the form

$$dA_t = \mu_t \, dt + \sigma A_t dW_t$$

for some general function μ, some constant σ and a Brownian motion W.

The value σ is then estimated from data. The initial choice of a log-normal model is one taken at the macro level and determines broad features of the model. Thus the primary modelling decision is made with reference to macro behaviour in the real world and the infinitesimals are merely a way to reproduce this behaviour. Indeed, whereas the resulting model may give a reasonable reflection of macro behaviour, it certainly gives a very poor reflection at the infinitesimal level – real-world price processes are piecewise constant with jumps, not continuous processes.

There is one further reason to avoid focusing too much on infinitesimals, and that is stability. As with ordinary differential equations, a small change in the form of an SDE can lead to large, and sometimes unrealistic, changes at the macro level.

9.2.2 Modelling via macro information

We have just seen why it can be dangerous to rely on infinitesimals when formulating a model for derivative pricing. An alternative is to use prices and null sets, the other things common to the real world and any equivalent measure.

The first of these, prices, are particularly relevant and useful. We shall see in Section 13.2.2, that if, for all strikes, we know the prices of call options on an asset then this gives us information about marginal distributions of that asset in an EMM. This macro information can and should be incorporated within the model. Furthermore, any derivative must be hedged with other instruments, so using this price information to the full also *calibrates* the model to the prices of hedge instruments. This is important because these hedge prices give the best measure of the real cost of replicating the derivative in practice. We shall rely heavily on this approach in the rest of this book.

The third invariant when changing measure is null sets. By ensuring that a model assigns zero probability in the real world to an event that we wish to exclude, we are also giving that event zero probability in an EMM. In practice we tend to push this a little further, insisting that an event that is very unlikely to occur in the real world should also be unlikely to occur in the EMM. This is sensible because we tend to model with continuous distributions and do not impose hard cut-off levels for distributions, even though we may feel they occur. Whereas this makes sense in practice, great care must be taken. Intuition about the real world can be misleading about the EMM because the change of measure can lead to large changes in distributions. Here is an example to illustrate.

Example 9.1 Consider an economy with three assets, the first being constant, $A^{(0)} \equiv 1$. Suppose that the other two are jointly log-normal with, for $i = 1, 2$,

$$dA_t^{(i)} = \mu^{(i)} A_t^{(i)} dt + \sigma A_t^{(i)} dW_t^{(i)}$$

where $\mu^{(1)}, \mu^{(2)}$ and σ are constants and $W = (W^{(1)}, W^{(2)})$ is a Brownian motion.

Suppose we set $\mu^{(1)} = 0$ and $A_0^{(1)} = A_0^{(2)} = A$. Then, for any K,

$$\mathbb{P}\big(A_T^{(1)} > K\big) = N(d_2)$$

where $N(\cdot)$ is the standard Gaussian distribution function and

$$d_2 = \frac{\log(A/K) - \frac{1}{2}\sigma^2 T}{\sigma\sqrt{T}}.$$

Fix A, T and σ and choose K such that this probability is small, say 10^{-6}. A similar calculation shows that

$$\mathbb{P}\big(A_T^{(2)} < K\big) = N\left(-d_2 - \frac{\mu^{(2)}}{\sigma}\sqrt{T}\right).$$

For $\mu^{(2)}$ sufficiently large this latter probability can be made arbitrarily small, say 10^{-6} once more.

Note that the economy above is arbitrage-free and complete. Consider pricing a call option on each of the assets $A^{(1)}$ and $A^{(2)}$ which matures at time T and has strike K. Having changed measure, they both have the same distribution and so the option prices are identical. Yet the real-world probability that the first option pays off is one in a million, whereas the real-world probability that the second does not pay off is also one in a million. Which option would you buy?

Fortunately the situation above, although theoretically possible for an arbitrage-free model, is not sustainable and would not occur in practice. The reader may wish to think about why. The point of the example is that the change of measure from the real world to the EMM may lead to very counter-intuitive results unless you are careful.

9.3 PRODUCT-BASED MODELLING

Suppose we wanted to price a derivative with payoff depending on a number of assets, n say. There are two ways we might first approach this problem. The first would be to look at historical time series for the prices of these assets and at the prices of other derivatives of these assets to build an understanding of the real market. Then, with this understanding, we might formulate a realistic model for the evolution of the n assets. Because of computational considerations this model would have to be of low dimension if we were to use it for derivative pricing. Finally, having formulated the model, we would apply it to price the derivative.

Following this approach will lead to a reasonable model for the economy but one which is not necessarily a good model when used to price the particular derivative in question. An alternative, therefore, is to understand the economy, as before, and then try to gain an understanding of what features of this economy will have the most impact on the derivative in question. Only then do we formulate the precise model.

This latter approach is what we call *product-based modelling*. In the rest of this section we illustrate its use. First, in Section 9.3.1 we show what can go wrong if we do not follow this approach. Then, in Section 9.3.2 we apply the technique for a product which is similar to one encountered in practice.

9.3.1 A warning on dimension reduction

When confronted with a high-dimensional set of data, such as a time series of the prices of a set of assets, a common and very powerful technique often applied is that of principal component analysis. Using this standard statistical tool it is possible to gain considerable understanding of the main structure of the asset price processes, vital knowledge for successful modelling of the economy (with a process of lower dimension). Here we consider a 'toy' pricing problem to illustrate some of the issues involved. The main point we are trying to communicate here is the importance of considering the derivative and the economy together: understanding both separately and also how they interact.

So consider an economy comprising $n + 1$ assets whose price processes $A^{(0)}$ and $A := (A^{(1)}, \ldots, A^{(n)})$ satisfy, for all $t \geq 0$,

$$A_t^{(0)} \equiv 1,$$
$$dA_t = \Sigma dW_t,$$

in the real-world measure $\mathbb{P}$, where W is an n-dimensional Brownian motion and Σ is some constant matrix of full rank n. Now consider the problem of pricing a European derivative which at time T has the quadratic payoff

$$V_T = A_T^T Q A_T.$$

We may here, without loss of generality (by considering $\frac{1}{2}(Q + Q^T)$ if necessary), assume that Q is symmetric. For illustrative purposes we shall also impose some extra structure, although this is not essential for the main points made below to hold. We assume that there exists some unitary matrix P (i.e. one for which $P^T P = I$) such that

$$P^T(\Sigma\Sigma^T)P = D_1 = \text{diag}(\sigma_i^2 : i = 1, 2, \ldots, n) \qquad (9.1)$$
$$P^T Q P = D_2 = \text{diag}(q_i : i = 1, \ldots, n).$$

This assumption corresponds exactly to saying that Q and $\Sigma\Sigma^T$ have the same eigenvectors, namely the columns p_i of P, with corresponding eigenvalues σ_i^2

and q_i respectively. Note that the representation (9.1) can always be achieved for a positive definite matrix Σ, and this is the basis of a principal component analysis. In that context the columns of P are referred to as the *principal components* and it is usual to order these columns such that $\sigma_1^2 \geq \ldots \geq \sigma_n^2$. We shall assume this ordering holds here.

The original measure $\mathbb{P}$ is in fact a martingale measure corresponding to the numeraire asset $A^{(0)}$, so the model is arbitrage-free. It further follows from the theoretical results of Chapter 7 that the market is complete, so this quadratic payoff can be replicated, and the derivative can be valued by taking expectations in the martingale measure $\mathbb{P}$. Hence the value at any time $t \leq T$ is given by

$$V_t = A_t^{(0)} \mathbb{E}_\mathbb{P}\left[\frac{A_T^T Q A_T}{A_T^{(0)}}\bigg|\mathcal{F}_t\right] = \mathbb{E}_\mathbb{P}\left[A_T^T Q A_T | \mathcal{F}_t\right],$$

where $\{\mathcal{F}_t\}$ is the augmented Brownian filtration. Evaluating this expectation is straightforward. First note that $M := D_1^{-1/2} P^T \Sigma$ satisfies $MM^T = I$, hence $\widetilde{W}_t := MW_t$ defines a standard Brownian motion. Thence

$$dA_t = \Sigma dW_t = \Sigma M^{-1} d\widetilde{W}_t = PD_1^{1/2} d\widetilde{W}_t$$

and so

$$A_t = A_0 + PD_1^{1/2} \widetilde{W}_t.$$

It follows from this representation and the independence of the Brownian increments and the filtration that

$$\mathbb{E}_\mathbb{P}\left[(A_T - A_t)^T Q(A_T - A_t)|\mathcal{F}_t\right]$$

$$= \mathbb{E}_\mathbb{P}\left[(A_T - A_t)^T Q(A_T - A_t)\right]$$

$$= \mathbb{E}_\mathbb{P}\left[(\widetilde{W}_T - \widetilde{W}_t)^T D_1^{1/2} P^T Q P D_1^{1/2}(\widetilde{W}_T - \widetilde{W}_t)\right]$$

$$= \mathbb{E}_\mathbb{P}\left[(\widetilde{W}_T - \widetilde{W}_t)^T D_1 D_2(\widetilde{W}_T - \widetilde{W}_t)\right]$$

$$= \sum_{i=1}^{n} \sigma_i^2 q_i(T - t).$$

This in turn yields the option value at time t,

$$V_t = \mathbb{E}_\mathbb{P}[A_T^T Q A_T \mid \mathcal{F}_t] = A_t^T Q A_t + (T - t)\sum_{i=1}^{n} \sigma_i^2 q_i.$$

This valuation formula depends on both the dynamics of the asset price movements, as summarized by the weightings σ_i^2 given to each component of the asset covariance matrix, and on the derivative payoff, as summarized by the corresponding weightings q_i. It is the product of the two which matters.

Suppose now that we need to approximate the process governing the asset price movements to reduce the dimension of the problem to m. The most natural choice of model, not taking into account the particulars of the derivative, would be to set to zero the smallest $n - m$ values of σ_i. This would then give an approximate value to the option of

$$\hat{V}_t = A_t^T Q A_t + (T - t) \sum_{i=1}^{m} \sigma_i^2 q_i.$$

In the situation when $q_1 = \cdots = q_m = 0, q_i \gg 0, i > n$, this would be a very poor approximation. The model value would be just the *intrinsic value*, $A_t^T Q A_t$, whereas the true value is $A_t^T Q A_t + (T-t) \sum_{i=m+1}^{n} \sigma_i^2 q_i$. This example illustrates the need to examine the derivative closely when choosing a model for the economy.

It appears, from the discussion above, that we could price any quadratic payoff with a low-dimensional model provided we were judicious in our model choice. However, this is not the case. Consider, for example, the situation when $q_i = 0$ for all $i < n$ and $q_n = 1$. A one-factor model might appear to be enough to find the right price. However, in practice this is not true as we would not know the covariance matrix Σ precisely. By data analysis or by examining the prices of other derivatives we would come up with some estimate $\hat{\Sigma}$. Our best estimate of the derivative price, even using a full n-factor model, would now be

$$\tilde{V}_t = A_t^T Q A_t + (T - t) q_n p_n^T \hat{\Sigma} p_n .$$

However, because the payoff is based on the last principal component of the matrix Σ, the variance of this estimate would typically be very large compared with the value itself. In other words, we should not even attempt to price this product.

9.3.2 Limit cap valuation

Here we present a further important example using the idea of choosing a model specifically for an application. Consider again the Gaussian economy of Section 9.3.1 but restrict to the case when $n = 2$. Assume that the covariance matrix Σ is known. Suppose we have some derivative to price whose payoff depends only on $A_{T_1}^{(1)}$ and $A_{T_2}^{(2)}$ for some $T_1 < T_2$. An example of this type of product would be one which pays $(A_{T_1}^{(1)} - K_1)_+$ if this is strictly positive, otherwise it pays $(A_{T_2}^{(2)} - K_2)_+$. This is the analogue for this 'toy' economy of the limit cap introduced in Section 18.2.3.

We could clearly value this product exactly using a two-factor model. But suppose, because of issues with the speed of a numerical implementation, that we wish instead to use a one-factor model. How should we do this?

One approach would be to use principal component analysis. That is, take the matrix $\Sigma\Sigma^T$ and calculate the first principal component p_1 and the corresponding eigenvalue σ_1^2. Use this component to define a one-factor approximation to the economy,

$$\hat{A}_t := A_0 + \sigma_1 p_1 W_t$$

for a one-dimensional Brownian motion W. Then calculate the option price within this model, which is given by

$$V_0 = \mathbb{E}_{\mathbb{P}}\left[\left(A_{T_1}^{(1)} - K_1\right)\mathbb{1}_{\{A_{T_1}^{(1)}>K_1\}} + \left(A_{T_2}^{(2)} - K_2\right)\mathbb{1}_{\{A_{T_1}^{(1)}\leq K_1,\, A_{T_2}^{(2)}>K_2\}}\right]. \quad (9.2)$$

This expectation can be evaluated in any of a number of standard ways. The accuracy of the model, and hence the price of the option, will depend on how dominant the first principal component is. See also the warnings in Section 9.3.1 above. But if the first component is dominant this model will work well for many options, provided the option being priced is not 'nearly orthogonal' to this first component.

An alternative approach involves examining the derivative more closely. Instead of using principal component analysis, we first note that the valuation formula for this option, equation (9.2), depends only on the joint distribution of the pair $(A_{T_1}^{(1)}, A_{T_2}^{(2)})$. Given the covariance matrix Σ,

$$\Sigma = \begin{pmatrix} \xi_1^2 & \xi_1\xi_2\rho \\ \xi_1\xi_2\rho & \xi_2^2 \end{pmatrix},$$

the vector $(A_{T_1}^{(1)}, A_{T_2}^{(2)})$ is a Gaussian vector with covariance matrix

$$\hat{\Sigma} = \begin{pmatrix} \xi_1^2 T_1 & \xi_1\xi_2\rho T_1 \\ \xi_1\xi_2\rho T_1 & \xi_2^2 T_2 \end{pmatrix}.$$

Now define the one-factor model for this real economy as

$$d\hat{A}_t^{(1)} := \xi_1 dW_t,$$

$$d\hat{A}_t^{(2)} := \left(\rho\mathbb{1}_{\{t<T_1\}} + \sqrt{\frac{T_2 - \rho^2 T_1}{T_2 - T_1}}\mathbb{1}_{\{t\geq T_1\}}\right)\xi_2 dW_t,$$

again for a one-dimensional Brownian motion W. The second model is also one-factor but the joint distribution $(\hat{A}_{T_1}^{(1)}, \hat{A}_{T_2}^{(2)})$ exactly matches that of $(A_{T_1}^{(1)}, A_{T_2}^{(2)})$. Thus the model will give an exact price for the option.

On this occasion the second approach gave the exact option price. It is, of course, not usually possible to achieve this. The cost of such a tailored approach to modelling is that the model produced will often be a very poor model for other derivatives – by fitting the joint distribution of $(A_{T_1}^{(1)}, A_{T_2}^{(2)})$ we have distorted other distributions in the economy quite severely. It is therefore imperative that one understands well the product in question before applying this technique.

9.4 LOCAL VERSUS GLOBAL CALIBRATION

Suppose, based on a general understanding of the market and a given product, we have selected a model for pricing purposes. Now we need to calibrate the model to market data. There are two distinct approaches to calibrating derivative pricing models to market data, the *global* and the *local* approaches. We prefer the latter for derivative pricing and hedging purposes, for reasons given below. However, the local approach is only better if you are very careful about how you apply the technique and it requires a deeper understanding of the product being priced and its relationship to the market more generally.

By the *global* approach to model calibration we mean one which uses a (large) number of existing vanilla market instruments' prices as inputs to a model and then fits the model to approximately price all these input prices. The model is then used to price the derivative to hand. Indeed it is likely that the same model would be used to price a number of different exotic derivatives.

By contrast to the above, the *local* approach fits a model to a much smaller number of market prices, ones which are highly relevant to the particular product being priced. The fit to these prices is exact. Every time a different product is considered a different set of calibrating instruments is chosen.

One can draw a simple analogue to each of the above techniques. Suppose we have a series of data points (x_i, y_i), $i = 1, \ldots, n$, and we know that

$$y_i = f(x_i)$$

where f is some unknown function. The analogue here of the derivative pricing problem is to find $f(\hat{x})$ for some given value $\hat{x}$.

A global approach to this problem would be to postulate some simple functional form for f, perhaps linear, quadratic, log-linear, ..., and then to perform a least squares fit to the points (x_i, y_i), as illustrated in Figure 9.1.

This would be the right thing to do if we did not know beforehand the value of $\hat{x}$, i.e. in the pricing context, if we did not know the derivative to be priced. The advantage of this approach is that it will give reasonable answers over a whole range of input values $\hat{x}$, i.e. in the pricing context it will price many products reasonably well. This approach is what an econometrician interested in general market structure would use.

The analogue of local calibration for this example is to fit the function f using the points close to the given $\hat{x}$ as shown in Figure 9.2.

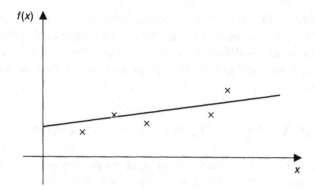

Figure 9.1 Least squares fit to data

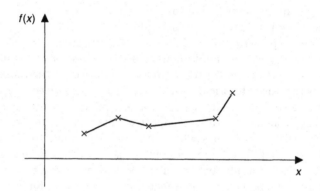

Figure 9.2 Local linear fit to data

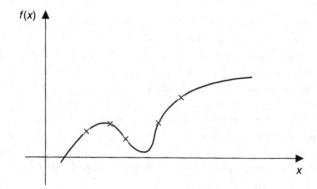

Figure 9.3 Overfitting to data

The main benefit of this second approach, the one that matters for derivative pricing, is that it exactly matches the prices of the most relevant instruments and those which will be used for the hedge. If relevant market prices are not the values one would expect based on historical experience and other market prices, to an extent this does not matter to us when trying to price an exotic. If they are out of line now, so should be the exotic's price. If, over time, the values come back into line, so will the value of the exotic.

One of the major dangers of the local approach is that perfect fitting has a cost. If we fit too many prices exactly this will give very poor results. In our simple analogue this would correspond to fitting a polynomial locally of too high an order, with obvious consequences as shown in Figure 9.3. An example of how this can arise in a derivative pricing context is discussed in Section 15.4.

10

Basic Instruments and Terminology

10.1 INTRODUCTION

To develop the theory and ideas of option pricing beyond the abstract setting of Part I of this book we need to introduce the basic assets that are traded in the financial markets. It is only with a thorough understanding of these assets, and equally importantly of how the financial markets view these assets, that we can start to formulate appropriate models for their evolution. Once we have done this we can start pricing derivatives.

The remainder of this book is motivated by examples and problems from the interest rate markets. Many of the techniques employed carry over to other markets, such as equity markets, and are often adaptations of techniques first employed in those other areas. However, interest rate derivatives have their own specific characteristics which make them particularly challenging to understand.

The purpose of this chapter is to introduce some of the basic and fundamental products that define the interest rate market. The products that we consider are all readily traded and will be the underlying assets on which other derivatives are based. We will also introduce standard market terminology which is important for the discussions that follow.

10.2 DEPOSITS

We begin with the most fundamental instrument in the market, the *deposit*. A deposit is an agreement between two parties in which one pays the other a cash amount and in return receives this money back at some *pre-agreed* future date, with a *pre-agreed* additional payment of interest. This interest is, of course, proportional to the amount of cash initially deposited.

Observe that in the above definition a deposit is made until a fixed date.

Financial Derivatives in Theory and Practice Revised Edition. P. J. Hunt and J. E. Kennedy
© 2004 John Wiley & Sons, Ltd ISBNs: 0-470-86358-7 (HB); 0-470-86359-5 (PB)

This would often in the retail market be referred to as a fixed term deposit. It is quite distinct from investing money in a deposit account such as those offered by UK building societies and from which money can be withdrawn at any time.

Deposits are available for a range of maturities, or terms. Only a small number of maturities, none greater than a year, are quoted as standard. Exactly which are quoted depends upon the currency, with typical values being overnight, 2 days, 7 days, 1, 2, 3, 6 and 12 months. The amount of interest paid at the end of the period is naturally larger for longer periods. The way this interest is quoted is via an *accrual factor* and an *interest rate*. It is important to understand exactly what these are as they play an important role in most of the derivatives that we shall encounter.

10.2.1 Accrual factors and LIBOR

LIBOR, sometimes called spot LIBOR, is the *London Interbank Offer Rate*. As the name implies, it is the rate of interest that one London bank will offer to pay on a deposit by another. There will, in general, be a different LIBOR for each of the standard deposit maturities.

The total amount of interest that the depositing bank will receive is calculated by multiplying the LIBOR by the amount of time, as a proportion of a year, for which this money has been on deposit. This amount of time is called the *accrual factor* or *daycount fraction*.

Accrual factors are calculated by dividing the number of days in the period by the number of days in a year. Different markets use different conventions to calculate these figures and the name given to the particular method is the *basis*. Two very common examples are actual/365 and actual/360. In both cases the number of days in the period is taken to be the exact number of calendar days between the payments, but the number of days in a year is calculated differently. For the actual/360 basis it is assumed to be 360 days; for the actual/365 basis it is assumed to be 365, or 366 in a leap year.

It will not be important for us to have a detailed knowledge of bases. All that we need to know is that the interest payable is calculated via an accrual factor which is a known number. Clearly it is also the case that any quotation of a LIBOR is for a specified maturity and basis.

In summary, if we denote the accrual factor by α and the corresponding LIBOR by L, a deposit of notional amount A will be repaid at maturity along with an interest payment of amount $A\alpha L$. Usually we will suppress the notional amount and assume it to be unity. These cashflows are illustrated in Figure 10.1.

We will often represent products in this way. The horizontal axis represents the time of cashflows; those above the axis are ones which we receive, while those below the line are ones that we pay.

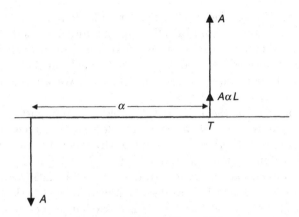

Figure 10.1 Deposit cashflows

10.3 FORWARD RATE AGREEMENTS

A *forward rate agreement*, or FRA, is an agreement between two counterparties to exchange cash payments at some specified date in the future. Associated with each FRA there is a notional amount and two dates, the *reset* date T and the *payment* date S. These two dates will be such that the period $[T, S]$ corresponds exactly to the accrual period for a standard deposit starting on date T, the most common lengths being 3 and 6 months. The reset date would usually be some multiple of a month out to two years.

Under an FRA, one counterparty agrees to pay to the other an amount $A\alpha K$, where A is the notional amount, α is the accrual factor for the period $[T, S]$, and K is the *fixed rate*. In return the second counterparty agrees to pay to the first an amount $A\alpha L_T[T, S]$, where $L_T[T, S]$ is the LIBOR for the period $[T, S]$ that sets on date T. Figure 10.2 summarizes this, with the convention that a 'wavy' cashflow is not known on the date that the agreement is made. We call this a *floating cashflow*.

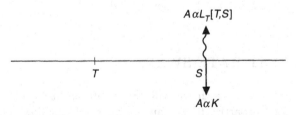

Figure 10.2 Cashflows for the holder of an FRA

When, at time t, two counterparties enter an FRA they do so at zero cost to both parties, and this is achieved by choosing an appropriate value for the fixed rate K. It is this breakeven value for K that is quoted in the FRA market and it is referred as the *forward* LIBOR, $L_t[T, S]$, for the period $[T, S]$, which we will tend to abbreviate to L_t when the context is clear. Note, however, that when an FRA is entered its fixed rate is determined but its value will then change through time as rates increase and decrease. We will see soon how to calculate this value.

One main purpose of the FRA is to enable a counterparty to fix today the amount of interest he will receive on a deposit he intends to make at some future date. It is worthwhile examining why this is the case. Suppose he wishes to invest a unit amount from time T until S but would like to fix today the amount of interest he will receive, rather than wait until T when market conditions may have changed to his disadvantage. To achieve this he *sells*, or *goes short*, an FRA today (although there is no exchange of cash today) at the current forward LIBOR $L_t[T, S]$. At time T he then invests his unit capital at the then spot LIBOR $L_T[T, S]$ and in return receives his capital back at time S along with interest $\alpha L_T[T, S]$. In addition, under the FRA he will receive $\alpha(L_t[T, S] - L_T[T, S])$, so his net income at time S will be his notional plus exactly $\alpha L_t[T, S]$, the interest amount he locked in at the time he entered the FRA.

Remark 10.1: In later chapters we will be using models for the forward LIBOR process $\{L_t[T, S] : t \leq T\}$. Note that the times T and S are held fixed as t varies. The *spot* LIBOR process, $\{L_t[t, t + \Delta] : t \geq 0\}$, for some fixed Δ, is something quite distinct which has different properties and which will not concern us. It is important to note the distinction. The same comments hold for the forward swap rate which we treat below.

Remark 10.2: Those familiar with the FRA market will realize that the definition we have just given is not strictly correct. In reality the FRA is settled at time T and not at time S, but the amount paid is exactly the value at T of what we have described above as an FRA. Consequently this could be invested at LIBOR to yield the cashflows we have described. We will, for ease of exposition, stick to our slightly altered FRA definition which is also more closely related to swaps, which we now introduce.

10.4 INTEREST RATE SWAPS

An *interest rate swap*, which we will abbreviate to *swap*, is an agreement between two counterparties to exchange a series of cashflows on pre-agreed dates in the future. This is illustrated in Figure 10.3.

We refer to all the payments we receive as one *leg* of the swap, and those we pay constitute the other leg. To specify the swap we must know its start

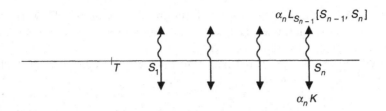

Figure 10.3 Payers interest rate swap

date (the start of the first accrual period), maturity date (the date of the last cashflow), and the payment frequency, for each of the legs. Typical maturities are 1–10, 12, 15, 20 and 30 years. Each leg can in general have a different payment frequency, but here we describe the case where it is the same for both legs.

Suppose there are a total of n cashflows in each leg, cashflow i occurring at time S_i. One of the legs will be a *fixed* leg, for which all the cashflows are known at the start of the swap. The amount of cashflow i is given by $\alpha_i K$ where α_i is the accrual factor for the period $[S_{i-1}, S_i]$ and K is the *fixed rate* of the swap. For the swap of Figure 10.3 the fixed leg is represented below the axis, so this is a pay-fixed, or *payers* swap; the contract with the cashflows reversed is a *receivers* swap.

The other leg is the *floating* leg, so called because the payment amounts will be set during the life of the swap. The amount of payment i is set at time S_{i-1} and is the accrual factor α_i multiplied by $L_{S_{i-1}}[S_{i-1}, S_i]$, the LIBOR then quoted for the period $[S_{i-1}, S_i]$.

Remark 10.3: In actual fact, the presence of holidays and weekends means it is very unusual for a swap to have an exact match between payment dates and the start of the next accrual period. The market allows for this fact when valuing swaps. However, when modelling more complex option-type products it usually simplifies the analysis to assume there is a perfect match. As long as one is careful, this does not cause any problems. In all the modelling in this book we will assume that there is a perfect match. However, for notational convenience we shall often refer to the start of the ith accrual period as T_i (rather than S_{i-1}). Thus the LIBOR appropriate to the ith period is $L_{T_i}[T_i, S_i]$.

Swaps are usually entered at zero initial cost to both counterparties. A swap with this property is called a *par* swap, and the value of the fixed rate K for which the swap has zero value is called the *par swap rate*. In the case when the swap start date is spot (i.e. the swap starts immediately), this is often abbreviated to just the *swap rate*, and it is these par swap rates that are quoted on trading screens in the financial markets. A swap for which the start date is not spot is, naturally enough, referred to as a *forward start swap*, and the corresponding par swap rate is the *forward swap rate*. Forward start swaps are less common than spot start swaps and forward swap rates are not quoted as standard on market screens.

We will denote the (forward) par swap rate, at time t for a swap starting on date T and making *fixed* payments on dates given by the vector $S = (S_1, \ldots, S_n)$, by $y_t[T, S]$ (often just y_t when the context is clear). Note that this definition is independent of the payment frequency in the floating leg. This is indeed the case, as we see when we learn how to value swaps in Section 10.6.4.

We can see from the above that a swap is just a series of FRAs, each having the same strike K. Note that in Figure 10.3 we have represented the first cashflow in the floating leg of the swap as an unknown floating payment, even though the amount of this cashflow will be set immediately the swap is agreed, based on spot LIBOR. This is a slight abuse of our previous notation but is convenient since, like all the other cashflows in the floating leg, its amount is based on LIBOR at the start of the corresponding accrual period.

10.5 ZERO COUPON BONDS

We have already met these in the theory of Part I. *Zero coupon bonds* (ZCBs), also known as pure discount bonds, are assets which entitle the holder to receive a cashflow at some future date T. The amount of this cashflow is part of the contract specification, although we will often assume it to be a unit amount.

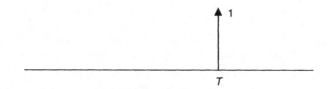

Figure 10.4 Zero coupon bond

ZCBs are not generally liquid market instruments. It is not possible, for example, to find their prices on standard quotation screens. However there are many in existence, and indeed a deposit, once the initial capital is handed over, is effectively a ZCB, so they certainly exist for all maturities less than a year. Furthermore, banks will quote prices for most maturities over a year as well.

The main value of introducing ZCBs comes from the fact that other products can be built up from them. In this sense they are fundamental.

As an example, the fixed leg of a swap consists of n known cashflows and so can be thought of as n ZCBs. Thus the value of the fixed leg will be the sum of the values of each of these component ZCBs. We shall shortly see how an FRA can also be related to ZCBs.

10.6 DISCOUNT FACTORS AND VALUATION

In this section we introduce discount factors, the most fundamental tool for summarizing the value of interest rate products. We will see how discount factors can be used to express the value of the basic instruments we have met and also how they relate to forward LIBOR and swap rates.

10.6.1 Discount factors

Suppose we are at time t. For any time $T \geq t$ there is a corresponding *discount factor*, denoted as D_{tT}, defined to be the value, at time t, of a ZCB paying a unit amount at time T. Note that, for all t, $D_{tt} = 1$ since a unit cashflow now is worth one unit.

We will usually assume that the initial *discount curve*, $\{D_{0T} : T \geq 0\}$ if today is time zero, is known. This is not strictly the case since the standard liquid market instruments (which do not include ZCBs of all maturities) only give us a limited amount of information. In practice banks take these liquid market instruments and construct a full discount curves consistent with their prices. There is no unique way to do this, and what is done often depends on the use to which the discount curve is to be put. We shall not discuss the techniques used to construct discount curves in detail and shall assume that this has been done as the starting point for the rest of our analyses.

10.6.2 Deposit valuation

A standard unit deposit at time T pays at maturity S an amount $1 + \alpha L_T[T, S]$. Since there are no further payments other than the initial deposit and the final redemption, these two cashflows must have the same value at time T. Thus it follows that

$$D_{TT} = (1 + \alpha L_T[T, S])D_{TS},$$
$$D_{TS} = (1 + \alpha L_T[T, S])^{-1},$$
$$L_T[T, S] = \frac{D_{TT} - D_{TS}}{\alpha D_{TS}}. \tag{10.1}$$

10.6.3 FRA valuation

This our first example of derivative pricing. Recall that the net payment under an FRA is given by $\alpha(L_T[T, S] - K)$. From (10.1) it can be seen that the final payment of an FRA depends on two ZCBs which mature at times T and S, thus the FRA is a derivative of these bonds. We can thus apply the theory developed in Chapters 7 and 8, but for this simple product we can use a more direct argument.

Suppose at time t we have an amount of cash

$$V_t = D_{tT} - (1 + \alpha K)D_{tS}. \tag{10.2}$$

Consider the following investment strategy. At time t we buy one unit of the ZCB maturing at time T and sell $1 + \alpha K$ units of the ZCB maturing at time S. When, at time T, we receive a unit payment from the maturing ZCB, we deposit this until time S (equivalently buy $(1 + \alpha L_T[T, S])^{-1}$ units of the other ZCB). At time S the deposit and ZCB mature, giving a net payment of

$$(1 + \alpha L_T[T, S]) - (1 + \alpha K) = \alpha(L_T[T, S] - K).$$

This is precisely the payment received under the FRA, and hence we have replicated the FRA by trading in the two ZCBs. It follows immediately, assuming there is no arbitrage in the economy, that V_t is the value of the FRA.

In general, of course, V_t will be non-zero. Recall that the value of K for which V_t is zero is defined to be the forward LIBOR for the period $[T, S]$ and, from (10.2), is given by

$$L_t[T, S] = \frac{D_{tT} - D_{tS}}{\alpha D_{tS}}.$$

We can use this to obtain an alternative representation for the valuation formula (10.2), namely

$$V_t = \alpha D_{tS}(L_t[T, S] - K). \tag{10.3}$$

Note that in this argument, unusually, we have not needed to specify a model for the evolution of the assets of the economy. This is because the hedging strategy is static up until the time T at which time all payments are known. We will see in Chapter 11 how to price the same product using the techniques of Chapter 7.

10.6.4 Swap valuation

A swap is a series of FRAs so it can be valued by adding together the values of the constituent FRAs. However, we will instead value the swap directly, considering each of the two legs in turn.

The fixed leg consists of a series of payments on dates $S_1, S_2, \ldots, S_n$, the payment at time S_i being $\alpha_i K$. Its value is therefore given by

$$V_t^{\text{FXD}} = K \sum_{j=1}^{n} \alpha_j D_{tS_j}$$

$$= K P_t[T, S],$$

where $S = (S_1, S_2, \ldots, S_n)$ and

$$P_t[T, S] := \sum_{j=1}^{n} \alpha_j D_{tS_j}.$$

We will refer to the expression $P_t[T, S]$ as the *present value of a basis point*, or PVBP, of the swap. It represents the value of the fixed leg of the swap if the fixed rate were unity. This is a slight abuse of notation since $P_t[T, S]$ as defined above is the present value of unity, but one which is common in the market and which we will always use. A *basis point* is actually 0.01%, or 10^{-4}, so P_t is $10,000$ times the true PVBP of the swap. The PVBP will play an important role in some of the products we examine later. When there can be no possible confusion we will usually abreviate $P_t[T, S]$ to P_t.

The floating leg of the swap is more difficult to value, comprising as it does a series of floating payments. However, just as we did for the FRA, we can find a simple trading strategy which replicates the swap's floating payments.

Suppose at time t we have an amount of cash

$$V_t^{\text{FLT}} = D_{tT} - D_{tS_n}. \tag{10.4}$$

Take this cash and buy one ZCB of maturity T and sell one of maturity S_n. At time T take the unit paid by the ZCB and deposit it at LIBOR until time S_1. At time S_1 ($= T_2$) we will receive $1 + \alpha_1 L_T[T, S_1]$. The term $\alpha_1 L_T[T, S_1]$ is the floating payment we need to replicate the swap, the extra unit of principal we deposit at LIBOR until time S_2. We repeat this at each floating payment date until at time S_n we receive $1 + \alpha_n L_{T_n}[T_n, S_n]$. The LIBOR part of this payment is what we need for the last floating payment of the swap, the notional pays the amount owed on the ZCB we sold at time t.

It follows that the net value of a payers swap is exactly

$$V_t = V_t^{\text{FLT}} - V_t^{\text{FXD}}$$

$$= D_{tT} - D_{tS_n} - K P_t[T, S]. \tag{10.5}$$

The forward swap rate $y_t[T, S]$ is the value of K which sets this value to zero. Substituting $V_t = 0$ into (10.5) then yields

$$y_t[T, S] = \frac{D_{tT} - D_{tS_n}}{P_t[T, S]}. \tag{10.6}$$

Substituting this back into (10.5), we obtain the more usual expression for the value of a payers swap,

$$V_t = P_t[T, S](y_t[T, S] - K). \tag{10.7}$$

The value of a receivers swap is, of course, just $-V_t$.

As a final point on swaps, recall that we allow the floating and fixed legs to have different payments dates. It is easy to see from the arguments above that the value of the floating leg, as given by (10.4), is independent of the floating payment frequency. It is therefore only the fixed payment frequency that matters when defining and valuing a forward starting swap.

11

Pricing Standard Market Derivatives

11.1 INTRODUCTION

In this chapter we apply the techniques of Part I to price some of the more common derivatives seen in the market. The models we present have been around for quite some time, but it is only recently that the theory has been understood well enough for those working in the area to realize that the prices derived are not just approximations but are actually theoretically exact, given a suitable initial model.

Most of the derivatives we price here are sufficiently common in the financial markets that models are no longer really used to derive the price from model inputs. Rather a model is selected and the market price is used to imply the parameters of the pricing model (such as volatility) which would give the observed market price. The model is then used to calculate the hedge for the derivative in terms of the underlying instruments. Furthermore, these derivatives are themselves often used as underlying instruments for other more complex derivatives. We shall see several examples of this kind in the chapters that follow.

11.2 FORWARD RATE AGREEMENTS AND SWAPS

We have already met the FRA in Chapter 10, where we derived the price via a simple replication argument in terms of zero coupon bonds. Recall that under an FRA one counterparty makes a fixed payment αK at some time S and in return receives an amount $\alpha L_T[T, S]$ on the same date. We saw that the value at time t is given by

$$V_t = D_{tT} - (1 + \alpha K)D_{tS}.$$

This is an unusual example in that we have not needed to specify a model for the evolution of asset prices in order to price the derivative. This is because

Financial Derivatives in Theory and Practice Revised Edition. P. J. Hunt and J. E. Kennedy
© 2004 John Wiley & Sons, Ltd ISBNs: 0-470-86358-7 (HB); 0-470-86359-5 (PB)

there is in this case a static replicating portfolio. It is instructive, however, to use the tools developed in Chapter 7 to price the FRA; the resulting price should be independent of the model we choose. So suppose we have some model for the evolution of discount bonds. Following the approach of Chapter 7, if $(N, \mathbb{N})$ is some numeraire pair and if $\{\mathcal{F}_t\}$ is the filtration generated by the assets in the economy, then the price of the FRA is given by

$$
\begin{aligned}
V_t &= N_t \mathbb{E}_{\mathbb{N}} \left[\alpha (L_T - K) N_S^{-1} | \mathcal{F}_t \right] \\
&= N_t \mathbb{E}_{\mathbb{N}} \left[\alpha (L_T - K) \mathbb{E}_{\mathbb{N}} \left[N_S^{-1} \,\middle|\, \mathcal{F}_T \right] \,\middle|\, \mathcal{F}_t \right] \\
&= N_t \mathbb{E}_{\mathbb{N}} \left[\alpha (L_T - K) D_{TS} N_T^{-1} | \mathcal{F}_t \right] \\
&= D_{tT} - (1 + \alpha K) D_{tS} \,,
\end{aligned}
$$

the last equality following by substituting for L_T using equation (10.1). A swap could be valued in exactly the same way, but as we saw in Chapter 10 it is just a linear combination of FRAs and pure discount bonds, and so its value follows immediately by summing the values of these FRAs and ZCBs.

11.3 CAPS AND FLOORS

It is often the case that a customer is either making or receiving a series of floating payments and does not wish to convert them into a series of fixed payments. This may be because he believes future rate moves will be in his favour. However, he is then exposed if rates move against him and would like to buy some protection against this without removing the benefits of the move in rates that he expects. The solution in this case is for the customer to buy a *cap* or a *floor*.

Caps and floors are similar to swaps in that they are made up of a series of payments on regularly spaced dates, S_j, $j = 1, 2, \ldots, n$. On date S_j the holder of a cap receives a payment of amount

$$
\alpha_j \max \left(L_{T_j} [T_j, S_j] - K, 0 \right),
$$

where $T_j := S_{j-1}$ is the setting time for the LIBOR which pays at time S_j. A floor is similar except the payment amount is given by

$$
\alpha_j \max \left(K - L_{T_j} [T_j, S_j], 0 \right).
$$

The constant K in these expressions is part of the contract specification and is known as the *strike* of the option. These payoff profiles are shown in Figures 11.1 and 11.2. A counterparty who is paying LIBOR and who buys a cap has ensured that he will never pay more that $\alpha_j K$ at time S_j; one who is receiving LIBOR and buys a floor will never receive less than $\alpha_j K$. Each

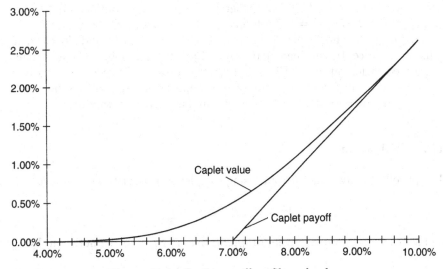

Figure 11.1. Caplet payoff profile and value

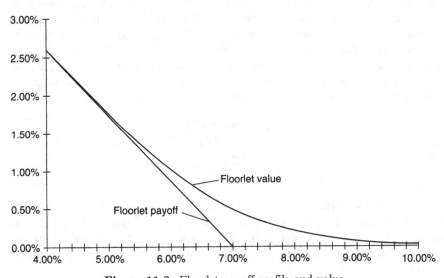

Figure 11.2. Floorlet payoff profile and value

individual payment is usually referred to as a *caplet* or *floorlet*. At any time $t \leq T$, the amount $\alpha_j(L_t[T_j, S_j] - K)_+$ is referred to as the *intrinsic value* of the caplet and, similarly, $\alpha_j(K - L_t[T_j, S_j])_+$ is referred to as the intrinsic value of the floorlet. In either case, if the intrinsic value is positive then the option is said to be *in the money*, whereas if it is zero the option is *out of the money*. When $L_t[T_j, S_j] = K$ both options are said to be *at the money*.

Since caps and floors are linear combinations of caplets and floorlets, it suffices to price these single payments. Caplets and floorlets are usually considered as derivatives of FRAs, but can of course also be seen as derivatives of ZCBs. Which is done depends on the application. We will consider the former here, and will derive the market standard pricing formula.

11.3.1 Valuation

The payoff at some time S from a single caplet setting at time T is just

$$V_S = \alpha(L_T[T, S] - K)_+$$

and so its time-t value is given, dropping the $[T, S]$ from the notation, by

$$V_t = N_t \mathbb{E}_{\mathbb{N}}[\alpha(L_T - K)_+ N_S^{-1} | \mathcal{F}_t], \tag{11.1}$$

for some numeraire pair $(N, \mathbb{N})$. To calculate this price we must choose a numeraire N and a suitable model for $L[T, S]$ in the measure $\mathbb{N}$. Suppose we choose $N_t = D_{tS}$, the discount bond maturing on the payment date S. The corresponding measure $\mathbb{N}$ is usually referred to as the *forward measure* and we denote it by $\mathbb{F}$. Using this measure has two important consequences. First, $N_S = D_{SS} = 1$ and so this disappears from equation 11.1. Secondly, the forward LIBOR L, which is of the form

$$L_t = \frac{D_{tT} - D_{tS}}{\alpha D_{tS}}$$

$$= \frac{D_{tT} - D_{tS}}{\alpha N_t},$$

is a ratio of asset prices over the numeraire and so must be a martingale. As long as we model L as a martingale (under $\mathbb{F}$) we will have a model that is arbitrage-free and the caplet value will be given by (11.1). Because interest rates are always assumed to be positive, we will model L as a log-normal martingale,

$$dL_t = \sigma_t L_t dW_t,$$

for some deterministic σ and a Brownian motion W. This is standard market practice and yields a solution

$$L_t = L_0 \exp\left(\int_0^t \sigma_u dW_u - \frac{1}{2}\int_0^t \sigma_u^2 du\right).$$

Substituting into (11.1) now yields

$$V_t = D_{tS}\mathbb{E}_{\mathbb{F}}\left[\alpha(L_T - K)_+|\mathcal{F}_t\right]$$

$$= D_{tS}\mathbb{E}_{\mathbb{F}}\left[\alpha\left(L_t\exp\left(\int_t^T \sigma_u dW_u - \tfrac{1}{2}\int_t^T \sigma_u^2 du\right) - K\right)_+\Bigg|\mathcal{F}_t\right]$$

$$= \alpha D_{tS}\left(L_t N(d_1) - K N(d_2)\right) \tag{11.2}$$

where

$$d_1 = \frac{\log(L_t/K)}{\tilde{\sigma}_t\sqrt{T-t}} + \tfrac{1}{2}\tilde{\sigma}_t\sqrt{T-t}, \tag{11.3}$$

$$d_2 = \frac{\log(L_t/K)}{\tilde{\sigma}_t\sqrt{T-t}} - \tfrac{1}{2}\tilde{\sigma}_t\sqrt{T-t},$$

$$\tilde{\sigma}_t^2 = \frac{1}{T-t}\int_t^T \sigma_u^2 du.$$

This is Black's formula, and was first published in Black (1976). The derivation at that time used a series of approximations. Criticisms of the assumptions made have been that the model admits arbitrage and that inconsistent assumptions are made about the evolution of the discount factor (taken to be deterministic) and the forward rate (taken to be stochastic). Whereas these criticisms may be valid for the thought process at the time, they are not valid for the above model and pricing formula, as the earlier analysis shows.

Pricing of floorlets is identical and yields the usual put option formula:

$$V_t = \alpha D_{tS}\mathbb{E}_{\mathbb{F}}\left[(K - L_T)_+|\mathcal{F}_t\right]$$

$$= \alpha D_{tS}\left(K N(-d_2) - L_t N(-d_1)\right). \tag{11.4}$$

Figures 11.1 and 11.2 show, for some choice of the model input parameters, the value at t of a caplet and floorlet as a function of the forward LIBOR L_t.

11.3.2 Put–call parity

There is a simple relationship between caplets, floorlets and FRAs which must always hold, namely

$$V_t^{CAP} - V_t^{FLOOR} = \alpha(L_t - K)D_{tS}. \tag{11.5}$$

This can be verified by substituting (11.2) and (11.4) into (11.5), but it must also hold more generally, irrespective of the model being used. This follows since

$$(L_T - K)_+ - (K - L_T)_+ = (L_T - K)$$

and thus, recalling the FRA valuation formula (10.3),

$$\mathbb{N}_t \mathbb{E}_{\mathbb{N}}[\alpha(L_T - K)_+ N_S^{-1}|\mathcal{F}_t] - N_t \mathbb{E}_{\mathbb{N}}[\alpha(K - L_T)_+ N_S^{-1}|\mathcal{F}_t]$$
$$= N_t \mathbb{E}_{\mathbb{N}}[\alpha(L_T - K)N_S^{-1}|\mathcal{F}_t]$$
$$= \alpha(L_t - K)D_{tS}.$$

11.4 VANILLA SWAPTIONS

Just as a caplet is an option on an FRA, a swaption is an option on a swap. Swaptions are also commonly traded in the market and are priced using Black's formula. There are two types of swaption: receivers swaptions (referred to as receivers) and payers swaptions (referred to as payers). In the first case, upon exercise the option holder enters a swap in which he receives a fixed rate K, the swaption strike, and pays the floating rate; a payers is the reverse. We refer to this as a vanilla swaption if the underlying swap is a plain vanilla interest rate swap.

We now price a payers swaption. Let $U_t[T, S]$ be the value at time t of a vanilla payers swap which starts on date T and makes fixed payments on the dates $S = (S_1, \ldots, S_n)$. The effective payoff to the option holder, who has the right but not the obligation to enter the swap, is given at the option expiry T by

$$V_T = \max(U_T, 0)$$

since he will only enter the swap if it is to his advantage to do so. The value at time t of the swaption is, by the usual valuation formula, given by

$$V_t = N_t \mathbb{E}_{\mathbb{N}}[\max(U_T, 0)N_T^{-1}|\mathcal{F}_t]$$
$$= N_t \mathbb{E}_{\mathbb{N}}[P_T(y_T - K)_+ N_T^{-1}|\mathcal{F}_t], \qquad (11.6)$$

for some suitable numeraire pair $(N, \mathbb{N})$. The final equality above follows by substituting (10.7) for the value of a swap, where P_T, y_T and K are, as usual, the PVBP and par swap rate at T and the fixed rate for the underlying swap. We usually refer to K as the *swaption strike*. At any time $t \leq T$ we use the terminology *in the money, out of the money* and *at the money* just as we did for caplets and floorlets according to the sign of $(y_t - K)$.

To evaluate (11.6) we follow a procedure similar to that for the FRA. On this occasion we choose P as the numeraire, in which case the corresponding martingale measure is called the *swaption measure* and will be denoted by $\mathbb{S}$. This reduces (11.6) to the form

$$V_t = P_t \mathbb{E}_{\mathbb{S}}[(y_T - K)_+|\mathcal{F}_t],$$

which should by now be quite familiar. We recall from (10.6) that the forward swap rate y_t is of the form

$$y_t = \frac{D_{tT} - D_{tS_n}}{P_t},$$

which is the ratio of asset prices over the numeraire, so must be a martingale under $\mathbb{S}$. Since we want interest rates to remain positive, we will model y to be a log-normal martingale,

$$dy_t = \sigma_t y_t dW_t, \tag{11.7}$$

for some deterministic function σ and a Brownian motion W. This puts us in precisely the same framework as when we priced caplets and floorlets and again yields Black's formula for the price,

$$V_t = P_t \Big(y_t N(d_1) - K N(d_2) \Big) \tag{11.8}$$

where

$$d_1 = \frac{\log(y_t/K)}{\tilde{\sigma}_t \sqrt{T-t}} + \tfrac{1}{2} \tilde{\sigma}_t \sqrt{T-t},$$

$$d_2 = \frac{\log(y_t/K)}{\tilde{\sigma}_t \sqrt{T-t}} - \tfrac{1}{2} \tilde{\sigma}_t \sqrt{T-t},$$

$$\tilde{\sigma}_t^2 = \frac{1}{T-t} \int_t^T \sigma_u^2 du.$$

The corresponding receivers valuation formula is just

$$V_t = P_t \Big(K N(-d_2) - y_t N(-d_1) \Big),$$

and the put–call parity relationship becomes

$$V_t^{\text{PAY}} - V_t^{\text{REC}} = P_t(y_t - K).$$

Remark 11.1: The choice of P as numeraire means that expression (11.7) is the only distributional assumption made, despite the fact that the underlying swap is made up of several ZCBs. For other applications we will need to make extra modelling assumptions, and this is the topic of the chapters that follow.

Remark 11.2: Precisely the same criticisms have been levelled against Black's formula for swaption pricing as for caplet and floorlet pricing. Again these criticisms have been shown to be invalid (Neuberger 1990; Jamshidian 1997), on this occasion by working in the swaption measure. This is the first example we have met where the numeraire is not the price of a single asset in the economy. Note that when the swap comprises a single payment date the swap reduces to an FRA and the swaption measure reduces to the forward measure.

11.5 DIGITAL OPTIONS

A digital option is one which pays either one or zero at some future date, depending on the level of some index rate. The two most common examples are digital caps and floors, and digital swaptions.

11.5.1 Digital caps and floors

Just as caps and floors are made up of a series of caplets and floorlets, so digital caps and floors are made up of digital caplets and floorlets. A digital caplet is an option which pays a unit amount at time S if at T the LIBOR for the period $[T, S]$ is above some strike level K. A floorlet pays a unit amount if the LIBOR is below the strike.

The value at S of a digital caplet is given by $V_S = 1_{\{L_T > K\}}$, and so it follows that

$$V_t = N_t \mathbb{E}_{\mathbb{N}}[1_{\{L_T > K\}} N_S^{-1} | \mathcal{F}_t].$$

Taking D_{tS} as numeraire, as we did for caps and floors, and using the same log-normal model, yields

$$
\begin{aligned}
V_t &= D_{tS} \mathbb{E}_{\mathbb{F}}[1_{\{L_T > K\}} | \mathcal{F}_t] \\
&= D_{tS} \mathbb{F}(L_T > K | \mathcal{F}_t) \\
&= D_{tS} N(d_2),
\end{aligned}
\tag{11.9}
$$

where d_2 is as defined in (11.3). The analogous digital floorlet pricing formula is then

$$
\begin{aligned}
V_t &= D_{tS} \mathbb{E}_{\mathbb{F}}[1_{\{L_T < K\}}] \\
&= D_{tS} N(-d_2),
\end{aligned}
\tag{11.10}
$$

and the put–call parity relationship is now

$$V_t^{\text{DCAP}} + V_t^{\text{DFLOOR}} = D_{tS}.$$

Remark 11.3: What we have just done in (11.9) and (11.10) is derive the price of a digital caplet and floorlet using the same log-normal model as we did to price a standard caplet and floorlet. Even when alternative models are used for the forward LIBOR there is still a fundamental relationship between the prices of caps and floors on the one hand and digital caps and digital floors on the other. This relationship essentially follows from the identity

$$\frac{d}{dK}(x - K)_+ = -1_{\{x > K\}}.$$

The more general relationship will be exploited further in Chapters 14 and 19. The reader may wish to derive it now.

11.5.2 Digital swaptions

Digital swaptions are less common than digital caps and floors, and they are also more difficult to price. There are, once again, two types of digital swaptions, payers and receivers. In the case of a digital payers swaption the option holder receives a unit amount at some date M if the index swap rate, y_T, is above the strike K on the setting date T. For a digital receivers the option holder receives the unit payment if the swap rate is below the strike K.

In the case of swaptions there is no obvious choice for the option payment date M relative to the rate setting date T. Common choices are $M = T$ or $M = S_1$, the first payment date of the underlying swap. Considering the payers digital, the value at time M is $\mathbb{1}_{\{y_T > K\}}$ and so the time-t value of the option is

$$V_t = N_t \mathbb{E}_{\mathbb{N}}[\mathbb{1}_{\{y_T > K\}} N_M^{-1} | \mathcal{F}_t]$$
$$= N_t \mathbb{E}_{\mathbb{N}}[\mathbb{1}_{\{y_T > K\}} D_{TM} N_T^{-1} | \mathcal{F}_t],$$

the last equality following by conditioning on $\mathcal{F}_T$. To value the digital consistently with standard swaptions, we would like to take the same model as in Section 11.4. Working in swaption measure, we have

$$V_t = P_t \mathbb{E}_{\mathbb{S}}[\mathbb{1}_{\{y_T > K\}} D_{TM} P_T^{-1} | \mathcal{F}_t]. \tag{11.11}$$

Evaluation of the expectation on the right-hand side of (11.11) presents us with a new problem. The expectation involves three distinct random terms, namely y_T, D_{TM} and P_T. We know that within our model y_T is a log-normal martingale, but we have not yet specified a suitable model for the other two terms. To price this product is thus beyond the scope of this current chapter. Chapters 13 and 14 show how this type of product can be treated.

12

Futures Contracts

12.1 INTRODUCTION

In Chapter 10 we met the forward rate agreement (FRA) and saw how it could be used to lock in today a rate of interest on a deposit or loan which is scheduled for some future date. This has the effect, therefore, of immunizing the buyer (or seller) from future changes in interest rates. One drawback with the FRA, however, is that it is a contract traded *over the counter*, directly between the two counterparties. If, before the settlement date of the contract (the date on which the cashflows occur), one of the counterparties were to default, the other would again be left exposed to interest rate moves that had occurred between the date when the FRA was entered and the date of the default.

The *Eurodollar futures contract* is a derivative designed to provide similar protection from interest rate moves as the FRA but without the accompanying counterparty risk. Other futures contracts fulfil a similar role in other markets, such as equity markets, and for other interest rate instruments, such as government bonds. Futures contracts generally are amongst the most important and liquidly traded instruments.

In this chapter we define and analyse the futures contract. We shall do this in some generality since the extra abstraction required to do so makes the analysis no more difficult but clarifies some of the underlying issues. As we shall see, any futures contract is defined with reference to some other financial instrument, commodity or derivative. We will derive the relationship between the *futures price* (which is in fact not a price at all) and the *forward price* or *forward rate* of the associated instrument.

12.2 FUTURES CONTRACT DEFINITION

Here, in Section 12.2.1, we describe the futures contract and how it works in practice. This is followed, in Section 12.2.2, by a brief discussion of how the

Financial Derivatives in Theory and Practice Revised Edition. P. J. Hunt and J. E. Kennedy
© 2004 John Wiley & Sons, Ltd ISBNs: 0-470-86358-7 (HB); 0-470-86359-5 (PB)

futures price is related to the price of the associated reference instrument, something that is not immediately clear from the contract definition. We then, in Section 12.2.3, give a mathematical formulation of the product and set notation.

12.2.1 Contract specification

The first important point to note about futures contracts, one which distinguishes them from the derivatives discussed so far, is that they are all traded on an exchange. When one counterparty *buys* a futures contract there is a second counterparty who simultaneously *sells* the same contract at the same *price*. However, neither counterparty knows who the other counterparty is and both counterparties, when buying and selling respectively, enter a legal agreement with the exchange rather than with each other. The exchange is a financially secure organization, backed by its members. It is very unlikely to default because of this backing and because the agreements it enters are set up in such a way that it does not become exposed to defaults by any trade counterparties.

Futures contracts are defined relative to some other reference instrument or commodity, examples being interest rates and stock prices. In the interest rate market this reference instrument is always a standard liquid market instrument, the FRA in the case of a Eurodollar futures contract. The aim of the future is to give the counterparties *roughly* the same exposure to market moves as if they had done the reference trade, but without the associated credit risk. How the contract defined here simultaneously achieves these two objectives is discussed in Section 12.2.2 below.

Associated with any futures contract there is a *settlement date* T and a *futures price process*, $\{\Phi_s : 0 \leq s \leq T\}$. A counterparty who buys the futures contract at time t, when the futures price is Φ_t, agrees to pay to the exchange the net amount $\Phi_t - \Phi_T$ over the time interval $[t, T]$. The precise payment dates and payment amounts, some of which may be positive and some negative, are determined by the evolution of the futures price process over the interval $[t, T]$. We will say more on this process and on its relationship to the price of the reference instrument later, but for now it is sufficient to know that the futures price is always available on the exchange (during opening hours) and that the settlement price Φ_T is derived, according to simple rules, from the price of the reference instrument. For example, in the case of the Eurodollar futures contract the final futures price is given by

$$\Phi_T := 100\big(1 - L_T[T, S]\big)$$

where, as usual, $L[T, S]$ is the forward LIBOR for the period $[T, S]$ (which is always a three-month period for the Eurodollar future).

To complete the definition of the future we need to specify the payment rules which determine precisely how the (possibly negative) net amount $\Phi_t - \Phi_T$

is paid. When a counterparty buys a futures contract at time t (similar rules apply to a sold future) he places with the exchange an *initial margin*. (This amount is set with reference to the volatility of the price of the reference instrument. The details of this will not concern us.) Over time the futures price will change from its initial value Φ_t, sometimes going up, sometimes going down. At the end of every day the exchange checks the closing futures price. If the price has gone up during the course of the day the exchange credits to the counterparty's margin account the amount of that increase; if it has gone down the exchange debits the amount of the decrease. Whenever the balance exceeds the initial margin level the counterparty has the right to (and therefore should) withdraw any excess credit amount above the initial margin from the account. On the other hand, whenever the balance drops below the *maintenance margin* level, a fixed level which is less than the initial margin amount, he must bring the balance back up to the initial margin level again. At the expiry of the contract he may close the account and remove all remaining funds. Thus the net payment made by the counterparty over the interval $[t, T]$ is indeed precisely the amount $\Phi_t - \Phi_T$.

12.2.2 Market risk without credit risk

As explained above, futures contracts are designed to allow a counterparty to gain exposure to or protection from various market movements without having to take on the accompanying counterparty risk. Here we explain why the contract just described has indeed achieved each of these objectives.

First consider the issue of counterparty risk. For the exchange to suffer a loss due to a credit default one of its customers must fail to make a margin payment when requested to do so. Consider the case when the counterparty bought the future. In the event of default, as soon as it becomes clear, at time s, that a counterparty is not going to honour his obligations and make the margin payment, the exchange cancels the contract with the counterparty and sells the same futures contract to some other counterparty at the then futures price Φ_s. By entering this new contract the exchange ensures that it will receive from this new counterparty an amount $\Phi_s - \Phi_T$ over the period $[s, T]$. On the other hand, at that same moment the total amount that the exchange has received from the defaulting counterparty is $\Phi_t - \Phi_s$ plus the balance in the margin account. Therefore as long as this balance is not negative the exchange has no exposure to the defaulting counterparty. The levels of the initial and maintenance margins are set in such a way that the likelihood of this balance being negative, as the result of a large market move between the default on a margin call and the close out of the position, is small.

This discussion demonstrates that the exchange has only a very small credit exposure to its counterparties. But this in turn means these counterparties have only a small credit exposure to the exchange. Because the exchange is not running credit risk and does not itself take market risk, and because it has the financial backing of its members, there is very little chance that it would ever default. And if it did the exposure of a counterparty to the exchange is limited to an amount which is approximately the size of the initial margin.

The above discussion demonstrates the effectiveness of the futures contract in removing credit risk. What is less clear from the original contract definition is that the future also gives the desired exposure to market moves. Here we will briefly explain why this is the case. To do this we will consider the specific example of the Eurodollar future.

Recall that the final settlement price of the Eurodollar futures contract is given by $\Phi_T = 100(1 - L_T[T, S])$ for some given period $[T, S]$. Here we argue that the futures price at time t is *approximately* given by

$$\Phi_t \simeq 100(1 - L_t[T, S]).\qquad(12.1)$$

If this is the case then a counterparty buying the future at time t will pay approximately the net amount

$$\Phi_t - \Phi_T \simeq 100(L_t[T, S] - L_T[T, S])$$

to the exchange over the period $[t, T]$. To see why (12.1) is roughly what one would expect, consider also the FRA on which the future is based.

Suppose that the relationship (12.1) were exact. Had the counterparty, instead of buying the future, bought a notional amount $100/\alpha$ of the corresponding FRA (α being the accrual factor for the period $[T, S]$) then, at time S, he would have paid the net amount $100(L_t[T, S] - L_T[T, S])$. By assumption this is exactly the same amount as for the future. But which is better? If the forward LIBOR rises over the interval $[t, T]$ the counterparty will in each case be making a net positive payment. Under the FRA this payment is all made at time $S > T$ so this would generally be the better contract to have entered (better to pay later than earlier). On the other hand, if the forward LIBOR dropped over the interval $[t, T]$ the counterparty will in each case receive a net positive payment and so would like to receive it early. So on this occasion the future would have been better.

The preceeding paragraph is intended to give a heuristic argument to justify why the futures price of the Eurodollar future should be approximately given by (12.1) and thus why the future gives a market exposure similar to the FRA. If (12.1) were true then both contracts make the same payment, with the timing sometimes favouring the future, sometimes favouring the FRA. The exact relationship between the futures price and the forward LIBOR is a little more subtle and depends, as one might guess, on the relationship between the price or rate on which the contract is based and the discount curve over the time period $[t, T]$. For the Eurodollar future, for example, the purchaser of the future pays money to the exchange when the reference LIBOR goes up. But when LIBOR goes up discount factors generally go down, which leads to an increased advantage to the FRA under which the payment is later. Furthermore, when the reference LIBOR goes down the purchaser receives money from the exchange and discount factors generally go up, which reduces the disadvantage to the FRA under which the payment is received later. These two effects combine and lead to a futures price which is actually less than that given by (12.1). Quantifying this relationship is the primary topic of the remainder of this chapter.

12.2.3 Mathematical formulation

In this section we give a more mathematical description of a futures contract. Throughout we shall suppose we are working on some filtered probability space $(\Omega, \{\mathcal{F}_t\}, \mathcal{F}, \mathbb{P})$ where $\{\mathcal{F}_t\}$ is the augmented natural filtration for some d-dimensional Brownian motion.

A futures *contract* is specified by the following:

(i)   the settlement date T;
(ii)   the settlement amount H, an $\{\mathcal{F}_T\}$-measurable random variable;
(iii) a resettlement arrangement, which determines when payments to and from the counterparties are to take place.

For a particular futures *trade*, entered at time t say, it is also important to know the entry time since this is required to determine the amount of the first resettlement payment.

We will denote by $\{\Phi_t^H : 0 \leq t \leq T\}$ the futures price process corresponding to this contract. The contract resettlement times will be denoted by $0 < T_1 < \ldots < T_n = T$.

In common with other authors we will make a simplifying assumption at this stage. We shall suppose that both the initial and variation margins are zero. If we expect movements in the futures price process over the life of the contract to be large compared with the size of the margin this will not be an unreasonable assumption. Note also that a non-zero margin in an arbitrage-free world is not possible (without a modification to the contract definition) since both the buyer and the seller must place margin with the exchange.

With this simplifying assumption regarding the margin, the payment made at time T_i, $1 \leq i \leq n$, by a counterparty who buys a futures contract at time t will be for the amount

$$\Phi_{T_i \vee t}^H - \Phi_{T_{i-1} \vee t}^H .$$

Note that this payment will be zero if $T_i < t$ and that the first non-zero payment, at T_{i^*} say, will be for the amount

$$\Phi_{T_{i^*}}^H - \Phi_t^H .$$

12.3 CHARACTERIZING THE FUTURES PRICE PROCESS

Throughout the rest of this chapter we will assume the existence of the risk-neutral measure $\mathbb{Q}$ equivalent to $\mathbb{P}$ and a short-rate process r such that if A is any asset in the economy then ζA is an $(\{\mathcal{F}_t\}, \mathbb{Q})$ martingale where

$$\zeta_t = \exp\left(-\int_0^t r_u \, du\right).$$

Recall this assumption implies that the economy is arbitrage-free, and we shall also assume it to be complete. One consequence of market completeness is that the cash account, ζ^{-1}, is a numeraire. We choose to use the language of numeraires throughout the remainder of this chapter rather than that of pricing kernels.

Our primary interest is in futures contracts in the interest rate markets, but under the assumptions above the discussion applies equally well to other markets.

12.3.1 Discrete resettlement

The futures contract specification of Section 12.2.3 completely defines the futures price process Φ^H, as we now show.

Let $(N, \mathbb{N})$ denote a numeraire pair for the economy. From the theory of Chapter 7, the total value of these payments at time t is given by

$$V_t = N_t \mathbb{E}_{\mathbb{N}}\left[\sum_{j=1}^n (\Phi^H_{T_j \vee t} - \Phi^H_{T_{j-1} \vee t}) N_{T_j}^{-1} \Big| \mathcal{F}_t\right]. \tag{12.2}$$

The property that a futures contract is entered at zero cost corresponds to $V_t = 0$ for all $t \in [0, T]$. The futures price process can thus be recovered from (12.2) by backwards induction as follows. First recall that

$$\Phi^H_{T_n} = H.$$

Now fix $i < n$ and assume that we have found $\Phi^H_{T_k}$ for all $i < k \leq n$. Then, setting $t = T_i$ in (12.2) and rearranging yields

$$\Phi^H_{T_i} = \frac{\mathbb{E}_{\mathbb{N}}\left[\Phi^H_{T_{i+1}} N_{T_{i+1}}^{-1} | \mathcal{F}_{T_i}\right] + \mathbb{E}_{\mathbb{N}}\left[\sum_{j=i+2}^n (\Phi^H_{T_j} - \Phi^H_{T_{j-1}}) N_{T_j}^{-1} | \mathcal{F}_{T_i}\right]}{\mathbb{E}_{\mathbb{N}}[N_{T_{i+1}}^{-1} | \mathcal{F}_{T_i}]}.$$

This defines the futures price $\Phi^H_{T_i}$ for $i = 0, 1, \dots, n$. For general $t \in (T_i, T_{i+1})$ we can now use (12.2) to obtain

$$\Phi^H_t = \frac{\mathbb{E}_{\mathbb{N}}\left[\Phi^H_{T_{i+1}} N_{T_{i+1}}^{-1} | \mathcal{F}_t\right] + \mathbb{E}_{\mathbb{N}}\left[\sum_{j=i+2}^n (\Phi^H_{T_j} - \Phi^H_{T_{j-1}}) N_{T_j}^{-1} | \mathcal{F}_t\right]}{\mathbb{E}_{\mathbb{N}}[N_{T_{i+1}}^{-1} | \mathcal{F}_t]}.$$

In principle the above completely characterizes the process Φ. However, a closed-form solution can only be found in very special cases, for example under Gaussian assumptions. A mathematically cleaner and more enlightening result can be obtained by considering the limit as the inter-resettlement times are taken to zero. This we do in the next section.

12.3.2 Continuous resettlement

Typically for a futures contract the settlement time T is large compared to the inter-resettlement time interval. In US dollars, for example, Eurodollar futures contracts are traded with settlement times up to ten years. From a modelling viewpoint it is therefore reasonable to consider the limiting situation as the inter-resettlement time period decreases to zero, i.e. continuous resettlement.

We have just seen in Section 12.3.1 that the futures price process is characterized by equation (12.2). The summation inside the expectation in (12.2) can be interpreted as the gain, associated with the numeraire N, between times t and T for a contract entered at t. That is, it is the number of units of the numeraire N which would be held at time T if all cash receipts between t and T were immediately invested in the numeraire. For a given numeraire N and an arbitrary partition $\Delta[t,T] = \{t = T_0^\Delta, T_1^\Delta, \dots, T_{m^\Delta}^\Delta = T\}$ of the interval $[t,T]$, define the associated gain to be

$$G_t^{\Delta[t,T]}(\Phi^H, N) = \sum_{j=1}^{m^\Delta}(\Phi_{T_j^\Delta}^H - \Phi_{T_{j-1}^\Delta}^H)N_{T_j^\Delta}^{-1}. \tag{12.3}$$

Assuming we are given a futures price process Φ, the following result identifies the gain process in the limit as the partition spacing converges to zero.

Theorem 12.1 *Let Φ and N be continuous semimartingales on $(\Omega, \{\mathcal{F}_t\}, \mathcal{F}, \mathbb{P})$ with $N > 0$ a.s. and let $\Delta^m[t,T]$, $m > 0$, be a nested sequence of partitions of the interval $[t,T]$ with partition spacing converging to zero as $m \to \infty$. Then*

$$G_t^{\Delta[t,T]}(\Phi, N) \to \int_t^T N_u^{-1}d\Phi_u + \int_t^T d[N^{-1}, \Phi]_u \tag{12.4}$$

in probability as $m \to \infty$.

Proof: First note that $G_t^{\Delta[t,T]}(\Phi, N)$, as defined by (12.3), can be rewritten as

$$G_t^{\Delta[t,T]}(\Phi, N) = \sum_{j=1}^{m^\Delta} N_{T_{j-1}^\Delta}^{-1}(\Phi_{T_j^\Delta} - \Phi_{T_{j-1}^\Delta}) + \sum_{j=1}^{m^\Delta}(N_{T_j^\Delta}^{-1} - N_{T_{j-1}^\Delta}^{-1})(\Phi_{T_j^\Delta} - \Phi_{T_{j-1}^\Delta}). \tag{12.5}$$

The first sum on the right-hand side of (12.5) is a Riemann sum approximation to the first (stochastic) integral on the right-hand side of (12.4). It can be shown that this sum converges in probability to this stochastic integral. The reader is referred to Proposition IV.2.13 of Revuz and Yor (1991). A proof that the second sum on the right-hand side of (12.5) converges in probability to the second integral on the right of (12.4) can also be found in Revuz and Yor (1991, p. 115).   □

Remark 12.2: Had we imposed further restrictions on the partitions allowed in the above result, we could have established almost sure convergence of the sums to the integrals. For the type of restrictions required and the method of proof, see for example Theorem 4.15 and Definition 3.71.

Remark 12.3: The second integral in (12.4) is significant. It has arisen because for a futures contract payment is made at the end of the time interval, once the price increment is known. This contrasts with the case of a trading strategy where the portfolio must be selected at the start of the interval, before the price increment is known. This latter case, of course, led to the stochastic integral.

Theorem 12.1 shows us how we should define the gain for a given futures price process corresponding to a contract with continuous resettlement. First define the continuous process $G(\Phi^H, N)$ by

$$G_t(\Phi^H, N) := \int_0^t N_u^{-1} d\Phi_u^H + \int_0^t d[N^{-1}, \Phi^H]_u. \tag{12.6}$$

The gain over the interval $[t, T]$ from holding the futures contract can be seen to be precisely

$$G_T(\Phi^H, N) - G_t(\Phi^H, N) = \int_t^T N_u^{-1} d\Phi_u^H + \int_t^T d[N^{-1}, \Phi^H]_u.$$

Equation (12.2), the requirement that a futures contract can at any time t be entered at zero cost, now becomes

$$0 = V_t = N_t \mathbb{E}_{\mathbb{N}} \left[G_T(\Phi^H, N) - G_t(\Phi^H, N) | \mathcal{F}_t \right]. \tag{12.7}$$

This is precisely the statement that $G(\Phi^H, N)$ is a martingale under the measure $\mathbb{N}$. Equation (12.2), the analogue of (12.7), characterized the futures price process in the discrete case. This is no longer true under continuous resettlement. What we do know from (12.6) is that

$$\Phi_t^H = \Phi_0^H + \int_0^t N_u dG_u(\Phi^H, N) + \int_0^t d[N, G(\Phi^H, N)]_u. \tag{12.8}$$

We will use this representation in the next section.

12.4 RECOVERING THE FUTURES PRICE PROCESS

To recover the futures price process from (12.8) we will now specialize to taking the cash account ζ^{-1} as numeraire. The reason for working in the risk-neutral measure is that ζ is a finite variation process. As such the quadratic variation term in (12.8) is identically zero and so Φ is a local martingale under $\mathbb{Q}$. This leads us to make the following definition.

Definition 12.4 *We say that Φ^H is a futures price process with settlement value H at time T if the process $G(\Phi^H, N)$ defined by (12.6) is a $\mathbb{Q}$ martingale. Note, in particular, that Φ^H is then a local martingale.*

If, additionally, Φ^H is a true martingale, and thus

$$\Phi_t^H = \mathbb{E}_{\mathbb{Q}}[\Phi_T^H | \mathcal{F}_t] = \mathbb{E}_{\mathbb{Q}}[H | \mathcal{F}_t], \tag{12.9}$$

we say that Φ^H is a martingale futures price process.

For a general futures price process, even when working with ζ^{-1} as numeraire, it is still unclear how to establish the existence of and how to recover the process Φ^H from H and T, the problem being the need to solve for Φ^H and $G(\Phi^H, \zeta^{-1})$ simultaneously. It is also difficult to check, and indeed need not be true in general, that Φ^H is uniquely specified by (12.7) and (12.8).

If we restrict to martingale futures price processes the situation is clearer. Recovering Φ^H from the settlement value H is immediate from the martingale property (12.9). Given any particular H, to establish that (12.9) defines a futures price process it is still necessary to check that the resulting gain process $G(\Phi^H, \zeta^{-1})$ is also a martingale. This can be done on a case by case basis, but see Theorem 12.6 below. That, for a given H, there is at most one corresponding martingale futures price process is easily established.

Theorem 12.5 *There exists at most one martingale futures price process corresponding to any settlement value H and settlement date T.*

Proof: If Φ^H and $\hat{\Phi}^H$ are two such processes then $M := \Phi^H - \hat{\Phi}^H$ is a martingale with $M_T = \Phi_T^H - \hat{\Phi}_T^H = 0$. Hence M is identically zero and $\Phi^H \equiv \hat{\Phi}^H$. □

Note that the above result has not in general ruled out the possibility of another (local martingale) futures price process.

We conclude this section with a result concerning the existence of a (martingale) futures price process under special restrictions on the random variable H and the cash account ζ^{-1}.

Theorem 12.6 *Suppose that the random variable H satisfies $\mathbb{E}_{\mathbb{Q}}[H^2] < \infty$ and that ζ is bounded above on the time interval $[0, T]$ (which holds, for example, if interest rates are positive). Then the martingale Φ^H defined by*

$$\Phi_t^H := \mathbb{E}_{\mathbb{Q}}[H | \mathcal{F}_t]$$

is a futures price process corresponding to H.

Proof: Since $\mathbb{E}_{\mathbb{Q}}\big[(\Phi_T^H)^2\big] = \mathbb{E}_{\mathbb{Q}}[H^2] < \infty$ the martingale Φ^H is a square-integrable martingale over the interval $[0, T]$. The process ζ, being bounded both above and below (by zero) lies in $\mathcal{L}^2(\Phi)$. Hence $G_t(\Phi^H, \zeta^{-1}) := \int_0^t \zeta_u d\Phi_u^H$ also defines a (square-integrable) martingale, by Corollary 4.14, and Φ^H is a futures price process with corresponding gain $G(\Phi^H, \zeta^{-1})$. $\square$

12.5 RELATIONSHIP BETWEEN FORWARDS AND FUTURES

A *forward contract* is a European derivative under which one counterparty pays to the other an unknown amount H at time S and in return receives a known amount on the same date. This differs from a standard European derivative only in that the 'premium' payment is delayed until the time of the derivative payoff. For the forward there is no other payment, in particular when the trade is done, so the forward contract has zero value on the trade date.

For a given payoff H and payment date S, the fixed payment F_t^H which gives the forward contract zero value at time t is called the *forward price*. We have already met an example of a forward contract, the FRA. In this case the forward price is, of course, referred to the forward rate or forward LIBOR.

It is important to understand the relationship between the forward price for the payoff H at time S and the corresponding futures price. Both products are liquidly traded and their relative pricing is therefore important. In general, as is the case for the FRA and Eurodollar futures contract, the future may expire at some time $T \leq S$ and we allow this possibility in the discussion.

The forward price is easily determined. For a forward contract agreed at time t, the net payment at time S is for the amount $H - F_t^H$. This has zero value at time t and thus, valuing in the risk-neutral measure,

$$0 = \zeta_t^{-1} \mathbb{E}_{\mathbb{Q}}\big[(H - F_t^H)\zeta_S | \mathcal{F}_t\big],$$

which yields

$$F_t^H = \frac{\zeta_t^{-1}\mathbb{E}_{\mathbb{Q}}[H\zeta_S|\mathcal{F}_t]}{\zeta_t^{-1}\mathbb{E}_{\mathbb{Q}}[\zeta_S|\mathcal{F}_t]} = \frac{\zeta_t^{-1}\mathbb{E}_{\mathbb{Q}}[H\zeta_S|\mathcal{F}_t]}{D_{tS}}.$$

On the other hand, we have just seen in Section 12.4 that the corresponding martingale futures price is given by

$$\Phi_t^H = \mathbb{E}_{\mathbb{Q}}[H|\mathcal{F}_t].$$

The difference between the two, which is often referred to as the *futures correction*, is thus given by

$$\Phi_t^H - F_t^H = \mathbb{E}_{\mathbb{Q}}[H|\mathcal{F}_t] - \frac{\zeta_t^{-1}\mathbb{E}_{\mathbb{Q}}[H\zeta_S|\mathcal{F}_t]}{\zeta_t^{-1}\mathbb{E}_{\mathbb{Q}}[\zeta_S|\mathcal{F}_t]}$$

$$= \frac{\zeta_t^{-1}\mathbb{E}_{\mathbb{Q}}[\zeta_S|\mathcal{F}_t]\mathbb{E}_{\mathbb{Q}}[H|\mathcal{F}_t] - \zeta_t^{-1}\mathbb{E}_{\mathbb{Q}}[H\zeta_S|\mathcal{F}_t]}{\zeta_t^{-1}\mathbb{E}_{\mathbb{Q}}[\zeta_S|\mathcal{F}_t]}$$

$$= -\frac{\zeta_t^{-1}\mathrm{cov}_{\mathbb{Q}}(H, \zeta_S|\mathcal{F}_t)}{\zeta_t^{-1}\mathbb{E}_{\mathbb{Q}}[\zeta_S|\mathcal{F}_t]}$$

$$= -\zeta_t^{-1}\frac{\mathrm{cov}_{\mathbb{Q}}(H, \zeta_S|\mathcal{F}_t)}{D_{tS}}.$$

In the particular case when $t = 0$ this reduces to

$$\Phi_0^H - F_0^H = \frac{\mathrm{cov}_{\mathbb{Q}}(H, \zeta_S)}{D_{0S}}. \tag{12.10}$$

Thus the futures correction is the forward value to the date S of the covariance of the payoff amount H with ζ_S. If they are independent there is no correction. For the Eurodollar future H and ζ_S will be positively correlated and so the futures price will be lower than the corresponding forward price $100(1 - L_0[T, S])$.

The futures correction (12.10) can be calculated analytically in special cases. Of particular interest is when $H = L_T[T, S]$ and $H = (L_T[T, S] - K)_+$, for some $K > 0$, under the assumption of a Gaussian short-rate model such as the Vasicek–Hull–White model described in Chapter 17. These calculations are left as exercises for the reader.

Orientation: Pricing Exotic European Derivatives

The next four chapters are concerned with the pricing of exotic European derivatives. By an exotic option we mean one which is not *vanilla*, i.e. not included in the set of liquid products discussed in Chapters 10–12. The lack of liquidity means that a bank trading an exotic derivative would have to quote a price based on its own pricing models and not by direct reference to the prices quoted by other market participants for the same product.

What distinguishes a European derivative from other path-dependent and American-type products is the fact that it makes payments which can be determined by observing the market on a single date T. This fact means that to model an exotic European derivative it is, in principle, only necessary to formulate a model for the economy (of pure discount bonds) on the single date T. This is clearly a simplification of the general situation but there is a price to pay. Most products traded are European, and most new exotic European derivatives are generalizations of those already in existence and commonly traded. It is essential when formulating a model to price a new product to ensure that the price which results is reasonable when compared to the prices of other similar products, and it must, for example, be worth more than any other product which is guaranteed to pay out less in all circumstances. Thus, although the problem is theoretically easier, more is demanded of the model. In particular, it is essential that the pricing model is *calibrated* to other similar but vanilla instruments. Indeed, we shall see in Chapter 14 that there are even products whose prices are completely determined by the prices of other vanilla products and, given these vanilla prices, their prices are independent of any model chosen.

With the above comments in mind, the process of developing a pricing model for an exotic European derivative can be split into two steps. The first step is to apply as much high-level insight as possible about the market and its relationship to this particular derivative to decide on general essential and desirable features of the pricing model. In particular, one should ask what factors in the market are going to have the most significant effect on the price and to which instruments it is most closely related. This determines the complexity of the model required (one factor or more?) and which vanilla instruments the model should be calibrated to. We shall see several examples of this approach in the coming chapters. The alternative approach, of trying to use one general model for all exotic derivatives, is dangerous and will not always work. For example, models we introduce later when discussing American and path-dependent products, such as the market models and

Markov-functional models, are well suited to those more complex products but often give a poor reflection of those features of the market which are relevant to pricing Europeans.

The second step in developing a pricing model, having identified the necessary properties, is to choose the specific model to be used, determine the inputs to this model, calibrate the model and finally apply the model to price the exotic derivative in question.

As an example of the above procedure, we will here briefly discuss the *constant maturity swap* (CMS). The CMS is described fully in Chapter 14 and a suitable pricing model is developed there. Like a vanilla swap, the CMS comprises a number of payments which can be priced independently so we shall discuss only one. Let $y[T, S]$ be a given forward swap rate which sets at time T. Then a CMS payment based on the index $y[T, S]$ is for the amount $y_T[T, S]$ and is made at some fixed time $M \geq T$. This is clearly a European product, its value at T being determined by D_{TM} and $y_T[T, S]$,

$$V_T = D_{TM} y_T[T, S].$$

Thus, in particular, it is only necessary to model the economy at this terminal time T and, by the usual valuation formula (presented in Corollary 7.34), the product's value at time zero is given by

$$V_0 = N_0 \mathbb{E}_{\mathbb{N}} \left[y_T[T, S] \frac{D_{TM}}{N_T} \right] \tag{$*$}$$

for some numeraire pair $(N, \mathbb{N})$.

We need now to select a numeraire N and then formulate a model to describe the law of $y_T[T, S]\frac{D_{TM}}{N_T}$ in the martingale measure $\mathbb{N}$ corresponding to this numeraire. The payout amount is determined by the swap rate $y_T[T, S]$ so the distribution of this rate will clearly have a major impact on the price of the CMS. Furthermore, there is a set of vanilla instruments whose payout also depends on this rate, namely the set of swaptions on the associated swap. These are obvious and natural calibrating instruments. It is then a natural choice to take $N = P[T, S]$ as numeraire, the PVBP for the underlying swap, and the valuation formula (*) becomes

$$V_0 = P_0 \mathbb{E}_{\mathbb{S}} \left[y_T[T, S] \frac{D_{TM}}{P_T} \right]. \tag{$**$}$$

It is in fact the case, and this follows from applying Remark 11.3 in the context of swaptions, that the prices for all strikes of the calibrating vanilla swaptions determine the law of $y_T[T, S]$ in the measure $\mathbb{S}$. But the valuation formula (**) also includes the term D_{TM}/P_T so this must also be modelled. Thus we need to develop a realistic model for the discount curve $\{D_{TS} : S \geq T\}$ which correctly values swaptions on the swap corresponding to the rate $y[T, S]$. We also need to ensure that the model is arbitrage-free, otherwise it is of little use for derivative pricing. Just such a model is developed in Chapters 13 and 14.

The next four chapters present ways to develop models for pricing European exotics. Chapter 13 provides a general framework, the class of *terminal swap-rate models*, and Chapter 14 applies these ideas to *convexity-related products* such as the CMS. The ideas of Chapter 13 are further extended in Chapter 15, and Chapter 16 shows how the ideas carry over to the multi-currency setting.

13

Terminal Swap-Rate Models

13.1 INTRODUCTION

The purpose of this chapter is to present a direct approach to modelling and pricing European interest rate derivatives. To do this we focus attention on the derivative product to be priced and develop models which explicitly capture precisely those properties of the market which are relevant to the product at hand. In particular, we insist that a model is calibrated to any existing market products which are related to the exotic derivative in question. The advantage of this approach over other techniques (for example, using a standard short-rate model such as one of those discussed later in Chapter 17) is that it is guaranteed to price the new product *accurately relative to existing products*; the model will, by construction, have realistic properties (for example, positive interest rates) and will be built upon a theoretically sound base. Furthermore, and at least as important, the characteristics of the model will usually be highly transparent and thus we will be able to understand the model's strengths and weaknesses as a tool for pricing any specific product.

13.2 TERMINAL TIME MODELLING

13.2.1 Model requirements

Suppose we wish to value a European product which we know at some future date T has value V_T, where V_T is some arbitrary function of the discount factors on date T. Typically (always in practice) the payoff function V_T will depend on only a finite number of discount factors, $V_T = V_T(D_{TM_1}, D_{TM_2}, \ldots, D_{TM_k})$. In particular, this payoff amount is independent of any modelling assumptions we might choose to make about the evolution of market rates.

The valuation of a European derivative with a payoff such as this was discussed in detail in Chapter 7. Given suitable assumptions about the model of the economy (that it be complete and arbitrage-free), the time-zero value

---

Financial Derivatives in Theory and Practice Revised Edition.   P. J. Hunt and J. E. Kennedy
© 2004 John Wiley & Sons, Ltd   ISBNs: 0-470-86358-7 (HB); 0-470-86359-5 (PB)

of the derivative is given by

$$V_0 = N_0 \mathbb{E}_{\mathbb{N}}[V_T N_T^{-1}], \qquad (13.1)$$

where $(N, \mathbb{N})$ is some numeraire pair for the economy. To evaluate (13.1) we need a model under the measure $\mathbb{N}$ for the pure discount bond prices D_{TM_i}, $i = 1, 2, \ldots, k$, and for the numeraire price N_T. We do not need to know any more about the distributions of these assets at earlier times since these other distributions are not relevant to the price of the exotic European product. We suppose that there is some (low-dimensional) random variable y_T such that the numeraire N_T and, for each $S \geq T$, the pure discount bond price D_{TS} can be written as a function of y_T:

$$D_{TS} = D_{TS}(y_T),$$
$$N_T = N_T(y_T).$$

If this is the case, then the derivative payoff can now also be written as $V_T(y_T)$ and V_0 can be evaluated from the distribution of y_T and the functional forms $V_T(y_T)$ and $N_T(y_T)$.

A model such as this, in which the economy can be modelled as a function of a low-dimensional random variable, is a special case of a *Markov-functional* model. These models are discussed more fully in Chapter 19.

Thus far the discussion has been very general. We have as yet discussed neither how to choose an appropriate random variable y_T, nor the choice of an appropriate functional form for $D_{TS}(y_T)$. These choices complete the specification of the model and will depend on the 'exotic' derivative that we are valuing. The first decision to be made is the choice of y_T, and it is important to ensure that this choice (including the dimension of y_T) is appropriate for the exotic product in question. Usually there will be some standard market products which are closely related to the exotic derivative, and it is essential that the model is 'calibrated' to these products. To achieve this we often choose y_T to be a natural characterizing variable associated with the vanilla products, usually some par swap rate. Given the choice of y_T there will usually be a natural choice of numeraire N such that under the corresponding measure $\mathbb{N}$ the distribution of y_T is easily inferred from the market prices of the vanilla products.

All that then remains to evaluate (13.1) is to specify the functional form $D_{TS}(y_T)$ and hence $V_T(y_T)$. This choice is important and different choices give different valuations. There are three key properties which are desirable for the resulting model:

(P1) (martingale property) $D_{.S} N^{-1}$ must be a martingale under the measure $\mathbb{N}$, otherwise the model will admit arbitrage;

(P2) (consistency property) any relevant relationships which hold in the market must also hold for the model;

(P3) (realism property) the model must have realistic properties.

(P1) is essential if the model is to be suitable for pricing derivative products since it ensures that the model does not admit arbitrage and is required for equation (13.1) to be valid. In many applications, such as calculating convexity corrections (discussed in the next chapter), this property is often not identified or acknowledged in practice. From a practical point of view it ensures, for example, that if one forward rate is today significantly higher than another then this will also be the case at maturity T. (P2) is required to ensure that any model we propose is internally consistent. An example of (P2), which we shall see shortly, arises when a model is parameterized in terms of a market interest rate, such as a swap rate; we have specified the discount factors in terms of the swap rate, which in turn is a function of discount factors – the two must agree. The need for (P3) is obvious.

If we can find a functional form obeying (P1)–(P3) then we have constructed a model which is well calibrated, realistic and arbitrage-free, and which can be used to value European derivatives. In practice the valuation will often be carried out by performing a numerical integration.

13.2.2 Terminal swap-rate models

In most situations the vanilla products most closely related to the derivative to be priced will be swaptions (or caplets, which are just single-period swaptions). This is, for example, the case for constant maturity swaps (discussed more fully in Chapter 14) and for zero coupon swaptions (discussed in Section 13.5). We will assume that this holds throughout this section and show how the ideas above can be applied.

Let $V_t^n(K)$ be the time-t value of the calibrating (payers) swaption with strike K, where n denotes the total number of payments in the underlying swap. Let $T = S_0$ be the swap start date, $S = (S_1, S_2, \ldots, S_n)$ be the payment dates in the swap, and define $\alpha_j = S_j - S_{j-1}, j = 1, 2, \ldots, n$, to be the associated accrual factors. As in Chapter 10, the forward par swap rate for this swap, $y_t[T, S]$, is given by

$$y_t[T, S] = \frac{D_{tT} - D_{tS_n}}{P_t[T, S]},$$

where

$$P_t[T, S] = \sum_{j=1}^{n} \alpha_j D_{tS_j}.$$

We will take the forward par swap rate $y_T[T, S]$ as the parameterizing random variable y_T and construct a model which is calibrated to the underlying vanilla swaption prices for all strikes.

We work in the measure $\mathbb{S}$ corresponding to the numeraire process P. The first thing we do is derive the law of the random variable y_T in this measure.

Recall from Chapter 11 that the value of a payers swaption with strike K is given by

$$V_0^n(K) = P_0 \mathbb{E}_{\mathbb{S}}[(y_T - K)_+] . \qquad (13.2)$$

Differentiating both sides of (13.2) with respect to the strike parameter K yields

$$\frac{\partial V_0^n(K)}{\partial K} = P_0 \mathbb{E}_{\mathbb{S}}[-\mathbb{1}_{\{y_T > K\}}] = -P_0 \mathbb{S}(y_T > K) ,$$

and thus the swaption prices have allowed us to recover the *implied* distribution for y_T under the measure $\mathbb{S}$. In the case when all the swaption prices are given by Black's formula this corresponds to y_T having a log-normal distribution, but the approach is more general and also incorporates *smile effects* (i.e. non-log-normal distributions).

All that it now remains to do is to specify the functional form $D_{TS}(y_T)$ and hence $V_T(y_T)$. Recall the three key properties identified earlier, which in this case become the following:

(P1) (martingale property) $D_{.S}P^{-1}$ must be a martingale under the measure $\mathbb{S}$;
(P2) (consistency property) $y_T = P_T^{-1}(D_{TT} - D_{TS_n})$;
(P3) (realism – monotonicity and limit properties) for all $S > T$, $D_{TS}(0) = 1$, $D_{TS}(\infty) = 0$ and $D_{TS}(y_T)$ should be decreasing in both y_T and in S.

If we can find a functional form obeying (P1)–(P3) then we have constructed a model which is well calibrated, realistic and arbitrage-free. There are many ways to do this, and we shall discuss three in the next section.

13.3 EXAMPLE TERMINAL SWAP-RATE MODELS

We shall now study three specific models in more detail: the exponential, the geometric and the linear swap-rate models. In what follows $S_1, \ldots, S_n$ denote the cashflow times for the calibrating swap.

13.3.1 The exponential swap-rate model

In practice, discount curves always exhibit an approximately exponential shape and so this is a natural functional form to consider. We will model discount factors as

$$D_{TS} = \exp(-C_S z_T)$$

for some random variable z_T and some C_S, $S \geq T$. The random variable z_T reflects the overall level of rates at T, whereas C_S gives the dependence on the maturity S. Note that, were the discount curve exactly exponential at T then we would have $C_S = S - T$. This relationship will approximately hold in our model, but the exact values C_S will be chosen to satisfy the martingale property (P1).

It remains to specify the distribution of z_T under $\mathbb{S}$. To do this note that, given the function C_S, z_T is defined implicitly from y_T via

$$y_T = \frac{D_{TT}(z_T) - D_{TS_n}(z_T)}{P_T(z_T)}.$$

This relationship yields both the distribution of z_T (from that of y_T) and also the (implicit) functional form $D_{TS}(y_T)$. Furthermore, the consistency property (P2) is automatically satisfied.

A simple algorithm can be used to solve for the values C_S needed for any given derivative. Suppose for now that $C_{S_1}, \ldots, C_{S_n}$ are known. Then the functional form for $P_T(z_T)$ is also known and, for the model to be consistent with the initial discount curve and the martingale property of P-rebased asset prices, any other C_S is the unique solution to the equation

$$\frac{D_{0S}}{P_0} = \mathbb{E}_{\mathbb{S}}\left[\frac{\exp(-C_S z_T)}{P_T(z_T)}\right], \qquad (13.3)$$

which can be found efficiently using any standard algorithm such as Brent's algorithm (see, for example, Press et al. (1988)). This has assumed knowledge of the C_{S_i} which we do not have. Indeed, we do not yet even know whether or not C_{S_i} exist which will generate an arbitrage-free model. That they do we establish in Section 13.4, where we also provide an algorithm to find the C_{S_i} that is guaranteed to converge.

13.3.2 The geometric swap-rate model

It is common in practice to model discount factors as geometrically decaying, based on some market rate. The corresponding model in the terminal swap-rate framework is

$$D_{TS_i} = \prod_{j=1}^{i}(1 + \hat{\alpha}_j z_T)^{-1}, \qquad i = 1, 2, \ldots, n,$$

$$D_{TS} = (1 + \hat{\alpha}_S z_T)^{-1} D_{TS_i}, \qquad S_i \leq S < S_{i+1},$$

$$y_T = \frac{D_{TT}(z_T) - D_{TS_n}(z_T)}{P_T(z_T)},$$

for some $\hat{\alpha}_i, \hat{\alpha}_S$. This model clearly satisfies (P2) and (P3), and (P1) can be made to hold by an appropriate choice for $\hat{\alpha}_i, \hat{\alpha}_S$, the fitting of which is done in a manner similar to that for the exponential model, solving first for the $\hat{\alpha}_i$ and then any other $\hat{\alpha}_S$ that are needed.

Note that under this model each forward LIBOR at T which corresponds to a payment period of the vanilla swap, i.e. $L_T[S_{i-1}, S_i]$ for $i = 1, \ldots, n$, is linear in z_T. The model could easily be adapted to make this hold for all forward LIBOR with the same accrual period $\Delta := S_i - S_{i-1}$. That is, the model can be adapted so that, for all $T \leq \hat{T} = \hat{S} - \Delta$, $L_T[\hat{T}, \hat{S}]$ is linear in z_T. Note also that y_T and z_T are closely related – if the initial curve is such that $\hat{\alpha}_i = \alpha_i$, then $y_T = z_T$.

13.3.3 The linear swap-rate model

Our third model is analytically the most tractable of the three and this can lead to insight for specific products (see Chapter 14 for a discussion of this point in the case of convexity-related products). Suppose, as in the examples above, that we have found a suitable (smooth) function $D_{TS}(y_T)$. Letting $\hat{D}_{TS} = D_{TS} P_T^{-1}$ and expanding $\hat{D}_{TS}(y_T)$ about y_0, we have

$$\hat{D}_{TS}(y_T) = \sum_{i=0}^{m} \frac{(y_T - y_0)^i}{i!} \hat{D}_{TS}^{(i)}(y_0) + \frac{(y_T - y_0)^{m+1}}{(m+1)!} \hat{D}_{TS}^{(m+1)}(\theta_T) \qquad (13.4)$$

for some $\theta_T \in [y_0, y_T]$, where $D_{TS}^{(i)}(y_T)$ is the ith derivative in y_T of the function $D_{TS}(y_T)$. We could truncate this at m terms and take this to be our model

$$\hat{D}_{TS}(y_T) = \sum_{i=0}^{m} a_i y_T^i.$$

The parameters a_i must now be chosen/adjusted from (13.4) to give the necessary martingale property for the numeraire-rebased discount factors. Without this adjustment the model will no longer correctly value zero coupon bonds, and the ability to do this is a highly desirable feature of any interest rate model.

For many applications a first-order model ($m = 1$) is adequate, and this is the one we develop further. (Most treatments of convexity are based on a first-order Taylor expansion but without parameter adjustments to get the necessary martingales and thus bond prices.) This yields a model of the form

$$\hat{D}_{TS}(y_T) = A + B_S y_T, \qquad (13.5)$$

for suitable A, B_S. We call this the *linear swap-rate model*. Note that since y is a martingale in the swaption measure $\mathbb{S}$ it follows from (13.5) and the martingale property of $\hat{D}_{.S}$ that

$$\hat{D}_{tS}(y_t) = A + B_S y_t,$$

for all $t \leq T$. Furthermore, again in the swaption measure $\mathbb{S}$, the discount factors $D_{TS}(y_T)$ have a rational log-normal distribution, *but they may become negative*. We discuss this latter rather undesirable feature further in the next section.

Parameter fitting for this model is straightforward. Substituting (13.5) into the definition for P_T yields

$$\sum_{i=1}^{n} \alpha_i \hat{D}_{TS_i} = \left(\sum_{i=1}^{n} \alpha_i\right) A + \left(\sum_{i=1}^{n} \alpha_i B_{S_i}\right) y_T = 1,$$

and thus, since this must hold for all values of the random variable y_T,

$$A = \left(\sum_{i=1}^{n} \alpha_i\right)^{-1}.$$

Now we apply the martingale property to find B_S:

$$\hat{D}_{0S} = \mathbb{E}_{\mathbb{S}}[\hat{D}_{TS}] = A + B_S y_0$$

and so

$$B_S = \left(\frac{\hat{D}_{0S} - A}{y_0}\right).$$

13.4 ARBITRAGE-FREE PROPERTY OF TERMINAL SWAP-RATE MODELS

The three terminal swap-rate models developed in Section 13.3 have been formulated to be arbitrage-free by ensuring that all PVBP-rebased assets are martingales in the swaption measure $\mathbb{S}$. We saw in Theorem 7.32 that this guarantees that the model is arbitrage-free. To complete the argument, however, there are two final points to check:

(i) In the case of the exponential and geometric models we have not yet established that there exist parameter values C_{S_i} and $\hat{\alpha}_i$, respectively, such that the martingale property holds for the (PVBP-rebased) bonds in the underlying calibrating swap, $D_{TS_0}, \ldots, D_{TS_n}$.

(ii) We have not established the existence of a full term structure economy $\{D_{tS} : 0 \leq t < S < \infty\}$ which is arbitrage-free and consistent with the model. That is, we have not extended the economy to exist over the time interval $[0, \infty)$.

This second step we do not need to carry out explicitly for applications, but we should check that it can be done to ensure that the model is theoretically valid.

13.4.1 Existence of calibrating parameters

To address the first of these questions, we present an algorithm which can be used to find the parameters $C_{S_i}, i = 1, \ldots, n$, in the exponential swap-rate model of Section 13.3.1. A similar algorithm can also be employed for the geometric model to find the $\hat{\alpha}_i$. The algorithm is iterative and guaranteed to converge. At each iteration the algorithm gets strictly closer to the solution. Note that the argument below also guarantees the existence of a solution to equation (13.3), i.e. the model is consistent with the initial term structure of interest rates and is arbitrage-free.

Let C_i^k denote the value of C_{S_i} at the kth iteration of the algorithm. For each k, take $C_0^k \equiv 0$ and, without loss of generality, $C_n^k \equiv 1$. Define

$$x_i = \frac{D_{0S_i}}{P_0}$$

and

$$E_i^k = \mathbb{E}_{\mathbb{S}}\left[\frac{\exp(-C_i^k z_T^k)}{P_T(z_T^k)}\right],$$

where

$$P_T(z_T^k) = \sum_{j=1}^n \alpha_j \exp(-C_j^k z_T^k)$$

and z_T^k solves

$$y_T = \frac{1 - \exp(-z_T^k)}{\sum_{j=1}^n \alpha_j \exp(-C_j^k z_T^k)}. \tag{13.6}$$

Note that for $C_j^k \geq 0$, $j = 1, \ldots, n$, (13.6) has a solution $z_T^k \geq 0$ for all $y_T \geq 0$. The x_i are the target values and our aim is to construct a sequence $C^k = (C_0^k, \ldots, C_n^k)$ such that $E_i^k \to x_i$, $i = 0, 1, \ldots, n$. The limit value of the C^k sequence is the solution to (13.3) that we require (for each i).

The algorithm which we present below has the following properties:

(i) $0 \leq C_i^{k+1} \leq C_i^k$, $i = 2, 3, \ldots, n-1$, for all k;
(ii) $E_i^k \leq x_i$, $i = 2, 3, \ldots, n-1$, for all k, with equality only if $C_i^k = 0$;
(iii) $E_0^k - E_n^k = x_0 - x_n = y_0$, for all k;
(iv) $\sum_{j=1}^n \alpha_j E_j^k = \sum_{j=1}^n \alpha_j x_j = 1$, for all k.

It follows from (i) that the algorithm converges to yield some limit values C and E. In the limit $E_i \leq x_i$ by (ii), but this will be an equality with $C_i > 0$ for all $i > 0$ (which is what we require), as we now show. For suppose $E_i < x_i$ for some i. Then property (ii) implies $C_i = 0$, in which case $E_i = E_0$. Further, by (iv), if (ii) holds and $E_i < x_i$ then $E_n > x_n$. Substituting these relationships into (iii) yields

$$x_0 = E_0 - E_n + x_n < E_0 = E_i < x_i,$$

which contradicts the property that discount factors are non-increasing in maturity.

Algorithm

Step 0: Define $C^0 = (0, A, \ldots, A, 1)$ for some A sufficiently large that property (ii) holds. That such A exists is clear (consider $A \to \infty$). At this first step, and all subsequent steps, property (iii) follows from (13.6) and the martingale property of y under $\mathbb{S}$. Property (iv) is immediate.

General step: Given C^k, define z_T^k via (13.6) and let C_i^{k+1}, $i = 2, 3, \ldots, n-1$, solve

$$x_i = \mathbb{E}_{\mathbb{S}} \left[\frac{\exp(-C_i^{k+1} z_T^k) y_T}{1 - \exp(-z_T^k)} \right]. \tag{13.7}$$

Note that if C_i^{k+1} is replaced by C_i^k in (13.7) then the right-hand side of (13.7) is just E_i^k, which is no greater than x_i by property (ii). Hence $C_i^{k+1} \le C_i^k$, $i = 2, 3, \ldots, n - 1$. To ensure that our argument as given is complete, we must at this point add the step that if $C^{k+1} < 0$ then set it to zero. Finally, note that $C^{k+1} \le C^k$ implies from (13.6) that $z_T^{k+1} \ge z_T^k$, for each y_T, and thus, from (13.7), $E_i^{k+1} \le x_i$.

Our earlier argument shows that $C_i^{k+1} > 0$, $i = 2, 3, \ldots, n-1$, and so there will never be an occasion on which we set any component of C^{k+1} equal to zero. Further, it is now clear that the inequalities in (i) and (ii) will always be strict.

13.4.2 Extension of model to $[0, \infty)$

As stated earlier, extending terminal swap-rate models to the time domain $[0, \infty)$ is not something we want to do in practice – the whole objective behind the approach is to model European derivatives for which the only relevant times are 0 and T. However, we would like to know whether or not an extension is possible and thus whether the model is consistent with a full arbitrage-free term structure model.

We carry out such an extension for the exponential model. That this approach applies equally well to any model for which the discount factors at time T are positive is easy to see. The extension is completed in three steps:

(1) Define a swap-rate *process* $\{y_t : 0 \le t \le T\}$ consistent with the distribution of the random variable y_T.
(2) Define the economy over $[0, T]$, $\{D_{tS} : 0 \le t \le T, S\}$.
(3) Extend the definition to also define the economy over $[T, \infty)$, $\{D_{tS} : T \le t \le S\}$.

Step 1: Consider a probability space $(\Omega, \{\mathcal{F}_t\}, \mathcal{F}, \mathbb{S})$ supporting a Brownian motion W. Let F be the probability distribution function for the random variable y_T (under the swaption measure $\mathbb{S}$),

$$F(x) = \mathbb{S}(y_T \le x),$$

and let Φ be the standard cumulative normal distribution. Note that F here can be general and is not restricted to the log-normal distribution. Now define the stochastic process $\{y_t : 0 \le t \le T\}$ by

$$y_t := \mathbb{E}_\mathbb{S}\left[F^{-1}\Phi(W_T/\sqrt{T})|\mathcal{F}_t\right]. \tag{13.8}$$

It is immediate from the tower property and equation (13.8) that y is an $(\{\mathcal{F}_t\}, \mathbb{S})$ martingale on $[0, T]$. Furthermore,

$$\mathbb{S}(y_T \le x) = \mathbb{S}\left(F^{-1}\Phi(W_T/\sqrt{T}) \le x\right)$$
$$= \mathbb{S}\left(W_T/\sqrt{T} \le \Phi^{-1}F(x)\right)$$
$$= F(x),$$

so y_T has the required distribution.

Step 2: We are given the functional form $D_{TS} = \exp(-C_S z_T)$ for $S \ge T$. We will use this, combined with the martingale property of y and all PVBP-rebased discount factors, to construct the model over $[0, T]$. However, the functional form of D_{TS} for $S \ge T$ tells us nothing about discount factors maturing before T. To model these discount factors and thus to be able to completely specify the model on $[0, T]$ we first introduce a set of 'pseudo-discount factors' $D_{TS} = \exp(-\hat{C}_S z_T)$ for $0 \le S \le T$. We have yet to fix the parameters $\hat{C}_S$. Clearly these are not real discount factors since all bonds with maturities $S \le T$ have expired by T, but, if one so wished, one could think of each D_{tS}, $t > S$, as a new asset which is bought with the proceeds of the bond maturing at S. We tie down the parameters $\hat{C}_S$ from the required martingale property of the PVBP-rebased pseudo-assets,

$$\frac{D_{0S}}{P_0} = \mathbb{E}_\mathbb{S}\left[\frac{\exp(-\hat{C}_S z_T)}{P_T(z_T)}\right],$$

and the complete model can now be recovered from the martingale property of the PVBP-rebased assets,

$$D_{tS} = \frac{D_{tS}/P_t}{D_{tt}/P_t} = \frac{\mathbb{E}_\mathbb{S}[D_{TS}P_T^{-1}|\mathcal{F}_t]}{\mathbb{E}_\mathbb{S}[D_{Tt}P_T^{-1}|\mathcal{F}_t]}.$$

Step 3: For the exponential and geometric models all discount factors at time T are positive. Extend the economy beyond T in a deterministic manner as follows: for $S \ge t \ge T$,

$$D_{tS} = \frac{D_{Tt}(z_T)}{D_{TS}(z_T)}.$$

That the resulting model is arbitrage-free follows from the fact that it is a numeraire model, as defined in Chapter 8; here we are taking the numeraire to be the PVBP until T and the unit-rolling numeraire from then on. The corresponding martingale measure is the swaption measure $\mathbb{S}$.

Remark 13.1: The derivative pricing theory presented in Chapter 7 required all numeraire-rebased assets to be martingales with respect to the *asset filtration* rather than with respect to the original filtration for the probability space. It is not difficult to see that for the models and examples presented in this chapter we can work with either filtration.

13.4.3 Arbitrage and the linear swap-rate model

One would like to apply the techniques of Section 13.4.2 to show that the linear swap-rate model can also be embedded in a full arbitrage-free term structure model. This cannot be done, the reason being the failure of step 3 above. In the linear swap-rate model discount factors at time T may, in general, be negative and it is not possible to have an arbitrage-free model with discount factors which are negative at time T but guaranteed to be positive, indeed unity, at a known future time.

This fact is not as inhibiting as it may at first appear. Recall the initial motivation for the linear swap-rate model as a first-order approximation to a more realistic and arbitrage-free model such as the exponential model – a model which can be extended. Also note that the martingale property *up until time T* ensures that there is no arbitrage up to this time, and for a European option it is only over the time interval $[0, T]$ that we need to consider the economy.

13.5 ZERO COUPON SWAPTIONS

Vanilla swaptions, as defined in Chapter 11, are widely traded in the interest rate markets. A related but less common product is the *zero coupon swaption*. In the case of a vanilla swaption the holder has the right to enter, at time T, a swap in which the fixed rate is some predetermined value K. The fixed leg then consists of a series of payments made on dates $S_1, S_2, \ldots, S_n$, payment j being of amount $\alpha_j K$. A zero coupon swaption gives the holder the right to enter a zero coupon swap at time T. A zero coupon swap is also an agreement between two counterparties to exchange cashflows, but in this case there is only one fixed payment which is made at the swap maturity time S_n. The amount of this fixed payment is given by $(1 + K)^{(S_n - S_0)} - 1$ where K is the 'zero coupon rate' of the swap. The floating leg is exactly the same as for a vanilla swap.

Here we present the results of pricing a zero coupon swaption using each of the exponential, geometric and linear swap-rate models, defined above, under various market scenarios. Three trades were considered, each being options to enter into payers zero coupon swaps: a one-year option into a nine-year swap; a five-year option into a five-year swap; and a seven-year option into a three-year swap. These trades were each valued in three different interest rate environments: a steeply upward-sloping forward curve; a flat forward curve;

and a downward-sloping forward curve. In each case the model was calibrated
to vanilla swaptions with matching option expiry and swap length. Vanilla
swaption volatilities of 5%, 12% and 20% were used, and the zero coupon
swaptions were valued for strikes ranging from 180 basis points (bp) out of
the money to 180 bp in the money. Tables 13.1–13.3 summarize these results,
showing the maximum price differences that were found between the different
models.

Table 13.1   Pricing differences for a one-year into nine-year trade

	Shape of forward curve		
	Upward-sloping	Flat	Downward-sloping
Geometric vs Exponential	7.61 bp at 20%	2.97 bp at 20%	3.89 bp at 20%
Geometric vs Linear	130.0 bp at 20%	6.34 bp at 20%	6.21 bp at 20%

Table 13.2   Pricing differences for a five-year into five-year trade

	Shape of forward curve		
	Upward-sloping	Flat	Downward-sloping
Geometric vs Exponential	1.03 bp at 20%	0.48 bp at 20%	0.29 bp at 20%
Geometric vs Linear	37.62 bp at 20%	2.90 bp at 20%	1.05 bp at 20%

Table 13.3   Pricing differences for a seven-year into three-year trade

	Shape of forward curve		
	Upward-sloping	Flat	Downward-sloping
Geometric vs Exponential	1.60 bp at 20%	0.59 bp at 20%	0.24 bp at 20%
Geometric vs Linear	8.20 bp at 20%	1.34 bp at 20%	0.42 bp at 20%

The exponential and geometric models compared well in most situations.
The greatest deviation of 7.61 bp was observed when pricing a one-year into
nine-year trade deep in the money, at high volatilities and in a steeply upward-
sloping curve. The price of the option in this case was 405.51 bp under the
exponential model, so this represents a 1.9% relative difference in value. In
general the differences were less than 1 bp.

The linear swap-rate model generally compares reasonably well with the geometric and exponential models, except in high-volatility and steep-curve environments. It consistently overpriced the options we considered when compared with the other two models, the divergence being most noticeable when pricing the one-year into nine-year trade, as one may expect, since in this example the zero coupon swaption is least similar to the calibrating vanilla swaption. The relative inaccuracy of the linear model is to be expected given that it is intended only as a first-order approximation to a more realistic model, such as the exponential or geometric.

In Figure 13.1 we examine the one-year into nine-year swaption in more detail. The figure shows the value of the zero coupon swaption for a range of strikes as calculated by each of the three models. The volatility of the calibrating swaption was 20% and the forward curve was upward-sloping, the situation in which the models differ most. To give the results some perspective the figure also shows the intrinsic value of the zero coupon swaption and, for each strike level K, the value of a vanilla swaption which has a strike $\hat{K}$ chosen such that the underlying vanilla and zero coupon swaps have the same value. Note how consistently the geometric and exponential perform relative to each other and the time value of the option which they are designed to capture.

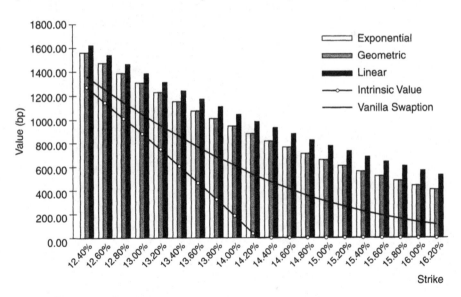

Figure 13.1 One-year into nine-year zero coupon swaption valuation

14

Convexity Corrections

14.1 INTRODUCTION

In this chapter we study the problem of how to correctly price products, relative to each other, which have in common the amount of payment made and differ only through the date of payment. As we shall see, by careful choice of model and numeraire for pricing we obtain a decomposition of the valuation formulae which has a natural and informative interpretation.

Two products of considerable importance in the interest rate derivatives market which fall into this category are the *constant maturity swap* (CMS) and the *LIBOR-in-arrears basis swap*. Each of these products shares the common property that there is some liquid product to which it is very closely related, and consequently each must be priced correctly relative to its more liquid counterpart.

The pricing of constant maturity and LIBOR-in-arrears swaps is an example of a more general pricing problem which is the subject of this chapter and which we now describe. We begin with a definition.

Definition 14.1 *Two derivative products are said to be convexity-related if there exists some time T such that the value at T of product i, $i = 1, 2$, is of the form*

$$V_T^{(i)} = \sum_{j=1}^{n_i} D_{TS_j^{(i)}} c_j^{(i)} F(D_{TS}, S \geq T).$$ (14.1)

Here $c_j^{(i)}$ are a set of known constants, $S_j^{(i)} \geq T$ are known times and F is some function of the discount curve at time T.

Each of a pair of convexity-related products promises a set of cashflows made at the times $S_j^{(i)}$. Modulo the constants $c_j^{(i)}$, the actual payments made for each product are for the same amount, hence the close relationship between them. However, where they differ is through the dates of the payments.

The result of this mismatch of payment dates is convexity. The value of both derivatives will change as the underlying index (curve) moves. However, they

Financial Derivatives in Theory and Practice Revised Edition. P. J. Hunt and J. E. Kennedy
© 2004 John Wiley & Sons, Ltd ISBNs: 0-470-86358-7 (HB); 0-470-86359-5 (PB)

will change by different amounts because of the differing payment dates. The word 'convexity' is used in this context because if the value at the maturity time T of one product is linear in some underlying market rate, the other is not, and in many common applications the value in this latter case is a convex function of the underlying market rate.

An important point to note about these products is the following. The value at time T takes the form of a product of some function of the market rates (the same in each case) with the value of a linear combination of pure discount bonds. In practice the value of the derivative is usually more sensitive to market rates through the function F than through the second 'discounting' term, and we assume this to be the case.

Mathematically, the key observation which lies at the heart of this chapter is that any linear combination of pure discount bonds can, assuming positivity, be taken as a numeraire in the valuation of options. This property, as we see in the next section, leads to an appealing decomposition of either of the convexity-related derivatives. Careful choice of modelling assumptions then allows a detailed comparison to be made between the products.

The layout of this chapter is as follows. In Section 14.2.1 we carry out the decomposition described above, appealing to the theory of option pricing via numeraires as described in Chapter 7. Applying these results in the context of the linear swap-rate model, we present some general and illuminating pricing formulae in Section 14.2.2. In Section 14.3 we apply these results to some explicit examples. We study the CMS and the LIBOR-in-arrears basis swap and present closed-form results. We also consider options on these products and are able to show why the *ad hoc* market standard approach works well and is, to first order, theoretically valid.

14.2 VALUATION OF 'CONVEXITY-RELATED' PRODUCTS

14.2.1 Affine decomposition of convexity products

Recall that the value at time t of a derivative that pays at T an amount V_T can be expressed in the form

$$V_t = N_t \mathbb{E}_\mathbb{N}[V_T N_T^{-1} | \mathcal{F}_t], \qquad (14.2)$$

where $(N, \mathbb{N})$ is some numeraire pair and $\{\mathcal{F}_t\}$ is the filtration generated by the underlying assets (in the application here these assets will be a finite collection of zero coupon bonds).

We can use (14.2) to express the value of convexity-related derivatives in a number of ways. Consider two European products having time-T value of the form (14.1). For $i = 1, 2$, let

$$P_t^{(i)} = \sum_{j=1}^{n_i} c_j^{(i)} D_{tS_j^{(i)}}.$$

If we assume that $P_t^{(i)} \neq 0, t \leq T$, a.s. we can take $P^{(i)}$ (or $-P^{(i)}$) as numeraire and apply (14.2) to obtain

$$V_t^{(i)} = P_t^{(i)} \mathbb{E}_{\mathbb{P}^{(i)}}[F|\mathcal{F}_t],$$

where $\mathbb{P}^{(i)}$ is a martingale measure corresponding to the numeraire $P^{(i)}$. This gives us formulae for each of the products independently, each as an expectation *of the same function* F, but, as the expectations are taken with respect to different measures, these formulae give little information about the relationship between the two products. To understand this relationship we extend the ideas above a little further.

Suppose we can find (and we always can) a set of numeraire pairs $(N^{(j)}, \mathbb{N}^{(j)})$ such that, for each $j = 1, \ldots, d$, $N^{(j)}$ is an a.s. non-zero linear combination of pure discount bonds and such that

$$P_t^{(i)} = \sum_{j=1}^{d} a_j^{(i)} N_t^{(j)}, \qquad (14.3)$$

for all t and some constants $a_j^{(i)}$. Then, for $i = 1, 2$, we have the representation

$$V_T^{(i)} = \Big(\sum_{j=1}^{d} a_j^{(i)} N_T^{(j)} \Big) F \qquad (14.4)$$

and thus

$$V_t^{(i)} = \sum_{j=1}^{d} a_j^{(i)} N_t^{(j)} \mathbb{E}_{\mathbb{N}^{(j)}}[F|\mathcal{F}_t].$$

This can be rewritten as

$$V_t^{(i)} = P_t^{(i)} \Big(\sum_{j=1}^{d} \omega_j^{(i)}(t) \mathbb{E}_{\mathbb{N}^{(j)}}[F|\mathcal{F}_t] \Big), \qquad (14.5)$$

where

$$\omega_j^{(i)}(t) = \frac{a_j^{(i)} N_t^{(j)}}{P_t^{(i)}}.$$

We have thus expressed both $V_t^{(1)}$ and $V_t^{(2)}$ as an affine combination of the same set of expected values of the single function F. The weightings, $w_j^{(i)}(t)$, depend *only* on the time-t discount factors D_{tS}, $S \geq T$, and not on any assumptions about the model of the economy – that only comes in via the expectations. Thus the effect of the mismatch of payment dates is manifested through these weighting terms.

What is the interpretation of these results, and why have we derived the valuation formulae in this way? This will become clear in the next section.

14.2.2 Convexity corrections using the linear swap-rate model

To understand better the relative pricing of convexity-related products we will employ the linear swap-rate model. The key to understanding the convexity lies in Section 14.2.1. In what follows we shall derive an affine decomposition similar to (14.4) which leads to pricing formulae that provide insight into how the values of convexity-related products are affected by the mismatch of payment dates.

When studing convexity-related products there is usually some market swap rate y closely associated with both products (as is the case in the examples of the next section). We assume this is indeed the case and take this rate as the underlying swap rate for the linear swap-rate model that we use for pricing. We denote, as usual, by P the PVBP of the fixed leg of the swap associated with the parameterizing swap rate y, and by α_j and S_j, $j = 1, 2, \ldots, n$, the corresponding accrual factors and payment dates.

Recall from Section 14.2.1 that the affine decomposition of products was done using suitable numeraire processes. In what follows we will perform a decomposition using the numeraires P and yP. That P is a numeraire is clear from the fact that it is the value of the fixed leg of a swap. Furthermore, since y_t is the fixed rate at t which ensures both fixed and floating legs of the swap have the same value, it follows that $y_t P_t$ is the time-t value of the floating leg of the same swap and is also suitable as a numeraire process.

Associated with the two numeraires P and yP are martingale measures, $\mathbb{S}$ and $\mathbb{Y}$ respectively, in which all numeraire-rebased assets are martingales. We have already met the swaption measure $\mathbb{S}$ in which P-rebased assets, and in particular y, are martingales. As is common market practice, we shall assume that, under $\mathbb{S}$, y satisfies an SDE of the form

$$dy_t = \sigma_t y_t dW_t^{\mathbb{S}}, \qquad (14.6)$$

where $W^{\mathbb{S}}$ is a standard Brownian motion under $\mathbb{S}$ and σ is some deterministic function of time. Recall that this yields Black's formula for swaption prices (Section 11.4). The measure $\mathbb{Y}$, which we call the *floating measure* (associated with the swap rate y), is defined from $\mathbb{S}$ by

$$\left. \frac{d\mathbb{Y}}{d\mathbb{S}} \right|_{\mathcal{F}_t} = \frac{y_t}{y_0}, \qquad t \leq T.$$

It follows from Girsanov's theorem (Theorem 5.24) that

$$dy_t = \sigma_t^2 y_t dt + \sigma_t y_t dW_t^{\mathbb{Y}}, \tag{14.7}$$

where $W^{\mathbb{Y}}$ is a standard Brownian motion under $\mathbb{Y}$. In particular, solving (14.6) and (14.7) yields

$$
\begin{aligned}
y_T &= y_0 \exp\left(\int_0^T \sigma_u dW_u^{\mathbb{S}} - \tfrac{1}{2} \int_0^T \sigma_u^2 \, du \right) \\
&= y_0 \exp\left(\int_0^T \sigma_u dW_u^{\mathbb{Y}} + \tfrac{1}{2} \int_0^T \sigma_u^2 \, du \right).
\end{aligned}
\tag{14.8}
$$

Returning to our convexity-related derivatives, recall that their values at T are given by $V_T^{(i)} = P_T^{(i)} F(D_{TS}, S \geq T)$. Under the assumptions of the linear swap-rate model we can decompose each of the terms $P_T^{(i)}$, $i = 1, 2$, as follows:

$$
\begin{aligned}
P_T^{(i)} &= \sum_{j=1}^{n_i} c_j^{(i)} D_{TS_j^{(i)}} \\
&= \sum_{j=1}^{n_i} c_j^{(i)} (A + B_{S_j^{(i)}} y_T) P_T \\
&= \left(\sum_{j=1}^{n_i} c_j^{(i)} A \right) P_T + \left(\sum_{j=1}^{n_i} c_j^{(i)} B_{S_j^{(i)}} \right) (y_T P_T).
\end{aligned}
\tag{14.9}
$$

This decomposition is of the form (14.3) and so we can apply (14.5), which yields

$$V_t^{(i)} = P_t^{(i)} \left(w_t^{(i)} \mathbb{E}_{\mathbb{S}}[F|\mathcal{F}_t] + (1 - w_t^{(i)}) \mathbb{E}_{\mathbb{Y}}[F|\mathcal{F}_t] \right) \tag{14.10}$$

where

$$w_t^{(i)} = \frac{P_t / \sum_{j=1}^{n} \alpha_j}{P_t^{(i)} / \sum_{j=1}^{n_i} c_j^{(i)}}. \tag{14.11}$$

Note the particular form of $w_t^{(i)}$, which depends only on the discount curve at time t and not on the model for the evolution of the pure discount bonds through time. Define the random time $A(P, t)$ via

$$P_t = D_{tA(P,t)} \left(\sum_{j=1}^{n} \alpha_j \right).$$

We call $A(P, t)$ the *PVBP-average time* corresponding to P, and $D_{tA(P,t)}$ the *PVBP-average discount factor* corresponding to P. With these definitions, the numerator in (14.11) is the PVBP-average discount factor for P; it is the amount by which a *single* future payment $\sum_{j=1}^{n} \alpha_j$ would have to be discounted if it is to have the same value as the actual payments in the PVBP P. Similarly, the denominator of (14.11) is the PVBP-average discount factor for the term $P^{(i)}$.

The decomposition at (14.9), with the coefficients multiplying P_T and $(y_T P_T)$ being independent of y_T, was only possible because of the particular form of the linear swap-rate model. This decomposition in terms of the floating and fixed legs is not the only one with this property, but it is the most useful. The reason for this is the fact that y_T is log-normally distributed under both $\mathbb{S}$ and $\mathbb{Y}$, as given by (14.8). As a consequence of this, defining

$$F_t(y_t) = \mathbb{E}_{\mathbb{S}}[F|\mathcal{F}_t],$$

we have

$$\mathbb{E}_{\mathbb{Y}}[F|\mathcal{F}_t] = F_t\left(y_t^*\right),$$

where

$$y_t^* = y_t \exp\left(\int_t^T \sigma_u^2 \, du\right).$$

Note that the dependence of $F_t(y_t)$ only on t and y_t follows from the facts that the economy at T is summarized by y_T and that (y_t, t) is a Markov process.

The valuation formula (14.10) can now be written as

$$\begin{aligned} V_t^{(i)} &= P_t^{(i)}\left(w_t^{(i)} F_t(y_t) + (1 - w_t^{(i)})F_t(y_t^*)\right) \\ &= P_t^{(i)}\left(F_t(y_t) + (1 - w_t^{(i)})\left(F_t(y_t^*) - F_t(y_t)\right)\right). \end{aligned} \tag{14.12}$$

The first term in (14.12) is what would result from naïvely valuing the product by taking the expected payoff in the swaption measure $\mathbb{S}$ and multiplying this by the PVBP term $P_t^{(i)}$. The second term inside the brackets in (14.12) is referred to as the *convexity correction*, the amount by which we must change the *expectation* term in the valuation formula in order to obtain the correct valuation. Practitioners have various *ad hoc* methods for obtaining this correction term but this one is particularly illuminating, as we explain in the next section.

14.3 EXAMPLES AND EXTENSIONS

The examples of greatest practical importance and interest are constant maturity swaps (CMS) and options on constant maturity swaps (CMS options), and these are the first two examples we discuss. The third example we consider is the LIBOR-in-arrears basis swap which is a special case of the

CMS. However, the LIBOR-in-arrears swap is simpler to analyse than the CMS and as a result much more can be said. In each of the three examples we discuss, one of the two convexity-related products is a standard *vanilla* product, and this is largely why it is so important to quantify this relationship correctly.

14.3.1 Constant maturity swaps

Recall that under the terms of a vanilla interest rate swap fixed and floating payments are made on a series of (approximately) regularly spaced dates $S_1, S_2, \ldots, S_n$. The amount of each floating payment is the (accrual factor multiplied by the) LIBOR that sets at the beginning of the period, whereas for the fixed leg the payment is the fixed rate K (multiplied by the accrual factor). A CMS is the same, with the exception that the payments in the floating leg are not set based on LIBOR, rather they are based on some other market swap rate. So, for example, the CMS may have a five-year maturity and make payments every six months, a total of ten payments per leg. The payments on the floating leg may be the two-year par swap rate that sets (usually but not always) at the beginning of each accrual period.

Valuation of the fixed leg is straightforward and need not concern us. To value the CMS floating leg we consider each cashflow in turn. A general floating payment in a CMS can be defined as follows. On some date T a market par swap rate is observed, y_T, and a payment of the amount y_T is made at time $M \geq T$.

This single-payment derivative is closely related to the vanilla swap corresponding to the rate y_T. The floating leg of this vanilla swap has value at T given by $P_T y_T$, which can be compared with the value of the CMS payment which is $D_{TM} y_T$. We can apply the results of Section 14.2 to value this CMS payment relative to the floating leg of the vanilla swap.

The value of the CMS payment at time zero is given, from (14.11) and (14.12), by

$$V_0 = D_{0M}(w y_0 + (1 - w) y_0^*)$$

where

$$w = \frac{\sum_{j=1}^n \alpha_j D_{0S_j} / \sum_{j=1}^n \alpha_j}{D_{0M}}, \tag{14.13}$$

$$y_0^* = y_0 e^{\sigma^2 T},$$

σ being the Black–Scholes volatility of a swaption (on the swap rate y_T) maturing at T. It follows that the 'convexity-adjusted forward rate', the known amount which, paid at time M, has the same value today as the CMS payment, is given by

$$\hat{y}_0 = w y_0 + (1 - w) y_0^*$$
$$= y_0 + (1 - w) y_0 (e^{\sigma^2 T} - 1). \tag{14.14}$$

From (14.13) and (14.14) we see that if the single CMS payment is made before the PVBP-average time for P, the correction term in (14.14) is positive, reflecting the fact that an increase in y_T will increase the value of the CMS payment by proportionately more than the amount by which the value of the vanilla swap floating leg will increase. On the other hand, if the CMS payment is made after the PVBP-average of the swap, the convexity correction is negative.

14.3.2 Options on constant maturity swaps

Having introduced the CMS, it is natural to consider options thereon, also common products in the market. A CMS cap is to a cap what the CMS is to a swap. It comprises a series of payments, the payment made at S_j being $\alpha_j(y_{S_{j-1}} - K)_+$ where α_j is the accrual factor, K is the strike and $y_{S_{j-1}}$ is some market swap-rate setting at S_{j-1}. The CMS floor is identical, but with the payment being $\alpha_j(K - y_{S_{j-1}})_+$. A general payment is therefore of the form $(\phi(y_T - K))_+$ and is made at time M, where $\phi \in \{-1, +1\}$ and $M \geq T$. Again following Section 14.2, the value of this payment at time zero is

$$V_0 = D_{0M}\Big(wBS(y_0, \sigma, K, T) + (1 - w)BS(y_0^*, \sigma, K, T) \Big)$$

where w and y_0^* are as above and BS is the market standard forward Black–Scholes option pricing formula,

$$BS(y, \sigma, K, T) = \phi y N(\phi d_1) - \phi K N(\phi d_2),$$
$$d_1 = \frac{\log(y/K)}{\sigma\sqrt{T}} + \tfrac{1}{2}\sigma\sqrt{T},$$
$$d_2 = \frac{\log(y/K)}{\sigma\sqrt{T}} - \tfrac{1}{2}\sigma\sqrt{T},$$

ϕ being $+1$ for a call and -1 for a put. The same comments made for the CMS also apply for the convexity corrections for CMS options.

It is common market practice when valuing CMS caps and floors first to calculate the convexity-adjusted forward swap rate, $\hat{y}_0$ in (14.14), and then claim that the value of the CMS option is given by

$$\widetilde{V}_0 = D_{0M}BS(\hat{y}_0, \sigma, K, T).$$

We can now see that this is justified, at least to first order, which is all any convexity correction of this nature achieves. For, given any payoff $F(y_T)$ at time M, we have

$$V_0^F = D_{0M}\Big(wF_0(y_0) + (1 - w)F_0(y_0^*) \Big) \tag{14.15}$$

where, recall from (14.12),

$$F_0(y_0) = \mathbb{E}_\mathbb{S}[F(y_T)].$$

Noting that F_0 is C^∞ we can apply a Taylor expansion to F_0 in (14.15) to obtain

$$V_0^F = D_{0M}\left(F_0(y_0) + (1 - w)\left(F_0(y_0^*) - F_0(y_0)\right)\right)$$

$$= D_{0M}\left(F_0(y_0) + (1 - w)(y_0^* - y_0)F_0'(y_0)\right) + O(y_0^* - y_0)^2$$

$$= D_{0M}F_0(\hat{y}_0) + O(y_0^* - y_0)^2.$$

As is often the case, *ad hoc* market practice turns out to have a theoretical justification.

14.3.3 LIBOR-in-arrears swaps

Unlike a vanilla interest rate swap, the LIBOR-in-arrears basis swap has two floating legs but no fixed leg. For notational convenience, define $L_t^j := L_t[S_{j-1}, S_j]$. The first floating leg is exactly the same as for a vanilla swap – that is, it comprises a sequence of payments made at times $S_1, S_2, \ldots, S_n$, the amount of the jth payment being given by $\alpha_j L_{S_{j-1}}^j$. The second floating leg makes payments on the same dates but the LIBOR is *set in arrears*, the amount of payment j being $\alpha_j L_{S_j}^{j+1}$ (with the obvious definition of S_{n+1} and L^{n+1}).

To calculate the value of the LIBOR-in-arrears leg we consider each cashflow in turn. It is easy to see that this is just a special case of the more general CMS above, taking, in the case of the jth payment, $M = T = S_j$ and $y_T = L_T^{j+1}$. However, to analyse this product we do not need even to formulate a model for the evolution of rates, as long as we know the prices of an appropriate set of market caplets. The technique we use here was first presented by Breeden and Litzenberger (1978).

The value at time S_j of the jth LIBOR-in-arrears payment is just the payment amount $\alpha_j L_{S_j}^{j+1}$. Observe that we could take this payment and immediately reinvest it until time S_{j+1}, at which time it will be worth

$$V_{S_{j+1}}^j(L_{S_j}^{j+1}) = \alpha_j L_{S_j}^{j+1}\left(1 + \alpha_{j+1}L_{S_j}^{j+1}\right)$$

$$= \alpha_j L_{S_j}^{j+1} + \alpha_j\alpha_{j+1}\left(L_{S_j}^{j+1}\right)^2. \tag{14.16}$$

This formula is smooth in the variable $L_{S_j}^{j+1}$, and for any smooth function $V(x)$, $x \geq 0$, we have the linear representation

$$V(x) = V(0) + xV'(0) + \int_0^\infty V''(K)(x - K)_+ \, dK. \tag{14.17}$$

Equation (14.17) represents the payoff $V(x)$ as a linear combination of caplet payoffs. But the value of a linear combination of derivatives is just the sum of their individual values. Substituting (14.16) into (14.17), we conclude that V_0^j, the value at time zero of the LIBOR-in-arrears payment, is given by

$$V_0^j = \alpha_j D_{0S_{j+1}} L_0^{j+1} + \int_0^\infty 2\alpha_j \alpha_{j+1} C^j(K)\,dK,$$

where $C^j(K)$ is the value at time zero of a caplet with strike K, setting at S_j and paying $(L_{S_j}^{j+1} - K)_+$ at S_{j+1}. Note that at no point have we assumed a model in this analysis.

15

Implied Interest Rate Pricing Models

15.1 INTRODUCTION

In Chapter 13 we introduced the general class of terminal swap-rate models and studied three examples: the linear, geometric and exponential swap-rate models. Each of these examples was constructed by postulating a functional form for the discount factors, at the terminal time T, and then using the distribution of one given market swap rate and the martingale property of numeraire-rebased asset prices to determine the remaining free parameters of the model. In this chapter we will consider a class of terminal swap-rate models constructed in a different way.

One of the major advantages of using the terminal swap-rate approach to develop a model for pricing European derivatives is the explicit control one can exercise over the functional form of the discount factors in terms of an underlying stochastic process. A second is the ability to calibrate this type of model accurately to the market-implied distribution of a given swap rate or LIBOR. A practical difficulty with the approach is that for the more realistic examples, such as the exponential swap-rate model, the calibration algorithm specified in Section 13.4.1 can be slow to converge when the calibrating swap has many cashflows (for example, quarterly for ten years is a problem). The algorithm given there can be adapted to give better convergence but the performance is still insufficient for a live trading environment. That is not to say that an efficient algorithm cannot be developed, and indeed this may be a simple matter, but at this point of time we are not aware of one. In the absence of a more efficient calibrating algorithm it is useful to develop an alternative terminal swap-rate model with similar properties to the exponential or geometric model but which calibrates much more efficiently. A special case of the *implied pricing models* presented in this chapter achieves precisely that.

The defining property of an implied interest rate pricing model is that it

Financial Derivatives in Theory and Practice Revised Edition. P. J. Hunt and J. E. Kennedy
© 2004 John Wiley & Sons, Ltd ISBNs: 0-470-86358-7 (HB); 0-470-86359-5 (PB)

is one which is constructed to correctly price a number of standard market products. The starting point is the assumption that the complete discount curve at the terminal time T is some (to be determined) function of a random variable r. The prices of the *calibrating* products which must be correctly valued are then used to *imply* what this functional form must be. To develop and use such a model in practice it is important to first identify which instruments will be used for the calibration (always a collection of caps and/or swaptions) and then to have a method for implying the functional form from these prices.

The rest of this chapter is as follows. Throughout we will consider the case when the random variable r is univariate. First, in Section 15.2, we show how the functional form of the discount curve at time T, $\{D_{TS} : S \geq T\}$, can be implied from the prices of the standard market swaptions which expire at time T. To do this we use the prices of swaptions of all strikes and, for the underlying swaps, of all (regularly spaced) maturities. The resulting model is, in fact, a terminal swap-rate model. In Section 15.3 we briefly consider the numerical implementation of an implied model, then in Section 15.4 we consider in detail the problem of pricing an *irregular swaption*. The feature of interest about this product is that it is not in general clear which standard market swaption is most closely related to this exotic product. The chapter concludes in Section 15.5 with a discussion of the relationship between these implied models and the geometric and exponential models we met in Chapter 13. It turns out that if the volatility of all the calibrating swaptions takes the same value then the implied model is almost indistinguishable from the equivalent exponential or geometric model.

15.2 IMPLYING THE FUNCTIONAL FORM D_{TS}

Consider a payers swap with fixed rate K which starts at time T and makes payments at times $S_1, S_2, \ldots, S_n$, and let y_t^n denote the forward par swap rate for this swap at time t. We know that the value of this swap can be written in the form

$$U_t^n = P_t^n (y_t^n - K)$$

and that the value at the maturity date T of a swaption written on this swap is then

$$V_T^n(K) = P_T^n (y_T^n - K)_+ .$$

We will suppose throughout this section that we are today, that is at time zero, given the prices of these swaptions for all $K \geq 0$, i.e. $\{V_0^n(K) : K \geq 0\}$. For each coupon date of this swap, S_i, $i = 1, 2, \ldots, n - 1$, one can also consider the swaps maturing at time S_i which make payments at the times $S_1, S_2, \ldots, S_i$. We will additionally suppose that we know the prices of swaptions for all strikes written on all these underlying swaps, $\{V_0^i(K) : K \geq$

$0, i = 1, \ldots, n-1\}$. What we shall do is develop a model which is consistent with all these prices simultaneously.

To develop this model we need to make the following assumptions:

(i) There exists some univariate random variable r which summarizes the state of the economy at time T and thus the discount curve at T can be written in the form $\{D_{TS}(r) : S \geq T\}$.

(ii) All the swap rates $y_T^i(r)$ (the exact functional form here being derived from the functional form of the underlying discount factors), are monotone (either all increasing or all decreasing) in the random variable r.

The first assumption means that the model is a one-factor model. The second assumption is a reasonable one to make since it means that the single factor being used is a measure of the overall level of interest rates. It is unlikely that one would want to develop a single-factor model for the discount curve in which the stochastic variable does not represent the level of rates.

We can now derive the functional form $\{D_{TS_i}(r) : i = 1, 2, \ldots, n\}$ from these two assumptions and the swaption prices $\{V_0^i(K) : K \geq 0, i = 1, \ldots, n\}$ (we shall provide the functional form for other S later in this section). To achieve this we shall, in Theorem 15.1, derive the functional forms $\{y_T^i(r) : i = 1, 2, \ldots, n\}$. That these are equivalent follows from the relationships

$$D_{TS_0} = D_{TT} = 1$$

and

$$y_T^i = \frac{D_{TT} - D_{TS_i}}{P_T^i}, \qquad i = 1, \ldots, n.$$

To obtain the functional forms $y_T^i(r)$ we make the following observation. Since we have assumed that there exists a single factor r which governs all discount bond prices, and since each y_T^i is monotone in r, then given K_1 there exist $K_2, \ldots, K_n$ such that

$$\{r : y_T^1(r) < K_1\} = \{r : y_T^i(r) < K_i\}. \tag{15.1}$$

If we can, in some measure, find the distribution of a single y_T^i and if we can also find, for each K_1, the values K_i in (15.1) then we have in effect derived the functional form that we desire. More precisely, we will have derived the functional forms $y_T^i(y_T^1)$, giving us the result we require in the case $r = y_T^1$. The result for any other random variable r can now be derived immediately from this.

The following theorem provides the desired link between each of the K_i.

Theorem 15.1   *Given any $\lambda \in (0, D_{0T})$, there is a unique solution to the equations (taking $V_0^0 = 0$)*

$$\frac{\partial V_0^{i-1}(K_{i-1})}{\partial K_{i-1}} = (1 + \alpha_i K_i) \frac{\partial V_0^i(K_i)}{\partial K_i} - \alpha_i (V_0^i(K_i) + \lambda). \tag{15.2}$$

Furthermore the solution $\{K_i(\lambda)\}$ satisfies

$$\lambda = D_{0T}\mathbb{F}(y_T^i < K_i(\lambda)) \tag{15.3}$$

for each i, where $\mathbb{F}$ is the forward measure to time T (the measure under which all $D_{.T}$-rebased discount factors are martingales), and

$$\{r : y_T^i < K_i(\lambda)\} = \{r : y_T^j < K_j(\lambda)\}$$

for all i and j.

Proof: We work in the forward measure $\mathbb{F}$, in which case

$$
\begin{aligned}
V_0^i(K_i) &= D_{0T}\mathbb{E}_{\mathbb{F}}[V_T^i(K_i)] \\
&= D_{0T}\int_{-\infty}^{\infty}(-D_{TT} + D_{TS_i} + K_i P_T^i)_+ \, d\mathbb{F}(r) \\
&= D_{0T}\int_{-\infty}^{r^*(K_i)}(-1 + D_{TS_i} + K_i P_T^i) \, d\mathbb{F}(r),
\end{aligned}
\tag{15.4}
$$

where $r^*(K_i)$ is the (possibly infinite) supremum value of r such that the integrand is positive. Differentiating with respect to K_i yields

$$\frac{\partial V_0^i(K_i)}{\partial K_i} = D_{0T}\int_{-\infty}^{r^*(K_i)} P_T^i \, d\mathbb{F}(r). \tag{15.5}$$

Substituting (15.4) and (15.5) into the right-hand side of (15.2) and rearranging gives

$$D_{0T}\int_{-\infty}^{r^*(K_i)} P_T^{i-1} \, d\mathbb{F}(r) - \alpha_i(\lambda - D_{0T}\mathbb{F}(r < r^*(K_i))). \tag{15.6}$$

Observe that this is increasing in K_i so, given λ, (15.2) will have at most one solution. Choosing $K_1, K_2, \ldots, K_n$ such that, for all i and j,

$$D_{0T}\mathbb{F}(y_T^1 < K_1) = \lambda,$$

$$\{r : y_T^i(r) < K_i\} = \{r : y_T^j(r) < K_j\},$$

we have

$$r^*(K_i) = r^*(K_j) = r^*,$$

$$D_{0T}\mathbb{F}(r < r^*) = \lambda.$$

With this choice, (15.6) now reduces to

$$D_{0T}\int_{-\infty}^{r^*(K_{i-1})} P_T^{i-1} \, d\mathbb{F}(r)$$

which is precisely the left-hand side of (15.2). It follows that this is our required solution.   □

The proof above is developed in the forward measure. When considering each swaption in isolation it is more natural to work with swap rates and to work in the appropriate swaption measure, as we did in Chapter 11. It is this property of the market that obscures the relationship between each of the swaptions – the proof is easy once we have the relationship (15.2) given to us.

We have derived an implied functional form at the swap payment dates S_i. It remains to specify the functional form at intermediate time points. There are many ways in which this could be done. One approach which has considerable practical advantages is to define the intermediate discount factors by some simple interpolation algorithm. For example, one could use a log-linear interpolation of discount factors,

$$
\log(D_{TS}) = \frac{S - S_i}{S_{i+1} - S_i} \log(D_{TS_{i+1}}) + \frac{S_{i+1} - S}{S_{i+1} - S_i} \log(D_{TS_i}), \qquad (15.7)
$$

where $S_i < S < S_{i+1}$. This has the advantage that it is exactly what many banks do in practice. At the maturity time T the liquid market instruments determine the discount factors D_{TS_i} and banks then calculate intermediate discount factors according to the formula (15.7), or some other similar simple interpolation routine. Theoretically, however, this model will admit arbitrage – numeraire-rebased discount bond prices for intermediate bonds will not be martingales. This effect is small enough that it is not relevant in practice.

If, despite the practical insignificance of the theoretical arbitrage caused by this assumption, one wants to generate a model that is exactly arbitrage-free then this can be done quite easily in a number of ways. Perhaps the simplest would be, for each intermediate time S, to find some constant w_S such that

$$
D_{0S} = w_S D_{0S_i} + (1 - w_S) D_{S_{i+1}}
$$

where $S_i < S < S_{i+1}$. It then immediately follows from the martingale property of the numeraire-rebased gridpoint bonds, D_{tS_i}, that the numeraire-rebased bond D_{tS} is also a martingale. If one wanted to mirror more closely the market practice of interpolating using (15.7), then an alternative would be to use the interpolation

$$
\log(D_{TS}) = w_S \log(D_{TS_{i+1}}) + (1 - w_S) \log(D_{TS_i}),
$$

or more generally

$$
f(D_{TS}, S) = w_S f(D_{TS_{i+1}}, S_{i+1}) + (1 - w_S) f(D_{TS_i}, S_i),
$$

for some function f. In these latter cases the values w_S would need to be calculated numerically so that the martingale property held. This would, however, be very efficient and could be done in exactly the same way as we solved for intermediate discount factors when we discussed the exponential swap-rate model in Section 13.3.1.

Now that we have derived, from market prices, the joint distribution of the swap rates at T, we can (numerically) price any derivative from market swaption prices. Of course, this will only be meaningful if a one-factor model is suitable for the particular product being considered. See Section 15.4 for further discussion of this point.

15.3 NUMERICAL IMPLEMENTATION

To implement an implied model in practice we must solve (15.2). For this we need to know the functional forms $V_0^i(K)$ and the first derivatives $\frac{\partial V_0^i(K)}{\partial K}$. All this information is available in the market. Note, in particular, that the swaptions market generally prices using Black's model. In this case the value at time zero of a receivers swaption of maturity T is given by

$$V_0^i(K_i) = P_0^i[K_i N(-d_2^i) - y_0^i N(-d_1^i)],$$

where

$$d_1^i = \frac{\log(y_0^i/K_i)}{\sigma_i \sqrt{T}} + \tfrac{1}{2}\sigma_i \sqrt{T},$$

$$d_2^i = d_1^i - \sigma_i \sqrt{T},$$

$N(\cdot)$ is the standard Gaussian distribution function and σ_i is the market volatility for this swaption.

To derive the joint distributions in this case we work in the forward measure again. Fixing $\lambda \in (0, D_{0T})$, equation (15.2) becomes

$$P_0^{i-1} N(-d_2^{i-1}) + \alpha_i \lambda = P_0^i[N(-d_2^i) + \alpha_i y_0^i N(-d_1^i)]. \qquad (15.8)$$

The form of (15.8) and, more generally, (15.2) is very simple and so it can be solved inductively in i using Newton–Raphson. This calculation is very quick because of the quadratic convergence of the Newton–Raphson algorithm (see Press et al. (1988)) and the well-behaved form of the functions involved. Furthermore, if we are solving (15.8) for a grid of λ values in $(0, D_{0T})$ then the solution for one λ will be a very good first guess at a solution for the next neighbouring λ. As a result convergence only takes one or two iterations of the algorithm.

One final decision to be made is what spacing should be used when choosing the values of λ at which to calculate the functional forms. One could space the values evenly over the interval $(0, D_{0T})$ which, given the interpretation of λ in (15.3) is certainly one reasonable choice. An alternative, the one we have usually adopted, is to choose instead an equal spacing for the first par swap rate y^1. For each value of y^1 the corresponding λ can be calculated from (15.8), thus saving one application of the Newton–Raphson routine.

15.4 IRREGULAR SWAPTIONS

We now consider the problem of pricing an irregular swaption and how an implied pricing model could be used to value such a product. We shall see that the problem is not as straightforward as one might at first expect.

Recall that under the terms of a standard swap one counterparty pays to another a series of floating cashflows and in return receives a series of fixed cashflows. Using the terminology of Chapter 10, the jth floating cashflow, paid at S_j, is for the amount $A\alpha_j L_{T_j}[T_j, S_j]$ (A being the notional amount) and the corresponding fixed cashflow is for $A\alpha_j K$, K being the fixed rate. By an *irregular* swap we mean an agreement in which, as for a vanilla swap, a series of floating and fixed cashflows are exchanged but for which the notional amount of each cashflow pair may differ between pairs. That is, the jth floating cashflow is for an amount $A_j \alpha_j L_{T_j}[T_j, S_j]$ and the jth fixed cashflow is for $A_j \alpha_j K$, where now the $A_j, j = 1, \ldots, n$, may differ *but must all be non-negative*.

This is a considerable generalization on a standard swaption. The restriction that all the A_j be non-negative ensures that this product is primarily sensitive to the overall level of interest rates and not to the shape of the yield curve. Thus a one-factor model will be sufficient. So suppose we now wish to use a terminal swap-rate model to price this product, first calibrating to some related vanilla swaption(s). If we adopt the approach of Chapter 13 and fit an exponential model calibrated to a single swaption, the question arises as to which swaption is appropriate. For example, if all the A_j are (very nearly) equal this is clearly a vanilla product and the calibrating swaption is obvious. On the other hand if all but the last A_j are equal and $A_n = \varepsilon$ which is small, then the correct calibrating swaption is one with an underlying swap of maturity S_{n-1}. When the A_j are linearly decreasing to zero the problem is altogether more difficult to decide. One alternative in this case is to use an implied model and calibrate to *all* the underlying swaption prices. This has a clear weakness, as the following theorem demonstrates.

Theorem 15.2 *Consider a linearly amortizing receivers swap, namely an irregular swap for which the notional amounts A_j, $j = 1, 2, \ldots, n$, are given by*

$$A_j = \frac{\sum_{l=j}^{n} \alpha_l}{\sum_{l=1}^{n} \alpha_l},$$

with the α_j being the usual accrual factors. Now let

$$U_t = \sum_{j=0}^{n} c_j D_{tS_j}$$

be the value at t of this swap (where, in fact, it follows from the FRA analysis of Section 10.6.3 that, taking $A_0 = A_{n+1} = 0$, $c_j = A_j(1 + \alpha_j K) - A_{j+1}$). For $i = 1, 2, \ldots, n$, let

$$U_t^i(K_i) = -D_{tS_0} + D_{tS_i} + K_i \sum_{j=1}^{i} \alpha_j D_{tS_j}$$

be the value at t of a set of vanilla (receivers) swaps with strikes K_i. Suppose that

$$U_t = \sum_{i=1}^{n} w_i U_t^i(K_i) \tag{15.9}$$

for some K_i and some positive w_i and consider now the amortizing and vanilla swaptions with payoffs $V_T = (U_T)_+$ and $V_T^i(K_i) = (U_T^i(K_i))_+$ respectively. Let V_t and $V_t^i(K_i)$ denote the values of these swaptions at time $t \leq T$. Then

$$V_t \leq \sum_{i=1}^{n} w_i V_t^i(K_i) \tag{15.10}$$

for all t.

Suppose now that we use a one-factor model to price each of these swaptions, with resultant prices $\hat{V}_t$ and $\hat{V}_t(K_i)$. Then, assuming all swap rates are increasing in the driving factor,

$$\hat{V}_t = \inf_{\mathcal{F}} \sum_{i=1}^{n} w_i \hat{V}_t^i(K_i) \tag{15.11}$$

where

$$\mathcal{F} = \left\{ \{(w_i, K_i), i = 1, 2, \ldots, n\} : w_i \geq 0, U_t = \sum_{i=1}^{n} w_i U_t^i(K_i) \right\}.$$

Remark 15.3: This theorem shows how decomposing the linearly amortizing swap into regular swaps, as in (15.9), gives a corresponding overhedge for the amortizing swaption in terms of vanilla swaptions, (15.10). The implications of the second part of the theorem for implied interest rate models is clarified in Corollary 15.4 below.

Proof: The inequality follows immediately since

$$V_T = \max(U_T, 0)$$

$$= \max\left(\sum_{i=1}^{n} w_i U_T^i(K_i), 0\right)$$

$$\leq \sum_{i=1}^{n} w_i \max(U_T^i(K_i), 0).$$

Thus the original option always pays an amount no greater than the sum of the vanilla swaptions and so its value at t must obey the same inequality. Note that this holds independently of any modelling assumptions.

The second part of the result relies heavily on the one-factor property and the monotonicity of each product in the driving random variable r. The argument is essentially the one-factor decomposition of Jamshidian (1989). Let $r^* \in (-\infty, \infty)$ be the value of r such that $U_T(r^*) = 0$. We deal with the degenerate case when no such r^* exists later. Choose each K_i such that $U_T^i(K_i, r^*) = 0$, namely

$$K_i = \frac{D_{TS_0}(r^*) - D_{TS_i}(r^*)}{\sum_{j=1}^{i} \alpha_j D_{TS_j}(r^*)}, \tag{15.12}$$

and let w_i, $i = 1, \ldots, n$, solve (15.9).

Telescoping back from c_n to c_1 shows that such w_i do exist and are given (with the convention $w_{n+1} = c_{n+1} = 0$), by

$$(1 + \alpha_i K_i)\frac{w_i}{\alpha_i} = \frac{w_{i+1}}{\alpha_{i+1}} + \frac{c_i}{\alpha_i} - \frac{c_{i+1}}{\alpha_{i+1}}. \tag{15.13}$$

The final identity that we require for these w_i to solve (15.9) is that

$$c_0 = -\sum_{i=1}^{n} w_i, \tag{15.14}$$

but this follows since, by construction, (15.9) holds at $t = T$, $r = r^*$.

We now make two observations. The first is that $w_i \geq 0$ for each i and so this solution is indeed feasible for our minimization problem. This follows inductively from (15.13) since the right-hand side is positive, given the form of the c_j for a linearly amortizing swaption, and also (15.12) ensures that $(1 + \alpha_i K_i) > 0$. The second observation is that U_T and $U_T^i(K_i)$ are both decreasing in r, which follows since all the cashflows except the first (whose value is independent of r) are positive and discount factors decrease in r.

To complete the proof note that, by construction, all the swaps have zero value when $r = r^*$ and so monotonicity in r implies

$$\hat{V}_T(r) = \sum_{i=1}^n w_i \hat{V}_T^i(r, K_i)$$

for all r. Hence (15.10) is an equality in this case and the infimum is attained.

The argument above started with the existence of a value r^* for which the amortizing swap had zero payout. In the absence of such an r^* the amortising swap value is either always positive or always negative. We deal with the former case, the latter being similar.

For $i = 2, 3, \ldots, n$, let $K_i = \sup_r K_i(r)$ where $K_i(r)$ is as in (15.12), and define w_i, as before, using (15.13). Now choose w_1 and K_1 such that (15.13) and (15.14) hold. Observe that each w_i is again non-negative and every swap has non-negative value for all r. The proof is complete.   □

Corollary 15.4 *A one-factor implied pricing model calibrated to all the market prices $V_0^i(K_i)$ will always yield an upper bound for the value of a linearly amortizing swaption.*

Proof: For the implied interest rate model, the model and market prices of the calibrating swaptions are equal by construction, $\hat{V}_0^i(K_i) = V_0^i(K_i)$ for all i and K_i. It thus follows from (15.11) and (15.10) that

$$V_t \le \hat{V}_t .$$

Thus the price from the implied model always bounds the true linearly amortizing swaption value from above.   □

This systematic overpricing of linearly amortizing swaptions is a direct result of the one-factor assumption and full calibration. It is worth spending a little time understanding why this is the case because the issue needs to be considered carefully when pricing any product with a single-factor model. The problem is that in the model all the forward LIBORs are perfectly correlated with each other, whereas in the real market they are not. The volatility of each calibrating swaption depends on the law of the underlying swap rate, and this in turn depends on a subset of the forward LIBOR and their interactions. In particular, the correlation structure of these forward LIBORs has a significant effect. On the other hand, in the one-factor model the correlation between the forward LIBORs is unity. In calibrating correctly to the swaption prices within a one-factor model, i.e. one with perfect correlations, the model must compensate by systematically adjusting the volatilities of the forward LIBORs. But this compensation, while it is ideal for the products to which the model is calibrated, is far from suitable for other products which have a different dependency on the underlying LIBORs.

To rectify this problem we must remember the purpose of developing and using a one-factor model and use general higher-level information before specializing to a specific model. In all the applications so far there has been a very clear market rate, with a known volatility, to which we should calibrate the model. We have then consistently developed a model for the whole term structure (at the terminal time T) so that we can capture some pricing effect, such as the convexity of a CMS. The situation for an irregular swaption is different since there is no clear liquid instrument against which to calibrate the model.

What we must do in this situation is first decide what it is in the market that has most effect on the price of the irregular swaption. In fact there is an obvious answer to this, the par swap rate for the underlying irregular swap, which we define shortly. As we shall see, this rate clearly captures the important features of the market, but it has the drawback that there is little direct market information about its volatility. But we can estimate this by various means.

To expand on this approach, consider a general irregular swap that comprises n cashflows and has a fixed rate K. Suppose that cashflow j occurs at time S_j and that the notional amount on which this cashflow is based is $A_j \geq 0$. Then the value at time t of the fixed leg is

$$\sum_{j=1}^{n} A_j \alpha_j K D_{tS_j}$$

and, using the FRA analysis of Chapter 10, the value of the floating leg is

$$\sum_{j=1}^{n} A_j (D_{tS_{j-1}} - D_{tS_j}).$$

For this swap structure we can define the corresponding PVBP, P, and par swap rate, y, via

$$P_t = \sum_{j=1}^{n} A_j \alpha_j D_{tS_j},$$

$$y_t = \frac{\sum_{j=1}^{n} A_j (D_{tS_{j-1}} - D_{tS_j})}{P_t}.$$

We now model and price the swaption very much as we did for vanilla swaptions. Take P as numeraire and y as a log-normal process in the corresponding swaption measure $\mathbb{S}$. Once more y is a martingale in this measure, being a linear combination of numeraire-rebased asset prices, and we obtain, for the payers swaption,

$$V_t = P_t \mathbb{E}_{\mathbb{S}}[(y_T - K)_+ | \mathcal{F}_t]$$
$$= P_t \Big(y_t N(d_1) - K N(d_2) \Big). \tag{15.15}$$

Equation (15.15) has exactly the same form as (11.8), the price of a vanilla swaption, and the definitions for d_1 and d_2 are exactly as given there:

$$d_1 = \frac{\log(y_t/K)}{\tilde{\sigma}_t\sqrt{T-t}} + \tfrac{1}{2}\tilde{\sigma}_t\sqrt{T-t}\,,$$

$$d_2 = \frac{\log(y_t/K)}{\tilde{\sigma}_t\sqrt{T-t}} - \tfrac{1}{2}\tilde{\sigma}_t\sqrt{T-t}\,,$$

$$\tilde{\sigma}_t^2 = \frac{1}{T-t}\int_t^T \sigma_u^2 du\,.$$

From the above it is clear that we do not, on this occasion, need to model the full discount curve at time T, $\{D_{TS} : S \geq T\}$ – everything we need to model is summarized by the irregular swap rate. However, we are left with the problem of how we should set the volatility for this irregular swap rate y. But it is now very clear that this is what matters and we are able to exploit to the full our knowledge of the true covariance structure between the various market LIBORs and swap rates. To do this one can use various degrees of sophistication based on the market swaption and cap floor volatilities.

The simplest reasonable approach would be the following. *Very* roughly speaking, any vanilla par swap rate can be considered to be a weighted average of the LIBORs which underlie the corresponding swap. To see this, write

$$y_t^i = \sum_{j=1}^{i} \beta_j L_t[T_j, S_j]\,, \tag{15.16}$$

where

$$\beta_j := \frac{\alpha_j D_{tS_j}}{P_t^i}\,.$$

Of course the β_j also depend on the forward LIBOR but, to simplify, we overlook this fact. If by some means (we shall not go into this, but one way is by historical analysis) we can estimate the correlation structure for the LIBORs then equations (15.16) allow us to derive the (approximate) LIBOR volatilities from the swaption volatilities. From these volatilities and correlations we can now calculate the volatility of the irregular swap rate,

$$y_t = \sum_{j=1}^{n} \gamma_j L_t[T_j, S_j]\,,$$

where

$$\gamma_j := \frac{A_j \alpha_j D_{tS_j}}{P_t}\,.$$

What we have done here is use full information about the multi-factor nature of the real world to choose the 'best' one-factor model for pricing the irregular swaption. We have not needed to use a full terminal swap-rate model because a single rate (the irregular par swap rate y) has summarized enough about the market to price the product.

As a final point on this product, note that if one were to price a CMS based on this irregular swap rate then it is possible, extending the approach of Chapter 13 in the obvious way, to develop a terminal swap-rate model based on this irregular swap rate. This product is unlikely to occur in practice because counterparties like payments to be determined from a readily available rate such as a LIBOR or a vanilla swap rate.

Returning to the implied pricing models, the subject of this chapter, the analysis above might suggest that they are of limited use. The fitting of a single factor to many swap-rate distributions has created a model with undesirable features. The problem was not one caused by the implied fitting idea itself, which is an efficient technique, rather it was caused by the choice of distributions to fit. However, if we use the technique in a slightly different way it turns out to be very powerful. We will demonstrate this in the next section by generating an arbitrage-free terminal swap-rate model with properties very similar to those of the exponential model defined in Chapter 13. This alternative model has the distinct advantage that it is much easier to calibrate because equations (15.2) can be solved sequentially rather than needing to be solved simultaneously as in the algorithm of Section 13.4.

15.5 NUMERICAL COMPARISON OF EXPONENTIAL AND IMPLIED SWAP-RATE MODELS

A single-factor implied interest rate model such as those derived in this chapter, or more generally the one-dimensional examples of Markov-functional models introduced in Chapter 19, can have undesirable properties and give misleading prices if used inappropriately. The problem, as discussed in the last section, is the reduction of a multi-factor world to a single-factor model and is one which is relevant to any one-factor model – you cannot get everything right so be careful to focus on the right quantities for the pricing problem to hand.

For the amortizing swaption, calibrating a one-factor implied model to the prices of all the underlying swaptions was not the right thing to do. So the idea of calibrating to all these underlying swaptions *as an end in itself* is not sensible. However, by careful choice of input prices it is possible to generate a one-factor implied pricing model which has relevant (to the problem at hand) properties and behaviour. This should not be surprising since, after all, any one-factor model is determined by its marginal swap-rate distributions and an implied model can be fitted to any desired marginals. We illustrate this idea and technique by formulating an implied pricing model which is very similar to a predefined exponential swap-rate model. Other behaviour can be imposed in a similar way.

The problem we consider is this. Suppose we are given an initial discount curve, $\{D_{0S}, 0 \leq S < \infty\}$, and we wish to generate an exponential swap-rate model for which the swap rate y^n is log-normal in its own swaption measure $\mathbb{S}^n$, i.e. a model which prices all swaptions of expiry time T on the swap rate y^n according to Black's formula. Calibration of this algorithm can be prohibitively slow, so we would like an alternative model which has very similar behaviour to the 'target' exponential model.

To achieve this, the procedure we follow is this. We will develop an implied model which values all the swaptions $\{V_0^i(K_i) : 1 \leq i \leq n\}$ according to Black's formula. Clearly we want the volatility of y^n to be σ_n, but what should we choose for the volatility σ_i of the swap rate y^i? We use the following heuristic. In the exponential model, we know that, for each $i = 1, \ldots, n$,

$$D_{TS_i}(z_T) = \exp(-C_{S_i} z_T), \tag{15.17}$$

$$y_T^i = \frac{1 - \exp(-C_{S_i} z_T)}{\sum_{j=1}^{i} \alpha_j \exp(-C_{S_j} z_T)}. \tag{15.18}$$

If we were to replace the random variable z_T in (15.17) and (15.18) by some constant $\hat{z}$, then we could conclude immediately, remembering that we have defined $C_{S_n} = 1$, that

$$\hat{z} = -\log(D_{TS_n}),$$

$$C_{S_i} = -\frac{\log(D_{TS_i})}{\hat{z}}, \qquad i = 1, \ldots, n-1,$$

and, furthermore,

$$\frac{dy_0^i}{d\hat{z}} = y_0^i \left(\frac{C_{S_i} D_{0S_i}}{1 - D_{0S_i}} + \frac{\sum_{j=1}^{i} C_{S_j} \alpha_j D_{0S_j}}{\sum_{j=1}^{i} \alpha_j D_{0S_j}} \right). \tag{15.19}$$

The sensitivity of the swap rate y^i to the variable $\hat{z}$ is given by (15.19). If, for each i, we use this as a proxy for the sensitivity of the swap rate y^i to a driving Brownian motion, then we are led to select the volatilities σ_i according to the rule

$$\frac{\sigma_i y_i}{\sigma_n y_n} := \frac{dy_0^i/d\hat{z}}{dy_0^n/d\hat{z}}.$$

So now we can build an implied model with these volatilities σ_i, $i = 1, \ldots, n$. The implied model will have each swap rate being precisely log-normally distributed in its own swaption measure. But how close is this to the 'target' exponential model? We can investigate this numerically and do so below. Note, however, that the point of this numerical analysis is to show how close the two models are. Thus we can be assured that the implied model has similar properties to the (more explicit) exponential model. However, there is no reason to believe the implied model is inferior to the exponential model which it attempts to emulate. Both are arbitrage-free, both calibrate perfectly to the swap rate y^n, both give rise to discount curves of approximately the same shape for each value of the random variable z_T (as guaranteed by the numerical comparison below). However, the major benefit of the implied model is that it can easily be calibrated efficiently, whereas the exponential model cannot.

To compare these two models we have considered a model with expiry time $T = 1$ calibrated to a nine-year swap which pays annual fixed coupons. The volatility of the one-year into nine-year swaption is taken to be 20% throughout and the corresponding forward swap rate is taken to be 7.00%. We then consider various interest rate curves, upward-sloping, downward-sloping and flat, and see how similar the models are.

The three different interest rate curves used are summarized in Table 15.1. For each of these curves and for each of the two models we calculated the following quantities:

(i) for each swap rate y^i and for a range of strikes K, the value of the associated payers swaption;

(ii) for each swap rate y^i and for a range of K, the probability in the corresponding swaption measure $\mathbb{S}^i$ that the swap rate is below the level K, $\mathbb{S}^i(y^i_T < K)$.

It turns out that the differences are greatest for the swap rate y^1, as one might expect given that this rate is least similar to the one-year into nine-year swap rate to which both models have been calibrated. The worst results were observed for the upward-sloping curve, and Figures 15.1 and 15.2 below present the results in this case. Figure 15.1 shows the difference in the PVBP-forward swaption values (i.e. swaption value/swap PVBP) for both models plotted against the strike. The maximum difference was 0.33 basis points (0.0033%). Figure 15.2 shows the difference in the cumulative probabilities for the two models as a function of the level K. In this case the maximum discrepancy was 0.00125.

Table 15.1   Interest rate curves

		Term/Years								
		1	2	3	4	5	6	7	8	9
Upward sloping	Swap rate	5.31%	5.54%	5.77%	5.99%	6.21%	6.42%	6.62%	6.81%	7.00%
	Fwd libor	5.31%	5.79%	6.27%	6.75%	7.23%	7.71%	8.18%	8.66%	9.12%
Flat	Swap rate	7.07%	7.06%	7.05%	7.04%	7.03%	7.02%	7.02%	7.01%	7.00%
	Fwd libor	7.07%	7.05%	7.03%	7.01%	6.99%	6.97%	6.95%	6.94%	6.92%
Downward sloping	Swap rate	8.76%	8.51%	8.28%	8.05%	7.83%	7.62%	7.41%	7.20%	7.00%
	Fwd Libor	8.76%	8.25%	7.75%	7.26%	6.78%	6.30%	5.82%	5.35%	4.88%

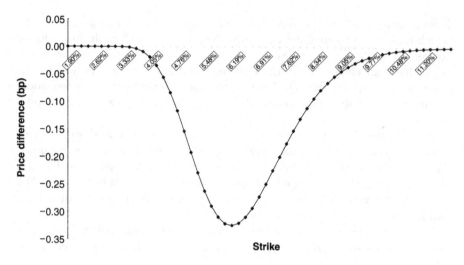

Figure 15.1 Difference in forward swaption prices

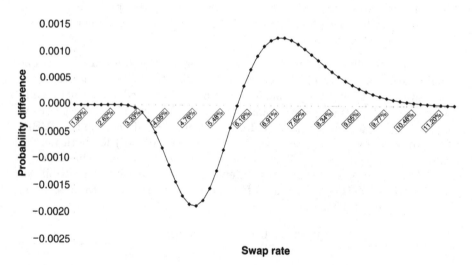

Figure 15.2 Difference in cumulative probabilities

16

Multi-Currency Terminal Swap-Rate Models

16.1 INTRODUCTION

The majority of exotic interest rate trades, even ones which involve several currencies, can be valued using single-currency models and thus fall within the scope of Chapters 13–15. The reason for this is the fact that the trades usually comprise two or more legs, one in each currency, and each can be valued independently of the others. However, there are still many trades which do genuinely require a model for the interest rate and foreign exchange (FX) rate evolution of several currencies simultaneously. Models of this type, again for application to European derivatives, are the topic of this chapter.

The terminal swap-rate models developed in Chapter 13 have many desirable features and are particularly suited to European derivatives. In developing a multi-currency model one would like to retain these properties of the models. Indeed, it would be nice to be able to develop a multi-currency analogue which, when considering a single currency, reduced to a standard terminal swap-rate model. This we can in fact do, as we shall see. As was the case for the single currency, the resulting models are arbitrage-free, have the advantage that they are parameterized in terms of standard market rates, and can be calibrated to appropriate prices in the interest rate market. Furthermore, within the model the resulting (PVBP-weighted) forward FX rate (defined precisely below) is log-normal and reduces to the standard market FX model when interest rates are taken to be deterministic. Another desirable feature is that the model is symmetric, in that it has the same general form when viewed from any currency.

Section 16.2 below provides the definition and derivation of a multi-currency terminal swap-rate model. These ideas are then applied, in Section 16.3, to value spread options and cross-currency swaptions. To do this we use the linear swap-rate model, and this results in convexity corrections which generalize those developed in Chapter 14. Of course, the more sophisticated exponential,

Financial Derivatives in Theory and Practice Revised Edition. P. J. Hunt and J. E. Kennedy
© 2004 John Wiley & Sons, Ltd ISBNs: 0-470-86358-7 (HB); 0-470-86359-5 (PB)

geometric or implied models could also be used for other products which warranted it.

16.2 MODEL CONSTRUCTION

As stated in Section 16.1, our objective in this chapter is to extend the class of terminal swap-rate models to the multi-currency setting. We want to do so in such a way that we retain all the desirable properties of a single-currency terminal swap-rate model. In particular, we want an arbitrage-free model (or more generally a modelling framework) which allows us to specify the functional form of discount factors in each currency. We now develop just such a model.

Suppose we wish to value some given product and to do so we must model a total of n currencies. Suppose further that in each of these currencies we have identified a market swap rate, the rate y^i in currency i, to which we want to calibrate the model. Extending the notation of Chapter 10 in the obvious way, we denote by P^i the PVBP corresponding to the calibrating swap in currency i and by $\{D^i_{tS} : 0 \le t \le S < \infty\}$ the discount factors in currency i.

The specification of the multi-currency model is carried out in two stages as follows.

Step 1: Consider each currency separately and in that currency define the functional form $D^i_{TS}(y^i_T)$, $S \ge T$.

Step 2: Fix a probability measure in which to work (usually the swaption measure for the calibrating swap in currency 1) and in this measure define the joint law of the calibrating swap rates and the spot exchange rates, $\{y^i, X^i : 1 \le i \le n, 0 \le t \le T\}$, X^i_t being the value at t in currency 1 of one unit of currency i.

Of course, in carrying out these two steps we must ensure that the resulting model is arbitrage-free and has all the properties intended at the outset of the modelling process.

We shall go through each of these steps in turn. For clarity we first consider the case when the calibrating swap rates y^i_T are log-normally distributed in their respective swaption measures (which we denote by $\mathbb{S}^i$). This corresponds to all the calibrating swaptions being priced by Black's formula and is the most common assumption made in practice – it would be unusual, with the possible exception of yen derivatives, for a bank to price an exotic option using a model which explicitly models a volatility smile (i.e. non-log-normal swap-rate distributions). It is the case, however, that the techniques presented here can be adapted to incorporate volatility smiles, and so we shall later show how this can be done.

16.2.1 Log-normal case

We carry out the two steps described above in turn.

Step 1: In each of the currencies $i = 1, \ldots, n$, we select the general functional form desired for the discount factors at the terminal time T as a function of the relevant calibrating swap rates. Usually this will be the same for all currencies but it need not be. Thus, for example, currency 1 could use the exponential swap-rate model whereas currency 2 could use the linear swap-rate model. We then follow exactly the procedure of Chapter 13 to determine the parameters of this functional form (the constants C_S for the exponential model, the constants A and B_S for the linear model). This step of the procedure requires that we know the volatilities of the respective swap rates but we do not actually set up the model on a probability space at this stage.

Step 2: To complete the model specification, we now define the law of all the swap rates (in a single measure) and of the various FX rates and show that these, together with the functional forms defined in step 1, combine to produce a full, arbitrage-free, multi-currency terminal swap-rate model.

Let $(\Omega, \{\mathcal{F}_t\}, \mathcal{F}, \mathbb{P})$ be a probability space supporting a correlated $2n$-dimensional Brownian motion $(W^i, \widehat{W}^i; 1 \leq i \leq n)$ with

$$dW_t^i dW_t^j = \rho_{ij} dt, \quad dW_t^i d\widehat{W}_t^j = \tilde{\rho}_{ij} dt, \quad d\widehat{W}_t^i d\widehat{W}_t^j = \hat{\rho}_{ij} dt.$$

As will become clear, the measure $\mathbb{P}$ will in fact be the swaption measure corresponding to the PVBP numeraire P^1. Define the vector process $(y^i, M^i; 1 \leq i \leq n)$ to be the solution to the SDE

$$dy_t^i = -\sigma_t^i \hat{\sigma}_t^i \tilde{\rho}_{ii} y_t^i \, dt + \sigma_t^i y_t^i dW_t^i,$$
$$dM_t^i = \hat{\sigma}_t^i M_t^i d\widehat{W}_t^i,$$

for deterministic functions σ^i and $\hat{\sigma}^i$. This SDE clearly has a unique strong solution and thus uniquely defines the process (y^i, M^i). Indeed it can be solved componentwise and yields explicitly the solution

$$y_t^i = y_0^i \exp\left(\int_0^t \sigma_u^i dW_u^i - \int_0^t \left(\tilde{\rho}_{ii} \sigma_u^i \hat{\sigma}_u^i + \tfrac{1}{2}(\sigma_u^i)^2 \right) du \right), \quad (16.1)$$

$$M_t^i = M_0^i \exp\left(\int_0^t \hat{\sigma}_u^i d\widehat{W}_u - \tfrac{1}{2} \int_0^t (\hat{\sigma}_u^i)^2 \, du \right).$$

The processes M^i will be used to define the FX rates and, accordingly, we will always set $\hat{\sigma}^1 \equiv 0$. Thus the Brownian motion $\widehat{W}^1$ is redundant and was only included to simplify notation (to save running the FX index from 2 to n rather than from 1 to n).

Denote by X_t^i the value in currency 1 at time t of one unit of currency i. The multi-currency model is now completed by defining all the discount factors and the FX rates in terms of the process $(y^i, M^i : i = 1, \ldots, n)$ as follows. First we use the functional forms derived in step 1 to define the discount factors at T from the swap-rate processes y^i via

$$D_{TS}^i := D_{TS}^i(y_T^i), \qquad i = 1, 2, \ldots, n.$$

Note that this automatically also defines the various PVBPs, P_T^i. Secondly, we define the FX rates at time T by

$$X_T^i := \frac{P_T^1 M_T^i}{P_T^i}.$$

This completely specifies the model at time T. To specify the model at any earlier time we use the martingale property of P^1-rebased assets, these assets being the pure discount bonds *quoted in units of currency 1*,

$$\frac{X_t^i D_{tS}^i}{P_t^1} = \mathbb{E}_{\mathbb{P}}\left[\frac{X_T^i D_{TS}^i}{P_T^1} \bigg| \mathcal{F}_t \right],$$

for all $t \leq T \leq S$. Note, in particular, that since each M^i is a martingale,

$$M_t^i = \frac{P_t^i X_t^i}{P_t^1},$$

for all $0 \leq t \leq T$.

Remark 16.1: Remark 13.1 regarding the filtration $\{\mathcal{F}_t\}$ also holds here. That is, all numeraire-rebased assets should be martingales with respect to the *asset filtration* rather than with respect to the original filtration for the probability space. Once again, for the models and examples presented here, we can work with either filtration.

It is straightforward to check that this completely determines all the asset price processes and the FX rate processes (although we will only need to consider them at the terminal time T in applications). This follows, as it did in Section 13.4 for the single-currency case, from the observation that $D_{tt}^i \equiv 1$ for all i and t and the fact that $X^1 \equiv 1$.

The steps above clearly uniquely define a multi-currency model. What is not yet clear is that it is arbitrage-free and that it has the kind of properties that we intended. This we now show.

To establish that the model in arbitrage-free, we show that all P^1-rebased assets are martingales. That is, for all i and S the process $\frac{D_{tS}^i X^i}{P^1}$ is an $(\{\mathcal{F}_t\}, \mathbb{P})$ martingale (and thus $(P^1, \mathbb{P})$ is a numeraire pair for the economy). To see this, fix i and note first that, for all t and S,

$$\frac{D_{tS}^i X_t^i}{P_t^1} = \frac{D_{tS}^i}{P_t^i} M_t^i.$$

It therefore follows, from Lemma 5.19, that $\frac{D_S^i X^i}{P^1}$ is an $(\{\mathcal{F}_t\}, \mathbb{P})$ martingale if and only if $\frac{D_S^i}{P^i}$ is an $(\{\mathcal{F}_t\}, \mathbb{Q})$ martingale where

$$\left.\frac{d\mathbb{Q}}{d\mathbb{P}}\right|_{\mathcal{F}_t} := \frac{M_t^i}{M_0^i}.$$

But, applying Girsanov's theorem for Brownian motion (Theorem 5.24) to W^i we see that (16.1) can be written as

$$y_t^i = y_0^i \exp\left(\int_0^t \sigma_u^i d\widetilde{W}_u^i - \tfrac{1}{2}\int_0^t (\sigma_u^i)^2\, du\right)$$

where $\widetilde{W}^i$ is an $(\{\mathcal{F}_t\}, \mathbb{Q})$ Brownian motion. But now, in currency i, we are in precisely the situation under which we set up the original single-currency terminal swap-rate model and, by construction, the process $\frac{D_S^i}{P^i}$ is an $(\{\mathcal{F}_t\}, \mathbb{Q})$ martingale, as required.

To conclude this section we make several observations about the properties of the model we have just defined:

(i) The measure $\mathbb{Q}$ just constructed is, in fact, the swaption measure $\mathbb{S}^i$ under which all P^i-rebased assets are martingales.

(ii) The *PVBP-weighted forward FX rates* M^i are log-normal martingales in the measure $\mathbb{P}$ $(= \mathbb{S}^1)$. One could also define the FX rates X^{ij}, where $X_t^{ij} := X_t^j / X_t^i$ is the value in currency i of one unit of currency j, and the corresponding PVBP-weighted forward FX rates, $M_t^{ij} := M_t^j / M_t^i$. It is left to the reader to check that, for each i and j, the process M^{ij} is an $\{\mathcal{F}_t\}$ martingale in the swaption measure $\mathbb{S}^i$ corresponding to the numeraire P^i. Furthermore, all the forward par swap rates y^i take a similar log-normal form in this measure and thus the model is symmetric – its general properties are not dependent on the particular currency used to set up the model initially.

(iii) Observe that the dynamics of any of the discount curves, currency i for example, depends only on the swap rate in that currency, y^i, and not on the FX rates or on the swap rates in other currencies. That is not to say the currencies are independent of each other: they are not. Furthermore, note that, if only one currency is considered, this multi-currency model reduces to the underlying single-currency terminal swap-rate model.

16.2.2 General case: volatility smiles

The treatment above was for the case when the swap rates were log-normal processes, this being the most common assumption made in practice. However, just as was the case for single-currency terminal swap-rate models, the analysis extends easily to allow for non-log-normal distributions for the swap rates y_T^i.

To cope with volatility smiles in the single-currency case, a model was set up, in Section 13.4.2, as follows. First we calculated a functional form $y_T(W_T)$ which would give y_T the required distribution, W here being a Brownian motion. Then the functional form $D_{TS}(y_T)$ was derived to be consistent with the martingale property of PVBP-rebased assets. Thus, indirectly, we calculated a functional form $D_{TS}(W_T)$.

In this multi-currency setting we apply these same ideas, along with the techniques introduced in Section 16.2.1 to extend to several currencies. First, as in step 1 above, we calculate the functional forms $D_{TS}^i(\widetilde{W}_T^i)$. We have not at this point defined the law of $\widetilde{W}^i$, but we know that if it were a Brownian motion in the measure $\mathbb{S}^i$ then currency i would, in isolation, define a single-currency terminal swap-rate model. We now carry out step 2 along similar lines to the log-normal case above. Take the same probability space, as before, supporting the correlated Brownian motion $(W^i, \widehat{W}^i; 1 \leq i \leq n)$, and now define the vector process $(z^i, M^i; 1 \leq i \leq n)$ to be the solution to the SDE

$$dz_t^i = -\sigma_t^i \hat{\sigma}_t^i \tilde{\rho}_{ii}\, dt + \sigma_t^i dW_t^i\,,$$
$$dM_t^i = \hat{\sigma}_t^i M_t^i d\widehat{W}_t^i\,.$$

The process $z = (z^1, \ldots, z^n)$ is, under the measure $\mathbb{S}^1$, a Brownian motion with drift, the drift being chosen such that, for each i, z^i is a (driftless) Brownian motion under the measure $\mathbb{S}^i$. Setting $D_{TS}^i := D_{TS}^i(z_T^i)$, and defining X^i from M^i, P^i and P^1 as in Section 16.2.1, now yields a multi-currency model with all the desired properties. The details are left as an exercise for the reader.

16.3 EXAMPLES

We consider two examples of particular practical importance: spread options and cross-currency swaptions. Henceforth, to simplify notation, we will drop the superscript [1] when the context is clear. We consider the case when the swap and FX rates are all log-normal processes with constant volatilities and use the linear swap-rate model in each currency. The constant volatility restriction is purely for notational convenience.

16.3.1 Spread options

Consider a cross-currency spread option, the payoff for which is made in currency 1 at time T and is for an amount

$$V_T = \left(\alpha y_T^{(2)} - \beta y_T^{(3)} - K\right)_+\,,$$

where α, β and K are positive constants. The index variables $y_T^{(2)}$ and $y_T^{(3)}$ are two predefined swap rates in currencies 2 and 3, and the payment currency, currency 1, may be the same as one of the index currencies.

Working in the measure $\mathbb{S}$, the time-t value for this option is given by

$$V_t = P_t \mathbb{E}_{\mathbb{S}}\left[\left(\alpha y_T^{(2)} - \beta y_T^{(3)} - K\right)_+ P_T^{-1} | \mathcal{F}_t\right]. \tag{16.2}$$

We can evaluate this expectation analytically in the case $K = 0$ or via a one-dimensional numerical integration when $K > 0$. However, rather than naïvely evaluating (16.2) by substituting in the appropriate distributions, it is instructive to examine the structure of the product in more detail. We do this by following the ideas and techniques introduced in Chapter 14 for treating convexity corrections.

We introduce some further notation. Let

$$M(A_1, A_2, \sigma_1, \sigma_2, \rho, K, T) = \mathbb{E}\left[\left(A_1 e^{N_1} - A_2 e^{N_2} - K\right)_+\right]$$

where (N_1, N_2) is a bivariate normal random variable such that

$$N_i \sim N\left(-\tfrac{1}{2}\sigma_i^2 T, \sigma_i^2 T\right)$$

and $\text{cov}(N_1, N_2) = \sigma_1 \sigma_2 \rho T$. In the case when $K = 0$ this reduces to Margrabe's exchange option formula (Margrabe (1978)) which is given by

$$M(A_1, A_2, \sigma_1, \sigma_2, \rho, 0, T) = A_1 N(d_1) - A_2 N(d_2),$$

where

$$d_1 = \frac{\log(A_1/A_2)}{\phi\sqrt{T}} + \tfrac{1}{2}\phi\sqrt{T},$$

$$d_2 = \frac{\log(A_1/A_2)}{\phi\sqrt{T}} - \tfrac{1}{2}\phi\sqrt{T},$$

$$\phi^2 = \sigma_1^2 + \sigma_2^2 - 2\rho\sigma_1\sigma_2.$$

In the case $K > 0$, M must be evaluated numerically, conditioning first on N_2 and evaluating the 'inner' integral explicitly.

Returning to the evaluation of (16.2) and writing

$$(\alpha u - \beta v - K)_+ = f(u, v),$$

the payoff can be expressed as the sum of two terms

$$V_T = f\left(y_T^{(2)}, y_T^{(3)}\right) = f\left(y_T^{(2)}, y_T^{(3)}\right)\left(A + B_T y_T\right) P_T,$$

the parameters A and B_T being as defined in Section 13.3.3 for the linear swap-rate model. Define $\mathbb{Y}$ to be the measure given by

$$\left.\frac{d\mathbb{Y}}{d\mathbb{S}}\right|_{\mathcal{F}_t} = \frac{y_t}{y_0}, \quad t \le T. \tag{16.3}$$

Then $\mathbb{Y}$ is a martingale measure corresponding to the numeraire $y_t P_t$, and it follows from Corollary 5.9 that

$$V_t = AP_t\mathbb{E}_{\mathbb{S}}\left[f\big(y_T^{(2)}, y_T^{(3)}\big)|\mathcal{F}_t\right] + B_T(y_t P_t)\mathbb{E}_{\mathbb{Y}}\left[f\big(y_T^{(2)}, y_T^{(3)}\big)|\mathcal{F}_t\right]$$
$$= wD_{tT}\mathbb{E}_{\mathbb{S}}\left[f\big(y_T^{(2)}, y_T^{(3)}\big)|\mathcal{F}_t\right] + (1-w)D_{tT}\mathbb{E}_{\mathbb{Y}}\left[f\big(y_T^{(2)}, y_T^{(3)}\big)|\mathcal{F}_t\right]$$

where

$$w = \frac{AP_t}{D_{tT}}.$$

We must now find the conditional distribution of $\big(y_T^{(2)}, y_T^{(3)}\big)$ given $\mathcal{F}_t$ in the measures $\mathbb{S}$ and $\mathbb{Y}$. We already know this for $\mathbb{S}$, the original measure under which the model was set up and under which the processes y^i satisfy (16.1). Under $\mathbb{S}$ the conditional distribution of $(\log y_T^{(2)}, \log y_T^{(3)})$ given $\mathcal{F}_t$ is bivariate normal. This joint distribution is specified via the mean and covariance structure which, recalling that we have dropped the time-dependency of the volatilities, is given by

$$\mathbb{E}_{\mathbb{S}}\left[\log y_T^i|\mathcal{F}_t\right] = \log y_t^i - \sigma^i \hat{\sigma}^i \tilde{\rho}_{ii}(T-t),$$

$$\mathrm{var}[\log y_T^i|\mathcal{F}_t] = (\sigma^i)^2(T-t),$$

and

$$\mathrm{corr}[\log y_T^{(2)}, \log y_T^{(3)}|\mathcal{F}_t] = \rho_{2,3}.$$

To find the corresponding law under $\mathbb{Y}$ we apply Girsanov's theorem (Theorem 5.24) to (16.1) (we know the Radon–Nikodým derivative is given by (16.3)), to establish that y^i is given by

$$y_t^i = y_0^i\exp\left(\sigma^i\widetilde{W}_t^i + (\sigma^1\sigma^i\rho_{1i} - \sigma^i\hat{\sigma}^i\tilde{\rho}_{ii} - \tfrac{1}{2}(\sigma^i)^2)t\right),$$

where $\widetilde{W}_t^i := W_t^i - \sigma^1\sigma^i\rho_{1i}t$ is an $(\{\mathcal{F}_t\}, \mathbb{Y})$ Brownian motion. Thus, under $\mathbb{Y}$, the conditional distribution of $\big(y_T^{(2)}, y_T^{(3)}\big)$, given $\mathcal{F}_t$, is again Gaussian, the covariance structure being as before but now having mean

$$\mathbb{E}_{\mathbb{Y}}\left[\log y_T^i|\mathcal{F}_t\right] = \log y_t^i + \sigma^i\sigma^1\rho_{1i}(T-t) - \sigma^i\hat{\sigma}^i\tilde{\rho}_{ii}(T-t).$$

It now follows that

$$V_t = w D_{tT} M^{\mathbb{S}} + (1-w) D_{tT} M^{\mathbb{Y}} \tag{16.4}$$

where

$$M^{\mathbb{N}} = M\left(\alpha\left(y_t^{(\mathbb{N};2)}\right)^*, \beta\left(y_t^{(\mathbb{N};3)}\right)^*, \sigma^{(2)}, \sigma^{(3)}, \rho_{23}, K, (T-t)\right)$$

and, for $i = 2, 3$,

$$\left(y_t^{(\mathbb{S};i)}\right)^* = y_t^i \exp\left(-\sigma^i \hat{\sigma}^i \tilde{\rho}_{ii}(T-t)\right),$$

$$\left(y_t^{(\mathbb{Y};i)}\right)^* = y_t^i \exp\left(\left(\sigma^1 \sigma^i \rho_{1i} - \sigma^i \hat{\sigma}^i \tilde{\rho}_{ii}\right)(T-t)\right).$$

The form of the valuation formula (16.4) is particularly striking. The value is expressed as an affine combination of two Margrabe-type terms, the weighting being dependent only on the domestic discount curve at time t. The Margrabe terms each have two 'convexity-adjusted' forward swap rates as input. In the first term the convexity adjustment is caused by the correlation between the respective swap rate and the corresponding FX rates. In the second term these forwards are additionally adjusted for the correlation between the swap rate and the domestic swap rate. This is a natural extension of the results presented in Chapter 14 for convexity in a single currency.

16.3.2 Cross-currency swaptions

A cross-currency swaption is an option to enter a cross-currency swap. There are several different types of cross-currency swap and we will consider four of them. The risks and pricing formulae for each of these are somewhat different, but all four fall within our framework.

With initial and final exchange on variable notional

This is one of the most common examples. On entering the swap on date T the two counterparties make an initial exchange of principal. One receives a unit amount of domestic currency (currency 1) and in return pays $(X_T^{(2)})^{-1}$ units of currency 2. This initial transaction has zero value. The counterparty who receives the unit domestic amount also pays, on dates S_j, $j = 1, 2, \ldots, n_1$, fixed amounts $\alpha_j K$, and finally he returns the initial unit payment at time S_{n_1}. In return for making these payments, this counterparty receives in currency 2 on dates $S_j^{(2)}$, $j = 1, \ldots, n_2$, a series of floating LIBOR payments plus a margin, m, on the notional $(X_T^{(2)})^{-1}$ (as in the floating leg of a standard interest rate swap). He also receives his initial principal $(X_T^{(2)})^{-1}$ back at time $S_{n_2}^{(2)}$.

To value this swaption, note first that in the case when $m = 0$ the net value at time T of all the payments in currency 2 is zero. Further note that the time-T value of the currency 1 principal exchange is $D_{TT} - D_{TS_{n_1}}$, the time-T value of the fixed payments made in currency 1 is $K \sum_{j=1}^{n_1} \alpha_j D_{TS_j}$ and the time-T value of the marginal payments received is $(X_T^{(2)})^{-1} m \sum_{j=1}^{n_2} \alpha_j^{(2)} D_{TS_j^{(2)}}^{(2)}$ in units of currency 2 or $m \sum_{j=1}^{n_2} \alpha_j^{(2)} D_{TS_j^{(2)}}^{(2)}$ expressed in units of currency 1. It then follows that for the swaption

$$
V_T = \left(D_{TT} - D_{TS_{n_1}} - K \sum_{j=1}^{n_1} \alpha_j D_{TS_j} + m \sum_{j=1}^{n_2} \alpha_j^{(2)} D_{TS_j^{(2)}}^{(2)} \right)_+
$$
$$
= \left(P_T(y_T - K) + m P_T^{(2)} \right)_+,
$$

and thus

$$
V_t = P_t \mathbb{E}_{\mathbb{S}} \left[\left((y_T - K) + m \frac{P_T^{(2)}}{P_T} \right)_+ \Big| \mathcal{F}_t \right]
$$
$$
= P_t \mathbb{E}_{\mathbb{S}} \left[\left((y_T - K) + m \left(\frac{A + B_T y_T}{A^{(2)} + B_T^{(2)} y_T^{(2)}} \right) \right)_+ \Big| \mathcal{F}_t \right].
$$

This can be evaluated by first conditioning on $y_T^{(2)}$ and then performing the remaining integral numerically.

Without initial and final exchange on variable notional

This is exactly as above except the initial and final principal payments are not made. Note that the time-T value of the floating LIBOR payments made in currency 2 is exactly $y_T^{(2)} P_T^{(2)}$ units of currency 1. We now have

$$
V_T = \left(y_T^{(2)} P_T^{(2)} - K P_T + m P_T^{(2)} \right)_+
$$

and so

$$
V_t = P_t \mathbb{E}_{\mathbb{S}} \left[\left(\left(y_T^{(2)} + m \right) \frac{P_T^{(2)}}{P_T} - K \right)_+ \Big| \mathcal{F}_t \right]
$$
$$
= P_t \mathbb{E}_{\mathbb{S}} \left[\left((A + B_T y_T) \left(\frac{y_T^{(2)} + m}{A^{(2)} + B_T^{(2)} y_T^{(2)}} \right) - K \right)_+ \Big| \mathcal{F}_t \right].
$$

Evaluation is carried out numerically as above.

With initial and final exchange on pre-agreed notionals

This is a variant of the first example. There the cashflows in currency 2 are all made based on a notional $(X_T^{(2)})^{-1}$. In this example they are made on some pre-agreed amount, C. Now we have

$$V_T = \left(P_T(y_T - K) + CmX_T^{(2)}P_T^{(2)}\right)_+$$

$$= P_T\left(y_T - K + CmM_T^{(2)}\right)_+,$$

and so

$$V_t = P_t\mathbb{E}_{\mathbb{S}}\left[\left(y_T - K + CmM_T^{(2)}\right)_+ \mid \mathcal{F}_t\right].$$

This expression is similar to those in the spread option example and evaluation is carried out in the same manner.

Without initial and final exchange on pre-agreed notionals

This is a variant of the second example. Earlier the floating payments were made on a notional amount $(X_T^{(2)})^{-1}$. In this example they are made on some pre-agreed amount, C. Now we have

$$V_T = \left(CX_T^{(2)}(y_T^{(2)} + m)P_T^{(2)} - KP_T\right)_+,$$

which yields the time-T value in currency 2 as

$$V_T^{(2)} = \left(C(y_T^{(2)} + m)P_T^{(2)} - K(X_T^{(2)})^{-1}P_T\right)_+,$$

and so, working in the measure $\mathbb{S}^{(2)}$,

$$V_t^{(2)} = P_t^{(2)}\mathbb{E}_{\mathbb{S}^{(2)}}\left[\left(C(y_T^{(2)} + m) - K(M_T^{(2)})^{-1}\right)_+ \mid \mathcal{F}_t\right].$$

This expression is again similar to those in the spread option example and evaluation is carried out in the same manner.

Orientation: Pricing Exotic American and Path-Dependent Derivatives

In the last three chapters of this book we consider the problem of how to price American and path-dependent products. By this we mean products which are not European, ones whose payoffs can only be determined by observing the market on several (more than one) distinct dates. If the product can be divided into simpler subproducts, it is assumed that at least one of the subproducts is not European. If the payoff can be determined completely by observing the market, with no decisions and actions being taken by the counterparties, we use the terminology *path-dependent*; if, in addition, the actions of one or both counterparties affect the payoff, we say the product is *American*. This latter class includes Bermudan swaptions which we introduce shortly.

The dependency of these products on the market on several dates means that to price them we must also model the market on all the relevant dates. This is certainly a generalization of the models considered in Chapters 13–16 where, although we proved that the models introduced extended to all dates, we only concentrated on the model's properties at a single date. This generalization clearly makes the modelling problem more difficult.

The general approach to modelling multi-temporal products is identical to that for European products. That is, the first step is to understand, at a high level, the relationship between the product in question and the market, in particular which features of the market are relevant to the product and must be modelled accurately, and which other products are already traded in the market relative to which this product must be priced. Once this has been done an appropriate model can be developed which reflects the required features of the market. Any model will only capture some properties of the real market and will be a (very) poor reflection of other features of the market more generally. Hence the need to understand the product in question well at the outset and to tailor the model to the product, just as we did for European products such as irregular swaptions in Section 15.4.

Understanding what in the market and within any given model has most effect on the price of a multi-temporal derivative is far from easy. There is still a shortage of research available on this problem. When modelling these products people often take the view that adding more factors (driving Brownian motions) to a model gives a more realistic model of the market and that this will therefore give a better price for the derivative. To an extent this is true. However, whenever an extra factor is added various parameters must be fitted. For many models it is often unclear how these should be chosen and what effect they will have on the derivative's price. And every time a new factor is added the computation problem of calculating a derivative price numerically gets much more difficult.

Recently, starting in 1995, there have been significant developments in interest rate modelling which have made the pricing of multi-temporal derivatives much easier. Prior to this date the primary models used for path-dependent and American products were short-rate models. The main benefits of these models are that they are arbitrage-free and easy to implement. However, they have the major drawback that they are parameterized in terms of a variable, the short rate, which is only distantly related to the product being priced. Consequently, it is usually difficult to assess how well a short-rate model captures those features of the market relevant to the pricing of any given product. We shall discuss models of this type in Chapter 17, in particular the Vasicek–Hull–White model which is highly tractable and easy to implement.

The breakthrough in 1995 was the introduction of the first *market models*. These models are formulated directly in terms of the dynamics of market rates such as LIBORs and swap rates. This has two major consequences. First, it is now possible to formulate models which reflect well the (implied) distributions of precisely those market rates relevant to any particular derivative product, and thus these models have much better calibration properties than short-rate models. The second major consequence is greater clarity and focus in the thought process around the pricing of multi-temporal derivatives. Now that we can model directly the law of a finite number of market rates on a finite number of dates, now that we realize (as long as we work in an appropriate probability measure) that it is only the distribution of these particular rates on these particular dates that affects the derivative price *and that given this distribution the infinitesimal behaviour of any process is irrelevant*, we can focus on what it is about this (high- but finite-dimensional) distribution that most affects the derivative price and which must be captured by the model. And now it is also much clearer what information in the real market we should use when pricing a particular derivative product. We shall discuss market models in Chapters 18 and a further development, *Markov-functional models*, in Chapter 19. This latter class of models shares with market models the ability to model market rates directly, but has improved implementation properties and also moves further away from the idea of specifying derivative pricing models via the infinitesimal behaviour of relevant stochastic processes.

An important example of this type of multi-temporal product which we shall meet repeatedly in the remainder of this book is the Bermudan swaption, or equivalently a *cancellable swap*. Let $T = S_0$ be the start date of a swap with payment dates $S_1 < \ldots < S_n$ and, as usual, let α_i, $i = 1, \ldots, n$, denote the corresponding accrual factors. If two counterparties enter a swap they are each obliged to make all the cashflows even if on some occasions it is not to their advantage. A cancellable swap is one in which one counterparty has the right, on any payment date, to cancel all the remaining cashflows. This he would clearly do if interest rates had moved against him. A *Bermudan swaption* is the option part of the cancellable swap. It gives the holder the right, on any of the swap payment dates, to enter the remaining swap. It is clear that a cancellable receive-fixed swap is just the combination of owning a Bermudan payers swaption and entering the underlying receive-fixed swap – if you exercise the swaption then the new pay-fixed swap will precisely offset all remaining obligations on the original receive-fixed swap.

What is it that we need to model to value a Bermudan swaption? Recall that for the vanilla swaption, analysed in Chapter 11, the only thing that we needed to model, as long as we worked in the relevant swaption measure, was the forward par swap rate. It is clear that for the Bermudan we must model *all* the discount factors relevant to the underlying swap, $\{D._{S_i} : 0 \leq i \leq n\}$. Indeed, working in the martingale measure corresponding to the numeraire $D._{S_n}$, it is straightforward to see (just write down the valuation formula) that all we need to model is the $\frac{1}{2}n(n+1)$-dimensional vector random variable $(D_{S_i S_j} : 0 \leq i < j \leq n)$. This is equivalent to modelling the $\frac{1}{2}n(n+1)$-dimensional swap-rate vector $(y_{T_i}^{ij} : 1 \leq i \leq j \leq n)$, where $T_i = S_{i-1}$ and

$$y_t^{ij} = \frac{D_{tT_i} - D_{tS_j}}{P_t^{ij}},$$

$$P_t^{ij} = \sum_{k=i}^{j} \alpha_k D_{tS_k}.$$

The market provides information, via swaption prices as described in Section 13.2.2, on the distribution of each $y_{T_i}^{ij}$ in its own swaption measure $\mathbb{S}^{ij}$. To develop a model which correctly captures all these marginal distributions would be overfitting. Even if this were a good idea, it would involve too many random factors to be implementable in practice. However, as we shall see in Chapter 19, it is possible to develop a practically implementable model (driven by a *univariate* Brownian motion) which correctly captures the market prices of swaptions corresponding to the par swap rates $y_{T_i}^{in}$, those corresponding to the swaps which could be entered if the Bermudan were exercised, and with enough free parameters to fit their joint distribution to a degree sufficient for Bermudan valuation.

17

Short-Rate Models

17.1 INTRODUCTION

This chapter is devoted to the study of short-rate models. We single out, in particular, the Vasicek–Hull–White model for more detailed study. At the theoretical level short-rate models have already been considered in Chapter 8, where they were related to other classes of models. What we actually mean here is the special case of arbitrage-free models of the terms structure for which the short rate $\{r_t, t \geq 0\}$ is a (time-inhomogeneous) Markov process in the risk-neutral measure $\mathbb{Q}$. Usually we will also restrict attention to a short-rate process driven by a univariate Brownian motion, but we need not make this restriction in general. This class of models has been of considerable importance historically, primarily because they are both arbitrage-free and can be implemented numerically. Recently their popularity has been decreasing amongst practitioners with the advent of market models which have better calibrating properties.

All the short-rate models in common use are specified via an SDE,

$$dr_t = \mu(r_t, t)dt + \sigma(r_t, t)\, dW_t,$$

where W is a Brownian motion in the risk-neutral measure $\mathbb{Q}$. We shall meet several examples in the coming sections. The functions μ and σ are chosen to give the model particular behaviour, tractability, and to make it arbitrage-free. For example, the Vasicek–Hull–White model satisfies an SDE of the form

$$dr_t = (\theta_t - a_t r_t)dt + \sigma_t\, dW_t, \tag{17.1}$$

where a, θ and σ are deterministic functions of time. The exact form of these functions is chosen to fit the model to initial bond prices (to make it arbitrage-free) and to option prices. We shall discuss this further later.

The Markovian property of the short rate, in the risk-neutral measure $\mathbb{Q}$, is essential for these models to be implementable in practice. To see why this

Financial Derivatives in Theory and Practice Revised Edition.   P. J. Hunt and J. E. Kennedy
© 2004 John Wiley & Sons, Ltd   ISBNs: 0-470-86358-7 (HB); 0-470-86359-5 (PB)

is the case, recall from Chapter 8 that the value D_{tT} of a pure discount bond at t which matures at T is given by (see equation (8.11))

$$D_{tT} = \mathbb{E}_{\mathbb{Q}}\left[\exp\left(-\int_t^T r_u\,du\right)\middle|\mathcal{F}_t\right],\qquad(17.2)$$

where $\{\mathcal{F}_t\}$ is the augmented natural filtration generated by the Brownian motion W. The Markov property of r ensures that, for all pairs $t \leq T$, (17.2) is a function of the triple (r_t, t, T). Thus the state of the market at t is completely summarized by the pair (r_t, t). It is this property that allows one to price a derivative using any of the standard numerical techniques such as numerical integration, simulation, trees or some finite-difference algorithm, as appropriate. We shall not discuss these standard techniques but refer the reader to Duffie (1996).

In this chapter we do the following. In Section 17.2 we give the defining SDE for a few of the best-known short-rate models. In each case there are parameters which need to be determined by fitting the model to market prices. This calibration must usually be done numerically and can often be onerous. One model for which the calibration is relatively straightforward is the Vasicek–Hull–White model which we study in detail in Sections 17.3 and 17.4. The form of the defining SDE (17.1) is such that the solution process is a Gaussian process and therefore much can be said about this model analytically. In particular, closed-form bond prices result. In Section 17.3 we show how to recover the Vasicek–Hull–White model parameters efficiently from market prices, and in Section 17.4 we consider the problem of pricing a Bermudan swaption using the Vasicek–Hull–White model. We provide an algorithm which is highly efficient and stable. The algorithm exploits the Gaussian property of the Vasicek–Hull–White model and can be applied to other Gaussian processes such as some of the example Markov-functional models introduced in Chapter 19.

17.2 WELL-KNOWN SHORT-RATE MODELS

Here we briefly describe a few of the best-known short-rate models. By far the most important for practitioners are the first two, the Vasicek–Hull–White model and the Black–Karasinski model.

17.2.1 Vasicek–Hull–White model

Vasicek (1977) introduced the first short-rate model, not with derivative pricing in mind but from an economic perspective. For his model, under $\mathbb{Q}$, the (mean reverting) short rate satisfies the SDE

$$dr_t = (\theta - ar_t)dt + \sigma dW_t$$

for some constants θ, a and σ. It is immediately clear that the Vasicek model does not have enough free parameters so that it can be calibrated to correctly price all pure discount bonds, i.e. we cannot in general choose a, σ and θ to simultaneously solve

$$D_{0T} = \mathbb{E}_{\mathbb{Q}}\left[\exp\left(-\int_0^T r_u\, du\right)\right]$$

for all maturities $T > 0$. This led Hull and White (1990) to extend this model by replacing these constants with deterministic functions,

$$dr_t = (\theta_t - a_t r_t)dt + \sigma_t dW_t \,. \tag{17.3}$$

It is apparent from (17.3) that the short rate r can take negative values in the Vasicek–Hull–White model. This is obviously an undesirable feature, one much criticized. However, in our view this criticism is overplayed since all the main short-rate models have their own particular weaknesses and, as long as the model is calibrated and used appropriately, this in itself should not usually present too many problems. Indeed, the Black–Karasinski model, which is often used instead of the Vasicek–Hull–White model precisely because it keeps interest rates positive, suffers from giving too high a probability to sample paths of r which spend large amounts of time at large positive values. This is just as problematic, leading, for example, to infinite futures prices!

The major advantage of the Vasicek–Hull–White model, and the primary reason why it is used so widely, is that it is highly tractable. This tractability is important in order to be able to calibrate the model efficiently to market prices. The SDE (17.3) is well known as one which yields a Gaussian process as a solution and it is easily solved. To do this, consider the process $y_t := \phi_t r_t$ where

$$\phi_t = \exp\left(\int_0^t a_u du\right).$$

Applying Itô's formula to y yields

$$dy_t = \phi_t \theta_t dt + \phi_t \sigma_t dW_t$$

and thus

$$y_t = y_0 + \int_0^t \phi_u \theta_u du + \int_0^t \phi_u \sigma_u dW_u$$

$$= y_0 + \int_0^t \phi_u \theta_u du + \widetilde{W}(\xi_t)\,, \tag{17.4}$$

where

$$\xi_t := \int_0^t \phi_u^2 \sigma_u^2 du$$

and $\widetilde{W}$ is a Brownian motion (adapted to the filtration $\{\tilde{\mathcal{F}}_t\} = \{\mathcal{F}_{\xi_t^{-1}}\}$). To see this last step just check that the process $\widetilde{W}$, defined by

$$\widetilde{W}_t := \int_0^{\xi^{-1}(t)} \phi_u \sigma_u dW_u,$$

satisfies the defining properties of a Brownian motion (Definition 2.4).

The representation for y_t given in (17.4) allows us to make all the necessary deductions and calculations for the short-rate process r_t ($= \phi_t^{-1} y_t$) that we need when calibrating the model to bond and option prices. In particular, given r_s, both r_t, $t \geq s$, and integrals of r_t against deterministic functions will be Gaussian. From this we obtain immediately the standard result for the value at time t of a pure discount bond paying unity at time T, namely,

$$D_{tT} = \mathbb{E}_{\mathbb{Q}}\left[\exp\left(-\int_t^T r_u du\right)\Big|\mathcal{F}_t\right] = A_{tT} e^{-B_{tT} r_t} \qquad (17.5)$$

for suitable A and B to be determined. These functions can be evaluated directly, using the properties of Brownian motion and log-normal random variables, to yield A and B in terms of the functions θ, a and σ, and the reader may choose to do so now. However, there is a more direct way to calculate these parameters, in terms of the initial discount curve $\{D_{0T} : 0 \leq T < \infty\}$. We shall do this in Section 17.3.

17.2.2 Log-normal short-rate models

Motivated by the log-normality of LIBORs and swap rates as given by Black's option pricing formula (see Sections 11.3.1 and 11.4), and by a desire to have a model with positive interest rates, one is led to postulate a short-rate model in which the short rate is a log-normal process,

$$d\log r_t = (\theta_t - a_t \log r_t)\, dt + \sigma_t dW_t. \qquad (17.6)$$

This model, usually referred to as the Black–Karasinski model (Black and Karasinski (1991)), and the Vasicek–Hull–White model are undoubtedly the most commonly used short-rate models in the markets. The Black–Karasinski model has the distinct advantage over the Vasicek–Hull–White model that it retains positive interest rates. However, as mentioned earlier, the cost, due to the log-normality, is heavy positive tails for the distribution of each r_t which leads to infinite values for futures prices (see Hogan and Weintraub (1993)). This illustrates quite clearly the dangers of a model which is parameterized in terms of a variable only remotely related to the relevant market rates. A second drawback is that fitting the model to market prices is time-consuming. If we knew the function σ, it would be a straightforward and quick matter to fit the remaining parameters. The way to proceed is using a tree algorithm and forward induction as described by Jamshidian (1991). (We shall not cover this topic.) However, when simultaneously fitting the model to option prices to determine σ one must resort to some search algorithm to solve for the function σ; fix σ, solve for θ, change σ, re-solve for θ, and so on.

The function a in the SDE (17.6), and also in (17.3), is extremely important. We shall defer discussion of this until Section 19.7. Some practitioners simultaneously fit all three functions θ, a and σ to the market, others fix a (as discussed in Section 19.7) and then fit just θ and σ. The latter is, in our view, the better approach. Note that if we were to set $a_t = -\dot{\sigma}_t/\sigma_t$ then

$$r_t = \exp(\alpha_t + \sigma_t W_t)$$

where

$$\alpha_t = \sigma_t \left(\frac{\log r_0}{\sigma_0} + \int_0^t \frac{\theta_u}{\sigma_u} \, du \right).$$

This special case is known as the Black–Derman–Toy model.

17.2.3 Cox–Ingersoll–Ross model

We include this model because it is extremely well known. However, it is not one that we have ever used for derivative pricing purposes.

The original Cox–Ingersoll–Ross model (Cox *et al.* (1985)) is defined via the short-rate SDE,

$$dr_t = (\theta - ar_t) \, dt + \sigma \sqrt{r_t} dW_t \,, \tag{17.7}$$

for constants θ, a and σ. Three features are immediately clear from the SDE (17.7). First, the short rate is mean-reverting, being pulled towards the level θ/a. Second, its diffusion term is decreasing with the level of the short rate according to $\sqrt{r_t}$. This is intermediate between the Vasicek–Hull–White model, in which the diffusion term does not depend on r_t and for which rates can go negative, and the log-normal models in which the diffusion term is proportional to r_t and rates are strictly positive. In fact, it can be shown that if $\theta < \frac{1}{2}\sigma^2$ the Cox–Ingersoll–Ross model will eventually hit zero; if $\theta \geq \frac{1}{2}\sigma^2$ it is always strictly positive.

The third property to note about the Cox–Ingersoll–Ross model is that there are not enough free parameters to fit an arbitrary initial term structure. This is easily rectified by modifying to the *generalized* Cox–Ingersoll–Ross model,

$$dr_t = (\theta_t - a_t r_t) \, dt + \sigma_t \sqrt{r_t} dW_t \,. \tag{17.8}$$

The SDE (17.8) is readily solved when $\theta_t \geq \frac{1}{2}\sigma_t^2$ for all t. For this generalized model there are, as was the case for the log-normal models, no closed-form bond pricing formulae and thus any implementation would again be relatively onerous to calibrate simultaneously to the initial discount curve and option prices.

17.2.4 Multidimensional short-rate models

This is not a topic we shall cover in much detail. In the financial literature there are a great many short-rate models which involve two factors but these are rarely used for derivative pricing in practice. See, for example, Musiela and Rutkowski (1997) and references therein. There are three primary reasons for this. The first is the difficulty with implementing such a model. For example, calibrating a tree, already onerous for all but the Vasicek–Hull–White model, will now be even more costly since the tree will be (at least) two-dimensional and there will be more parameters to fit. The second problem is one of understanding how these factors and associated model parameters affect the properties of the model and consequently option prices. The third reason is the fact that a one-factor model is usually enough for path-dependent and American-type products, as long as the model is fitted with the product to be priced in mind, a point discussed in Section 15.4 in the context of European amortizing swaptions.

Having given all these caveats, two-factor models can be developed. From an implementation viewpoint, one needs the resulting model to be of the form

$$r_t = f(X_t, Y_t, t) \tag{17.9}$$

for some (time-inhomogeneous) Markov process (X, Y). Typically the process (X, Y) would be driven by a (two-dimensional) Brownian motion. A special case of (17.9) is when the process (X, Y) takes the form

$$X_t = \mu_t^X + \int_0^t \sigma_u^X \, dW_u^X,$$

$$Y_t = \mu_t^Y + \int_0^t \sigma_u^Y \, dW_u^Y,$$

for deterministic functions μ^X, μ^Y, σ^X and σ^Y and for a Brownian motion (W^X, W^Y). In this special case one knows explicitly, for any $t \geq s$, the distribution of (X_t, Y_t) (hence of r_t) given (X_s, Y_s). When the function f at (17.9) is linear in (X, Y) we are once more in the situation where bond prices are available in closed form and this considerably simplifies the calibration of the model to market data (but the effect of each of the two factors is still difficult to appreciate).

17.3 PARAMETER FITTING WITHIN THE VASICEK–HULL–WHITE MODEL

We now detail how all the parameters of the Vasicek–Hull–White model can be calculated from market prices. Recall that the short rate is driven by the SDE

$$dr_t = (\theta_t - a_t r_t)dt + \sigma_t dW_t \qquad (17.10)$$

in the risk-neutral measure $\mathbb{Q}$, and furthermore (equation (17.5)), bond prices in the model are of the form

$$D_{tT} = A_{tT} e^{-B_{tT} r_t} .$$

We shall show how to derive each of the functions a, θ, σ, A and B. Note, however, that for derivative valuation explicit knowledge of each of these parameters is not necessary. The reason for this, as we shall shortly see, is that for the Vasicek–Hull–White model it is more convenient to work not in the risk-neutral measure but in the forward measure corresponding to taking some fixed bond $D_{.T}$ as numeraire. If we do this the only parameters we need to explicitly derive are (various integrals of) the functions σ and a.

Throughout this section we assume only that the initial discount curve, $\{D_{0T} : T \geq 0\}$, is right-differentiable with continuous right derivative almost everywhere (the usual case in practice). We comment on the implications of the discount curve not having a continuous derivative in Section 17.3.3.

The market prices to which the model will be calibrated are bond prices (i.e. the initial discount curve) and option prices, either caps and floors or swaptions as appropriate to the pricing application. We shall assume that the *mean reversion* function a has been prespecified. This means that it is only possible to calibrate the model to correctly price at most one option, with a given strike, per maturity date. One could fit the function a as well, and doing so would allow us to simultaneously fit to the prices of two different options per maturity date (such as a swaption and a caplet). However, in Section 15.4 we discovered the dangers of fitting a one-factor model to the price of more than one option per expiry date. Furthermore, as discussed in Section 19.7, the mean reversion has such a significant effect on the joint distribution of market swap rates that we should take care to ensure that any fitting algorithm fixes this parameter in a way most appropriate to the exotic option being priced. With the two remaining degrees of freedom (θ and σ) one could, in principle (as long as the prices are consistent), calibrate the model to the full initial discount curve and one option price for each option maturity date T. However, in applications it is only necessary to calibrate to options with a finite number of maturity dates and thus it is common to fix a functional form for σ between the relevant dates T_i.

We also suppose that the prices of the market options that we use to calibrate the model are consistent with the initial assumptions about the form of a, i.e. given our choice for a there exist θ and σ consistent with the market prices.

For future reference we make the following definitions, all relating to integrals of the original model parameters.

$$\phi_t = \exp\left(\int_0^t a_u du\right), \quad \psi_t = \int_0^t \phi_u^{-1} du,$$

$$\xi_t = \int_0^t \phi_u^2 \sigma_u^2 du, \quad \zeta_t = \int_0^t \psi_u \phi_u^2 \sigma_u^2 du, \quad \eta_t = \int_0^t \psi_u^2 \phi_u^2 \sigma_u^2 du,$$

$$\mu_t = \int_0^t \phi_u \theta_u du, \quad \lambda_t = \int_0^t \psi_u \phi_u \theta_u du.$$

It is only actually ψ and ξ that we shall need for the pricing algorithm described in Section 17.4 below.

17.3.1 Derivation of ϕ, ψ and $B._T$

Given that a is a predefined user input (for current purposes) the evaluation of ϕ and ψ is straightforward. To calculate B_{tT}, $0 \le t \le T$, we note that, for each T,

$$M_{tT} := \exp\left(-\int_0^t r_u du\right) D_{tT} = \exp\left(-\int_0^t r_u du\right) A_{tT} e^{-B_{tT} r_t}$$

defines a martingale $\{M._T, t \le T\}$ in the risk-neutral measure, i.e. the measure in which (17.10) holds. Applying Itô's formula to $M._T$ yields

$$\frac{dM_{tT}}{M_{tT}} = -r_t dt + \frac{dD_{tT}}{D_{tT}}$$

$$= \left[-r_t + (A'_{tT}/A_{tT}) - r_t B'_{tT} + \tfrac{1}{2}\sigma_t^2 B_{tT}^2 - B_{tT}(\theta_t - a_t r_t)\right] dt$$
$$\quad - \sigma_t B_{tT} dW_t$$

(A' and B' representing the derivative with respect to the first parameter t). Since $M._T$ is a martingale the finite variation term is identically zero for all values of r_t and t. It thus follows that

$$-1 - B'_{tT} + a_t B_{tT} = 0, \qquad (17.11)$$
$$(A'_{tT}/A_{tT}) + \tfrac{1}{2}\sigma_t^2 B_{tT}^2 - B_{tT}\theta_t = 0. \qquad (17.12)$$

Solving (17.11) subject to the condition $B_{TT} = 0$ (which is required since $D_{TT} \equiv 1$) yields

$$B_{tT} = \phi_t \int_t^T \phi_u^{-1} du = \phi_t(\psi_T - \psi_t).$$

17.3.2 Derivation of ξ, ζ and η

We already know all the terms in each of the respective integrands except σ. It suffices to calculate $\xi_t, t \geq 0$, since this implicitly defines σ and thus ζ and η. In practice we will only have a finite set of market option prices to calibrate to and so will calculate ξ_t for a finite set of times and assume a (simple) functional form for σ at intermediate points.

Suppose we have available the price V_0 of an option expiring at T and paying

$$\left(\sum_{j \leq J} c_j D_{TS_j} - K \right)_+ .$$

This general form includes all standard caps, floors and swaptions. Taking $D_{\cdot T}$ as numeraire, it follows that

$$V_0 = D_{0T} \mathbb{E}_{\mathbb{F}} \left[\left(\sum_{j \leq J} c_j D_{TS_j} - K \right)_+ \right],$$

where the measure $\mathbb{F}$ is equivalent to the original risk-neutral measure and is such that $\frac{D_{\cdot S}}{D_{\cdot T}}$ is a martingale for all $S > 0$. Indeed, $\mathbb{F}$ and $\mathbb{Q}$ are related by

$$\left. \frac{d\mathbb{F}}{d\mathbb{Q}} \right|_{\mathcal{F}_t} = \exp \left(- \int_0^t r_u du \right) \frac{D_{tT}}{D_{0T}} .$$

Applying Itô's formula to (the martingale) $\frac{D_{\cdot S_j}}{D_{\cdot T}}$ and changing measure to $\mathbb{F}$ yields

$$d \left(\frac{D_{tS_j}}{D_{tT}} \right) = \frac{D_{tS_j}}{D_{tT}} (B_{tT} - B_{tS_j}) \sigma_t d\bar{W}_t$$

$$= \frac{D_{tS_j}}{D_{tT}} (\psi_T - \psi_{S_j}) \phi_t \sigma_t d\bar{W}_t,$$

for an $(\{\mathcal{F}_t\}, \mathbb{F})$ Brownian motion $\bar{W}$. This SDE can be solved explicitly and yields the unique solution (see Chapter 6)

$$\frac{D_{tS_j}}{D_{tT}} = \frac{D_{0S_j}}{D_{0T}} \exp \left((\psi_T - \psi_{S_j}) \int_0^t \phi_u \sigma_u d\bar{W}_u - \frac{1}{2} (\psi_T - \psi_{S_j})^2 \xi_t \right)$$

$$= \frac{D_{0S_j}}{D_{0T}} \exp \left((\psi_T - \psi_{S_j}) \hat{W}(\xi_t) - \frac{1}{2} (\psi_T - \psi_{S_j})^2 \xi_t \right) \qquad (17.13)$$

where $\hat{W}$ is an $(\{\mathcal{F}_{\xi_t^{-1}}\}, \mathbb{F})$ Brownian motion, and thus

$$V_0 = \mathbb{E}_{\mathbb{F}} \left[\left(\sum_{j \leq J} c_j D_{0S_j} e^{(\psi_T - \psi_{S_j}) \hat{W}(\xi_T) - \frac{1}{2}(\psi_T - \psi_{S_j})^2 \xi_T} - K D_{0T} \right)_+ \right]. \qquad (17.14)$$

Suppose for now that we already know the value of ξ_T. Here we show how then to calculate $V_0(\xi_T)$, the option value for this particular ξ_T. Solving for ξ_T can then be completed by iterating over ξ_T with either Brent or Newton-Raphson to match the value $V_0(\xi_T)$ to the true market option value V_0.

Given a value for ξ_T, evaluation of (17.14) is performed using the decomposition first introduced by Jamshidian (1989). A requirement for this approach to (be guaranteed to) work is that the payoff must be monotone in the realized value of $\hat{W}$. This is always the case for standard options, including those with amortizing and forward start features.

The first step in this approach is to find the value $\hat{\omega}$ for $\hat{W}(\xi_T)$ which sets the term inside the expectation (17.14) to be exactly zero. A safe Newton-Raphson algorithm (Press *et al.* (1988)) will do this quickly. We can then rewrite (17.14) as

$$V_0 = \mathbb{E}_{\mathbb{F}}\left[\sum_{j \leq J} c_j D_{0S_j} \left(e^{(\psi_T - \psi_{S_j})\hat{W}(\xi_T) - \frac{1}{2}(\psi_T - \psi_{S_j})^2 \xi_T} - K_j \right)_{\pm} \right],$$

where

$$K_j = \exp\left((\psi_T - \psi_{S_j})\hat{\omega} - \tfrac{1}{2}(\psi_T - \psi_{S_j})^2 \xi_T \right)$$

and the $\pm$ in the expectation is $+$ if the payoff is increasing in $\hat{W}$ and $-$ if it is decreasing. With this decomposition the expectation can now be easily evaluated.

Before moving on we note that should an alternative model be required in which a is not chosen a priori but is instead a function of σ then the starting point for the algorithm would be this section, solving for a and σ simultaneously.

17.3.3 Derivation of μ, λ and $A._T$

Substituting for B_{tT} in (17.12) and integrating yields

$$\log(A_{tT}) = \log(A_{0T}) + \psi_T \mu_t - \lambda_t - \tfrac{1}{2}[\psi_T^2 \xi_t - 2\psi_T \zeta_t + \eta_t]. \qquad (17.15)$$

We know A_{0T} from the values D_{0T}, namely $A_{0T} = D_{0T} e^{B_{0T} r_0}$. Substituting $t = T$ into (17.15), differentiating with respect to T and noting that $A_{TT} \equiv 1$ yields

$$\mu_T = \psi_T \xi_T - \zeta_T - \phi_T \frac{\partial}{\partial T} \log(A_{0T}) \qquad (17.16)$$

and thence

$$\lambda_T = \log(A_{0T}) - \psi_T \phi_T \frac{\partial}{\partial T} \log(A_{0T}) + \tfrac{1}{2}(\psi_T^2 \xi_T - \eta_T).$$

Substituting back into (17.15) now gives the function A_{tT}.

Remark 17.1: Observe from equations (17.16) that if the initial discount curve does not have a continuous first derivative everywhere then there will be time points at which the function μ is also discontinuous. This corresponds to the function θ having a delta function component at these time points. As a result the underlying short-rate process is continuous whenever the initial discount curve has a continuous derivative but undergoes jumps at the time points where the discount curve has a discontinuous derivative.

In the theory of SDEs as discussed in Chapter 6 we only considered SDEs for continuous processes. However, the jump in the function μ is of a deterministic amount at a deterministic time and this corresponds to a deterministic jump in the short-rate process r at a deterministic time. Thus, although we have not strictly covered such processes, it is easy to see that the theory developed is sufficient by inductively applying the continuous theory between jump times, explicitly incorporating the jumps, and piecing together the solutions.

17.4 BERMUDAN SWAPTIONS VIA VASICEK–HULL–WHITE

We defined the Bermudan swaption in the preceding Orientation section on multi-temporal products. Here we consider it again and describe an efficient algorithm to value Bermudan swaptions within the Vasicek–Hull–White model.

The Bermudan we consider here is more general than that described in the Orientation. There we restricted to a Bermudan which could be exercised only on payment dates of a vanilla swap. Furthermore, the swaps on which the constituent options were based were all themselves vanilla swaps. Here we allow the exercise dates, on which the holder has the right but not the obligation to enter a swap, to be an arbitrary but finite sequence of times $T_1, \ldots, T_K$. If the option is exercised the underlying swap which the holder enters may be irregular, with irregular payment dates and irregular coupon amounts. For the swap which can be entered at the exercise time T_i, we denote by S_{i0} the swap start date and by $S_{ij}, 1 \leq j \leq J_i$, the payment dates. We saw in Chapter 10 that the value of any swap can be written as a linear combination of discount factors. Thus if the option is exercised at T_i the value of the swap entered is

$$I_i := \Big(\sum_{j \leq J_i} c_{ij} D_{T_i S_{ij}} \Big)_+ \tag{17.17}$$

for some known constants c_{ij}. Where there is no risk of confusion in what follows we will tend to drop some of the subscripts.

In the following sections we describe in detail an algorithm for pricing a Bermudan swaption such as this. Bermudan pricing is usually done by building a tree to carry out a dynamic programming calculation via backward induction and is standard. The algorithm described below also uses backward induction but exploits the Gaussian structure to gain extra efficiencies. We refer to the set of (r_t, t) pairs used in the algorithm as a *tree* but we do not explicitly assign probabilities to transitions within the tree as is usual for standard binomial and trinomial pricing trees. The steps we need to go through to implement the algorithm are the following:

(i) Choose a probability measure within which to work and use a set of market instruments to calibrate the Vasicek–Hull–White model. If we were to work in the risk-neutral measure $\mathbb{Q}$ this would determine (various integrals of) the functions a, θ and σ in (17.10). We shall see that not all of these are needed explicitly.

(ii) Specify the locations of the nodes in the tree and how the intrinsic value of the Bermudan will be calculated at each of these nodes.

(iii) Calculate at each node the discounted future value of the Bermudan, and thus calculate the Bermudan value at this node by taking the maximum of this with the intrinsic value at the node.

We address each of these in turn.

17.4.1 Model calibration

The model needs to be calibrated so that it correctly prices a set of market instruments. These instruments will typically be (pure discount) bonds and either swaptions or caps and floors, as appropriate.

Calibrating a Vasicek–Hull–White model to these products is, because of the one-factor Gaussian structure, a straightforward task. We will for our current purposes assume that the (critical) mean reversion function a has been chosen and fixed beforehand, although these techniques also apply to the more general situation when a depends on σ.

For our algorithm we shall work in the forward measure corresponding to taking as numeraire the bond $D_{\cdot T_K}$, that is the bond which matures on the exercise date of the last option in the Bermudan swaption. We are given the initial discount curve, $\{D_{0T}, 0 \le T < \infty\}$, and a set of option prices, one maturing on each date T_i. The only parameters we shall need to imply are the underlying bond prices $D_{0S_{ij}}$, which can be read off from the discount curve, the values ψ_{T_i} and $\psi_{S_{ij}}$, a user input since a is predefined, and the values ξ_{T_i}. Recall that we saw how to derive the function ξ, using a safe Newton–Raphson or Brent algorithm (Press *et al.* (1988)), in Section 17.3.2.

17.4.2 Specifying the 'tree'

The first step is to identify the random process that will be modelled by the 'tree' and then allocate the positions of the tree nodes.

We saw in Section 17.3.2 that, if we work in the forward measure $\mathbb{F}$

corresponding to the numeraire bond $D._{T_K}$, then all discount factors could be written in the form

$$\frac{D_{tS}}{D_{tT_K}} = \frac{D_{0S}}{D_{0T_K}} \exp\Big((\psi_{T_K} - \psi_S) X_t - \tfrac{1}{2} (\psi_{T_K} - \psi_S)^2 \xi_t \Big), \qquad (17.18)$$

where X is a (deterministically) time-changed Brownian motion,

$$X_t := \hat{W}(\xi_t),$$

and $\hat{W}$ is as defined in (17.13). Note that (17.18), an equation for ratios of discount factors, defines all discount factors since we can recover D_{tT_K} by substituting $S = t$ into (17.18). We shall take the Gaussian process X as the one we model with the tree.

The most striking property of the tree is that we *only* position nodes at times when decisions must be made, namely $T_1, \ldots, T_K$. At intermediate times no decisions need to be made and the value function V is well behaved. We are thus able to calculate any intermediate properties explicitly.

At each grid-time T_i we position nodes to be equally spaced in the variable X_{T_i}. This 'vertical' spacing between nodes is a user input, depending on the accuracy required. We will describe shortly the location of the top and bottom nodes in any column.

The importance of being able to choose the times of nodes should not be overlooked. We have positioned the nodes at precisely those times at which we need to know the option's value and so have not lost accuracy caused by nodes missing decision points.

The choice of the locations of the extreme nodes in each column depends on the required accuracy. To understand the effect of this choice, note that $V_0(X_0)$, the Bermudan value at zero, can be written as

$$V_0(X_0) = D_{0T_K} \mathbb{E}_{\mathbb{F}} \left[V_\tau(X_\tau) D_{\tau T_K}^{-1} \right]$$

$$= D_{0T_K} \mathbb{E}_{\mathbb{F}} \left[V_\tau(X_\tau) D_{\tau T_K}^{-1} \mid X_{T_i} < m_i \right] \mathbb{P}(X_{T_i} < m_i)$$

$$+ D_{0T_K} \mathbb{E}_{\mathbb{F}} \left[V_\tau(X_\tau) D_{\tau T_K}^{-1} \mid X_{T_i} \geq m_i \right] \mathbb{P}(X_{T_i} \geq m_i) \, (17.19)$$

where τ is the random time at which the option is exercised and $V_t(X_t)$ is its value at time t. By truncating column i of the tree at m_i we are ignoring the second term in (17.19) and also the probability in the first term. Given that the process X is Gaussian, the first probability differs from one by a term $O(\exp(-m_i^2/2\xi_{T_i}))$. The second term decays to zero at the same rate since the payoff is at most $O(\exp(cm_i))$ for some constant c. This argument extends to include all the exercise times T_i and shows that errors resulting from truncation are below polynomial in order of magnitude.

Should we wish to choose the limits to guarantee a certain level of accuracy we can use the above argument to do so. What is important is the number of standard deviations of the Gaussian distribution for X_{T_i} covered by the nodes. The limits should thus be set at a fixed number of standard deviations above and below zero, the mean for X_{T_i}. In practice, because the error decays so quickly, the exact level is not important and a conservative value can be chosen without adversely affecting the algorithm's speed.

The precise algorithm for placing nodes is as follows. First select your node spread, m. Locate the top node in each column at the value $m\sqrt{\text{var}[X_{T_i}]} = m\sqrt{\xi_{T_i}}$. Then locate nodes at a regular spacing Δ, chosen depending on the accuracy required (as described in Section 17.4.5), down from this node until the first node lower than $-m\sqrt{\xi_{T_i}}$. We do not specify any further tree structure at this point. Strictly speaking, the algorithm is not a standard tree algorithm, and we do not assign arcs and probabilities as in a tree.

17.4.3 Valuation through the tree

To value the Bermudan swaption we work back in time from the final exercise date assigning option values at each node in the tree. On the final exercise date, if we have not already exercised, the option value is its intrinsic value which can be calculated as a closed-form expression from (17.17) and (17.18).

To calculate the value of the option at any other node in the tree, suppose first that we have calculated the value at all later nodes. Let t be the time of this node and T be the time of the next set of nodes. Let $V_t(X_t)$ be the value at this node, $E_t(X_t)$ be the 'discounted future value' of the option at this node (if we choose not to exercise), and $I_t(X_t)$ be the intrinsic value at the node. Then

$$V_t(X_t) = \max\big(E_t(X_t), I_t(X_t)\big) \tag{17.20}$$

$$\frac{E_t(X_t)}{D_{tT_K}} = \mathbb{E}_\mathbb{F}\left[V_T(X_T)D_{TT_K}^{-1} \mid X_t\right]$$

$$= \frac{D_{0T}}{D_{0T_K}}e^{-\frac{1}{2}(\psi_{T_K}-\psi_T)^2\xi_T}\mathbb{E}_\mathbb{F}\left[V_T(X_T)\exp\Big((\psi_{T_K}-\psi_T)X_T\Big) \mid X_t\right]. \tag{17.21}$$

All that remains is to evaluate (17.21), and thus $V_t(X_t)$, efficiently.

17.4.4 Evaluation of expected future value

There are many ways to evaluate an expectation for a univariate random variable. There are two important considerations for us. The first is that the integrand in (17.21), $V_T(X_T)\exp\big((\psi_{T_K}-\psi_T)X_T\big)$, is piecewise C^∞, but not in itself C^∞. The problem is (see equation (17.20)) that V takes two distinct functional forms, one below an exercise boundary X_T^*, the other above it. We must first estimate the level X_T^* from the values of $E_T(X_T)$ and $I_T(X_T)$ at the nodes. This can be done by fitting a polynomial to E_T and I_T about the exercise point and solving, as illustrated in Figure 17.1.

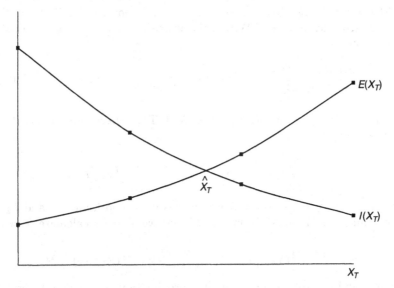

Figure 17.1. Changing functional form in the integration

Having calculated $\hat{X}_T$, our estimate of X_T^*, we now perform the integration on each side of $\hat{X}_T$, separately. To do this efficiently, over each interval $[x, x + \Delta]$, approximate $V_T(X_T)$ by a polynomial P, of order n, using neighbouring points. Then approximate the integral of V_T over this interval by instead calculating the integral of P analytically (we know the integral of the polynomial against a Gaussian explicitly; see Appendix 3). This procedure is very closely related to that of Gaussian quadrature (Press *et al.* (1988)).

One further point on efficiency should be mentioned. This issue does not affect the asymptotic order of convergence, rather it affects convergence only through the constant term. The point to consider is that we do not just wish to integrate for one value of X_t but for a whole set of values. Also one of the most expensive calculations in the whole algorithm is to calculate the exponential function which is needed both as part of the integrand and also in the evaluation of the normal density function which appears in the expressions for the integrals of a Gaussian against a polynomial. We do not go into the details here, but it is a straightforward matter to exploit the fact that the values X_t and X_T are equally spaced to considerably reduce the number of calls to the exponential function. Effectively as much of the calculation as is similar for differing nodes is carried out only once and cached.

17.4.5 Error analysis

There are three (and only three) distinct sources of error within the algorithm. The first, which we have already discussed in Section 17.4.2, is that caused by 'clipping' the tree. This is of negligible order compared with the other two.

All the errors come about as part of the integration routine. Let $t < T$ be two successive exercise times. Within the algorithm we are aiming to calculate the function

$$\frac{E_t(X_t)}{D_{tT_K}} = \frac{D_{0T}}{D_{0T_K}} e^{-\frac{1}{2}(\psi_{T_K} - \psi_T)^2 \xi_T} \mathbb{E}_\mathbb{F}\Big[F(X_T) \mid X_t\Big] \qquad (17.22)$$

where $F(X_T) = V_T(X_T)\exp((\psi_{T_K} - \psi_T)X_T)$. The expectation in (17.22) can be expressed as

$$\mathbb{E}_\mathbb{F}\Big[F(X_T) \mid X_t\Big] = \mathbb{E}_\mathbb{F}\Big[F_1(X_T)\mathbb{1}_{\{X_T > X_T^*\}} \mid X_t\Big] + \mathbb{E}_\mathbb{F}\Big[F_2(X_T)\mathbb{1}_{\{X_T \leq X_T^*\}} \mid X_t\Big]$$

where X_T^* is the exercise point at which F is non-smooth and F_1 and F_2 are smooth functions (E_T and I_T in fact). What we actually calculate (exactly) is

$$\mathbb{E}_\mathbb{F}\Big[G_1(X_T)\mathbb{1}_{\{X_T > \hat{X}_T\}} \mid X_t\Big] + \mathbb{E}_\mathbb{F}\Big[G_2(X_T)\mathbb{1}_{\{X_T \leq \hat{X}_T\}} \mid X_t\Big]$$

where G_1 and G_2 are our polynomial approximations to F_1 and F_2 and $\hat{X}_T$ is our estimate of X_T^*. It follows that the error, the difference between these two, is

$$\mathbb{E}_\mathbb{F}\Big[(F_1 - G_1)(X_T)\mathbb{1}_{\{X_T > \hat{X}_T\}} \mid X_t\Big] + \mathbb{E}_\mathbb{F}\Big[(F_2 - G_2)(X_T)\mathbb{1}_{\{X_T \leq \hat{X}_T\}} \mid X_t\Big]$$
$$+ \mathbb{E}_\mathbb{F}\Big[(F_1 - F_2)(X_T)(\mathbb{1}_{\{X_T > X_T^*\}} - \mathbb{1}_{\{X_T > \hat{X}_T\}}) \mid X_t\Big].$$

Since G is a local polynomial approximation of order n to F, it follows that the magnitude of the first two terms is $O(\Delta^{n+1})$, where Δ is, as before, the spacing between neighbouring nodes at time T. The remaining term is the expectation of a function taking the value zero outside an interval of size $|\hat{X}_T - X_T^*|$ which is $O(\Delta^{n+1})$ if we calculate $\hat{X}_T$ with a polynomial of order n. Over this interval the integrand $F_1 - F_2$ is also $O(\Delta^{n+1})$, so the net error contribution is $O(\Delta^{2(n+1)})$. Finally, the function F is not dependent on Δ and thus the net overall error is $O(\Delta^{n+1})$.

In terms of the time to run the algorithm, there are K columns of nodes corresponding to the K decision times, each containing $O(m/\Delta)$ nodes, where m is the range covered by the columns. Each node performs a calculation involving all the nodes at the next column, so the total number of calculations is $O(m^2/\Delta^2)$, which is thus the order of the time to run the algorithm. Now we must increase m as Δ decreases to ensure that the clipping error is not significant. Suppose we increase m like $\Delta^{-\varepsilon}$ for some $\varepsilon > 0$. The clipping error will then decrease faster than any polynomial order in Δ. Thus the run time is $O(\Delta^{-2(1-\varepsilon)})$, and the error is $o(\Delta^{n+1-\eta})$, for any η. This finally yields an error which is $o(T^{-\frac{1}{2}(n+1)+\varepsilon})$ for any $\varepsilon > 0$. This compares with a standard Vasicek–Hull–White tree where the error is $O(T^{-\frac{1}{2}})$.

18

Market Models

18.1 INTRODUCTION

The short-rate models of Chapter 17 have been around for many years. They
have the virtue of being easy to understand (although they are not usually
transparent in their properties) and, in the case of the Vasicek–Hull–White
model, can be very efficient to implement. Where they are less ideal is in their
calibration properties. For example, when pricing a Bermudan swaption it is
only possible to calibrate a Vasicek–Hull–White model, for each swaption
maturity date, to correctly price one swaption of a given strike (two if we
also fit the mean reversion parameter). But which one should it be? That
is not at all clear, and given that the Vasicek–Hull–White model yields a
Gaussian short-rate process whereas the market does not, the precise choice
will in general be important.

Recently, to overcome this calibration problem, a new class of models has
been developed, the class of *market models*. The approach was pioneered by
Miltersen *et al.* (1997) and Brace *et al.* (1997) in the context of LIBOR-
based models and later extended to the swap-rate context by Jamshidian
(1997). The essential new and defining feature of these models is that they
are parameterized in terms of standard market rates such as LIBORs and
swap rates. Furthermore, it is easy, in the case when option prices are given
by Black's formula and thus the link between the SDE governing the evolution
of the appropriate market interest rates and the terminal distributions of these
rates is clear, to define these models so that they exactly match market prices.

The major breakthrough that came with the introduction of market models
was a new way of thinking, one which focuses directly on the *market* interest
rates to be modelled. In this chapter we shall present these models. We provide
the necessary defining equations and prove that models do exist with these
properties. Importantly, we do not prove that they fall within the (HJM) class,
defined fully in Chapter 8, rather we show that they are numeraire models
(N). The extra burden of proving the models to be (HJM) is unnecessary,
as discussed in Chapter 8. We shall also prove that the models are strong

---

Financial Derivatives in Theory and Practice Revised Edition. P. J. Hunt and J. E. Kennedy
© 2004 John Wiley & Sons, Ltd ISBNs: 0-470-86358-7 (HB); 0-470-86359-5 (PB)

Markov, an important property if they are to be used for pricing in practice.

Despite all their benefits, the market models defined in this chapter have a significant drawback. Although the models are strong Markov, the dimension of the Markov process is the same as the number of LIBORs or swap rates being modelled. So, for example, to price a Bermudan swaption which gives the right to cancel a ten-year swap paying quarterly coupons, the underlying Markov process would be 39-dimensional (there are 39 unknown cashflows in the swap). It is infeasible, in practice, to implement such a model except by simulation or by approximation. Certainly the techniques introduced in Chapter 17 could not be employed. We shall meet an alternative class of models in the next chapter that overcomes this problem.

Since they were first introduced in 1995, market models have generated considerable interest and there is already an extensive literature on the topic. Of particular note are the original articles, mentioned above, and a comprehensive survey article by Rutkowski (2001). For a discussion of implementation issues and market models with smiles, see Glasserman and Zhao (2000) and Andersen and Andreasen (2000) respectively.

In this chapter we shall describe the two basic models, one for LIBOR, the other for swap rates. These models are ideal for pricing path-dependent LIBOR and swap derivatives for which the price can be obtained by simulation. Because of the implementation problems, Bermudan swaptions are not easily treated using market models. We also introduce an alternative swap-rate model, the *reverse swap-rate model*. This latter model is intended for pricing European swap and LIBOR derivatives, but again prices must be obtained via simulation. We develop each of these models in turn in the next three sections, first deriving the form of the SDE they satisfy (which is what we need in practice), then establishing that the SDE does admit a (strong Markov) solution and thus that such a model does indeed exist.

18.2 LIBOR MARKET MODELS

The first market models studied were LIBOR market models. The idea was to develop a model directly in terms of a set of (contiguous) LIBORs, and in particular to develop a model under which the caplets corresponding to these LIBORs were all priced using the market standard Black formula. More generally, however, the objective was to (partially) define a term structure model via an SDE for the LIBORs.

There are several steps required to define a model in this way, and these steps are common to the swap-rate models developed in Sections 18.3 and 18.4 below. These are as follows.

(i) Specify an SDE for the forward LIBOR (swap rates in Sections 18.3 and 18.4) and determine the necessary relationship between the drift and diffusion terms for any corresponding term structure model to be arbitrage-free.

(ii) Prove that a solution exists to the SDE defined in step (i).

(iii) Check that, for this solution, there exists a numeraire pair $(N, \mathbb{N})$ for which all the numeraire-rebased bonds defined by the SDE (only a finite number at this stage) are martingales under the measure $\mathbb{N}$.

(iv) Demonstrate that this partially defined model can be extended to a model for the whole term structure in such a way that the extended model also admits a numeraire pair.

Here we shall carry out steps (i)–(iv) for the LIBOR market model. So, let $S_0 < S_1 < \ldots < S_n$ be a sequence of dates and, for $i = 1, \ldots, n$, we define the corresponding forward LIBORs,

$$L_t^i := \frac{D_{tS_{i-1}} - D_{tS_i}}{\alpha_i D_{tS_i}}. \tag{18.1}$$

We will develop the model working in the forward measure $\mathbb{F}$ corresponding to taking the bond $D_{\cdot S_n}$ as numeraire, and with this in mind we also make the definitions, for $0 \le i \le n$,

$$\hat{D}_t^i := \frac{D_{tS_i}}{D_{tS_n}}, \tag{18.2}$$

$$\Pi_t^i := \prod_{j=1}^{i} \left(1 + \alpha_j L_t^j\right),$$

where in the second equation the product over the empty set is unity, i.e. $\Pi^0 \equiv 1$. It will also be convenient to define $\hat{D}^{n+1} \equiv 1$ and $L^{n+1} \equiv 0$.

Note immediately from (18.1) and (18.2) that, for $i = 0, \ldots, n - 1$,

$$\hat{D}_t^i = (1 + \alpha_{i+1} L_t^{i+1}) \hat{D}_t^{i+1}, \tag{18.3}$$

a relationship that will be used inductively to derive the form of the SDE for the forward LIBORs, and further that

$$\hat{D}_t^i = \prod_{j=i+1}^{n} \left(1 + \alpha_j L_t^j\right) = \frac{\Pi_t^n}{\Pi_t^i}. \tag{18.4}$$

18.2.1 Determining the drift

Suppose that we wish to develop a model for the term structure of interest rates for which, for each $i = 1, \ldots, n$, the forward LIBORs L^i solve an SDE of the form

$$dL_t^i = \mu_t^i(L_t) L_t^i dt + \sigma_t^i(L_t) L_t^i dW_t^i \tag{18.5}$$

where W is a (correlated) Brownian motion with $dW_t^i dW_t^j = \rho_{ij} dt$ and $L = (L^1, \ldots, L^n)$. As stated earlier, we work in the forward measure $\mathbb{F}$ corresponding to taking $D_{\cdot S_n}$ as numeraire. Then, if this model is to be arbitrage-free, each of the $\hat{D}^i$ must be a martingale under this measure. This imposes a relationship between the diffusion coefficients σ^i and the drifts μ^i in the SDE (18.5). Here we derive that relationship. Note that, in this section, we only provide a necessary relationship between the drift and diffusion terms. We do not yet show that a suitable process exists which satisfies (18.5), i.e. we only carry out step (i) described above. Steps (ii)–(iv) are treated in the next section.

So suppose each of the $\hat{D}^i$ is a martingale under the measure $\mathbb{F}$. Then, applying Itô's formula to (18.3), we obtain

$$d\hat{D}_t^i = (1 + \alpha_{i+1} L_t^{i+1}) d\hat{D}_t^{i+1} + \alpha_{i+1} \hat{D}_t^{i+1} dL_t^{i+1} + \alpha_{i+1} dL_t^{i+1} d\hat{D}_t^{i+1} \, . \quad (18.6)$$

Substituting from (18.5) and equating local martingale parts in (18.6) yields

$$d\hat{D}_t^i = (1 + \alpha_{i+1} L_t^{i+1}) d\hat{D}_t^{i+1} + \hat{D}_t^{i+1} \alpha_{i+1} L_t^{i+1} \sigma_t^{i+1} \, dW_t^{i+1} \, ,$$

whence

$$\Pi_t^i d\hat{D}_t^i = \Pi_t^{i+1} d\hat{D}_t^{i+1} + \Pi_t^{i+1} \hat{D}_t^{i+1} \left(\frac{\alpha_{i+1} L_t^{i+1}}{1 + \alpha_{i+1} L_t^{i+1}} \right) \sigma_t^{i+1} dW_t^{i+1} \, .$$

(We have here suppressed the dependence of σ_t on L_t. We shall do this later also, for both σ_t and μ_t.) It follows by backward induction, down from $i = n$, that for $i = 0, \ldots, n - 1$,

$$\Pi_t^i d\hat{D}_t^i = \sum_{j=i+1}^{n} \Pi_t^j \hat{D}_t^j \left(\frac{\alpha_j L_t^j}{1 + \alpha_j L_t^j} \right) \sigma_t^j \, dW_t^j \, ,$$

and thus

$$\begin{aligned}
d\hat{D}_t^i &= \hat{D}_t^i \sum_{j=i+1}^{n} \frac{\Pi_t^j \hat{D}_t^j}{\Pi_t^i \hat{D}_t^i} \left(\frac{\alpha_j L_t^j}{1 + \alpha_j L_t^j} \right) \sigma_t^j dW_t^j \\
&= \hat{D}_t^i \sum_{j=i+1}^{n} \left(\frac{\alpha_j L_t^j}{1 + \alpha_j L_t^j} \right) \sigma_t^j dW_t^j \, .
\end{aligned} \quad (18.7)$$

Now we can also equate the finite variation terms in (18.6) to obtain, for $i = 0, \ldots, n - 1$,

$$\alpha_{i+1} \hat{D}_t^{i+1} \mu_t^{i+1} L_t^{i+1} dt + \alpha_{i+1} \sigma_t^{i+1} L_t^{i+1} dW_t^{i+1} d\hat{D}_t^{i+1} = 0 \, ,$$

and thus, again by backward induction, it follows that

$$\mu_t^i(L_t) = - \sum_{j=i+1}^{n} \left(\frac{\alpha_j L_t^j}{1 + \alpha_j L_t^j} \right) \sigma_t^i(L_t) \sigma_t^j(L_t) \rho_{ij} \,.$$

Finally, the original SDE (18.5) becomes

$$dL_t^i = - \left(\sum_{j=i+1}^{n} \left(\frac{\alpha_j L_t^j}{1 + \alpha_j L_t^j} \right) \sigma_t^i(L_t) \sigma_t^j(L_t) \rho_{ij} \right) L_t^i \, dt + \sigma_t^i(L_t) L_t^i \, dW_t^i \,. \quad (18.8)$$

18.2.2 Existence of a consistent arbitrage-free term structure model

The results of Section 18.2.1 are all that we need in practice when using a market model. However, we need to check that the model given there is consistent with a full arbitrage-free term structure model (steps (ii)–(iv) as defined above), and this we now do.

To begin, we must show that the SDE (18.8) for the forward LIBORs has a solution. In general it will not since the SDE given there is of a very general form. However, in the important case when σ^i is bounded over any time interval $[0, t]$ a solution does exist. This includes the special case of most practical relevance (and which leads to Black's formula for option prices) when $\sigma_t^i(L_t) = \sigma_t^i$, a locally bounded function of time which does not depend on the LIBORs L^i, $i = 1, \dots, n$.

Establishing existence of a solution to the SDE in the above cases follows from the general theory for SDEs as developed in Chapter 6.

Theorem 18.1 *Suppose that, for $i = 1, \dots, n$, the functions $\sigma^i : \mathbb{R}^n \times \mathbb{R}^+ \to \mathbb{R}$ are bounded on any time interval $[0, t]$. Then strong existence and pathwise uniqueness hold for the SDE (18.8). Furthermore, the solution process L is a strong Markov process.*

Proof: First consider the modified SDE

$$dL_t^i = \hat{\mu}_t^i(L_t) L_t^i \, dt + \hat{\sigma}_t^i(L_t) L_t^i \, dW_t^i \quad (18.9)$$

where

$$\hat{L}^i = L^i \mathbf{1}_{\{L^i > 0\}},$$
$$\hat{\mu}_t^i(L) = \mu_t^i(\hat{L}),$$
$$\hat{\sigma}_t^i(L) = \sigma_t^i(\hat{L})$$

and $\hat{L} = (\hat{L}^1, \dots, \hat{L}^n)$. For any fixed $t > 0$, this modified SDE is globally Lipschitz over the time interval $[0, t]$. This follows since both $\hat{\mu}$ and $\hat{\sigma}$ are bounded on $[0, t]$. It thus follows from Theorem 6.27 that strong existence and pathwise uniqueness hold for the SDE $(\hat{\sigma}, \hat{\mu})$. (Note that all the components of the driving d-dimensional Brownian motion in Theorem 6.27 are independent, whereas W in (18.9) is a correlated Brownian motion. Casting (18.9) in that form is a straightforward matter.)

Let $\tilde{L}$ now be a solution process for the SDE $(\hat{\sigma}, \hat{\mu})$. By applying Itô's formula to obtain an SDE for $\log \tilde{L}$ it is clear that the process $\tilde{L}$ can be written in the form

$$\tilde{L}_t^i = \tilde{L}_0^i \exp\left(\int_0^t \hat{\sigma}_u^i(\tilde{L}_u)\, dW_u^i + \int_0^t \left(\hat{\mu}_u^i(\tilde{L}_u) - \tfrac{1}{2}(\hat{\sigma}_u^i(\tilde{L}_u))^2 \right) du \right).$$

Since $\hat{\sigma}$ and $\hat{\mu}$ are bounded, the process $\tilde{L}$ is a.s. strictly positive and thus is also a solution for the original SDE (σ, μ) (which only differs from $(\hat{\sigma}, \hat{\mu})$ when the solution process takes negative values).

The strong Markov property follows immediately (for $\tilde{L}$ and thus for L) from Theorem 6.36.   □

We have just shown that there is indeed a process L satisfying the SDE (18.8). This was necessary for the model to be arbitrage-free. Of course it is not sufficient. We must additionally check that all $D_{\cdot S_n}$-rebased assets are martingales in the measure $\mathbb{F}$ (step (iii)). But, from (18.7), $\hat{D}^i$ can be written as a Dolean exponential,

$$\hat{D}_t^i = \hat{D}_0^i \mathcal{E}\left(\int_0^t \sum_{j=i+1}^n \left(\frac{\alpha_j L_u^j}{1 + \alpha_j L_u^j} \right) \sigma_u^j dW_u^j \right).$$

Observing that the exponentiated term has bounded quadratic variation over any time interval $[0, t]$, it follows from Novikov's condition (Theorem 5.16) that $\hat{D}^i$ is indeed a true martingale.

It remains to show that the economy can be extended in such a way that all numeraire-rebased assets for the whole economy are martingales. This is straightforward. Up until time S_n we have taken $D_{\cdot S_n}$ as numeraire but have not defined its law fully as yet. What we have done is define the law of the processes L^i. However, since $D_{S_i S_i} \equiv 1$, we can recover the functional form $D_{S_i S_n}(L_{S_i})$ from (18.4). Extend the model as follows. For any $S_i \le t \le S_{i+1}$, define

$$D_{t S_n}(L_t) := \frac{S_{i+1} - t}{S_{i+1} - S_i} D_{S_i S_n}(L_t) + \frac{t - S_i}{S_{i+1} - S_i} D_{S_{i+1} S_n}(L_t).$$

This is a continuous semimartingale and its law can be found from that of the process L and the defining functional forms. We use this to define a numeraire model (as defined in Chapter 8): for $0 \le t \le S \le S_n$, define

$$D_{tS} := D_{t S_n}(L_t) \mathbb{E}_{\mathbb{F}}\left[D_{S S_n}^{-1}(L_S) | \mathcal{F}_t \right].$$

It is readily seen that this is consistent with the assumptions made above. To complete the model, for $t \leq S_n < S$ take

$$D_{tS} = D_{tS_n} \frac{D_{0S}}{D_{0S_n}},$$

and for $S_n < t < S$

$$D_{tS} = \frac{D_{0S}}{D_{0t}}.$$

For the corresponding numeraire, take $D._{S_n}$ until time S_n and then the unit rolling numeraire from then on.

18.2.3 Example application

One important exotic product which is relatively popular is the *limit cap*. This product is so closely related to the underlying vanilla cap that accurate calibration is essential, and thus the LIBOR market model is ideally suited to valuing this product.

A *limit cap* is identical to a vanilla cap except there is a limit imposed on the total number of caplets that can be exercised. As per a cap, the limit cap thus comprises a number of payment dates S_i, $i = 1, 2, \ldots, n$, and a set of strikes K_i, $i = 1, 2, \ldots, n$. There is also a limit number m. The holder of a limit cap receives the first m caplets that set in the money. Note that the option holder does not have the right to refuse to exercise any particular caplet that sets in the money (perhaps because it is only just in the money) and thus the product is path-dependent but not American. This means that pricing can be carried out via simulation of the LIBOR process and option payout, and the high dimensionality of the Markov LIBOR process does not cause major difficulties.

One could choose to implement a multi-factor LIBOR market model to price this product. Because for each reset date there is only one market rate used to determine the option payoff, in practice we ourselves have always restricted to a single-factor model (see the discussion in Chapter 9). One can, by introducing a mean reversion parameter, as discussed in Section 19.7, reflect the relevant market distributions even within the one-factor setting.

18.3 REGULAR SWAP-MARKET MODELS

For path-dependent products based on swaps and swap rates, such as barrier swaptions (discussed in Section 18.3.3), we would like to develop swap-rate models analogous to the LIBOR models just introduced. The procedure for doing this is identical to that for LIBOR models, with only the notation and algebra being more difficult.

We make the following definitions relating to a series of swaps all maturing on the date S_n. Taking for convenience $\alpha_0 = 1$, we define, for $i = 0, \ldots, n$,

$$P_t^i := \sum_{j=i}^{n} \alpha_j D_{tS_j},$$

$$\hat{P}_t^i := \frac{P_t^i}{D_{tS_n}} = \sum_{j=i}^{n} \alpha_j \frac{D_{tS_j}}{D_{tS_n}}, \tag{18.10}$$

and, for $1 \leq i \leq n$,

$$y_t^i := \frac{D_{tS_{i-1}} - D_{tS_n}}{P_t^i}, \tag{18.11}$$

$$\Psi_t^i := \prod_{j=1}^{i} \left(1 + \alpha_j y_t^{j+1}\right).$$

We also take $\hat{P}^{n+1} \equiv y^{n+1} \equiv 0$ and $\Psi^0 \equiv 1$, $\Psi^{-1} \equiv (\Psi^1)^{-1}$. It follows from (18.10) and (18.11) that, for $i = 0, \ldots, n$,

$$\hat{P}_t^i = \alpha_i + (1 + \alpha_i y_t^{i+1})\hat{P}_t^{i+1}, \tag{18.12}$$

whence, for $i = 0, \ldots, n$,

$$\Psi_t^{i-1} \hat{P}_t^i = \alpha_i \Psi_t^{i-1} + \Psi_t^i \hat{P}_t^{i+1}.$$

It then follows inductively that

$$\hat{P}_t^i = \frac{\sum_{j=i}^{n} \alpha_j \Psi_t^{j-1}}{\Psi_t^{i-1}}. \tag{18.13}$$

We follow the same four steps in developing the swap-market model as we did in Section 18.2 for the LIBOR market model, starting with the derivation of the relationship between the diffusion and drift of the market swap rates.

18.3.1 Determining the drift

We want to develop a model for the term structure of interest rates for which, for each $i = 1, \ldots, n$, the forward par swap rates y^i solve an SDE of the form

$$dy_t^i = \mu_t^i(y_t) y_t^i \, dt + \sigma_t^i(y_t) y_t^i \, dW_t^i, \tag{18.14}$$

where W is a (correlated) Brownian motion with $dW_t^i dW_t^j = \rho_{ij} dt$ and $y = (y^1, \ldots, y^n)$. We again work in the forward measure $\mathbb{F}$ corresponding to taking $D.S_n$ as numeraire. Then, if the model is to be arbitrage-free, each of the $\hat{P}^i$, $i = 0, \ldots, n$, must be a martingale under this measure. This again imposes a relationship between the diffusion coefficients σ^i and the drifts μ^i in the SDE (18.14), which we now derive.

Suppose each of the $\hat{P}^i$ is a martingale under the measure $\mathbb{F}$ and apply Itô's formula to (18.12) to obtain

$$d\hat{P}_t^i = (1 + \alpha_i y_t^{i+1})d\hat{P}_t^{i+1} + \alpha_i \hat{P}_t^{i+1}dy_t^{i+1} + \alpha_i dy_t^{i+1}d\hat{P}_t^{i+1} \, . \tag{18.15}$$

Suppressing the dependence of σ^i (and later of μ^i) on the vector y, equating local martingale parts yields

$$d\hat{P}_t^i = (1 + \alpha_i y_t^{i+1})d\hat{P}_t^{i+1} + \hat{P}_t^{i+1}\alpha_i y_t^{i+1}\sigma_t^{i+1}dW_t^{i+1} \, ,$$

whence

$$\Psi_t^{i-1}d\hat{P}_t^i = \Psi_t^i d\hat{P}_t^{i+1} + \Psi_t^i \hat{P}_t^{i+1}\left(\frac{\alpha_i y_t^{i+1}}{1 + \alpha_i y_t^{i+1}}\right)\sigma_t^{i+1}dW_t^{i+1} \, .$$

It now follows by backward induction, down from $i = n$, that for $i = 0, \ldots, n-1$,

$$\Psi_t^{i-1}d\hat{P}_t^i = \sum_{j=i+1}^{n} \Psi_t^{j-1}\hat{P}_t^j\left(\frac{\alpha_{j-1} y_t^j}{1 + \alpha_{j-1} y_t^j}\right)\sigma_t^j dW_t^j \, ,$$

and so

$$d\hat{P}_t^i = \hat{P}_t^i \sum_{j=i+1}^{n} \frac{\Psi_t^{j-1}\hat{P}_t^j}{\Psi_t^{i-1}\hat{P}_t^i}\left(\frac{\alpha_{j-1} y_t^j}{1 + \alpha_{j-1} y_t^j}\right)\sigma_t^j dW_t^j \, . \tag{18.16}$$

Now we can also equate the finite variation terms in (18.15) to obtain, for $i = 0, \ldots, n-1$,

$$\alpha_i \hat{P}_t^{i+1}\mu_t^{i+1}y_t^{i+1}dt + \alpha_i\sigma_t^{i+1}y_t^{i+1}dW_t^{i+1}d\hat{P}_t^{i+1} = 0 \, ,$$

and thus, for $i = 1, \ldots, n$,

$$\mu_t^i(y_t) = -\sum_{j=i+1}^{n} \frac{\Psi_t^{j-1}\hat{P}_t^j}{\Psi_t^{i-1}\hat{P}_t^i}\left(\frac{\alpha_{j-1} y_t^j}{1 + \alpha_{j-1} y_t^j}\right)\sigma_t^i(y_t)\sigma_t^j(y_t)\rho_{ij} \, .$$

Finally, the original SDE (18.14) becomes

$$dy_t^i = -\left(\sum_{j=i+1}^{n} \frac{\Psi_t^{j-1}\hat{P}_t^j}{\Psi_t^{i-1}\hat{P}_t^i}\left(\frac{\alpha_{j-1} y_t^j}{1 + \alpha_{j-1} y_t^j}\right)\sigma_t^i(y_t)\sigma_t^j(y_t)\rho_{ij}\right)y_t^i dt + \sigma_t^i(y_t)y_t^i \, dW_t^i \, . \tag{18.17}$$

18.3.2 Existence of a consistent arbitrage-free term structure model

It remains for us to prove that the SDE (18.17) has a solution and that there is a full arbitrage-free term structure model consistent with this SDE. This is precisely steps (ii)–(iv) of Section 18.2. Step (ii) is carried out as for the LIBOR market model and leads to the following theorem. Note that this again includes the important special case when $\sigma_t^i(y_t) = \sigma_t^i$, a locally bounded deterministic function of time.

Theorem 18.2 *Suppose that, for $i = 1, \ldots, n$, the functions $\sigma^i : \mathbb{R}^n \times \mathbb{R}^+ \to \mathbb{R}$ are bounded on any time interval $[0, t]$. Then strong existence and pathwise uniqueness hold for the SDE (18.17). Furthermore, the solution process y is a strong Markov process.*

Proof: The proof is identical to that for Theorem 18.1 once we have established that the functions μ^i are bounded over the time interval $[0, t]$ when $y^i \geq 0$ for all i. This is immediate once we observe, from (18.13), that $\Psi_t^{i-1}\hat{P}_t$ is decreasing in i.   □

Now that we know that (18.17) admits a solution for suitable σ^i, we must carry out step (iii) and show that for this solution and for each $i = 0, \ldots, n$, the processes $D_{\cdot S_i}/D_{\cdot S_n}$ are martingales in the measure $\mathbb{F}$. Equivalently, we need to show that the processes $\hat{P}^i$ are martingales. That they are local martingales follows from the form of (18.16). But, from (18.16), $\hat{P}^i$ can be written as a Dolean exponential,

$$\hat{P}_t^i = \hat{P}_0^i \mathcal{E}\left(\int_0^t \sum_{j=i+1}^n \frac{\Psi_u^{j-1}\hat{P}_u^j}{\Psi_u^{i-1}\hat{P}_u^i} \left(\frac{\alpha_{j-1}y_u^j}{1 + \alpha_{j-1}y_u^j} \right) \sigma_u^j \, dW_u^j \right).$$

The integrand is bounded over any finite time interval $[0, t]$ and thus $\hat{P}^i$ is a martingale by Novikov's condition. The model is completed by carrying out step (iv) and extending to all bonds and to the time interval $[0, \infty)$. This last step is precisely the same as for the LIBOR model of Section 18.2.

18.3.3 Example application

The most important swap-based multi-temporal product in the market is undoubtedly the Bermudan swaption. Whereas the models described in this section are theoretically well suited to pricing this product, the high dimensionality of the underlying Markov process, combined with the American nature of the Bermudan swaption, means it is very difficult to use market swaption models for the Bermudan.

One product for which a market swap-rate model is suited is the *discrete barrier swaption*. This product is similar to a vanilla swap with one modification. If, on any of the swap reset dates $T_i = S_{i-1}$, the par swap rate y^i is above a given barrier level H_i then the rest of the swap is cancelled and no further cashflows take place. This is an *up and out barrier swaption*. There are also the obvious variants, the up and in, down and out, and down and in swaps.

The discrete barrier swaption is path-dependent and can be valued within a swap-market model using simulation. One factor is a natural choice and a mean reversion parameter (see Section 19.7) should be included to create the desired joint distributions for the underlying swap rates.

18.4 REVERSE SWAP-MARKET MODELS

The model described in Section 18.3 above is suitable for path-dependent products such as barrier swaptions where the exotic product to be priced is related to a collection of swap rates which set on different dates but which have a common maturity date. There are other situations where several swaps need to be modelled and they have in common not their maturity date but their start date. One example of this was the spread option which was discussed in Section 16.3.1. Here we return to the problem of pricing European derivatives and present a variant of the model described in Section 18.3 which is suitable for such products. As will become clear, most of the procedure for developing the model is similar to that in Section 18.3, with the exception of step (iv) in the construction which uses techniques developed for terminal swap-rate models.

Now we make the following definitions relating to a series of swaps all starting on date $T = S_0$ but with maturity dates $S_1, \ldots, S_n$. Note that these differ from those of Section 18.3. We define, for $i = 1, \ldots, n$,

$$P_t^i := \sum_{j=1}^{i} \alpha_j D_{tS_j},$$

$$y_t^i := \frac{D_{tT} - D_{tS_i}}{P_t^i}, \tag{18.18}$$

$$\hat{P}_t^i := \frac{P_t^i}{D_{tT}} = \sum_{j=1}^{i} \alpha_j \frac{D_{tS_j}}{D_{tT}} \tag{18.19}$$

and

$$\Pi_t^i := \prod_{j=1}^{i} \left(1 + \alpha_j y_t^j\right).$$

It follows from (18.18) and (18.19) that, for $i = 1, \ldots, n$,

$$(1 + \alpha_i y_t^i)\hat{P}_t^i = \alpha_i + \hat{P}_t^{i-1}, \tag{18.20}$$

$$\Pi_t^i \hat{P}_t^i = \alpha_i \Pi_t^{i-1} + \Pi_t^{i-1} \hat{P}_t^{i-1} \,,$$

where $\hat{P}^0 \equiv 0$ and $\Pi^0 \equiv 1$. Induction now yields

$$\hat{P}_t^i = \frac{\sum_{j=1}^i \alpha_j \Pi_t^j}{\Pi_t^i} \,. \tag{18.21}$$

18.4.1 Determining the drift

Once more, we wish to develop a model for the term structure of interest rates for which, for each $i = 1, \dots, n$, the forward par swap rates y^i solve an SDE of the form

$$dy_t^i = \mu_t^i(y_t) y_t^i \, dt + \sigma_t^i(y_t) y_t^i \, dW_t^i \tag{18.22}$$

where, as before, W is a (correlated) Brownian motion with $dW_t^i dW_t^j = \rho_{ij} dt$. For this model we work in the measure $\mathbb{F}$ corresponding to taking $D._T$ as numeraire. Applying Itô's formula to (18.20) yields, for $i = 1, \dots, n$,

$$d\hat{P}_t^{i-1} = (1 + \alpha_i y_t^i) d\hat{P}_t^i + \alpha_i \hat{P}_t^i dy_t^i + \alpha_i dy_t^i d\hat{P}_t^i \,. \tag{18.23}$$

Since we require each $\hat{P}^i$ to be a martingale under the measure $\mathbb{F}$, we can substitute (18.22) into (18.23) and equate local martingale parts to obtain

$$d\hat{P}_t^{i-1} = (1 + \alpha_i y_t^i) d\hat{P}_t^i + \hat{P}_t^i \alpha_i y_t^i \sigma_t^i \, dW_t^i \,,$$

$$\Pi_t^{i-1} d\hat{P}_t^{i-1} = \Pi_t^i d\hat{P}_t^i + \Pi_t^i \hat{P}_t^i \left(\frac{\alpha_i y_t^i}{1 + \alpha_i y_t^i} \right) \sigma_t^i dW_t^i \,,$$

and thus, inductively,

$$\Pi_t^i d\hat{P}_t^i = -\sum_{j=1}^i \Pi_t^j \hat{P}_t^j \left(\frac{\alpha_j y_t^j}{1 + \alpha_j y_t^j} \right) \sigma_t^j dW_t^j \,,$$

$$d\hat{P}_t^i = -\hat{P}_t^i \sum_{j=1}^i \frac{\Pi_t^j \hat{P}_t^j}{\Pi_t^i \hat{P}_t^i} \left(\frac{\alpha_j y_t^j}{1 + \alpha_j y_t^j} \right) \sigma_t^j dW_t^j \,.$$

Equating the finite variation terms in (18.23) now yields

$$\hat{P}_t^i \mu_t^i(y_t) \, dt + \sigma_t^i(y_t) \, dW_t^i d\hat{P}_t^i = 0 \,,$$

and thus

$$\mu_t^i(y_t) = -\sum_{j=1}^i \frac{\Pi_t^j \hat{P}_t^j}{\Pi_t^i \hat{P}_t^i} \left(\frac{\alpha_j y_t^j}{1 + \alpha_j y_t^j} \right) \sigma_t^i(y_t) \sigma_t^j(y_t) \rho_{ij} \,.$$

Thus the SDE (18.22) can finally be written in the form

$$dy_t^i = -\left(\sum_{j=1}^{i} \frac{\Pi_t^j \hat{P}_t^j}{\Pi_t^i \hat{P}_t^i} \left(\frac{\alpha_j y_t^j}{1 + \alpha_j y_t^j}\right) \sigma_t^i(y_t)\sigma_t^j(y_t)\rho_{ij}\right) y_t^i \, dt + \sigma_t^i(y_t)y_t^i \, dW_t^i \, .$$

$$(18.24)$$

18.4.2 Existence of a consistent arbitrage-free term structure model

Again we must complete the model by carrying out steps (ii)–(iv) of Section 18.2. Observing from (18.21) that $\Pi_t^i \hat{P}_t^i$ is increasing in i, exactly the same proof as used for Theorem 18.2 yields the analogous result here.

Theorem 18.3 *Suppose that, for $i = 1, \ldots, n$, the functions $\sigma^i : \mathbb{R}^n \times \mathbb{R}^+ \to \mathbb{R}$ are bounded on any time interval $[0, t]$. Then strong existence and pathwise uniqueness hold for the SDE (18.24). Furthermore, the solution process y is a strong Markov process.*

The proof of this result also establishes that the solution process y has all its components strictly positive. It then follows from the definition of y_t^i (equation (18.18)) that, for each i, $D_{tS_i} < D_{tT}$ and thus $\hat{P}_t^i < \sum_{j=1}^{i} \alpha_j$. So the continuous local martingale $\hat{P}^i$ is bounded, hence a true martingale and we have completed step (iii).

Thus far we have only defined the dynamics of a finite number of bonds and so we must extend the model. There is, as was the case for the terminal swap-rate models of Chapter 13, no unique way to extend the economy, and in practice we do not carry out this extension, we just need to know that it can be done. We can do this in a very similar way to that used in Section 13.4.2 to extend a terminal swap-rate model.

To carry out this extension, observe that at time T the discount factors D_{TS_i} can all be written as a function of the strong Markov process y, indeed as a function of y_T (see equation (18.21)). For intermediate maturities S, define D_{TS} by

$$D_{TS} = \frac{S_{i+1} - S}{S_{i+1} - S_i} D_{TS_i} + \frac{S - S_i}{S_{i+1} - S_i} D_{TS_{i+1}}$$

where $S_i \leq S \leq S_{i+1}$, and for $S < T$ define the 'pseudo'-discount factors $D_{TS} = \exp(-\hat{C}_S y_T^1)$ for $\hat{C}_S$ being the unique solution to

$$\frac{D_{0S}}{D_{0T}} = \mathbb{E}_{\mathbb{F}}\left[\exp(-\hat{C}_S y_T^1)\right].$$

We then define discount factors at any earlier time $t \leq T$ from the (required) martingale property,

$$D_{tS} = \frac{D_{tS}/D_{tT}}{D_{tt}/D_{tT}} = \frac{\mathbb{E}_{\mathbb{F}}[D_{TS}|\mathcal{F}_t]}{\mathbb{E}_{\mathbb{F}}[D_{Tt}|\mathcal{F}_t]} \, .$$

Finally, we extend the model beyond T in a deterministic manner by setting

$$D_{tS} = \frac{D_{Tt}(y_T)}{D_{TS}(y_T)}\,,$$

for $T \le t \le S$.

18.4.3 Example application

As mentioned above, this model is designed for pricing European swap-rate derivatives. One important application is the single-currency spread option which pays an amount $(\alpha \hat{y}_T^1 - \beta \hat{y}_T^2 - K)_+$ for two par swap rates $\hat{y}^1$ and $\hat{y}^2$ and for constants α, β and K.

This product needs a two-factor model because there are two market rates *setting on the same date* required to determine the payoff. Recall that in Chapter 16 we priced spread options based on the spread between two market rates in different currencies but could not work in one currency since there we only developed terminal swap-rate models with one factor per currency.

The path-dependent nature of the SDE (18.24) means that pricing must be done by simulation, even though the underlying product is European.

19

Markov-Functional Modelling

19.1 INTRODUCTION

In this final chapter we consider models that can fit the observed prices of liquid instruments in a similar fashion to the market models, but which also have the advantage that derivative prices can be calculated just as efficiently as in the Vasicek–Hull–White model, the most tractable short-rate model. To achieve this we consider the general class of *Markov-functional interest rate models*, the defining characteristic of which is that pure discount bond prices are at any time a function of some low-dimensional process which is Markovian in some martingale measure. This ensures that implementation is efficient since it is only necessary to track the driving Markov process, something that is particularly important for American products such as Bermudan swaptions. Market models do not possess this property (for a *low*-dimensional Markov process) and this is the impediment to their efficient implementation. The freedom to choose the functional form is what permits accurate calibration of Markov-functional models to relevant market prices, a property not possessed by short-rate models. The remaining freedom to specify the law of the driving Markov process means the model can be formulated to capture well those features of the real market relevant to a given exotic product.

19.2 MARKOV-FUNCTIONAL MODELS

In Chapters 13–16 we developed a range of models to price exotic European derivatives. We did this by focusing on the specific product to be priced and on those features of the market relevant to that product. Because we were studying European products we were able to focus on market distributions at one particular time. We explicitly modelled the distribution of one appropriate market rate and then fitted a suitable functional form to the full discount curve on that date. What resulted was a model with the following important properties:

Financial Derivatives in Theory and Practice Revised Edition. P. J. Hunt and J. E. Kennedy
© 2004 John Wiley & Sons, Ltd   ISBNs: 0-470-86358-7 (HB); 0-470-86359-5 (PB)

(a) it was arbitrage-free;
(b) it was well calibrated, correctly pricing relevant liquid instruments
     without being overfitted;
(c) it was realistic and transparent in its properties;
(d) it allowed an efficient implementation.

Here we develop models which also have the four properties above but which
are specifically designed for the pricing of multi-temporal derivatives such as
Bermudan swaptions. In particular, we shall develop models which satisfy the
following definition.

Definition 19.1 *An interest rate model is said to be Markov-functional if
there exists some numeraire pair $(N, \mathbb{N})$ and some process x such that:*

*(P.i)   the process x is a (time-inhomogeneous) Markov process under the
measure $\mathbb{N}$;*

(P.ii) the pure discount bond prices are of the form

$$D_{tS} = D_{tS}(x_t), \qquad 0 \leq t \leq \partial_S \leq S,$$

for some boundary curve $\partial_S : [0, \partial^*] \to [0, \partial^*]$ and some constant ∂^*;
(P.iii) the numeraire N, itself a price process, is of the form

$$N_t = N_t(x_t) \qquad 0 \leq t \leq \partial^*.$$

Remark 19.2: The constant ∂^* and the *boundary curve* $\partial_S : [0, \partial^*] \to [0, \partial^*]$
are introduced so that the model does not need to be defined over the whole
time domain $0 \leq t \leq S < \infty$. In the applications we have encountered it has
always been of the form

$$\partial_S = \begin{cases} S, & \text{if } S \leq T, \\ T, & \text{if } S > T, \end{cases} \tag{19.1}$$

for some constant T.

All the models discussed in Chapters 13–16 were Markov-functional. In
fact all models of practical use are Markov-functional, including the short-
rate models of Chapter 17 and the market models of Chapter 18. However,
each of these fails to some degree to satisfy properties (a)–(d). By focusing
on the Markov-functional property we will be able to develop models which
satisfy all four criteria and can be used to price multi-temporal products.

To satisfy property (d) and to be able to implement a Markov-functional
model in practice the driving Markov process must be of dimension one or at
most two. This is an ambitious requirement given that the product to which
the model is being applied is multi-temporal. The payoff, and thus also the

valuation, of any multi-temporal derivative depends on (the joint distribution of) a set of forward swap rates or forward LIBORs. The model must capture (certain aspects of) the joint distribution of these rates.

Suppose we are aiming to price a product whose payoff depends (primarily) on the swap rates (or LIBORs) $\{y^1, \ldots, y^n\}$ on their respective setting dates $T_1, \ldots, T_n$. We assume, for now, that for each distinct setting date T_i there is only one associated swap rate y^i relevant to the product. (This, therefore, explicitly excludes products such as Bermudan spread options for which a two-dimensional driving Markov process would be appropriate. We discuss how to develop a Markov-functional model for products such as this in Section 19.5.) The question arises as to whether it is possible to specify a Markov-functional model for which the driving Markov process is only one-dimensional but which nonetheless accurately captures the important properties of the joint distribution of the y^i, $i = 1, \ldots, n$. In particular:

(Q1) Given a one-dimensional Markov process x and a set of market swap rates (or LIBORs), can we find functional forms $D_{tS}(x_t)$ such that the model accurately reflects the prices of other *vanilla* market instruments (i.e. swaptions) with payoffs dependent on each of the $y^i_{T_i}$? Equivalently (as we saw in Section 13.2.2), can we find functional forms such that the *marginal distributions* of the $y^i_{T_i}$ in the associated swaption measure $\mathbb{S}^i$ agree with those implied from the market?

(Q2) If we can do this and capture the relevant marginal distributions, is there still enough freedom in the model so that the full *joint* distribution of the vector $\{y^i_{T_i} : 1 \leq i \leq n\}$ is adequately captured, or will satisfying (Q1) lead to a model that is overfitted and thus a poor model for the application to hand? We saw a case of the latter problem when discussing irregular swaptions in Section 15.4.

The answer to the first question is yes, and Section 19.3 is devoted to showing how to do this.

Turning to the second question, for all the applications that we shall consider it turns out that, given the Markov process x, the functional forms $y^i_{T_i}(x_{T_i})$ are uniquely determined by the associated swaption prices. Thus the only way to affect the joint distribution of these rates is then through the choice of the Markov process x. But x is arbitrary and so there is considerable freedom remaining in the selection of x to affect this joint distribution. There is enough flexibility remaining even if we restrict x to be a Gaussian process, one for which numerical implementation is particularly efficient (as we saw when pricing Bermudans with the Vasicek–Hull–White model in Section 17.4). In order to be able to choose x to give a good model we must have some understanding of how the dynamics of x affects the relevant joint distributions. Much insight can be gained by a close examination of the Vasicek–Hull–White model because of its tractability. We do this in Section 19.7.

19.3 FITTING A ONE-DIMENSIONAL MARKOV-FUNCTIONAL MODEL TO SWAPTION PRICES

Throughout this section and the next we shall restrict attention to developing and understanding Markov-functional models for which the driving Markov process is one-dimensional. We extend to the more general case in Section 19.5. We shall suppose that we have already selected the martingale measure $\mathbb{N}$ in which to work and have specified the driving Markov process x. We consider the problem of how to choose the functional forms for the numeraire and discount factors so that the model calibrates accurately to the prices of a set of swaptions, those corresponding to the rates $y^1, \ldots, y^n$ setting on dates $T_1 < \ldots < T_n$.

Defining the payment dates for the swap (FRA in the case when y^i is a LIBOR) associated with the rate y^i by S^i_j, $j = 1, 2, \ldots, m_i$, we assume in the discussions below that, for all i, j, either $S^i_j > T_n$ or $S^i_j = T_k$, for some $k > i$. The assumption makes the ideas below easier to develop but is not strictly necessary for the results that follow. It does hold for the most important products encountered in practice, such as a Bermudan swaption associated with a cancellable swap. If the assumption does not hold one can always make it hold by introducing auxillary swap rates y^k as necessary.

To specify the model fully we must define the functional forms $D_{tS}(x_t)$ for all $0 \le t \le S \le S^n_{m_n}$, $0 \le t \le T_n$, and also the functional form $N_t(x_t)$. Note from Definition 19.1 that, given the law of the process x under $\mathbb{N}$, to completely define a general Markov-functional model it is sufficient to specify:

(P.ii') the functional form of the discount factors on the boundary ∂_S, i.e. $D_{\partial_S S}(x_{\partial_S})$ for $S \in [0, \partial^*]$;

(P.iii') the functional form of the numeraire $N_t(x_t)$ for $0 \le t \le \partial^*$.

That is, we do not need to explicitly specify the functional form of discount factors on the interior of the region bounded by ∂_S. This follows because these interior discount factors can be recovered via the martingale property for numeraire-rebased assets under $\mathbb{N}$,

$$D_{tS}(x_t) = N_t(x_t)\mathbb{E}_{\mathbb{N}}\left[\frac{D_{\partial_S S}(x_{\partial_S})}{N_{\partial_S}(x_{\partial_S})}\bigg|\mathcal{F}_t\right]. \tag{19.2}$$

We shall always consider Markov-functional models for which the boundary is as given by (19.1). In this case the terminal time is the last swap start date T_n. For $t < T_n$ we know the form of the discount factors on the boundary, $D_{tt}(x_t) \equiv 1$. Thus, all we need to know is the functional form of discount factors at T_n, $D_{T_n S}(x_{T_n})$, $S \ge T_n$, and the functional form of the numeraire for $t \le T_n$, $N_t(x_t)$.

Before deriving these functional forms, we make one further important observation. For the multi-temporal product in question the value depends only on discount factors at the times $T_1, \ldots, T_n$ and for maturities S_j^i, $i = 1, \ldots, n$, $j = 1, \ldots, m_i$. Thus we are interested in the functional forms only on the grid shown in Figure 19.1.

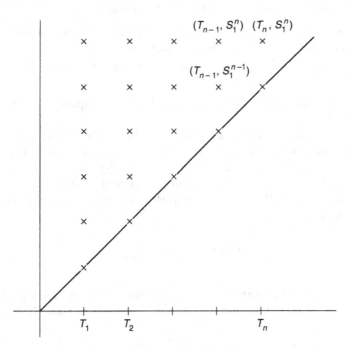

Figure 19.1 Model gridpoints

Since we need only specify the model on the boundary, it follows that all we need to define to recover the grid are the functional forms $D_{T_n S_j^n}(x_{T_n})$, $j = 1, \ldots, m_n$, and the functional form $N_{T_i}(x_{T_i})$, $1 \leq i \leq n$.

19.3.1 Deriving the numeraire on a grid

We now show how to construct these functional forms inductively in such a way that the resulting model correctly prices options on the swaps associated with the rates y^i. We work backwards from T_n to T_1 constructing the functional forms on the grid of Figure 19.1. Completing the specification of the model at other times is not necessary for the application. However, just as was the case for our discussion of terminal swap-rate models and market models, we need to know that it can be done in such a way that the resulting model is arbitrage-free. We do this in Section 19.3.2.

As a first step we must specify the functional forms $D_{T_n S_j^n}(x_{T_n})$, $j = 1, \ldots, m_n$, to be consistent with swaptions associated with the swap rate

y^n. We have seen how to do this in Chapter 13 in our discussion of terminal swap-rate models so we shall assume we have already carried out this step.

It remains to specify the functional form of the numeraire at the times $T_1, \ldots, T_n$. We will make two further assumptions:

(A.i) the choice of numeraire is such that $N_{T_n}(x_{T_n})$ can be inferred from the functional forms of the discount factors at T_n;

(A.ii) the ith forward par swap rate at time T_i, $y_{T_i}^i$, is a monotonic increasing function of the variable x_{T_i}.

The latter assumption is natural if x is intended to capture the overall level of interest rates.

In the remainder of this subsection we show how market prices of the calibrating vanilla swaptions can be used to imply, numerically at least, the functional forms $N_{T_i}(x_{T_i})$ for $i = 1, \ldots, n-1$. Equivalent to calibrating to vanilla swaption prices is calibrating to the inferred market prices of $PVBP$-$digital\ swaptions$. A PVBP-digital swaption corresponding to y^i having strike K has payoff at time T_i of

$$\widetilde{V}_{T_i}^i(K) = P_{T_i}^i \mathbb{1}_{\{y_{T_i}^i > K\}},$$

where P^i is the PVBP of the swap corresponding to the swap rate y^i. Its value at time zero, applying the usual valuation formula (Corollary 7.34), is given by

$$\widetilde{V}_0^i(K) = N_0(x_0)\mathbb{E}_{\mathbb{N}}\left[\hat{P}_{T_i}^i(x_{T_i})\mathbb{1}_{\{y_{T_i}^i(x_{T_i}) > K\}}\right] \qquad (19.3)$$

where

$$\hat{P}_{T_i}^i(x_{T_i}) = \frac{P_{T_i}^i(x_{T_i})}{N_{T_i}(x_{T_i})}.$$

That knowledge of $\widetilde{V}_0^i(K)$ for all K is equivalent to knowlege of the vanilla swaption prices $V_0^i(K)$, again for all K, follows as in Section 13.2.2 from the identity $\widetilde{V}_0^i(K) = -\frac{\partial}{\partial K}V_0^i(K)$.

To determine the functional forms of the numeraire $N_{T_i}(x_{T_i})$ we work back iteratively from the terminal time T_n. Consider the ith step in this procedure. Assume that $N_{T_k}(x_{T_k}), k = i+1, \ldots, n$, have already been determined. We can also assume that

$$\hat{D}_{T_i S}(x_{T_i}) := \frac{D_{T_i S}(x_{T_i})}{N_{T_i}(x_{T_i})}$$

is known for relevant grid-times $S > T_i$, having been determined using (19.2) and the known (conditional) distributions of x_{T_k}, $k = i, \ldots, n$. Note that this implies that $\hat{P}_{T_i}^i$ is also known.

Now consider $y_{T_i}^i$, which can be written as

$$
\begin{aligned}
y_{T_i}^i &= \frac{D_{T_i T_i} - D_{T_i S_{m_i}^i}}{P_{T_i}^i} \\
&= \frac{N_{T_i}^{-1} - D_{T_i S_{m_i}^i} N_{T_i}^{-1}}{P_{T_i}^i N_{T_i}^{-1}}.
\end{aligned}
\tag{19.4}
$$

Rearranging equation (19.4), we have that

$$
N_{T_i}(x_{T_i}) = \frac{1}{\hat{P}_{T_i}^i(x_{T_i}) y_{T_i}^i(x_{T_i}) + \hat{D}_{T_i S_{m_i}^i}(x_{T_i})}.
\tag{19.5}
$$

Thus to determine $N_{T_i}(x_{T_i})$ it is sufficient to find the functional form $y_{T_i}^i(x_{T_i})$.

By assumption (A.ii) there exists a unique value of K, say $K^i(x^*)$, such that the set identity

$$
\{x_{T_i} > x^*\} = \{y_{T_i}^i > K^i(x^*)\}
\tag{19.6}
$$

holds. Now define

$$
J_0^i(x^*) = N_0(x_0) \mathbb{E}_{\mathbb{N}} \left[\hat{P}_{T_i}^i(x_{T_i}) \mathbb{1}_{\{x_{T_i} > x^*\}} \right].
\tag{19.7}
$$

For any given x^* we can calculate the value of $J_0^i(x^*)$ using the known distribution of x_{T_i} under N. Furthermore, using market prices we can then find the value of K such that

$$
J_0^i(x^*) = \tilde{V}_0^i(K).
\tag{19.8}
$$

Comparing (19.3) and (19.7), we see that the value of K satisfying (19.8) is precisely $K^i(x^*)$. Clearly, from (19.6), knowing $K^i(x^*)$ for any x^* is equivalent to knowing the functional form $y_{T_i}^i(x_{T_i})$, and we are done.

It is common market practice to use Black's formula to determine the swaption prices $V_0^i(K)$. Observe, however, that the techniques here apply more generally. In particular, if the currency concerned is one for which there is a large volatility skew, meaning the volatility used as input to Black's formula is highly dependent on the level of the strike K, these techniques can still be applied. This is particularly important in currencies such as yen in which it is not reasonable to model rates via a log-normal process. The ability of Markov-functional models to 'adapt' to all different markets is one of their major strengths.

19.3.2 Existence of a consistent arbitrage-free term structure model

As we did for the terminal swap-rate models in Chapter 13, and for the market models in Chapter 18, we need to ensure that there exists a full arbitrage-free term structure model consistent with this Markov-functional model. The approach we adopt here is very similar to those used earlier.

What we have already done in Section 19.3.1 is to derive a functional form for the numeraire N at the grid-times T_i. By definition each of the bonds $D_{.T_i}$ when divided by the numeraire is a martingale,

$$\frac{D_{tT_i}}{N_t} := \mathbb{E}_{\mathbb{N}}\left[\frac{D_{T_iT_i}}{N_{T_i}}\bigg|\mathcal{F}_t\right].$$

It remains to show that we can develop a model such that this also holds for maturities $T \notin \{T_1,\ldots,T_n\}$. We extend to other maturities as follows. For any $t \leq T_n$, define

$$N_t(x_t) := \frac{T_{i+1}-t}{T_{i+1}-T_i}N_{T_i}(x_t) + \frac{t-T_i}{T_{i+1}-T_i}N_{T_{i+1}}(x_t), \qquad (19.9)$$

where $T_0 := 0$ and $T_i \leq t \leq T_{i+1}$. This is a continuous semimartingale and its law can be found from that of the process x and the defining functional forms $N_{T_i}(x)$. Use this to define a numeraire model (as introduced in Chapter 8) as follows. For $0 \leq t \leq S \leq T_n$ define

$$D_{tS}(x_t) := N_t(x_t)\mathbb{E}_{\mathbb{N}}[N_S^{-1}(x_S)|\mathcal{F}_t],$$

and for $0 \leq t \leq T_n \leq S$ define

$$D_{tS}(x_t) := N_t(x_t)\mathbb{E}_{\mathbb{N}}\left[\frac{D_{T_nS}(x_{T_n})}{N_{T_n}(x_{T_n})}\bigg|\mathcal{F}_t\right].$$

It is readily seen that this is consistent with the assumptions made above.

To complete the model, extend the model in a deterministic fashion from time T_n:

$$D_{tS} = \frac{D_{T_n t}(x_{T_n})}{D_{T_n S}(x_{T_n})},$$

for $T_n < t \leq S$.

For the corresponding numeraire, take N until time S_n and then the unit rolling numeraire from then on.

Remark 19.3: Note we could also have taken

$$N_t^{-1}(x_t) := \frac{D_{0T}-D_{0T_{i+1}}}{D_{0T_i}-D_{0T_{i+1}}}N_{T_i}^{-1}(x_t) + \frac{D_{0T_i}-D_{0T}}{D_{0T_i}-D_{0T_{i+1}}}N_{T_{i+1}}^{-1}(x_t)$$

instead of (19.9). If we wanted to actually build such a general model this latter interpolation is more convenient since it is N^{-1} that appears in valuation formulae, and it is also consistent with a general initial discount curve $\{D_{0t} : t \geq 0\}$.

19.4 EXAMPLE MODELS

In this section we introduce two example models, both one-dimensional, which can be used to price LIBOR- and swap-based interest rate derivatives respectively.

19.4.1 LIBOR model

The model we now describe, the LIBOR Markov-functional model, is designed to price path-dependent and American products which depend explicitly on a contiguous set of forward LIBORs. We denote this set of forward LIBORs by L^i for $i = 1, 2, \ldots, n$. Denote by T_i the start of the ith LIBOR period, by S_i the end, and let $\alpha_i = S_i - T_i$ be the corresponding accrual factor. So, in particular, $S_i = T_{i+1}$ for $i = 1, \ldots, n-1$ and, using the notation of Chapter 10, $L_t^i = L_t[T_i, S_i]$. Further, write $T_{n+1} := S_n$.

An example of the type of product we might wish to price with this model is the *flexible cap*. A flexible cap is defined by the sequence L^i, a corresponding set of strikes K_i, and a limit number m. The holder of a flexible cap has the right to exercise (or not, as he may decide) up to a maximum of m caplets. This contrasts with the limit cap, introduced in Chapter 18, for which the option holder was forced to exercise any caplet setting in the money, up to a maximum of m. It is the added American feature of the flexible cap which makes it hard to price using the market models of Chapter 18.

As described in Sections 19.2 and 19.3, a Markov-functional model can be specified by properties (P.i), (P.ii') and (P.iii'). We work in the swaption measure $\mathbb{S}^n$ corresponding to the numeraire $D._{S_n}$. Here we will be consistent with Black's formula for caplets on L^n if we assume that L^n is a log-normal martingale under $\mathbb{S}^n$, i.e.

$$dL_t^n = \sigma_t^n L_t^n dW_t \tag{19.10}$$

where W is a standard Brownian motion under $\mathbb{S}^n$ and σ^n is some deterministic function. We will often take $\sigma_t^n = \sigma e^{at}$, for some $\sigma > 0$ and some *mean reversion parameter* a, for reasons explained in Section 19.7. Note, however, that this form is too simplistic to use in practice as it does not dynamically fit an appropriate implied market correlation structure.

It follows from (19.10) that we may write

$$L_t^n = L_0^n \exp\left(-\frac{1}{2}\int_0^t (\sigma_u^n)^2 du + x_t\right),$$

where x, a deterministic time-change of a Brownian motion, satisfies

$$dx_t = \sigma_t^n dW_t. \tag{19.11}$$

We take x as the driving Markov process for our model, and this completes the specification of (P.i).

The boundary curve ∂_S for this problem is exactly that of (19.1). For this application the only functional forms needed on the boundary are $D_{T_iT_i}(x_{T_i})$ for $i = 1, 2, \ldots, n$, trivially the unit map, and $D_{T_nS_n}(x_{T_n})$. This latter form follows from the relationship

$$D_{T_nS_n} = \frac{1}{1 + \alpha_n L_{T_n}^n},$$

which yields

$$D_{T_nS_n} = \frac{1}{1 + \alpha_n L_0^n \exp(-\frac{1}{2}\int_0^{T_n}(\sigma_u^n)^2\,du + x_{T_n})}.$$

This completes (P.ii').

It remains to find the functional form $N_{T_i}(x_{T_i})$, in our case $D_{T_iS_n}(x_{T_i})$, and for this we apply the techniques of Section 19.3.1. Take as calibrating instruments the caplets on the forward LIBORs L^i. The value of the ith corresponding digital caplet of strike K is given by

$$\widetilde{V}_0^i(K) = D_{0S_n}(x_0)\mathbb{E}_{\mathbb{S}^n}\left[\frac{D_{T_iT_{i+1}}(x_{T_i})}{D_{T_iS_n}(x_{T_i})}\mathbb{1}_{\{L_{T_i}^i(x_{T_i})>K\}}\right].$$

If we assume the market value is given by Black's formula with volatility $\tilde{\sigma}^i$, the price at time zero for this digital caplet is

$$\widetilde{V}_0^i(K) = D_{0T_{i+1}}(x_0)\Phi(d_2^i), \tag{19.12}$$

where

$$d_2^i = \frac{\log(L_0^i/K)}{\tilde{\sigma}^i\sqrt{T_i}} - \frac{1}{2}\tilde{\sigma}^i\sqrt{T_i}$$

and Φ denotes the cumulative normal distribution function.

To determine the functional form $D_{T_iS_n}(x_{T_i})$ for $i < n$, we proceed as in Section 19.3.1. Suppose we choose some $x^* \in \mathbb{R}$. Evaluate by numerical integration

$$\begin{aligned}
J_0^i(x^*) &= D_{0S_n}(x_0)\mathbb{E}_{\mathbb{S}^n}\left[\frac{D_{T_iT_{i+1}}(x_{T_i})}{D_{T_iS_n}(x_{T_i})}\mathbb{1}_{\{x_{T_i}>x^*\}}\right] \\
&= D_{0S_n}(x_0)\mathbb{E}_{\mathbb{S}^n}\left[\mathbb{E}_{\mathbb{S}^n}\left[\frac{D_{T_{i+1}T_{i+1}}(x_{T_{i+1}})}{D_{T_{i+1}S_n}(x_{T_{i+1}})}\bigg|\mathcal{F}_{T_i}\right]\mathbb{1}_{\{x_{T_i}>x^*\}}\right] \\
&= D_{0S_n}(x_0)\int_{x^*}^{\infty}\left[\int_{-\infty}^{\infty}\frac{1}{D_{T_{i+1}S_n}(u)}\phi_{x_{T_{i+1}}|x_{T_i}}(u)\,du\right]\phi_{x_{T_i}}(v)\,dv
\end{aligned}$$
$$\tag{19.13}$$

where $\phi_{x_{T_i}}$ denotes the transition density function of x_{T_i} and $\phi_{x_{T_{i+1}}|x_{T_i}}$ the density of $x_{T_{i+1}}$ given x_{T_i}. Note from (19.11) that $\phi_{x_{T_{i+1}}|x_{T_i}}$ is a normal density function with mean x_{T_i} and variance $\int_{T_i}^{T_{i+1}} (\sigma_u^n)^2 \, du$. Finally, note the integrand in (19.13) only depends on $D_{T_{i+1} S_n}(x_{T_{i+1}})$ which has already been determined in the previous iteration at T_{i+1}.

The value of $D_{T_i S_n}(x^*)$ can now be determined as follows. Recall from (19.5) that to determine $D_{T_i S_n}(x^*)$ it is sufficient to find the functional form $L_{T_i}^i(x^*)$. From (19.6) and (19.8),

$$L_{T_i}^i(x^*) = K^i(x^*)$$

where $K^i(x^*)$ solves

$$J_0^i(x^*) = \widetilde{V}_0^i(K^i(x^*)) . \tag{19.14}$$

We have just evaluated the left-hand side of (19.14) numerically and $K^i(x^*)$ can thus be found from (19.12) using some simple algorithm. Formally,

$$L_{T_i}^i(x^*) = L_0^i \exp\left[-\tfrac{1}{2}(\tilde{\sigma}^i)^2 T_i - \tilde{\sigma}^i \sqrt{T_i} \Phi^{-1}\left(\frac{J_0^i(x^*)}{D_{0 T_{i+1}}(x_0)} \right) \right] .$$

Finally, to obtain the value of $D_{T_i S_n}(x^*)$ we use (19.5).

19.4.2 Swap model

Here we construct a *swap Markov-functional model* suitable for pricing swap-based products such as the Bermudan swaption. The model described here is suitable for the special case of pricing a Bermudan swaption which corresponds to a cancellable swap. We restrict to this case to keep notation simple. Generalizations are straightforward.

We have already met Bermudans and cancellable swaps on several occasions, in particular in the preceding Orientation on multi-temporal products and in the discussion of the Vasicek–Hull–White model in Chapter 17. Here we suppose that the cancellable swap starts on date T_1 and has payment dates $S_1, \ldots, S_n$. As previously, we refer to the exercise times of a Bermudan swaption as $T_1, \ldots, T_n$ and denote by α_i the accrual factor for the period $[T_i, S_i]$.

For this model the ith forward par swap rate y^i, which sets on date T_i, has coupons precisely at the dates $S_i, \ldots, S_n$, and is given by

$$y_t^i = \frac{D_{t T_i} - D_{t S_n}}{P_t^i},$$

where

$$P_t^i := \sum_{j=i}^{n} \alpha_j D_{t S_j} .$$

Note that the last par swap rate y^n is just the forward LIBOR, L^n, for the period $[T_n, S_n]$. As in the LIBOR example above, we take $D_{\cdot S_n}$ as our numeraire and work in the swaption measure $\mathbb{S}^n$. Furthermore, we specify properties (P.i), (P.ii$'$) and $D_{T_n S_n}(x_{T_n})$ exactly as for the LIBOR model. However, (P.iii$'$), the functional form for the numeraire $D._{S_n}$ at times T_i, $i = 1, \dots, n - 1$, will need to be determined.

For this new model the value of the calibrating PVBP-digital swaption (as defined in Section 19.3.1) having strike K and corresponding to y^i is given by

$$\widetilde{V}_0^i(K) = D_{0S_n}(x_0)\mathbb{E}_{\mathbb{S}^n}\left[\frac{P_{T_i}^i(x_{T_i})}{D_{T_i S_n}(x_{T_i})}\mathbb{1}_{\{y_{T_i}^i(x_{T_i})>K\}}\right].$$

If we assume that the market value is given by Black's formula then the price at time zero of this PVBP-digital swaption has the form

$$\widetilde{V}_0^i(K) = P_0^i(x_0)\Phi(d_2^i) \qquad (19.15)$$

where

$$d_2^i = \frac{\log(y_0^i/K)}{\tilde{\sigma}^i\sqrt{T_i}} - \tfrac{1}{2}\tilde{\sigma}^i\sqrt{T_i}\,.$$

Here Black's formula implies that the marginal distribution of the $y_{T_i}^i$ are log-normally distributed in their respective swaption measures.

Next suppose we choose some $x^* \in \mathbb{R}$ and, for $i < n$, evaluate by numerical integration

$$
\begin{aligned}
J_0^i(x^*) &= D_{0S_n}(x_0)\mathbb{E}_{\mathbb{S}^n}\left[\frac{P_{T_i}^i(x_{T_i})}{D_{T_i S_n}(x_{T_i})}\mathbb{1}_{\{x_{T_i}>x^*\}}\right] \\
&= D_{0S_n}(x_0)\mathbb{E}_{\mathbb{S}^n}\left[\mathbb{E}_{\mathbb{S}^n}\left[\frac{P_{T_{i+1}}^i(x_{T_{i+1}})}{D_{T_{i+1} S_n}(x_{T_{i+1}})}\bigg|\mathcal{F}_{T_i}\right]\mathbb{1}_{\{x_{T_i}>x^*\}}\right] \\
&= D_{0S_n}(x_0)\int_{x^*}^{\infty}\left[\int_{-\infty}^{\infty}\frac{P_{T_{i+1}}^i(u)}{D_{T_{i+1} S_n}(u)}\phi_{x_{T_{i+1}}|x_{T_i}}(u)\,du\right]\phi_{x_{T_i}}(v)\,dv
\end{aligned}
$$

Note that to calculate a value for $J_0^i(x^*)$ we need to know $D_{T_{i+1}T_j}(x_{T_{i+1}})$, $j > i$. These will have already been determined in the previous iteration.

Now, as in Section 19.3.1,

$$y_{T_i}^i(x^*) = K^i(x^*)$$

where $K^i(x^*)$ solves

$$J_0^i(x^*) = \widetilde{V}_0^i(K^i(x^*))\,. \qquad (19.16)$$

Having evaluated the left-hand side of (19.16) numerically, $K^i(x^*)$ can be recovered from (19.15). Formally, we have

$$y_{T_i}^i(x^*) = y_0^i\exp\left[-\tfrac{1}{2}(\tilde{\sigma}^i)^2 T_i - \tilde{\sigma}^i\sqrt{T_i}\,\Phi^{-1}\left(\frac{J_0^i(x^*)}{P_0^i(x_0)}\right)\right]\,.$$

The value of $D_{T_i S_n}(x^*)$ can now be calculated using (19.5).

19.5 MULTIDIMENSIONAL MARKOV-FUNCTIONAL MODELS

All the discussion so far has focused on Markov-functional models for which the driving Markov process is one-dimensional. This is sufficient for many practical applications, but not all. One product not covered is the Bermudan callable spread option. Another important example is as follows. Consider a swap for which one leg is a standard floating leg, with payments being based on LIBOR, and the other is a CMS leg, with the amount of each payment being based on some swap rate. Now suppose we wish to value this swap when, additionally, the counterparty paying LIBOR has the right, on any payment date, to cancel the rest of the swap. We call this product a *Bermudan callable CMS*. A two-factor model is needed to price this product accurately. The reason for this is the fact that the value of the underlying swap is insensitive to the overall level of interest rates (both legs increase in value as interest rates go up), but it is highly sensitive to the shape of the forward interest rate curve, i.e. to the relative level of different forward rates.

The techniques developed in Sections 19.3 and 19.4 relied heavily on the one-dimensional property of the Markov process x. In particular, we assumed in (A.ii) that the swap rates to which we calibrate are monotone functions of the process x. This and the one-dimensional property of x resulted in the set identity (19.6) which was crucial for the efficient implementation. This property would be lost if x were taken to be of higher dimension.

The solution to this problem, of retaining the set identity (19.6) while increasing the dimensionality of the Markov process, is to generalize x to be a one-dimensional process which is not itself Markovian but which is a deterministic function of some higher-dimensional process which is Markovian:

$$x_t = f(z_t, t),$$

where z is a (time-inhomogeneous) Markov process of dimension d and $f : \mathbb{R}^d \times \mathbb{R} \to \mathbb{R}$. As in the one-dimensional case, in practice we usually take z to be a time-changed Brownian motion.

If we adopt this approach, all the techniques developed in Sections 19.3 and 19.4 still apply. Of course, the algorithm takes longer to run since one-dimensional integrals have been replaced by d-dimensional integrals, but that is always the case for a genuine d-dimensional model. What remains is to choose the Markov process z and the function f. By careful choice of these it is possible to generate a wide range of models which calibrate to selected marginal and joint distributions. Work in this area is at a very early stage, but here we outline one of the simplest choices that can be made.

19.5.1 Log-normally driven Markov-functional models

The high dimensionality of the (log-normal) market models in Chapter 18 is a severe impediment to their use for pricing and hedging American-type products. A considerable literature has developed which attempts to overcome this, sometimes by clever use of simulation, sometimes by approximating the market model by some process of lower dimension. In this latter case the most common approach used is to replace the state-dependent drift in the market model by some simpler drift. For example, we could replace the state-dependent drift $\mu_t^i(L_t)$ in equation (18.5) by the corresponding drift with time t replaced by time zero throughout. The resultant LIBOR process is then jointly log-normal and Markov-functional, with the dimension of the driving Markov process being equal to that of the underlying driving Brownian motion in equation (18.5) (which will usually be much less than the total number of LIBORs being modelled).

More sophisticated approximations along similar lines, allowing for (deterministic) time-dependent drifts and accurate convexity adjustments, are also available. All admit arbitrage but they do give yield curve distributions similar to those of the market model being approximated.

A simple log-normal approximation to a log-normal market model makes a good basis for selecting the function f in the Markov-functional approach. For example, what follows is, in outline, a model designed for pricing a Bermudan callable CMS. Suppose we have in mind a two-factor Markov-functional model which prices exactly all swaptions based on the rates $y_{T_i}^i, i = 1, \ldots, n$, on which the constant maturity leg payments are based. Start by defining an approximate LIBOR market model using equation (18.5) but with the drift term $\mu_t^i(L_t)$ replaced by $\mu_0^i(L_0)$. (We use a LIBOR market model in preference to a swap market model because LIBOR is the other relevant index for the Bermudan callable CMS.) It follows from equation (18.8) that the approximate LIBOR process $\hat{L}$ for this approximating model now satisfies the SDE

$$d\hat{L}_t^i = -\left(\sum_{j=i+1}^{n} \left(\frac{\alpha_j L_0^j}{1 + \alpha_j L_0^j} \right) \sigma_t^i \sigma_t^j \rho_{ij} \right) \hat{L}_t^i dt + \sigma_t^i \hat{L}_t^i dW_t^i,$$

which in turn admits the solution

$$\hat{L}_t^i = \hat{L}_0^i \exp\left(\mu^i t - \tfrac{1}{2} \int_0^t (\sigma_u^i)^2 du + \int_0^t \sigma_u^i dW_u^i \right), \tag{19.17}$$

where

$$\mu^i = -\sum_{j=i+1}^{n} \left(\frac{\alpha_j L_0^j}{1 + \alpha_j L_0^j} \right) \sigma_t^i \sigma_t^j \rho_{ij}.$$

In the simplest case one might choose the volatilities σ^i to be constant in time and consistent with caplet prices and choose the correlations ρ_{ij} so that, within this approximate LIBOR market model, the swap rates $\hat{y}_{T_i}^i$ (which are a known function of the LIBORs) have the correct market-implied variance. The time-independent volatility structure can be generalized to incorporate mean reversion, but we omit that possibility here for ease of exposition.

Now take the driving Brownian motion W to be of dimension 2, meaning it can be written in the form $W_t = M z_t$ for some $n \times 2$ matrix M and a standard two-dimensional Brownian motion z. This and the assumption of time-independent volatility mean that (19.17) can be rewritten as

$$\hat{L}_t^i = \hat{L}_0^i \exp(\mu^i t - \tfrac{1}{2}(\sigma^i)^2 t + M_{i1} z_t^1 + M_{i2} z_t^2),$$

which is of Markov-functional form,

$$\hat{L}_t^i \equiv \hat{L}_t^i(z_t).$$

In particular

$$\hat{L}_{T_i}^i \equiv \hat{L}_{T_i}^i(z_{T_i})$$

Suppose now that we are constructing a Markov-functional model and have done so at times $T_{i+1}, \ldots, T_n$. Applying the martingale property of numeraire-rebased assets, we can recover the functional forms $\{L_{T_i}^j(z_{T_i}), j > i\}$. We now write the swap rate $\hat{y}^i$ as

$$\hat{y}_{T_i}^i = \hat{y}_{T_i}^i(\hat{L}_{T_i}^i, L_{T_i}^j, j > i) = f^i(z_{T_i}, T_i).$$

Using the above approximation to a market model as a guide, we take x as any one-dimensional process such that, for each $i = 1, \ldots, n, x(z_{T_i}, T_i) = f^i(z_{T_i}, T_i)$. There are many ways to define x at intermediate times but this has no effect on the model so we shall not specify it further. Having chosen x, we now proceed as in Section 19.3.

Following the above procedure leads to a model which is arbitrage-free, calibrated exactly to the swap rates $y_{T_i}^i, i = 1, \ldots, n$, and which has the desired correlation structure for each pair $(L_{T_i}^i, y_{T_i}^i)$. The intertemporal distributions of these rates can be controlled by the introduction of, and appropriate choice of, mean reversion parameters. Section 19.7 discusses this in the one-dimensional case.

19.6 RELATIONSHIP TO MARKET MODELS

An obvious question which arises is how the Markov-functional models of this chapter relate to the market models introduced in Chapter 18. The class of Markov-functional models is strictly larger. That it contains all the market models described in Chapter 18 follows from Theorems 18.1, 18.2 and 18.3 which showed that the market models given there were all Markovian. That the Markov-functional class is strictly larger follows since the driving Markov process could be taken to have dimension greater than the number of swap rates being modelled. In practice, of course, one would not choose to do this – a primary motivation for the Markov-functional formulation is to reduce the dimensionality.

One could, as is common, define market models more generally than we did in Chapter 18 by allowing the coefficients of the driving SDE to depend on the whole sample path and not just the current value of the market rates. Clearly for this more general case the resultant model will not be Markovian and, therefore, not Markov-functional. Indeed, in this generality the class of market models is equivalent to the class (PDB) described in Chapter 8 and thus is the most general class of continuous interest rate models that we have considered. In particular, it is more general than the (HJM) class.

Often when people refer to market models they mean the special case when the forward par swap rates are assumed to be log-normal martingales in their associated swaption measures. This yields Black's formula for the prices of the associated swaptions. A further question, therefore, is whether it is possible in this case to write a market model as a function of some *low*-dimensional Markov process. In fact it is not. The reason for this is that the assumption of log-normal *processes* in a market model is much stronger and more restrictive than the assumptions made when fitting a Markov-functional model to swaption prices given by Black's formula. In the latter case we only assume that the swap rates, in their associated swaption measures, satisfy the martingale property and have a log-normal distribution *on their respective fixing dates*. The additional restriction for log-normal market models is precisely what makes those models difficult to use in practice because they cannot be usefully characterized by a low-dimensional Markov process. The following result, which can be extended to include the swap-market models, makes this statement precise for the log-normal LIBOR-based market models.

Theorem 19.4 *Let $L = (L^1, \ldots, L^n)$, $n > 1$, be a non-trivial log-normal market model, where the L^i are a set of contiguous forward LIBORs. Then there exists no one-dimensional process x such that:*

(i) $L_t^i = L_t^i(x_t) \in C^{2,1}(\mathbb{R}, \mathbb{R}^+)$ for $i = 1, 2, \ldots, n$;
(ii) each $L_t^i(x_t)$ is strictly monotone in x_t.

That is, L is not a one-dimensional Markov-functional model.

Remark: We believe this result extends to hold for any process x of dimension less than n and any functional forms $L_t^i(x_t)$.

Proof: Suppose such a process x exists. Then it follows from the invertibility of the map $x_t \to L_t^i(x_t)$ that we can write

$$L_t^i = L_t^i(L_t^n), \qquad i = 1, 2, \ldots, n. \tag{19.18}$$

Since L is a log-normal market model, it follows from (18.8) that it satisfies the SDE

$$dL_t^i = -\left(\sum_{j=i+1}^{n} \left(\frac{\alpha_j L_t^j}{1 + \alpha_j L_t^j} \right) \sigma_t^i \sigma_t^j \rho_{ij} \right) L_t^i \, dt + \sigma_t^i L_t^i \, dW_t^i, \tag{19.19}$$

under the measure $\mathbb{S}^n$ (corresponding to taking $D._{S_n}$ as numeraire). Here $W = (W^1, \ldots, W^n)$ is a (correlated) n-dimensional Brownian motion with $dW_t^i dW_t^j = \rho_{ij} dt$, $\sigma^1, \ldots, \sigma^n$ are deterministic functions of time and the α_j are, as usual, accrual factors. On the other hand, if we apply Itô's formula to (19.18), we obtain

$$dL_t^i = \left[\frac{\partial L_t^i}{\partial t} + \tfrac{1}{2}(\sigma_t^n L_t^n)^2 \frac{\partial^2 L_t^i}{\partial (L_t^n)^2} \right] dt + \frac{\partial L_t^i}{\partial L_t^n} dL_t^n$$

$$= \left[\frac{\partial L_t^i}{\partial t} + \tfrac{1}{2}(\sigma_t^n L_t^n)^2 \frac{\partial^2 L_t^i}{\partial (L_t^n)^2} \right] dt + \frac{\partial L_t^i}{\partial L_t^n} \sigma_t^n L_t^n dW_t^n . \qquad (19.20)$$

Equating the local martingale terms in (19.19) and (19.20) yields $W^i \equiv W^n$, for all i, and

$$\frac{\partial L_t^i}{\partial L_t^n} = \frac{\sigma_t^i L_t^i}{\sigma_t^n L_t^n} . \qquad (19.21)$$

Solving (19.21), we find

$$L_t^i = c_i(t)(L_t^n)^{\beta_i(t)} \qquad (19.22)$$

where $c_i(t)$ is some function of t and $\beta_i(t) = \sigma_t^i / \sigma_t^n$.
Having concluded that $W^i = W^n$, for all i, (19.19) reduces to the form

$$dL_t^i = -\left(\sum_{j=i+1}^n \left(\frac{\alpha_j L_t^j}{1 + \alpha_j L_t^j} \right) \sigma_t^i \sigma_t^j \right) L_t^i \, dt + \sigma_t^i L_t^i \, dW_t^n. \qquad (19.23)$$

Substituting (19.22) back into (19.20) and equating finite variation terms in (19.20) and (19.23) now gives (after some rearrangement)

$$\frac{1}{\sigma_t^i} \left(\frac{c_i'(t)}{c_i(t)} + \beta_i'(t) \log L_t^i \right) + \tfrac{1}{2}\sigma_t^i(\beta_i(t) - 1) = - \sum_{j=i+1}^n \frac{\alpha_j \sigma_t^j L_t^j}{1 + \alpha_j L_t^j} .$$

In particular, taking $i = n-1$ yields $\sigma_t^n = 0$ and thus the model is degenerate.
□

19.7 MEAN REVERSION, FORWARD VOLATILITIES AND CORRELATION

19.7.1 Mean reversion and correlation

Mean reversion of interest rates is considered a desirable property of a model because it is perceived that interest rates tend to trade within a fairly tightly defined range. This is indeed true, but when pricing exotic derivatives it is the

effect of mean reversion on the correlation of interest rates at different times
that is more important. We illustrate this for the Vasicek–Hull–White model
because it is particularly tractable. The insights carry over to the market
models of Chapter 18 and the Markov-functional models of this chapter.

In the standard Vasicek–Hull–White model the short-rate process r solves
the SDE

$$dr_t = (\theta_t - a_t r_t)dt + \sigma_t dW_t, \tag{19.24}$$

under the risk-neutral measure $\mathbb{Q}$, where a, θ and σ are deterministic functions
and W is a standard Brownian motion. For the purposes of this discussion we
take $a_t \equiv a$, for some constant value.

Suppose we fit the Vasicek–Hull–White model to market bond and caplet
prices. The best we can do is to fit the model so that it correctly values
one caplet for each date T_i, the one having strike K_i say. It turns out, and
the reader can verify this from equations (17.4) and (17.14), that, given the
initial discount curve $\{D_{0S}, S > 0\}$ and the mean reversion parameter a, the
correlation structure for the r_{T_i} is given by

$$\mathrm{corr}(r_{T_i}, r_{T_j}) = e^{-a(T_j - T_i)} \sqrt{\frac{v_i}{v_j}},$$

where $v_i = \mathrm{var}(r_{T_i})$. Furthermore, given the market cap prices, the ratio
v_i/v_j is independent of a. Thus increasing a has the effect of reducing
the correlation between the short rate at different times, hence also the
covariance of spot LIBOR at different times. This is important for pricing
path-dependent and American options whose values depend on the joint
distribution $(r_{T_i}, i = 1, 2, \dots, n)$.

For other models, including Markov-functional models, the analytic
formulae above will not hold, of course, but the general principal does. Given
the marginal distributions for a set of spot interest rates $\{y_{T_i}^i, i = 1, 2, \dots, n\}$,
a higher mean reversion for spot interest rates leads to a lower correlation
between the $\{y_{T_i}^i, i = 1, 2, \dots, n\}$.

19.7.2 Mean reversion and forward volatilities

In the models of Section 19.4 we have not explicitly presented an SDE for
spot LIBOR or spot par swap rates. We therefore need to work a little harder
to understand how to introduce mean reversion within these models. To do
this we consider the Vasicek–Hull–White model once again.

We have in (19.10) parameterized our Markov-functional example in terms
of a *forward* LIBOR process L^n. If we can understand the effect of mean
reversion within a model such as Vasicek–Hull–White on L^n we can apply the
same principles to a more general Markov-functional model.

An analysis of the Vasicek–Hull–White model defined via (19.24) shows
that

$$L_t^n = X_t - \alpha_n^{-1}$$

where

$$dX_t = \left(\frac{e^{-aT_n} - e^{-aS_n}}{a} \right) \sigma e^{at} X_t d\widetilde{W}_t,$$

$\widetilde{W}$ being a standard Brownian motion under the measure $\mathbb{S}^n$. The forward LIBOR is a log-normal martingale minus a constant. Notice the dependence on time t of the diffusion coefficient of the martingale term: the volatility is of the form $constant \times X_t \times e^{at}$. This is the motivation for the volatility structure we chose for L^n in the Markov-functional model definition in (19.10). But notice also that, for a more realistic correlation structure, one would at a minimum need to make σ a function of time and calibrate (19.24) to market prices. This is not a topic we shall cover in any detail here. What is important is not so much the exact functional form as the fact that the volatility increases through time. The faster the increase, the lower the correlation between spot interest rates which set at different times.

19.7.3 Mean reversion within the Markov-functional LIBOR model

To conclude this discussion of mean reversion we return to the Markov-functional LIBOR model of Section 19.4.1 and show how taking $\sigma_t^n = \sigma e^{at}$ in (19.10) leads to mean reversion of spot LIBOR. Suppose the market cap prices are such that the implied volatilities for L^1 and L^n are the same, $\hat{\sigma}$ say, and all initial forward values are the same, $L_0^i = L_0$, for all i.

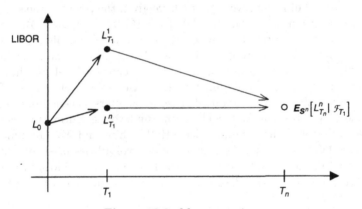

Figure 19.2. Mean reversion

Figure 19.2 shows the evolution of L^1 and L^n in the situation when the driving Markov process x has increased (significantly), $x_{T_1} > x_0$. Both L^1 and L^n have increased over the interval $[0, T_1]$ but L^1 has increased by more. The reason for this is as follows. Over $[0, T_1]$, L^1 has (root mean square) volatility $\hat{\sigma}$. By comparison, L^n has (root mean square) volatility $\hat{\sigma}$ *over the whole interval* $[0, T_n]$, but its volatility is increasing exponentially (in the case when $\sigma_t \equiv \sigma$, a constant) and thus its (root mean square) volatility over $[0, T_1]$ is less than $\hat{\sigma}$. Since L^n is a martingale under $\mathbb{S}^n$, it follows that

$$\mathbb{E}_{\mathbb{S}^n}\left[L^n_{T_n} \big| \mathcal{F}_{T_1}\right] = L^n_{T_1} < L^1_{T_1} \, .$$

That is, in Figure 19.2 when spot LIBOR has moved up from its initial value L_0 at time zero to its value at T_1, $L^1_{T_1}$, the expected value of spot LIBOR at T_n is less than $L^1_{T_1}$. Conversely, when spot LIBOR moves down by time T_1 ($x_{T_1} < x_0$) the expected value of spot LIBOR at T_n is greater than its value at T_1. This is mean reversion.

19.8 SOME NUMERICAL RESULTS

We conclude with some numerical results comparing the prices of Bermudan swaptions derived using some of the models discussed in this book. For a 30-year Deutschmark Bermudan, which is exercisable every five years, we have compared in Table 19.1, for different levels of mean reversion, the prices calculated by three different models: Black–Karasinski, one-dimensional Markov-functional and the Vasicek–Hull–White model. Note that our implementation of the Vasicek–Hull–White model prohibits negative mean reversions so we have not been able to include these results.

For every level of mean reversion, we have given the prices of the embedded European swaptions (5×25, 10×20, 15×15, 20×10, 25×5). Since all three models are calibrated to these prices, all models should agree exactly on them. The differences reported in the table are due to numerical errors. We adopted the following approach. First we chose a level for the Black–Karasinski mean reversion parameter and reasonable levels for the Black–Karasinski volatilities. We then used these parameters to generate the prices, using the Black–Karasinski model, of the underlying European swaptions. We then calibrated the other, Vasicek–Hull–White and Markov-functional, models to these prices. The resultant implied volatilities used for each case are reported in the second column. At the bottom of each block in the table, we report the value of the Bermudan swaption as calculated by each of the three models.

From the table we see that the mean reversion parameter has a significant impact on the price of the Bermudan swaption. It is intuitively clear why this is the case. The reason a Bermudan option has more value than the maximum of the embedded European option prices is the freedom it offers to delay or advance the exercise decision of the underlying swap during the life of the contract to a date when it is most profitable. The relative value between exercising 'now' or 'later' depends very much on the correlation of the underlying swap rates between different time points. This correlation structure is exactly what is being controlled by the mean reversion parameter.

By contrast, the effect of changing the model is considerably less, and what difference there is will be due in part to the fact that the mean reversion parameter has a slightly different meaning for each model. The precise (marginal) distributional assumptions made have a secondary role in determining prices for exotic options (relative to the joint distributions as captured by the mean reversion parameter).

Table 19.1 Value of 30-year Bermudan, exercisable every five years

Strike: 6.24%, Currency: DEM, Valuation Date: 11-feb-98

Mean Reversion = −0.05							
European		Receivers			Payers		
Mat	ImVol	BK	MF	HW	BK	MF	HW
05 × 25	8.17	447.7	445.9	–	456.6	456.6	–
10 × 20	7.74	365.2	363.8	–	448.2	448.6	–
15 × 15	7.90	285.8	284.7	–	350.8	351.4	–
20 × 10	8.29	191.8	190.8	–	241.6	242.1	–
25 × 05	8.68	91.0	90.9	–	126.2	126.1	–
Bermudan		510.0	502.7	–	572.2	566.9	–

Mean Reversion = 0.06							
European		Receivers			Payers		
Mat	ImVol	BK	MF	HW	BK	MF	HW
05 × 25	8.46	463.7	462.8	464.2	472.6	472.0	473.3
10 × 20	7.91	374.0	373.5	374.2	457.0	456.9	457.5
15 × 15	7.80	281.8	281.5	281.9	346.8	346.8	347.1
20 × 10	7.81	179.5	179.4	179.6	229.4	229.3	229.5
25 × 05	8.16	84.8	84.7	84.7	119.9	120.0	120.1
Bermudan		606.1	602.9	608.2	743.6	717.7	727.3

Mean Reversion = 0.20							
European		Receivers			Payers		
Mat	ImVol	BK	MF	HW	BK	MF	HW
05 × 25	8.37	458.7	457.6	459.3	467.5	466.8	468.0
10 × 20	6.99	326.2	325.4	326.2	409.1	408.8	409.3
15 × 15	6.67	237.1	236.4	237.3	302.1	301.7	302.4
20 × 10	7.32	166.7	166.4	166.6	216.6	216.4	216.6
25 × 05	9.40	99.8	99.5	99.6	134.9	134.8	134.8
Bermudan		647.4	665.8	673.9	885.5	814.7	813.6

Mat denotes maturity and tenor of embedded European option.
ImVol denotes implied volatility of embedded European option.
BK are prices calculated with Black–Karasinksi model.
MF are prices calculated with Markov-functional model.
HW are prices calculated with Hull–White model.

20

Exercises and Solutions

Exercises

Exercise 1 Let M be a non-negative local martingale defined on the filtered space $(\Omega, \{\mathcal{F}_t\}, \mathcal{F}, \mathbb{P})$ such that $\mathbb{E}[M_0] < \infty$. Show that M is a supermartingale.

Exercise 2 Let $(\Omega, \{\mathcal{F}_t\}, \mathcal{F}, \mathbb{P})$ be a filtered probability space supporting a one-dimensional Brownian motion W. Let X denote a local martingale of the form

$$X_t := \int_0^t H_u \, dW_u \,,$$

where $H \in \Pi(W)$ is a deterministic function of time. Suppose that $[X]_t \uparrow \infty$ as $t \uparrow \infty$ and, for $t \geq 0$, define

$$\tau_t := \inf \{ u > 0 : [X]_u > t \}.$$

Show that X can be written in the form

$$X_t = \widetilde{W}\big([X]_t \big),$$

where $\widetilde{W}$ is a Brownian motion adapted to $\{\tilde{\mathcal{F}}_t\} := \{\mathcal{F}_{\tau_t}\}$.
Remark: The result above is sufficient for the application in Section 17.2.1. For the more general result due to Dubins and Schwarz, where X is any continuous local martingale, see Section IV.34 of Rogers and Williams (1987).

Exercise 3 Consider the SDE

$$dX_t = -aX_t dt + \sigma dW_t,$$

where W is a one-dimensional Brownian motion and a and σ are constants. By applying Itô's formula to the process $Y_t := e^{at} X_t$, show that the unique solution to this SDE is given by

$$X_t = e^{-at} \left(X_0 + \int_0^t e^{as} \sigma \, dW_s \right).$$

Assuming $X_0 = x_0$, a constant, specify the distribution of X_t.

Financial Derivatives in Theory and Practice Revised Edition.   P. J. Hunt and J. E. Kennedy
© 2004 John Wiley & Sons, Ltd   ISBNs: 0-470-86358-7 (HB); 0-470-86359-5 (PB)

Further show that for $t_1 < t_2$,

$$\operatorname{corr}(X_{t_1}, X_{t_2}) = e^{-a(t_2-t_1)}\sqrt{\frac{v_1}{v_2}},$$

where $v_i = \operatorname{var}(X_{t_i})$.

Exercise 4 (i) Let $\mathbb{P}$ and $\mathbb{Q}$ be equivalent probability measures with respect to $\mathcal{F}$ on the filtered space $(\Omega, \{\mathcal{F}_t\}, \mathcal{F})$ and let $X \in \mathcal{L}^1(\Omega, \mathcal{F}_{t^*}, \mathbb{Q})$ for some fixed t^*. For $s \le t^*$, express the conditional expectation

$$\mathbb{E}_{\mathbb{Q}}[X \mid \mathcal{F}_s]$$

in terms of the Radon–Nikodým process

$$\rho_s := \left.\frac{d\mathbb{Q}}{d\mathbb{P}}\right|_{\mathcal{F}_s}, \qquad s \le t^*.$$

(ii) Show that M is an $(\{\mathcal{F}_t\}, \mathbb{Q})$ martingale if and only if ρM is an $(\{\mathcal{F}_t\}, \mathbb{P})$ martingale. (This is just Lemma 5.19(i).)

In Exercises 5 to 9 and 11, $(\Omega, \mathcal{F}, \mathbb{P})$ is a probability space supporting a one-dimensional Brownian motion W and $\{\mathcal{F}_t^W\}$ denotes the augmented natural filtration generated by W. We will also use the notation A' to denote the transpose of the vector A.

The next exercise can be attempted after reading Chapter 5.

Exercise 5 Let μ, $r > 0$ and $\sigma > 0$ be constants and consider a process $A' = (A^{(1)}, A^{(2)})$ satisfying

$$dA_t = \begin{pmatrix} rA_t^{(1)} \\ \mu A_t^{(2)} \end{pmatrix} dt + \begin{pmatrix} 0 \\ \sigma A_t^{(2)} \end{pmatrix} dW_t$$

$$A_0' = (e^{-rT}, 1).$$

For fixed $T < \infty$, define a measure $\mathbb{Q} \sim \mathbb{P}$ on $\mathcal{F}_T^W$ via

$$\left.\frac{d\mathbb{Q}}{d\mathbb{P}}\right|_{\mathcal{F}_T^W} = \exp\big(\phi W_T - \tfrac{1}{2}\phi^2 T\big),$$

where $\phi = \sigma^{-1}(r - \mu)$. Further, for some $V_T \in \mathcal{L}^1(\Omega, \mathcal{F}_T^W, \mathbb{Q})$, define

$$V_0 := e^{-rT}\mathbb{E}_{\mathbb{Q}}[V_T].$$

Suppose that there exists an $\{\mathcal{F}_t^W\}$ semimartingale X and $\{\mathcal{F}_t^W\}$-predictable processes $\alpha^{(1)}$ and $\alpha^{(2)}$ such that

$$dX_t = \alpha_t^{(1)} dA_t^{(1)} + \alpha_t^{(2)} dA_t^{(2)},$$

$$X_t = \alpha_t^{(1)} A_t^{(1)} + \alpha_t^{(2)} A_t^{(2)},$$

$$X_t \geq 0 \quad \text{for } t \in [0, T],$$

$$X_T = V_T.$$

(i) Show that $X_0 \geq V_0$.

Hint: Consider the process $\hat{X} := X/A^{(1)}$ under $\mathbb{Q}$.

(ii) Further show that we can find processes X, $\alpha^{(1)}$ and $\alpha^{(2)}$ with the properties described above and such that $X_0 = V_0$.

Hint: Define $M_t := \mathbb{E}_{\mathbb{Q}}[V_T \mid \mathcal{F}_t^W]$ and show we can take $X_t = A_t^{(1)} M_t$.

Exercise 6 *Black–Scholes economy*

Consider the Black–Scholes economy as defined in Example 7.1: for the finite time interval $0 \leq t \leq T < \infty$, the price process $A' = (D, S)$ satisfies

$$dD_t = r D_t \, dt \qquad (D_T = 1, \, r > 0),$$

$$dS_t = \mu S_t \, dt + \sigma S_t \, dW_t \qquad (S_0 \text{ a constant}).$$

(i) Writing $S_t^D = D_t^{-1} S_t$, explain how we could use Girsanov's theorem to find a solution to the SDE

$$dS_t^D = (\mu - r) S_t^D \, dt + \sigma S_t^D \, dW_t.$$

Remark: The easiest way to solve this equation is, of course, to consider $\log S_t^D$.

(ii) Set $\phi = \sigma^{-1}(r - \mu)$. Use the solution you found in (i) to show that

$$S_t^D \exp\left(\phi W_t - \tfrac{1}{2}\phi^2 t\right)$$

is an $\{\mathcal{F}_t^W\}$ martingale under $\mathbb{P}$.

(iii) Let $\mathbb{Q}$ be a measure equivalent to $\mathbb{P}$ on $\mathcal{F}_T^W$ defined via

$$\left.\frac{d\mathbb{Q}}{d\mathbb{P}}\right|_{\mathcal{F}_T^W} := \exp\left(\phi W_T - \tfrac{1}{2}\phi^2 T\right).$$

Deduce from (ii) that S^D is an $\left(\{\mathcal{F}_t^W\}, \mathbb{Q}\right)$ martingale.

Next define a measure $\hat{\mathbb{Q}} \sim \mathbb{Q}$ on $\mathcal{F}_T^W$ via

$$\left.\frac{d\hat{\mathbb{Q}}}{d\mathbb{Q}}\right|_{\mathcal{F}_T^W} = \frac{e^{-rT}S_T}{S_0}.$$

Show that, for $0 \le u \le t \le T$,

$$\mathbb{E}_{\hat{\mathbb{Q}}}\left[\frac{D_t}{S_t}\middle|\mathcal{F}_u^W\right] = \frac{D_u}{S_u}.$$

(iv) Show that the time-zero value of a derivative paying $(S_T - K)_+$ at time T is given by
$$V_0 = S_0\,\hat{\mathbb{Q}}(S_T > K) - e^{-rt}K\,\mathbb{Q}(S_T > K).$$

Exercise 7   For the Black–Scholes economy described in Exercise 6, let $\mathbb{N}$ denote an EMM corresponding to the numeraire $N = D + S$.

(i) Find the SDE satisfied by (A^N, N) under the measure $\mathbb{N}$.

(ii) Use your working in (i) and Lemma 5.17 to show that

$$\left.\frac{d\mathbb{N}}{d\mathbb{P}}\right|_{\mathcal{F}_t^W} = \frac{N_t/N_0}{D_t/D_0}\,\exp\left(\phi W_t - \tfrac{1}{2}\phi^2 t\right),$$

where $\phi = \sigma^{-1}(r - \mu)$.

Exercise 8   Let $\{\sigma_t, t \ge 0\}$ be an $\{\mathcal{F}_t^W\}$-predictable process bounded above and below by positive constants. Consider the Black–Scholes economy described in Exercise 6 but where now

$$dS_t = \mu S_t\,dt + \sigma_t S_t\,dW_t \qquad (S_0 = 1).$$

(i) Find the SDE satisfied by $D^S := S^{-1}D$ under $\mathbb{P}$.

(ii) Show that there exists a unique EMM $\mathbb{Q}^S$ corresponding to the numeraire S.

(iii) Find the solution to the SDE in (i) using Girsanov's theorem.

Exercise 9   *Example of a path-dependent option*

(i) Consider the Black–Scholes economy as defined in Exercise 6. Define

$$m_t^S := \min_{u \in [0,t]} S_u$$

and

$$M_{t,T}^X := \max_{u \in [t,T]}(X_u - X_t),$$

where

$$X_t := -\sigma W_t + \left(\tfrac{1}{2}\sigma^2 - \mu\right)t$$

and W is a one-dimensional Brownian motion with respect to $\left(\Omega, \{\mathcal{F}_t^W\}, \mathcal{F}, \mathbb{P}\right)$. Show that

$$m_T^S = \min\left\{m_t^S, S_t e^{-M_{t,T}^X}\right\}.$$

(ii) A *lookback call option* has payoff at time T given by

$$V_T = \left(S_T - m_T^S\right)_+ = S_T - m_T^S.$$

Use Girsanov's theorem and the result of Example 2.17 to show that the value at time $t \in [0, T]$ of this option is given by

$$V_t = s N(h_1) - m e^{-r\tau} N(h_2)$$

$$- \frac{s\sigma^2}{2r} N(-h_1) + e^{-r\tau} \frac{s\sigma^2}{2r} \left(\frac{m}{s}\right)^{2r\sigma^{-2}} N\left(-h_2 + 2\frac{r_2}{\sigma}\sqrt{\tau}\right),$$

where

$$h_1 := \frac{\log(s/m)}{\sigma\sqrt{\tau}} + \frac{r_1}{\sigma}\sqrt{\tau},$$

$$h_2 := \frac{\log(s/m)}{\sigma\sqrt{\tau}} + \frac{r_2}{\sigma}\sqrt{\tau},$$

and $s := S_t$, $m := m_t^S$, $\tau = T - t$, $r_1 = r + \tfrac{1}{2}\sigma^2$ and $r_2 = r - \tfrac{1}{2}\sigma^2$.

Exercise 10 Let W be a two-dimensional Brownian motion defined on the probability space $(\Omega, \mathcal{F}, \mathbb{P})$ and let $\{\mathcal{F}_t^W\}$ be the augmented natural filtration generated by W.

Consider an economy defined for the finite time interval $0 \le t \le T < \infty$ and composed of three assets having price process $A' = \left(D, S^{(1)}, S^{(2)}\right)$ satisfying

$$dD_t = r D_t \, dt \qquad\qquad (D_T = 1, \, r > 0),$$

$$dS_t^{(i)} = S_t^{(i)}\left(\mu_i \, dt + dM_t^{(i)}\right), \qquad i = 1, 2,$$

where $M^{(1)}$ and $M^{(2)}$ are continuous $\{\mathcal{F}_t^W\}$ martingales such that

$$\left[M^{(i)}, M^{(j)}\right]_t = \xi_{ij} \, dt, \qquad i = 1, 2,$$

for constants ξ_{ij} with $\xi_{ii} > 0$, and

$$\rho := \frac{\xi_{12}}{\sqrt{\xi_{11}}\sqrt{\xi_{22}}} \in (-1, 1).$$

(i) Use Girsanov's theorem to show that there exists an EMM $\mathbb{Q}$ corresponding to the numeraire $S^{(2)}$.

Hence, setting $X_t := S_t^{(1)}/S_t^{(2)}$, show that

$$dX_t = X_t \, dY_t,$$

where

$$Y_t = \left(\sqrt{\xi_{11}} - \sqrt{\xi_{22}}\rho\right)\widetilde{W}_t^{(1)} - \sqrt{\xi_{22}(1-\rho^2)}\,\widetilde{W}_t^{(2)}$$

and $\widetilde{W}' = \left(\widetilde{W}^{(1)}, \widetilde{W}^{(2)}\right)$ is a Brownian motion under $\mathbb{Q}$.

(ii) Show that the value at time zero of the attainable derivative paying $\left(S_T^{(1)} - S_T^{(2)}\right)_+$ at time T is given by

$$V_0 = S_0^{(1)} N(d_1) - S_0^{(2)} N(d_2),$$

where

$$d_1 = \frac{\log(X_0)}{\sigma_0 \sqrt{T}} + \tfrac{1}{2}\sigma_0\sqrt{T},$$

$$d_2 = \frac{\log(X_0)}{\sigma_0 \sqrt{T}} - \tfrac{1}{2}\sigma_0\sqrt{T}$$

and

$$\sigma_0^2 = \xi_{11} - 2\xi_{12} + \xi_{22}.$$

Exercise 11 *Completeness for the Black–Scholes economy*

Remark: Completeness of the Black–Scholes economy follows immediately from Theorem 7.45. This exercise works through a direct approach.

Consider the Black–Scholes economy as described in Exercise 6. Let $\mathbb{Q}^S$ denote the *unique* EMM corresponding to the numeraire S. (That $\mathbb{Q}^S$ is unique is a special case of Exercise 8(ii).)

(i) Apply the martingale representation theorem for Brownian motion (Theorem 5.49) to show that, for any $\mathcal{F}_T^W$-measurable random variable X satisfying

$$\mathbb{E}_{\mathbb{Q}^S}\big[|X/S_T|\big] < \infty,$$

there exists an $\{\mathcal{F}_t^W\}$-predictable self-financing trading strategy ϕ such that

$$X = \phi_0 \cdot A_0 + \int_0^T \phi_u \cdot dA_u.$$

(ii) Suppose $(N, \mathbb{N})$ is some other numeraire pair for the economy. Show that, for $0 \le t \le T$,

$$\left. \frac{d\mathbb{Q}^S}{d\mathbb{N}} \right|_{\mathcal{F}_t^W} = \frac{S_t/S_0}{N_t/N_0}.$$

(iii) Show that the trading strategy ϕ of part (i) is admissible and hence that the Black–Scholes economy is complete for the set

$$S = \left\{ X : X \text{ is an } \mathcal{F}_T^W\text{-measurable random variable such that } \mathbb{E}_{\mathbb{Q}}[|X|] < \infty \right\},$$

where $\mathbb{Q}$ is the unique EMM corresponding to numeraire D.

Exercise 12 (i) Let $\mathcal{E}$ be an economy defined on the probability space $(\Omega, \{\mathcal{F}_t\}, \mathcal{F}, \mathbb{P})$. Suppose $(N, \mathbb{N})$ is a numeraire pair for $\mathcal{E}$ and let V_T be some attainable contingent claim. Starting from the martingale valuation formula, equation (7.15), show that

$$V_t = Z_t^{-1} \mathbb{E}_{\mathbb{P}}\big[V_T Z_T \mid \mathcal{F}_t^A\big]$$

for some pricing kernel Z.

(ii) Find a pricing kernel for the economy described in Exercise 10. Is this economy complete? Write down the Radon–Nikodým derivative

$$\left. \frac{d\mathbb{Q}^D}{d\mathbb{P}} \right|_{\mathcal{F}_t^A}, \qquad 0 \le t \le T,$$

where $\mathbb{Q}^D$ denotes an EMM corresponding to the numeraire D.

Exercise 13 Recall that a rational log-normal model is defined by

$$D_{tT} = \frac{A_T + B_T M_t}{A_t + B_t M_t},$$

where M is a log-normal martingale under $\mathbb{Z}$, a measure locally equivalent to the 'real world' measure $\mathbb{P}$, $M_0 = 1$, and A and B are deterministic positive absolutely continuous functions decreasing to zero at infinity.

(i) Show that a rational log-normal model is in the class (FH). (See Chapter 8, page 190, for a definition of the class (FH).)

(ii) In the definition of an (FH) model suppose that

$$M_t = \exp\big(\sigma W_t - \tfrac{1}{2}\sigma^2 t\big),$$

where W is a one-dimensional Brownian motion under $\mathbb{Z}$. Use this rational log-normal model to find the time-zero value of a payers swaption having cashflows at times $S_1 < S_2 < \ldots < S_n$, start date $T = S_0 < S_1$, and fixed rate K.

Exercise 14 Let $(\Omega, \{\mathcal{F}_t\}, \mathcal{F}, \mathbb{P})$ be a probability space supporting a one-dimensional Brownian motion W.

(i) For a short-rate model having short-rate process r and risk-neutral measure $\mathbb{Q}$, show that the time-t value of any attainable derivative with value $V_T \in \mathcal{F}_T^A$ at maturity time T can be expressed as

$$V_t = \mathbb{E}_\mathbb{Q}\left[\exp\left(- \int_t^T r_u \, du \right) V_T \,\middle|\, \mathcal{F}_t^A \right].$$

Further, show that we can write

$$V_t = D_{tT} \, \mathbb{E}_\mathbb{F}\left[V_T \,\middle|\, \mathcal{F}_t^A \right],$$

where $\mathbb{F}$ is a measure equivalent to $\mathbb{Q}$ on $\mathcal{F}_T^A$ which you should specify.

(ii) *Ho–Lee model*
Suppose that under $\mathbb{Q}$ the short-rate process r satisfies the SDE

$$dr_t = \theta_t \, dt + \sigma \, dW_t,$$

where θ_t is a deterministic function of time and σ is a constant.
Assuming that the model is calibrated to the initial discount curve, show that

$$\theta_t = -\frac{\partial^2 \log D_{0t}}{\partial t^2} + \sigma^2 t.$$

Further, show that under $\mathbb{F}$ the short-rate process r satisfies

$$dr_t = \left(\theta_t - \sigma^2(T - t) \right) dt + \sigma \, d\widetilde{W}_t,$$

where $\widetilde{W}$ is a standard Brownian motion under $\mathbb{F}$.

Exercise 15 Consider an economy involving three currencies. For $i = 1, 2, 3$, let D_{tT}^i denote the value at time t of a zero coupon bond that pays a unit amount at T in currency i. Further, for $i, j = 1, 2, 3$, $i \neq j$, let X_t^{ij} denote the value at time t in currency i of one unit of currency j, and let M_{tT}^{ij} denote the corresponding forward FX rate at time t for maturity at T. Recall that

$$M_{tT}^{ij} = \frac{D_{tT}^j}{D_{tT}^i} X_t^{ij}.$$

Suppose that a complete arbitrage-free model has been specified for the economy for the finite time interval $0 \le t \le T < \infty$ and let $\mathbb{Q}^1$ denote the EMM corresponding to numeraire $D^1_{\cdot T}$.

(i) Show that the time-zero value, stated in currency 2, of a call option that pays $(X^{23}_T - K)_+$ in currency 2 at time T can be expressed as

$$V^{23}_0(K) = X^{21}_0 D^1_{0T} \mathbb{E}_{\mathbb{Q}^1}\left[(X^{13}_T - X^{12}_T K)_+\right].$$

(ii) Suppose we are given the prices for all strikes of call options on X^{21}_T and X^{31}_T. If the model is calibrated to these prices explain, using the expression in (i), what these prices tell us about the distributional information needed to calculate call prices on X^{23}_T.

(iii) Let W^* be a two-dimensional Brownian motion on the probability space $(\Omega, \mathcal{F}, \mathbb{Q}^1)$ and let $\{\mathcal{F}_t\}$ be the augmented natural filtration generated by W^*. Now suppose that under $\mathbb{Q}^1$ the processes $M^{12}_{\cdot T}$ and $M^{13}_{\cdot T}$ satisfy

$$dM^{12}_{tT} = \sigma^{(1)} e^{at} M^{12}_{tT} dW^{(1)}_t,$$

$$dM^{13}_{tT} = \sigma^{(2)} e^{at} M^{13}_{tT} dW^{(2)}_t,$$

where $\sigma^{(1)}$, $\sigma^{(2)}$ and a are positive constants and

$$W^{(1)}_t = \sqrt{1 - \rho^2}\, W^{*(1)}_t + \rho W^{*(2)}_t, \quad W^{(2)}_t = W^{*(2)}_t.$$

If $\mathbb{Q}^2$ denotes the EMM corresponding to numeraire $D^2_{\cdot T}$ show that, under $\mathbb{Q}^2$,

$$dM^{23}_{tT} = \sigma^{(3)}_t M^{23}_{tT} d\widehat{W}_t,$$

where $\widehat{W}$ is an $(\{\mathcal{F}_t\}, \mathbb{Q}^2)$ Brownian motion and

$$\sigma^{(3)}_t := e^{at} \sqrt{(\sigma^{(1)})^2 - 2\rho\sigma^{(1)}\sigma^{(2)} + (\sigma^{(2)})^2}.$$

For this model derive an expression for $V^{23}_0(K)$, the time-zero value of the vanilla FX call option described in (i).

Exercise 16 (i) Consider a Hull–White model in which the short-rate process r satisfies the SDE

$$dr_t = (\theta_t - ar_t)dt + \sigma dW_t$$

under the risk-neutral measure $\mathbb{Q}$, where θ is a deterministic function of time and σ and a are constants.

Let $0 < S_0 < S_1 < \ldots < S_{n-1} < S_n$ be a sequence of regularly spaced times and set $T_i = S_{i-1}$, $\alpha_i = S_i - T_i$. Show that the forward LIBOR $L_t^i = L_t[T_i, S_i]$ satisfies the SDE

$$dL_t^i = \left(\frac{e^{-aT_i} - e^{-aS_i}}{a} \right) e^{at} \left(L_t^i + \alpha_i^{-1} \right) \sigma d\widetilde{W}_t^i,$$

$\widetilde{W}^i$ being a standard Brownian motion under the EMM $\mathbb{F}^i$ corresponding to taking $D._{S_i}$ as numeraire.

(ii) Let $V_t^i(K)$ denote the value at time t of a European put option on a discount bond having payoff $(K - D_{T_i S_i})_+$ at time T_i (not the more usual payment time S_i).

Show that the value at time t of a cap having payment dates $S_1, \ldots, S_n$, and strike K can be written in the form

$$V_t^{\text{cap}}(K) = \sum_{i=1}^{n} (1 + \alpha_i K) V_t^i \left(\frac{1}{1 + \alpha_i K} \right)$$

and find a closed-form formula for $V_t^i(K)$ for the Hull–White model described in (i) above.

Exercise 17 (i) Consider a payers swaption on a swap with cashflows at times $S_1 < S_2 < \ldots < S_n$, start date $T = S_0 < S_1$ and fixed rate K. According to Black's formula, the value at time zero of this swaption is given by

$$V_0 = P_0 \left[y_0 N(d_1) - K N(d_2) \right],$$

where

$$d_1 = \frac{\log(y_0/K)}{\tilde{\sigma}_0 \sqrt{T}} + \tfrac{1}{2} \tilde{\sigma}_0 \sqrt{T},$$

$$d_2 = \frac{\log(y_0/K)}{\tilde{\sigma}_0 \sqrt{T}} - \tfrac{1}{2} \tilde{\sigma}_0 \sqrt{T},$$

$\tilde{\sigma}_0$ is a positive constant and $N(\cdot)$ denotes the standard cumulative normal distribution.

Suppose an arbitrage-free term structure model has been defined for which the price of this swaption is consistent with Black's formula as stated above with volatility $\tilde{\sigma}_0$ for all strikes. Show that for this model the distribution of y_T is log-normal under an EMM corresponding to numeraire $P = \sum_{i=1}^{n} \alpha_i D._{S_i}$ where $\alpha_i = S_i - S_{i-1}$.

(ii) Let $(\Omega, \mathcal{F}, \mathbb{P})$ be a probability space supporting a one-dimensional Brownian motion W. Let σ_t be a deterministic function of time t bounded above and below by positive constants and let μ be a constant.

Now suppose an arbitrage-free term structure model has been defined for the finite time interval $0 \le t \le T < \infty$ such that, under $\mathbb{P}$, the forward par swap rate y satisfies the SDE

$$dy_t = \mu y_t \, dt + \sigma_t y_t \, dW_t.$$

Show that the model will be consistent with Black's formula for the swaption as given in (i), provided

$$\tilde{\sigma}_0^2 = \frac{1}{T} \int_0^T \sigma_u^2 \, du.$$

In Exercises 18 to 21, $0 < S_0 < S_1 < \ldots < S_{n-1} < S_n$ denotes a sequence of regularly spaced times with $T_i := S_{i-1}$ for $i = 1, \ldots, n+1$. Further, we let $L_t^i = L_t[T_i, S_i]$ denote the forward LIBOR at time t for the period $[T_i, S_i]$.

Exercise 18 (i) For each $i = 1, \ldots, n$, consider a caplet with start date T_i, cashflow at time S_i and strike K. According to Black's formula, the value at time $t = 0$ of the ith caplet is given by

$$V_0^i = \alpha_i D_{0S_i} \big(L_0^i N(d_1^i) - K N(d_2^i) \big),$$

where $\alpha_i = S_i - T_i$, σ^i is a positive constant and

$$d_1^i = \frac{\log(L_0^i/K)}{\sigma^i \sqrt{T_i}} + \tfrac{1}{2}\sigma^i \sqrt{T_i},$$

$$d_2^i = \frac{\log(L_0^i/K)}{\sigma^i \sqrt{T_i}} - \tfrac{1}{2}\sigma^i \sqrt{T_i}.$$

Suppose an arbitrage-free term structure model has been defined which is consistent with Black's formula at time $t = 0$ for all strikes for each of these caplets. Find the distribution of $L_{T_i}^i$ under an EMM corresponding to the numeraire $D_{\cdot S_i}$.

(ii) In a one-factor LIBOR market model, working in the measure $\mathbb{F}$ corresponding to the numeraire $D_{\cdot S_n}$, suppose $L' := (L^1, \ldots, L^n)$ satisfies an SDE of the form

$$dL_t^i = \mu_t^i \, dt + \sigma^i L_t^i \, dW_t, \qquad i = 1, \ldots, n,$$

for constants σ^i, W a one-dimensional Brownian motion, $\mu_t^n = 0$ for all $t \ge 0$ and, for $1 \le i < n$,

$$\mu_t^i = -\sum_{j=i+1}^{n} \frac{\alpha_j L_t^j}{1 + \alpha_j L_t^j} \sigma^j \sigma^i L_t^i.$$

Show that this model is consistent with Black's formula as given in (i) for each of the caplets.

Exercise 19 An alternative model to the LIBOR market model given in Exercise 18(ii) is proposed as follows.

Suppose now that, under $\mathbb{F}$, for $i = 1, \ldots, n$,

$$L_t^i = L_0^i \exp\left(\sigma^i W_t + \eta_t^i\right),$$

where W and σ^i are as in Exercise 18(ii) and η_t^i is a deterministic function of $t \leq T_i$ chosen to satisfy

$$\mathbb{E}_{\mathbb{F}}[\hat{D}_t^{(i-1)}] = \hat{D}_0^{(i-1)},$$

where

$$\hat{D}_t^{(i-1)} := \frac{D_{tT_i}}{D_{tS_n}}.$$

Show that, for $i = 1, \ldots, n$,

$$\eta_t^i = \log \hat{D}_0^i - \log \mathbb{E}_{\mathbb{F}}[\exp(\sigma^i W_t)\hat{D}_t^i].$$

Is this model arbitrage-free? Would this model be easy to implement in practice?

Exercise 20 Fix k, where $1 < k < n$. Working in the measure $\mathbb{F}^k$ corresponding to the numeraire $D_{\cdot S_k}$, suppose that, in a LIBOR market model,

$$L' := (L^1, \ldots, L^n)$$

satisfies an SDE of the form

$$dL_t^i = \mu_t^i \, dt + \sigma^i L_t^i \, dW_t^i, \qquad i = 1, \ldots, n.$$

Here each σ^i is a constant, each μ_t^i is some general process, to be determined, and $W' = (W^1, \ldots, W^n)$ is an n-dimensional Brownian motion having

$$dW_t^i \, dW_t^j = \rho_{ij} \, dt$$

for constants ρ_{ij} with $\rho_{ii} = 1$ for all i.

(i) Why is $\mu_t^k = 0$?

(ii) Show that, for $1 \leq i < k$,

$$\mu_t^i = -\sum_{j=i+1}^{k} \frac{\alpha_j L_t^j}{1 + \alpha_j L_t^j} \sigma^j \sigma^i \rho_{ij} L_t^i,$$

where $\alpha_j = S_j - T_j$ for each j.

(iii) Show that, for $k < i \le n$,

$$\mu_t^i = \sum_{j=k+1}^{i} \frac{\alpha_j L_t^j}{1 + \alpha_j L_t^j} \sigma^j \sigma^i \rho_{ij} L_t^i.$$

Exercise 21 Consider a LIBOR Markov-functional model specified using the numeraire pair $(D._{S_n}, \mathbb{S}^n)$ and having a driving Markov process x satisfying

$$dx_t = \sigma e^{at} dW_t,$$

where W is a standard Brownian motion under $\mathbb{S}^n$ and $\sigma > 0$ and a are constants. Show that if this model is calibrated to caplet prices as given by the Hull–White model discussed in Exercise 16 then this Markov-functional model coincides with the Hull–White model on the 'grid', i.e. the functional forms

$$\left\{ D_{T_i S_j}(x_{T_i}), \ 1 \le i \le j \le n \right\}$$

are as for the Hull–White model.

Solutions

Solution 1 We need to check that conditions (i)–(iii) of Definition 3.90 hold. Clearly M is adapted by the definition of a local martingale and thus (i) is satisfied. Condition (iii) requires us to show that $M_s \ge \mathbb{E}[M_t \,|\, \mathcal{F}_s]$ for $s < t$. Letting $\{T_n\}$ be a reducing sequence for $M - M_0$, we have

$$\mathbb{E}\left[M_t^{T_n} \,|\, \mathcal{F}_s\right] = M_s^{T_n},$$

where $M_t^{T_n} = M_{t \wedge T_n}$. Now

$$M_s = \lim_{n \to \infty} M_s^{T_n} = \lim_{n \to \infty} \mathbb{E}\left[M_t^{T_n} \,|\, \mathcal{F}_s\right] \ge \mathbb{E}\left[M_t \,|\, \mathcal{F}_s\right]. \tag{20.1}$$

This last inequality follows from the conditional form of Fatou's lemma: if $\{Y_n\}$ is a sequence of random variables in $\mathcal{L}^1(\Omega, \mathcal{F}, \mathbb{P})$ with $Y_n \ge 0$, and if $\mathcal{G}$ is a sub-σ-algebra of $\mathcal{F}$, then

$$\mathbb{E}\left[\liminf Y_n \,|\, \mathcal{G}\right] \le \liminf \mathbb{E}\left[Y_n \,|\, \mathcal{G}\right].$$

Finally, to establish (ii), take expectations of both sides of inequality (20.1) and set $s = 0$. This yields, for all $t \ge 0$,

$$\mathbb{E}\left[|M_t|\right] = \mathbb{E}[M_t] \le \mathbb{E}[M_0] < \infty$$

and we are done.

Solution 2   Recall, from Section 4.5.1, that we have

$$[X]_t = \int_0^t H_u^2 \, du$$

and, since H is deterministic, τ_t is an increasing deterministic function of time. Setting

$$\widetilde{W}_t := X_{\tau_t} = \int_0^{\tau_t} H_u \, dW_u \, ,$$

we have that

$$\widetilde{W}\big([X]_t\big) = X_{\tau_{[X]_t}} = X_t$$

and so we will have established the result if we can prove that $\widetilde{W}$ is a Brownian motion adapted to $\{\mathcal{F}_{\tau_t}\}$. We do this by an application of Lévy's theorem. Thus we must show that $\widetilde{W}$ is a continuous local martingale adapted to $\{\mathcal{F}_{\tau_t}\}$ and that $\big[\widetilde{W}\big]_t = t$.

That $\widetilde{W}$ is adapted follows from the adaptedness of the stochastic integral. Further, since X is constant on any interval of constancy of $[X]$, it follows by the continuity of the stochastic integral that $\widetilde{W}$ is continuous. Next we find a reducing sequence for $\widetilde{W}$ in order to show that it is a local martingale. Define

$$S_n := \inf\big\{t > 0 : |X_t| > n\big\}$$

and

$$T_n := \tau_{S_n}^{-1} = [X]_{S_n}.$$

Since τ is increasing it follows, for each fixed t, that

$$\tau_{t \wedge T_n} = S_n \wedge \tau_t$$

and

$$\{T_n \le t\} = \{S_n \le \tau_t\} \in \mathcal{F}_{\tau_t}.$$

Thus

$$\widetilde{W}_t^{T_n} := \widetilde{W}_{t \wedge T_n} = X_{\tau_t}^{S_n}$$

and, since $\{S_n\}$ is a reducing sequence for X, it follows that $\widetilde{W}_t^{T_n}$ is a martingale with respect to $\{\mathcal{F}_{\tau_t}\}$. We thus see that $\{T_n\}$ is a reducing sequence for the continuous local martingale $\widetilde{W}$.

To apply Lévy's theorem and conclude that $\widetilde{W}$ is an $\{\mathcal{F}_{\tau_t}\}$ Brownian motion, it remains only to show that $[\widetilde{W}]_t = t$. But it is easy to check using Theorem 4.18 and localization that

$$[\widetilde{W}]_t = \int_0^{\tau_t} H_u^2\, du = [X]_{\tau_t} = t\,.$$

Solution 3 Strong existence and pathwise uniqueness for the SDE follow immediately from Theorem 6.27. By Itô's formula,

$$
\begin{aligned}
dY_t &= d(e^{at} X_t)\\
&= a e^{at} X_t\, dt + e^{at}\, dX_t\\
&= a Y_t\, dt - a Y_t\, dt + \sigma e^{at}\, dW_t\\
&= \sigma e^{at}\, dW_t\,,
\end{aligned}
$$

whence

$$Y_t = Y_0 + \int_0^t \sigma e^{as}\, dW_s$$

and the required equation for X_t follows. An application of Lévy's theorem, as in Solution 2 above, establishes that

$$\widetilde{W}_t := \int_0^{\xi_t^{-1}} \sigma e^{as}\, dW_s$$

is a Brownian motion where $\xi_t := \int_0^t \sigma^2 e^{2au}\, du$, and we can write

$$X_t = e^{-at}\left(X_0 + \widetilde{W}(\xi_t)\right).$$

Thus, when $X_0 = x_0$, X_t has a normal distribution with mean $e^{at}x_0$ and variance

$$e^{-2at}\xi_t = e^{-2at}\sigma^2 \int_0^t e^{2au}\, du = \sigma^2 \frac{1 - e^{-2at}}{2a}\,.$$

Finally, observe that

$$
\begin{aligned}
\operatorname{cov}(X_{t_1}, X_{t_2}) &= e^{-a(t_1+t_2)} \operatorname{cov}\left(\widetilde{W}(\xi_{t_1}), \widetilde{W}(\xi_{t_2})\right)\\
&= e^{-a(t_1+t_2)}\xi_{t_1}
\end{aligned}
$$

and

$$\operatorname{corr}(X_{t_1}, X_{t_2}) = \frac{e^{-a(t_1+t_2)}\xi_{t_1}}{\sqrt{e^{-2at_1}\xi_{t_1}}\,\sqrt{e^{-2at_2}\xi_{t_2}}} = e^{-a(t_2-t_1)}\frac{\sqrt{e^{-2at_1}\xi_{t_1}}}{\sqrt{e^{-2at_2}\xi_{t_2}}}$$

as required.

Solution 4 (i) If $\mathbb{P} \sim \mathbb{Q}$ with respect to $\mathcal{F}$, then $\mathbb{P} \sim \mathbb{Q}$ with respect to $\mathcal{F}_t$. By Corollary 5.9 applied to the σ-algebras $\mathcal{F}_s \subseteq \mathcal{F}_t$ we have

$$\mathbb{E}_{\mathbb{Q}}[X \mid \mathcal{F}_s] = \frac{\mathbb{E}_{\mathbb{P}}[\rho_t X \mid \mathcal{F}_s]}{\mathbb{E}_{\mathbb{P}}[\rho_t \mid \mathcal{F}_s]} = \frac{\mathbb{E}_{\mathbb{P}}[\rho_t X \mid \mathcal{F}_s]}{\rho_s}.$$

(ii) Suppose ρM is an $(\{\mathcal{F}_t\}, \mathbb{P})$ martingale. To show that M is an $(\{\mathcal{F}_t\}, \mathbb{Q})$ martingale we must check that conditions (M.i)–(M.iii) of Definition 3.1 are satisfied. Adaptedness of M follows from that of ρ and ρM. Since ρM is integrable with respect to $\mathbb{P}$ (property (M.i) applied to the martingale ρM) we have

$$\mathbb{E}_{\mathbb{Q}}\big[|M_t|\big] = \mathbb{E}_{\mathbb{P}}\big[|\rho_t M_t|\big] < \infty.$$

Finally, using part (i) and the martingale property of ρM, we have

$$\mathbb{E}_{\mathbb{Q}}[M_t \mid \mathcal{F}_s] = \rho_s^{-1} \mathbb{E}_{\mathbb{P}}[\rho_t M_t \mid \mathcal{F}_s] = M_s.$$

The converse implication can be proved similarly.

Solution 5 (i) Write $\hat{X}_t = X_t / A_t^{(1)}$ and $\hat{A}_t^{(2)} = A_t^{(2)} / A_t^{(1)}$. An application of Itô's formula shows that

$$d\hat{A}_t^{(2)} = (\mu - r)\hat{A}_t^{(2)}\, dt + \sigma \hat{A}_t^{(2)}\, dW_t \tag{20.2}$$

and that

$$\hat{X}_t = \alpha_t^{(1)} + \alpha_t^{(2)} \hat{A}_t^{(2)} = \hat{X}_0 + \int_0^t \alpha_u^{(2)}\, d\hat{A}_u^{(2)}. \tag{20.3}$$

By Girsanov's theorem

$$\widetilde{W}_t = W_t - \phi t$$

is an $(\{\mathcal{F}_t^W\}, \mathbb{Q})$ Brownian motion. Substituting for dW_t in (20.2), we have

$$d\hat{A}_t^{(2)} = \sigma \hat{A}_t^{(2)}\, d\widetilde{W}_t$$

and so

$$\hat{X}_t = \hat{X}_0 + \int_0^t \alpha_u^{(2)} \sigma \hat{A}_u^{(2)}\, d\widetilde{W}_u$$

and thus $\hat{X}$ is a (positive) local martingale under $\mathbb{Q}$. Since $\mathcal{F}_0^W$ is trivial we can use the result of Exercise 1 to conclude that $\hat{X}$ is a supermartingale. In particular, noting that $A_0^{(1)} = e^{-rT}$, we have

$$\hat{X}_0 \geq \mathbb{E}_{\mathbb{Q}}[\hat{X}_T]$$

and so $X_0 \geq V_0$ as required.

(ii) Consider the $\left(\{\mathcal{F}_t^W\}, \mathbb{Q}\right)$ martingale

$$M_t = \mathbb{E}_{\mathbb{Q}}\left[V_T \mid \mathcal{F}_t^W\right].$$

It follows from the martingale representation theorem for Brownian motion (Theorem 5.49) that there exists some $\{\mathcal{F}_t^W\}$ predictable $H \in \Pi(W)$ such that

$$M_t := \mathbb{E}_{\mathbb{Q}}[V_T] + \int_0^t H_u \, d\widetilde{W}_u.$$

From (i), $d\widetilde{W}_t = \sigma^{-1}\big(\hat{A}_t^{(2)}\big)^{-1} d\hat{A}_t^{(2)}$ and so

$$M_t = \mathbb{E}_{\mathbb{Q}}[V_T] + \int_0^t H_u \sigma^{-1}\big(\hat{A}_u^{(2)}\big)^{-1} d\hat{A}_u^{(2)}. \tag{20.4}$$

Suppose we set $X_t = A_t^{(1)} M_t$. We can now use (20.3) and (20.4) to form the required $\big(\alpha_t^{(1)}, \alpha_t^{(2)}\big)$. Take

$$\alpha_t^{(2)} = \sigma^{-1}\big(\hat{A}_t^{(2)}\big)^{-1} H_t.$$

Then by (20.3) we see that we need to choose

$$\alpha_t^{(1)} = \hat{X}_t - \alpha_t^{(2)} \hat{A}_t^{(2)} = M_t - \alpha_t^{(2)} \hat{A}_t^{(2)}.$$

It follows by Itô's formula that $dX_t = \alpha_t^{(1)} \, dA_t^{(1)} + \alpha_t^{(2)} \, dA_t^{(2)}$ and further, since V_T is $\mathcal{F}_T^W$-measurable and $A_T^{(1)} = 1$, we have $X_T = A_T^{(1)} M_T = V_T$, as required. Since M_t is the conditional expectation of a positive random variable we have $M_t \geq 0$. Noting that $A_t^{(1)} = e^{-r(T-t)}$, it follows that $X_t \geq 0$ for $t \in [0, T]$. Finally, observe $X_0 = A_0^{(1)} M_0 = e^{-rT} \mathbb{E}_{\mathbb{Q}}[V_T] = V_0$.

Solution 6 (i) Using Wald's martingale (Example 3.3) define

$$\left.\frac{d\mathbb{Q}}{d\mathbb{P}}\right|_{\mathcal{F}_t^W} := \exp\left(\phi W_t - \tfrac{1}{2}\phi^2 t\right), \qquad t \leq T,$$

where $\phi = \sigma^{-1}(r - \mu)$. By Girsanov's theorem

$$dS_t^D = \sigma S_t^D \, d\widetilde{W}_t,$$

where $\widetilde{W}_t := W_t - \phi t$ is an $\left(\{\mathcal{F}_t^W\}, \mathbb{Q}\right)$ Brownian motion. Using Doléan's exponential (Example 4.36) we have

$$\begin{aligned}
S_t^D &= S_0^D \exp\left(\sigma \widetilde{W}_t - \tfrac{1}{2}\sigma^2 t\right) \\
&= S_0^D \exp\left(\sigma W_t - \tfrac{1}{2}\sigma^2 t - (r - \mu)t\right).
\end{aligned}$$

(ii) Substituting for S_t^D, we have

$$S_t^D \exp\left(\phi W_t - \tfrac{1}{2}\phi^2 t\right) = S_0^D \exp\left((\sigma + \phi)W_t - \tfrac{1}{2}(\sigma + \phi)^2 t\right)$$

which is of the form of Wald's martingale with $\lambda = \phi + \sigma$.

(iii) Observe that

$$\rho_t := \left.\frac{d\mathbb{Q}}{d\mathbb{P}}\right|_{\mathcal{F}_t^W} = \mathbb{E}_\mathbb{P}\left[\left(\left.\frac{d\mathbb{Q}}{d\mathbb{P}}\right|_{\mathcal{F}_T^W}\right)\middle| \mathcal{F}_t^W\right] = \exp\left(\phi W_t - \tfrac{1}{2}\phi^2 t\right).$$

By Lemma 5.19(i), since ρS^D is an $(\{\mathcal{F}_t^W\}, \mathbb{P})$ martingale, S^D is an $(\{\mathcal{F}_t^W\}, \mathbb{Q})$ martingale. Note this result is also immediate from the working for (i).

Finally, using the fact that S^D is an $(\{\mathcal{F}_t^W\}, \mathbb{Q})$ martingale, define

$$\hat{\rho}_t := \left.\frac{d\hat{\mathbb{Q}}}{d\mathbb{Q}}\right|_{\mathcal{F}_t^W} := S_0^{-1}\mathbb{E}_\mathbb{Q}\left[e^{-rT}S_T \,\middle|\, \mathcal{F}_t^W\right] = \frac{e^{-rT}S_t^D}{S_0}.$$

Again we can apply Lemma 5.19(i) to deduce that, since $D_t^S\hat{\rho}_t = S_0^{-1}e^{-rT}$ is a constant (hence an $(\{\mathcal{F}_t^W\}, \mathbb{Q})$ martingale), it follows that D_t^S is an $(\{\mathcal{F}_t^W\}, \hat{\mathbb{Q}})$ martingale and we are done.

Alternatively, a direct proof of the identity follows easily using Exercise 4(i).

(iv) Write

$$V_T = (S_T - K)_+ = S_T \mathbb{1}_{\{S_T > K\}} - K\mathbb{1}_{\{S_T > K\}}.$$

Then, applying the martingale valuation formula, equation (7.15), using numeraire pairs $(N^1, \mathbb{N}^1)$ and $(N^2, \mathbb{N}^2)$, we have

$$V_0 = N_0^1 \mathbb{E}_{\mathbb{N}^1}\left[\frac{S_T \mathbb{1}_{\{S_T > K\}}}{N_T^1}\right] - N_0^2 \mathbb{E}_{\mathbb{N}^2}\left[\frac{K\mathbb{1}_{\{S_T > K\}}}{N_T^2}\right].$$

From our working in (iii) it is clear we can take $(N^1, \mathbb{N}^1) = (S, \hat{\mathbb{Q}})$ and $(N^2, \mathbb{N}^2) = (D, \mathbb{Q})$. Thus

$$V_0 = S_0 \mathbb{E}_{\hat{\mathbb{Q}}}\left[\mathbb{1}_{\{S_T > K\}}\right] - e^{-rT}K\mathbb{E}_\mathbb{Q}\left[\mathbb{1}_{\{S_T > K\}}\right]$$

as required.

Solution 7   (i) Define $Y_t := N_t/D_t$ and observe that

$$d\left(\frac{N_t}{D_t}\right) = d\left(\frac{D_t + S_t}{D_t}\right) = dS_t^D = (\mu - r)S_t^D + \sigma S_t^D dW_t.$$

Thus

$$
\begin{aligned}
dD_t^N = dY_t^{-1} &= -Y_t^{-2}dY_t + Y_t^{-3}(dY_t)^2 \\
&= -\left(D_t^N\right)^2\left((\mu - r)S_t^D dt + \sigma S_t^D dW_t\right) + \left(D_t^N\right)^3\sigma^2\left(S_t^D\right)^2 dt \\
&= \left(-(\mu - r) + \sigma^2 S_t^N\right)D_t^N S_t^N dt - D_t^N S_t^N \sigma dW_t \\
&= -\sigma D_t^N S_t^N\left(dW_t - \frac{-(\mu - r) + \sigma^2 S_t^N}{\sigma}dt\right).
\end{aligned}
\tag{20.5}
$$

Noting that $S_t^N = 1 - D_t^N$, we now also have $dS_t^N = -dD_t^N$.

Since $\mathbb{N}$ is equivalent to $\mathbb{P}$ with respect to $\mathcal{F}_T^W$, by Girsanov's theorem (Theorem 5.24) there exists an $\{\mathcal{F}_t^W\}$-predictable process C such that

$$
\left.\frac{d\mathbb{N}}{d\mathbb{P}}\right|_{\mathcal{F}_t^W} = \exp\left(\int_0^t C_u dW_u - \tfrac{1}{2}\int_0^t C_u^2\, du\right)
\tag{20.6}
$$

and

$$
\widetilde{W}_t = W_t - \int_0^t C_u\, du
$$

is an $\{\mathcal{F}_t^W\}$ Brownian motion under $\mathbb{N}$.

Further, since $(N, \mathbb{N})$ is a numeraire pair, we must have, from (20.5), that

$$
C_t = \frac{\left(-(\mu - r) + \sigma^2 S_t^N\right)D_t^N S_t^N}{\sigma D_t^N S_t^N} = \sigma S_t^N - \sigma^{-1}(\mu - r).
$$

It then follows that

$$
dA_t^N = d\begin{pmatrix} D_t^N \\ S_t^N \end{pmatrix} = \begin{pmatrix} -\sigma D_t^N S_t^N \\ \sigma D_t^N S_t^N \end{pmatrix}d\widetilde{W}_t.
$$

Finally, we have

$$
\begin{aligned}
dN_t &= dS_t + dD_t \\
&= (\mu S_t + rD_t)dt + \sigma S_t dW_t \\
&= (\mu S_t + rD_t + \sigma S_t C_t)dt + \sigma S_t d\widetilde{W}_t \\
&= \left(rN_t + \sigma^2\frac{S_t^2}{N_t}\right)dt + \sigma S_t d\widetilde{W}_t.
\end{aligned}
$$

(ii) Substituting for C_t in equation (20.6), we have

$$\left.\frac{d\mathbb{N}}{d\mathbb{P}}\right|_{\mathcal{F}_t^W} = \exp\left(\int_0^t \sigma S_u^N dW_u - \frac{1}{2}\int_0^t \left(\sigma^2 \left(S_u^N\right)^2 + 2\sigma\phi S_u^N\right) du\right) \exp\left(\phi W_t - \frac{1}{2}\phi^2 t\right).$$

It remains to show that this first exponential is just $\dfrac{N_t/N_0}{D_t/D_0}$.

Let $\mathbb{Q}$ be the equivalent martingale measure corresponding to numeraire D. Recall that

$$\left.\frac{d\mathbb{Q}}{d\mathbb{P}}\right|_{\mathcal{F}_t^W} = \exp\left(\phi W_t - \frac{1}{2}\phi^2 t\right),$$

where $\phi = \sigma^{-1}(r - \mu)$, and that, under $\mathbb{Q}$, $\rho_t := \dfrac{N_t/N_0}{D_t/D_0}$ is a martingale with

$$d\rho_t = d\left(\frac{N_t/N_0}{D_t/D_0}\right) = \frac{D_0}{N_0}dS_t^D = \frac{D_0}{N_0}\sigma S_t^D dW_t^*,$$

where $W_t^* = W_t - \phi t$ is a Brownian motion under $\mathbb{Q}$.

Using ρ as a Radon–Nikodým derivative, it follows from Lemma 5.17 that

$$\frac{N_t/N_0}{D_t/D_0} = \exp\left(\int_0^t \frac{D_u}{N_u}\sigma S_u^D dW_u^* - \frac{1}{2}\int_0^t \left(\frac{D_u}{N_u}\sigma S_u^D\right)^2 du\right)$$

$$= \exp\left(\int_0^t \sigma S_u^N dW_u - \int_0^t \sigma\phi S_u^N du - \frac{1}{2}\int_0^t \sigma^2 \left(S_u^N\right)^2 du\right)$$

as required.

Remark: The identity in part (ii) of the question also follows by solving the SDE for D_t^N found in part (i).

Solution 8 (i) Observe

$$dD_t^S = S_t^{-1}dD_t + D_t dS_t^{-1} + dD_t dS_t^{-1}$$

and

$$dS_t^{-1} = -S_t^{-2}dS_t + S_t^{-3}(dS_t)^2$$
$$= (\sigma_t^2 - \mu)S_t^{-1}dt - \sigma_t S_t^{-1}dW_t,$$

whence

$$dD_t^S = (\sigma_t^2 + (r - \mu))D_t^S dt - \sigma_t D_t^S dW_t. \tag{20.7}$$

(ii) If $\mathbb{Q}^S$ is an EMM corresponding to numeraire S then, under $\mathbb{Q}^S$, D^S is a martingale. Using the SDE for D^S from part (i), Girsanov's theorem (Theorem 5.24) suggests we take

$$\rho_t := \frac{d\mathbb{Q}^S}{d\mathbb{P}}\bigg|_{\mathcal{F}_t^W} = \exp\left(\int_0^t \frac{\sigma_u^2 + (r-\mu)}{\sigma_u} dW_u - \tfrac{1}{2}\int_0^t \left(\frac{\sigma_u^2 + (r-\mu)}{\sigma_u}\right)^2 du\right).$$

(20.8)

Since σ_t is bounded above and below by a positive constant, clearly

$$\mathbb{E}\left[\exp\left(\tfrac{1}{2}\int_0^t \left(\frac{\sigma_u^2 + (r-\mu)}{\sigma_u}\right)^2 du\right)\right] < e^{ct} < \infty$$

for some constant c, and so, by Novikov (Theorem 5.16), ρ is an $(\{\mathcal{F}_t^W\}, \mathbb{P})$ martingale with $\rho_0 = 1$. Thus equation (20.8) with $t = T$ does indeed define a measure $\mathbb{Q}^S \sim \mathbb{P}$ on $\mathcal{F}_T^W$.

Girsanov's theorem tells us that

$$\widetilde{W}_t = W_t - \int_0^t \left(\frac{\sigma_u^2 + (r-\mu)}{\sigma_u}\right) du$$

(20.9)

is an $(\{\mathcal{F}_t^W\}, \mathbb{Q}^S)$ Brownian motion and, substituting (20.9) into (20.7),

$$dD_t^S = -\sigma_t D_t^S d\widetilde{W}_t.$$

Using Doléan's exponential, this SDE has the unique solution

$$D_t^S = e^{-rT}\exp\left(-\int_0^t \sigma_u d\widetilde{W}_u - \tfrac{1}{2}\int_0^t \sigma_u^2 du\right)$$

(20.10)

and it follows by Novikov that D^S is an $(\{\mathcal{F}_t^W\}, \mathbb{Q}^S)$ martingale. Since $S^S \equiv 1$ is trivially a martingale we have established that $\mathbb{Q}^S$ is an EMM corresponding to the numeraire S.

To show that $\mathbb{Q}^S$ is unique, let $\mathbb{Q}^*$ be some other EMM for the numeraire S. Then by Girsanov's theorem for Brownian motion (Theorem 5.24) there exists an $\{\mathcal{F}_t^W\}$-predictable process γ such that

$$W_t^* = \widetilde{W}_t - \int_0^t \gamma_u\, du$$

is an $(\{\mathcal{F}_t^W\}, \mathbb{Q}^*)$ Brownian motion. But then

$$dD_t^S = -\sigma_t D_t^S(dW_t^* + \gamma_t\, dt).$$

For D^S to be a martingale under $\mathbb{Q}^*$ we must have $\int_0^t \sigma_u D_u^S \gamma_u \, du = 0$ for all $0 \le t \le T$. Noting that $\sigma_t > 0$ and $D_t^S > 0$ for $0 \le t \le T$, it follows that $\gamma \equiv 0$. Thus by Girsanov's theorem,

$$\left. \frac{d\mathbb{Q}^*}{d\mathbb{Q}^S} \right|_{\mathcal{F}_T^W} = 1$$

and hence $\mathbb{Q}^* = \mathbb{Q}^S$ on $\mathcal{F}_T^W$.

(iii) Substituting for $\widetilde{W}$ in (20.10) yields

$$D_t^S = e^{-rT} \exp\left(-\int_0^t \sigma_u \, dW_u + \tfrac{1}{2} \int_0^t \sigma_u^2 \, du + (r - \mu)t \right).$$

Solution 9 (i) Observe that

$$m_T^S = \min\left\{ m_t^S, \min_{u \in [t,T]} S_u \right\}.$$

From Solution 6(i) we have that

$$S_t = S_0 \exp\left(\sigma W_t + (\mu - \tfrac{1}{2}\sigma^2)t \right)$$

and so, for $u > t$, we may write

$$S_u = S_t \exp\left(-(X_u - X_t) \right).$$

Noting that

$$\min_{u \in [t,T]} S_u = S_t e^{-M_{t,T}^X},$$

the result follows.

(ii) Working with the numeraire pair $(D, \mathbb{Q})$, we have

$$V_t = D_t \mathbb{E}_{\mathbb{Q}} \left[S_T - m_T^S \mid \mathcal{F}_t^W \right]$$

$$= D_t S_t^D - D_t \mathbb{E}_{\mathbb{Q}} \left[\min\left\{ m_t^S, S_t e^{-M_{t,T}^X} \right\} \mid \mathcal{F}_t^W \right].$$

Now set $\tau = T - t$ and $M_\tau^X := \max_{u \in [0,\tau]} X_u$. Noting that both m_t^S and S_t are $\mathcal{F}_t^W$-measurable random variables and that the random variable $M_{t,T}^X$ is independent of $\mathcal{F}_t^W$, we have

$$\mathbb{E}_{\mathbb{Q}} \left[\min\left\{ m_t^S, S_t e^{-M_{t,T}^X} \right\} \mid \mathcal{F}_t^W \right] = L\left(S_t, m_t^S \right),$$

where

$$L(s, m) := \mathbb{E}_{\mathbb{Q}}\left[\min\left\{m, se^{-M_\tau^X}\right\}\right]$$

for $s \geq m > 0$. It remains to calculate a closed-form expression for $L(s, m)$. Recall that, under $\mathbb{Q}$, $X_t = -\sigma\widetilde{W}_t + \left(\frac{1}{2}\sigma^2 - r\right)t$ where $\widetilde{W}$ is a one-dimensional Brownian motion with respect to $\left(\{\mathcal{F}_t^W\}, \mathbb{Q}\right)$. Define a measure $\mathbb{Q}^* \sim \mathbb{Q}$ on $\mathcal{F}_T^W$ via

$$\frac{d\mathbb{Q}^*}{d\mathbb{Q}}\bigg|_{\mathcal{F}_T^W} := \exp\left(-\frac{r_2}{\sigma}\widetilde{W}_T - \frac{1}{2}\frac{r_2^2}{\sigma^2}T\right),$$

where $r_2 = r - \frac{1}{2}\sigma^2$. Then, by Girsanov's theorem,

$$W_t^* := \widetilde{W}_t + \frac{r_2}{\sigma}t$$

is an $\left(\{\mathcal{F}_t^W\}, \mathbb{Q}^*\right)$ Brownian motion and $X_t = -\sigma W_t^*$.

By symmetry

$$\mathbb{Q}^*\left(-W_\tau^* \in dx, M_\tau^{X/\sigma} \in da\right) = \mathbb{Q}^*\left(W_\tau^* \in dx, M_\tau^{W^*} \in da\right)$$
$$= f(x, a)\, dx\, da$$

where, using Example 2.17,

$$f(x, a) = \begin{cases} \dfrac{\sqrt{2}(2a - x)}{\sqrt{\pi\tau^3}}\exp\left(-\dfrac{(2a - x)^2}{2\tau}\right)\, dx\, da & \text{if } \max\{0, x\} \leq a, \\ 0 & \text{otherwise.} \end{cases}$$

Thus

$$L(s, m) - m = \mathbb{E}_{\mathbb{Q}}\left[\left(se^{-M_\tau^X} - m\right)\mathbf{1}_{\left\{M_\tau^X \geq -\log\left(\frac{m}{s}\right)\right\}}\right]$$

$$= \mathbb{E}_{\mathbb{Q}^*}\left[\frac{d\mathbb{Q}}{d\mathbb{Q}^*}\bigg|_{\mathcal{F}_\tau^W}\left(se^{-M_\tau^X} - m\right)\mathbf{1}_{\left\{M_\tau^X \geq -\log\left(\frac{m}{s}\right)\right\}}\right]$$

$$= \exp\left(-\frac{1}{2}\frac{r_2^2}{\sigma^2}\tau\right)\mathbb{E}_{\mathbb{Q}^*}\left[e^{\frac{r_2}{\sigma}W_\tau^*}\left(se^{-\sigma M_\tau^{X/\sigma}} - m\right)\mathbf{1}_{\left\{M_\tau^{X/\sigma} \geq -\frac{1}{\sigma}\left(\log\frac{m}{s}\right)\right\}}\right]$$

$$= \exp\left(-\frac{1}{2}\frac{r_2^2}{\sigma^2}\tau\right)$$

$$\times \int_0^\infty \int_{-\infty}^a e^{-\frac{r_2}{\sigma}x}\left(se^{-\sigma a} - m\right)\mathbf{1}_{\left\{a \geq -\frac{1}{\sigma}\log\left(\frac{m}{s}\right)\right\}}f(x, a)\, dx\, da.$$

Evaluating the double integral (which is a little tedious but straightforward) and substituting the resulting expression for $L(s, m)$ leads to the required result.

Solution 10 (i) In order to be able to apply Girsanov's theorem, we first define a Brownian SDE for the $S^{(2)}$-rebased assets. An application of Itô's formula yields

$$dX_t = S_t^{(1)} d\big(S_t^{(2)}\big)^{-1} + \big(S_t^{(2)}\big)^{-1} dS_t^{(1)} + dS_t^{(1)} d\big(S_t^{(2)}\big)^{-1}$$

$$= X_t\big(dM_t^{(1)} - dM_t^{(2)}\big) + X_t(\mu_1 - \mu_2 + \xi_{22} - \xi_{12})\, dt.$$

Similarly, setting $U_t := D_t/S_t^{(2)}$, we find that

$$dU_t := -U_t dM_t^{(2)} + U_t(r - \mu_2 + \xi_{22})\, dt.$$

By Lévy's theorem we have

$$\begin{pmatrix} M^{(1)} \\ M^{(2)} \end{pmatrix} = \begin{pmatrix} \sqrt{\xi_{11}} & 0 \\ \rho\sqrt{\xi_{22}} & \sqrt{\xi_{22}(1-\rho)^2} \end{pmatrix} \begin{pmatrix} \widehat{W}_t^{(1)} \\ \widehat{W}_t^{(2)} \end{pmatrix},$$

where $\widehat{W}' = (\widehat{W}^{(1)}, \widehat{W}^{(2)})$ is an $\{\mathcal{F}_t^W\}$ Brownian motion, and it then follows that

$$d\begin{pmatrix} U_t \\ X_t \end{pmatrix} = \begin{pmatrix} U_t & 0 \\ 0 & X_t \end{pmatrix} \begin{pmatrix} r - \mu_2 + \xi_{22} \\ \mu_1 - \mu_2 + \xi_{22} - \xi_{12} \end{pmatrix} dt$$

$$+ \begin{pmatrix} U_t & 0 \\ 0 & X_t \end{pmatrix} \begin{pmatrix} 0 & -1 \\ 1 & -1 \end{pmatrix} \begin{pmatrix} \sqrt{\xi_{11}} & 0 \\ \rho\sqrt{\xi_{22}} & \sqrt{\xi_{22}(1-\rho^2)} \end{pmatrix} \begin{pmatrix} d\widehat{W}_t^{(1)} \\ d\widehat{W}_t^{(2)} \end{pmatrix},$$

which is our desired SDE.

Observe that $\{\mathcal{F}_t^A\} = \{\mathcal{F}_t^{\widehat{W}}\}$, where $\{\mathcal{F}_t^{\widehat{W}}\}$ is the natural augmented filtration generated by $\widehat{W}$. In order to define an EMM $\mathbb{Q}$ corresponding to numeraire $S^{(2)}$, Girsanov's theorem (Theorem 5.24) suggests we take

$$C_t := -\begin{pmatrix} \sqrt{\xi_{11}} & 0 \\ \rho\sqrt{\xi_{22}} & \sqrt{\xi_{22}(1-\rho^2)} \end{pmatrix}^{-1} \begin{pmatrix} 0 & -1 \\ 1 & -1 \end{pmatrix}^{-1} \begin{pmatrix} r - \mu_2 + \xi_{22} \\ \mu_1 - \mu_2 + \xi_{22} - \xi_{12} \end{pmatrix}$$

$$=: \begin{pmatrix} c_1 \\ c_2 \end{pmatrix}.$$

We can define $\mathbb{Q}$ via

$$\frac{d\mathbb{Q}}{d\mathbb{P}}\bigg|_{\mathcal{F}_t^{\widehat{W}}} = \exp\Big(c_1\widehat{W}_t^{(1)} + c_2\widehat{W}_t^{(2)} - \tfrac{1}{2}(c_1^2 + c_2^2)t\Big),$$

which is clearly a martingale by Novikov's condition.

It then follows by Girsanov's theorem that

$$d \begin{pmatrix} U_t \\ X_t \end{pmatrix} = \begin{pmatrix} U_t & 0 \\ 0 & X_t \end{pmatrix} \begin{pmatrix} -\rho\sqrt{\xi_{22}} & -\sqrt{\xi_{22}(1-\rho^2)} \\ \sqrt{\xi_{11}} - \rho\sqrt{\xi_{22}} & -\sqrt{\xi_{22}(1-\rho^2)} \end{pmatrix} d \begin{pmatrix} \widetilde{W}_t^{(1)} \\ \widetilde{W}_t^{(2)} \end{pmatrix},$$

where $\widetilde{W}' = (\widetilde{W}^{(1)}, \widetilde{W}^{(2)})$ is a Brownian motion under $\mathbb{Q}$.

To establish that $\mathbb{Q}$ is an EMM it remains to check that the solution to the above SDE is a martingale. Solving for X, we have that

$$X_t = X_0 \exp\left(Y_t - \tfrac{1}{2}[Y]_t\right),$$

where Y_t is defined as in the question and

$$[Y]_t = (\xi_{11} - 2\xi_{12} + \xi_{22})t =: \sigma_0^2 t.$$

Again it follows by Novikov's condition that X is a martingale. That U is a martingale follows similarly.

(ii) Taking $(N, \mathbb{N}) = (S^{(2)}, \mathbb{Q})$, we have that

$$V_0 = S_0^{(2)} \mathbb{E}_{\mathbb{Q}}\left[(X_T - 1)_+\right].$$

Further, from (i),

$$X_T \sim X_0 e^{-\frac{1}{2}\sigma_0^2 t + Z}$$

where $Z \sim N(0, \sigma_0^2 t)$ under $\mathbb{Q}$, and thus the required formula follows immediately from the Black–Scholes formula.

Solution 11 (i) By numeraire invariance (Theorem 7.13) the trading strategy $\phi' = (\phi^{(1)}, \phi^{(2)})$ is self-financing with respect to A if and only if it is self-financing with respect to A^S. Thus we seek ϕ satisfying

$$\frac{X}{S_T} = \phi_0^{(1)} D_0^S + \phi_0^{(2)} + \int_0^T \phi_u^{(1)} dD_u^S = \phi_T^{(1)} D_T^S + \phi_T^{(2)}.$$

By Solution 8(ii), equation (20.9), it follows that $\widetilde{W}_t := W_t - \sigma^{-1}(\sigma^2 + r - \mu)t$ is a Brownian motion under $\mathbb{Q}^S$. Further, note that $\mathcal{F}_t^{\widetilde{W}} = \mathcal{F}_t^W$.
Define

$$M_t := \mathbb{E}_{\mathbb{Q}^S}\left[\frac{X}{S_T}\,\Big|\,\mathcal{F}_t^W\right], \qquad 0 \le t \le T.$$

Then M is an $\left(\{\mathcal{F}_t^W\}, \mathbb{Q}^S\right)$ martingale and by Theorem 5.49 we can find some $\{\mathcal{F}_t^W\}$-predictable H such that

$$M_t = M_0 + \int_0^t H_u \, d\widetilde{W}_u \,.$$

In particular, taking $t = T$,

$$\frac{X}{S_T} = \mathbb{E}_{\mathbb{Q}^S}\left[\frac{X}{S_T}\right] + \int_0^T H_u \, d\widetilde{W}_u \,.$$

Noting that under $\mathbb{Q}^S$ (see Exercise 8(ii))

$$d\widetilde{W}_t = -\sigma^{-1}\left(D_t^S\right)^{-1} dD_t^S,$$

we have

$$M_t = \mathbb{E}_{\mathbb{Q}^S}\left[\frac{X}{S_T}\right] - \int_0^T H_u \sigma^{-1}\left(D_u^S\right)^{-1} dD_u^S. \qquad (20.11)$$

Using (20.11), we find ϕ satisfying

$$M_t = \mathbb{E}_{\mathbb{Q}^S}\left[\frac{X}{S_T}\right] + \int_0^t \phi_u^{(1)} dD_u^S + \int_0^t \phi_u^{(2)} dS_u^S$$

$$= \phi_t^{(1)} D_t^S + \phi_t^{(2)}.$$

Take

$$\phi_t^{(1)} = -H_t \sigma^{-1}\left(D_t^S\right)^{-1}$$

and

$$\phi_t^{(2)} = M_t - \phi_t^{(1)} D_t^S.$$

Note that for replication it does not matter what $\phi_t^{(2)}$ is taken to be. We must make this choice for the self-financing property to hold.

(ii) Let $(N, \mathbb{N})$ be some other numeraire pair. Then, in particular, S_t/N_t is a martingale under $\mathbb{N}$. Use this fact to form an equivalent measure $\hat{\mathbb{Q}}$ via

$$\left.\frac{d\hat{\mathbb{Q}}}{d\mathbb{N}}\right|_{\mathcal{F}_t^W} = \frac{S_t/S_0}{N_t/N_0} =: \rho_t.$$

We show that $\hat{\mathbb{Q}} = \mathbb{Q}^S$, which establishes the result. Noting that D_t/N_t is an $\left(\{\mathcal{F}_t^W\}, \mathbb{N}\right)$ martingale, it follows from part (i) of Lemma 5.19 that

$$\frac{D_t}{N_t} \cdot \frac{N_t}{S_t} = \frac{D_t}{S_t}$$

is an $\left(\{\mathcal{F}_t^W\}, \hat{\mathbb{Q}}\right)$ martingale. Trivially S^S is also a $\left(\{\mathcal{F}_t^W\}, \hat{\mathbb{Q}}\right)$ martingale and so $\hat{\mathbb{Q}}$ is an EMM for the numeraire S. But the measure for which D^S is a martingale with respect to $\{\mathcal{F}_t^W\}$ is unique. Thus $\hat{\mathbb{Q}} = \mathbb{Q}$ on $\mathcal{F}_T^W$ and so

$$\left. \frac{d\mathbb{Q}^S}{d\mathbb{N}} \right|_{\mathcal{F}_t^W} = \rho_t$$

as required.

(iii) Recall that an $\{\mathcal{F}_t^W\}$-predictable process ϕ is admissible for the Black–Scholes economy if it is self-financing and if, for all numeraire pairs $(N, \mathbb{N})$, the numeraire-rebased gain process G^ϕ/N is an $\left(\{\mathcal{F}_t^W\}, \mathbb{N}\right)$ martingale (see Definition 7.25).

Using the result of Exercise 4(i), observe that, for any numeraire pair $(N, \mathbb{N})$, for $0 \leq u \leq t$ we have

$$\mathbb{E}_{\mathbb{N}}\left[\phi_t \cdot A_t^N \mid \mathcal{F}_u^W \right] = \mathbb{E}_{\mathbb{Q}^S}\left[\phi_t \cdot \frac{A_t}{N_t} \cdot \frac{N_t}{S_t} \,\bigg|\, \mathcal{F}_u^W \right] \frac{S_u}{N_u}$$

$$= \frac{S_u}{N_u} \phi_u \cdot A_u^S$$

$$= \phi_u \cdot A_u^N.$$

The second equality above follows since, by part (i), $M_t = \phi_t \cdot A_t^S$ is a martingale. That $\phi_t \cdot A_t^N$ is adapted is clear, and integrability follows from that of $\phi_t \cdot A_t^S$ under $\mathbb{Q}^S$. Thus $G_t^\phi/N_t = \phi_t \cdot A_t^N - \phi_0 \cdot A_0^N$ is an $\left(\{\mathcal{F}_t^W\}, \mathbb{N}\right)$ martingale as required.

We have shown that the Black–Scholes economy is complete for the set

$$S' = \left\{ X : X \text{ is an } \mathcal{F}_T^W\text{-measurable random variable with } \mathbb{E}_{\mathbb{Q}^S}\left[\left| \frac{X}{S_T} \right| \right] < \infty \right\}.$$

Observe by (ii) that

$$\mathbb{E}_{\mathbb{Q}}\left[|X| \right] = \mathbb{E}_{\mathbb{Q}^S}\left[|X| \left. \frac{d\mathbb{Q}}{d\mathbb{Q}^S} \right|_{\mathcal{F}_T^W} \right]$$

$$= \frac{S_0}{D_0} \mathbb{E}_{\mathbb{Q}^S}\left[|X| \frac{D_T}{S_T} \right]$$

and so $S = S'$ and we are done.

Solution 12   (i) Set $Z_t = N_t^{-1} \frac{d\mathbb{N}}{d\mathbb{P}}\big|_{\mathcal{F}_t^A}$. Using the result of Exercise 4(i), it is straightforward to check that Z is a pricing kernel (see the proof of part (i) of Theorem 7.48). By equation (7.15) and Exercise 4(i) we have

$$V_t = N_t \mathbb{E}_\mathbb{N}\left[V_T/N_T \mid \mathcal{F}_t^A \right]$$

$$= N_t \frac{\mathbb{E}_\mathbb{P}\left[V_T/N_T \frac{d\mathbb{N}}{d\mathbb{P}}\big|_{\mathcal{F}_T^A} \mid \mathcal{F}_t^A \right]}{\left(\frac{d\mathbb{N}}{d\mathbb{P}}\big|_{\mathcal{F}_t^A} \right)}$$

$$= Z_t^{-1} \mathbb{E}_\mathbb{P}\left[V_T Z_T \mid \mathcal{F}_t^A \right]$$

as required.

(ii) Using part (i) above, we can set

$$Z_t = \left(S_t^{(2)} \right)^{-1} \frac{d\mathbb{Q}}{d\mathbb{P}}\bigg|_{\mathcal{F}_t^A}.$$

The working in Solution 10 leads to an explicit expression for this pricing kernel, as we now show.

Observe that the SDE for $S^{(i)}$ can be written in exponential form

$$dS_t^{(i)} = S_t^{(i)} dX_t^{(i)}$$

with $X_t^{(i)} = \mu_i t + M_t^{(i)}$ and, by Example 4.36, this has the unique solution $S_0^{(i)} \mathcal{E}\left(X^{(i)} \right)_t$. In particular, we have

$$S_t^{(2)} = S_0^{(2)} \exp\left(M_t^{(2)} + (\mu_2 - \tfrac{1}{2}\xi_{22})t \right). \tag{20.12}$$

Set

$$R := \begin{pmatrix} \sqrt{\xi_{11}} & 0 \\ \rho\sqrt{\xi_{22}} & \sqrt{\xi_{22}(1-\rho^2)} \end{pmatrix}$$

and recall from Solution 10 that

$$M = R\widehat{W}, \tag{20.13}$$

where $M' = \left(M^{(1)}, M^{(2)} \right)$ and $\widehat{W}' = \left(\widehat{W}^{(1)}, \widehat{W}^{(2)} \right)$. Let $\mathbb{Q}$ be an EMM corresponding to numeraire $S^{(2)}$. Then, again using Solution 10, we have that

$$\frac{d\mathbb{Q}}{d\mathbb{P}}\bigg|_{\mathcal{F}_t^A} = \exp\left(C^T R^{-1} M_t - \tfrac{1}{2}|C|^2 t \right) \tag{20.14}$$

where $C' = (c_1, c_2)$, the components of the vector C being defined as in Solution 10.

Finally, combining (20.12) and (20.14), we have

$$Z_t = \left(S_0^{(2)}\right)^{-1} \exp(CR^{-1}M_t - M_t^{(2)} - \left(\tfrac{1}{2}|C|^2 + \mu_2 - \tfrac{1}{2}\xi_{22}\right)t).$$

To establish completeness, first observe that the SDE for $A' = (D, S^{(1)}, S^{(2)})$ can, using (20.13), be written in the form (7.1). That pathwise uniqueness and hence weak uniqueness holds for this SDE follows from the uniqueness of the solution stated above. Since the economy has a finite variation asset D, it then follows immediately from Theorem 7.43 that the economy is complete.

By Theorem 7.41 and part (iii) of Theorem 7.48, the pricing kernel Z is unique, and so if $(M, \mathbb{M})$ is some other numeraire pair we have

$$M_t^{-1} \frac{d\mathbb{M}}{d\mathbb{P}}\bigg|_{\mathcal{F}_t^A} = Z_t.$$

Thus, taking $(M, \mathbb{M}) = (D, \mathbb{Q}^D)$, we have

$$\frac{d\mathbb{Q}^D}{d\mathbb{P}}\bigg|_{\mathcal{F}_t^A} = D_t Z_t$$

where Z_t is as given above.

Solution 13 (i) Since A and B are absolutely continuous and decreasing to zero, we may write

$$A_t = \int_0^t a_u \, du + A_0, \qquad B_t = \int_0^t b_u \, du + B_0$$

where $A_\infty = B_\infty = 0$.

To see that a rational log-normal model is in (FH), define

$$M_{tS} := -(a_S + b_S M_t).$$

Then

$$\int_t^\infty M_{tS} dS = -\int_t^\infty (a_S + b_S M_t) \, dS$$

$$= A_t - A_\infty + (B_t - B_\infty) M_t$$

$$= A_t + B_t M_T$$

and

$$D_{tT} = \frac{\int_T^\infty M_{tS}\,dS}{\int_t^\infty M_{tS}\,dS}$$

as required.

(ii) From Theorem 8.25 it follows that for a model in (FH) we can take as pricing kernel

$$Z_t := \int_t^\infty M_{tS}\,dS.$$

Thus for a rational log-normal model we have the valuation formula

$$V_0 = Z_0^{-1}\mathbb{E}_{\mathbb{Z}}[Z_T V_T]$$
$$= (A_0 + B_0)^{-1}\mathbb{E}_{\mathbb{Z}}\big[(A_T + B_T M_T)V_T\big].$$

Noting that the value of the swaption at T is

$$V_T = \left(1 - D_{TS_n} - K\sum_{j=1}^n \alpha_j D_{TS_j}\right)_+$$

where $\alpha_j = S_j - S_{j-1}$, and that under the log-normal model

$$D_{TS_j} = \frac{A_{S_j} + B_{S_j}M_T}{A_T + B_T M_T},$$

we have that the time-zero value of the swaption under the log-normal model is given by

$$V_0 = (A_0 + B_0)^{-1}\mathbb{E}_{\mathbb{Z}}\left[\left[\left(A_T - \left(A_{S_n} + K\sum_{j=1}^n \alpha_j A_{S_j}\right)\right)\right.\right.$$
$$\left.\left. + \left(B_T - \left(B_{S_n} + K\sum_{j=1}^n \alpha_j B_{S_j}\right)\right)M_T\right]_+\right].$$

Assuming M_T is of the form given, this reduces to

$$V_0 = (A_0 + B_0)^{-1}\left[\left(B_T - \left(B_{S_n} + K\sum_{j=1}^n \alpha_j B_{S_j}\right)\right)N(d_1)\right.$$
$$\left. + \left(A_T - \left(A_{S_n} + K\sum_{j=1}^n \alpha_j A_{S_j}\right)\right)N(d_2)\right],$$

where

$$d_1 = \frac{\log \left(\dfrac{B_T - (B_{S_n} + K \sum_{j=1}^n \alpha_j B_{S_j})}{A_T - (A_{S_n} + K \sum_{j=1}^n \alpha_j A_{S_j})} \right)}{\sigma \sqrt{T}} + \tfrac{1}{2}\sigma\sqrt{T}$$

and

$$d_2 = \frac{\log \left(\dfrac{B_T - (B_{S_n} + K \sum_{j=1}^n \alpha_j B_{S_j})}{A_T - (A_{S_n} + K \sum_{j=1}^n \alpha_j A_{S_j})} \right)}{\sigma \sqrt{T}} - \tfrac{1}{2}\sigma\sqrt{T}.$$

Solution 14 (i) From the definition of an (SR) model it follows immediately that the risk-neutral measure $\mathbb{Q}$ is an EMM corresponding to the numeraire $\exp(\int_0^t r_u\, du)$. The first valuation formula is just the martingale valuation formula, equation (7.15), using this numeraire.

For $0 \le t \le T$ define a measure $\mathbb{F}$ equivalent to $\mathbb{Q}$ via

$$\left. \frac{d\mathbb{F}}{d\mathbb{Q}} \right|_{\mathcal{F}_t^A} = \exp\left(-\int_0^t r_u du \right) \frac{D_{tT}}{D_{0T}} = D_{0T}^{-1} \mathbb{E}_{\mathbb{Q}}\left[\exp\left(-\int_0^T r_u du \right) \bigg| \mathcal{F}_t^A \right].$$

(20.15)

Setting $\rho_t := \left. \frac{d\mathbb{Q}}{d\mathbb{F}} \right|_{\mathcal{F}_t^A}$ and using the result of Exercise 4, we have

$$V_t = \rho_t^{-1} \mathbb{E}_{\mathbb{F}}\left[\exp\left(-\int_t^T r_u du \right) V_T \rho_T \bigg| \mathcal{F}_t^A \right]$$

$$= \exp\left(-\int_0^t r_u du \right) D_{tT}\, \mathbb{E}_{\mathbb{F}}\left[V_T \exp\left(\int_0^t r_u du \right) \bigg| \mathcal{F}_t^A \right]$$

$$= D_{tT}\, \mathbb{E}_{\mathbb{F}}\left[V_T \,|\, \mathcal{F}_t^A \right].$$

Note that $D_{.S}/D_{.T}$ is a martingale under $\mathbb{F}$ and thus we have formed the Radon–Nikodým derivative from the ratio of numeraires.

(ii) For the Ho–Lee model observe that

$$\int_0^T r_u du = r_0 T + \int_0^T \int_0^u \theta_s\, ds\, du + \int_0^T \sigma W_u\, du.$$

(20.16)

Applying Itô's formula to tW_t and integrating, we obtain

$$T W_T = \int_0^T W_u\, du + \int_0^T u\, dW_u.$$

Substituting into (20.16) and interchanging the order of integration, we have

$$\int_0^T r_u \, du = r_0 T + \int_0^T \theta_s(T-s) \, ds + \int_0^T (T-u)\sigma dW_u. \qquad (20.17)$$

Now, from Exercise 2 it follows that, under $\mathbb{Q}$, $\int_0^T (T-u)\sigma \, dW_u$ has a normal distribution with mean zero and variance $\int_0^T \sigma^2(T-u)^2 du = \frac{1}{3}\sigma^2 T^3$. Thence, if the model is calibrated to the initial discount curve, we have

$$D_{0T} = \mathbb{E}_{\mathbb{Q}}\left[\exp\left(-\int_0^T r_u du \right) \right]$$

$$= \exp\left(-r_0 T - \int_0^T \theta_s(T-s)ds + \frac{1}{6}\sigma^2 T^3 \right). \qquad (20.18)$$

Taking logarithms and differentiating twice with respect to T yields the required equation for θ_t.

Next observe from (20.15), (20.17) and (20.18) that

$$\left. \frac{d\mathbb{F}}{d\mathbb{Q}} \right|_{\mathcal{F}_T^A} = \exp\left(-\frac{1}{6}\sigma^2 T^3 - \int_0^T (T-u)\sigma \, dW_u \right).$$

It then follows from Girsanov's theorem (check!) that, under $\mathbb{F}$,

$$\widetilde{W}_t := W_t - \int_0^t -(T-u)\sigma \, du \qquad (20.19)$$

is a Brownian motion. Substituting (20.19) into the SDE for r given in the question yields the required result.

Solution 15   (i) Using the numeraire $D_{\cdot T}^2$, the corresponding EMM $\mathbb{Q}^2$ and noting that

$$\left. \frac{d\mathbb{Q}^2}{d\mathbb{Q}^1} \right|_{\mathcal{F}_t} = \frac{M_{tT}^{12}}{M_{0T}^{12}},$$

we have

$$V_0^{23}(K) = D_{0T}^2 \mathbb{E}_{\mathbb{Q}^2}\left[(X_T^{23} - K)_+ \right]$$

$$= D_{0T}^2 \mathbb{E}_{\mathbb{Q}^1}\left[\frac{X_T^{12} D_{0T}^1}{X_0^{12} D_{0T}^2} (X_T^{23} - K)_+ \right]$$

$$= X_0^{21} D_{0T}^1 \mathbb{E}_{\mathbb{Q}^1}\left[X_T^{12} (X_T^{23} - K)_+ \right].$$

Noting that $X_T^{12} = X_T^{13}/X_T^{23}$ yields the required result.

(ii) Observe that

$$
\begin{aligned}
V_0^{21} &= D_{0T}^2 \mathbb{E}_{\mathbb{Q}^2}\left[\left(X_T^{21} - K\right)_+\right] \\
&= X_0^{21} D_{0T}^1 \mathbb{E}_{\mathbb{Q}^1}\left[X_T^{12}\left(X_T^{21} - K\right)_+\right] \\
&= K X_0^{21} D_{0T}^1 \mathbb{E}_{\mathbb{Q}^1}\left[\left(K^{-1} - X_T^{12}\right)_+\right].
\end{aligned}
$$

Thus from the prices of call options on X_T^{21} we can recover the prices of put options on X_T^{12} and hence, via put–call parity, the prices of call options on X_T^{12}. This means we can recover the implied marginal distribution of X_T^{12} under $\mathbb{Q}^1$. Similarly, we can recover the implied marginal distribution of X_T^{13} under $\mathbb{Q}^1$. However, no information concerning the dependence structure between X_T^{12} and X_T^{13} can be inferred from these prices.

(iii) Noting that the unique solution to the SDE for M_{tT}^{12} is

$$
M_{tT}^{12} = M_{0T}^{12}\exp\left(\int_0^t \sigma^{(1)}e^{au}dW_u^{(1)} - \tfrac{1}{2}\left(\sigma^{(1)}\right)^2 \int_0^t e^{2au}du\right),
$$

we have

$$
\left.\frac{d\mathbb{Q}^2}{d\mathbb{Q}^1}\right|_{\mathcal{F}_t} = \exp\left(\int_0^t C_u \cdot dW_u^* - \tfrac{1}{2}\int_0^t |C_u|^2\, du\right),
$$

where

$$
C_t = \sigma^{(1)}e^{at}\begin{pmatrix}\sqrt{1-\rho^2}\\ \rho\end{pmatrix}.
$$

It then follows by Girsanov's theorem that

$$
\widetilde{W}_t^* := \begin{pmatrix}\widetilde{W}_t^{*(1)}\\ \widetilde{W}_t^{*(2)}\end{pmatrix} = W_t^* - \left(\int_0^t \sigma^{(1)}e^{au}\, du\right)\begin{pmatrix}\sqrt{1-\rho^2}\\ \rho\end{pmatrix} \tag{20.20}
$$

is an $\{\mathcal{F}_t\}$ Brownian motion under $\mathbb{Q}^2$.

We now find the SDE for M_T^{23} under $\mathbb{Q}^2$. By integration by parts,

$$
dM_{tT}^{23} = d\left(\frac{M_{tT}^{13}}{M_{tT}^{12}}\right) = M_{tT}^{13}d\left(M_{tT}^{12}\right)^{-1} + \left(M_{tT}^{12}\right)^{-1}dM_{tT}^{13} + dM_{tT}^{13}d\left(M_{tT}^{12}\right)^{-1}.
$$

By Itô's formula,

$$
d\left(M_{tT}^{12}\right)^{-1} = -\sigma^{(1)}e^{at}\left(M_{tT}^{12}\right)^{-1}dW_t^{(1)} + \left(\sigma^{(1)}\right)^2 e^{2at}\left(M_{tT}^{12}\right)^{-1} dt
$$

and thus

$$dM_{tT}^{23} = M_{tT}^{23}\left(e^{2at}\left(\left(\sigma^{(1)}\right)^2 - \sigma^{(1)}\sigma^{(2)}\rho\right)dt + e^{at}\left(\sigma^{(2)}dW_t^{(2)} - \sigma^{(1)}dW_t^{(1)}\right)\right).$$

Substituting for $dW_t^{(1)}$ and $dW_t^{(2)}$ and using (20.20), we obtain

$$dM_{tT}^{23} = M_{tT}^{23}dZ_t,$$

where

$$dZ_t := \left(\sigma^{(2)} - \sigma^{(1)}\rho\right)e^{at}d\widetilde{W}_t^{*(2)} - e^{at}\sigma^{(1)}\sqrt{1-\rho^2}\,d\widetilde{W}_t^{*(1)}.$$

Observe that

$$d[Z]_t = \left(\sigma_t^{(3)}\right)^2 dt.$$

Setting

$$d\widehat{W}_t := \left(\sigma_t^{(3)}\right)^{-1}dZ_t,$$

it follows by Lévy's theorem that $\widehat{W}$ is an $\left(\{\mathcal{F}_t\}, \mathbb{Q}^2\right)$ Brownian motion. It can now easily be seen that $M_{\cdot T}^{23}$ satisfies the given SDE.

Finally, noting that

$$X_T^{23} = M_{TT}^{23} = M_{0T}^{23}\exp\left(\int_0^T \sigma_t^{(3)}d\widehat{W}_t - \tfrac{1}{2}\int_0^T \left(\sigma_t^{(3)}\right)^2 dt\right),$$

routine calculation yields

$$V_0^{23}(K) = X_0^{23}D_{0T}^3 N(d_1) - KD_{0T}^2 N(d_2),$$

where

$$d_1 := \frac{\log(D_{0T}^3 X_0^{23}/D_{0T}^2 K)}{\tilde{\sigma}_0} + \tfrac{1}{2}\tilde{\sigma}_0,$$

$$d_2 := \frac{\log(D_{0T}^3 X_0^{23}/D_{0T}^2 K)}{\tilde{\sigma}_0} - \tfrac{1}{2}\tilde{\sigma}_0,$$

and

$$\tilde{\sigma}_0^2 := \frac{e^{2at} - 1}{2a}\left(\left(\sigma^{(1)}\right)^2 - 2\rho\sigma^{(1)}\sigma^{(2)} + \left(\sigma^{(2)}\right)^2\right).$$

Solution 16 (i) Observe that

$$L_t^i := L_t[T_i, S_i] = \alpha_i^{-1}\frac{D_{tT_i}}{D_{tS_i}} - \alpha_i^{-1}$$

will be a martingale under $\mathbb{F}^i$. From (17.5), for the Hull–White model we have that the value at time t of a pure discount bond is of the form

$$D_{tT} = A_{tT}e^{-B_{tT}r_t},$$

where here

$$B_{tT} = e^{at} \int_t^T e^{-au}du = \frac{1}{a}\left(1 - e^{-a(T-t)}\right).$$

Thence

$$dL_t^i = \alpha_i^{-1}d\left[\frac{A_{tT_i}}{A_{tS_i}}\exp\left((B_{tS_i} - B_{tT_i})r_t\right)\right].$$

Applying Itô's formula to the $\mathbb{C}^{1,2}$ function $L^i(t, r_t)$ yields

$$dL_t^i = \frac{\partial}{\partial t}L^i(t, r_t)\,dt + \frac{\partial}{\partial r_t}L^i(t, r_t)\,dr_t + \frac{1}{2}\frac{\partial^2}{\partial r_t^2}L^i(t, r_t)\,(dr_t)^2.$$

Working in the measure $\mathbb{F}^i$, we can ignore finite variation terms (which must be zero since L^i is a martingale) and so

$$dL_t^i = \alpha_i^{-1}\frac{A_{tT_i}}{A_{tS_i}}(B_{tS_i} - B_{tT_i})e^{(B_{tS_i} - B_{tT_i})r_t}\sigma d\widetilde{W}_t^i$$

$$= \alpha_i^{-1}(B_{tS_i} - B_{tT_i})\frac{D_{tT_i}}{D_{tS_i}}\sigma d\widetilde{W}_t^i.$$

The result now follows by noting that

$$B_{tS_i} - B_{tT_i} = \frac{1}{a}\left[e^{-a(T_i-t)} - e^{-a(S_i-t)}\right] = \frac{e^{at}}{a}\left(e^{-aT_i} - e^{-aS_i}\right).$$

(ii) The caplet which pays at S_i has value $\widetilde{V}_{S_i}^i(K) = \alpha_i(L_{T_i}^i - K)_+$ at S_i and value $\widetilde{V}_{T_i}^i(K) = D_{T_iS_i}\widetilde{V}_{S_i}^i(K)$ at T_i. Recalling

$$D_{T_iS_i} = \frac{1}{1 + \alpha_i L_{T_i}^i},$$

it easily follows that

$$\widetilde{V}_{T_i}^i(K) = \alpha_i D_{T_iS_i}(L_{T_i}^i - K)_+$$

$$= (1 - D_{T_iS_i} - \alpha_i K D_{T_iS_i})_+$$

$$= (1 + \alpha_i K)\left(\frac{1}{1 + \alpha_i K} - D_{T_iS_i}\right)_+.$$

Thus a caplet having payment date S_i has the same value at time T_i as $(1 + \alpha_i K)$ times the value of a put option on the discount bond $D_{T_i S_i}$ with strike $(1 + \alpha_i K)^{-1}$. The required result now follows by summing over the values of the individual caplets at time $t < T_1$.

From the discussion and the martingale valuation formula, equation (7.15), it also follows that

$$V_t^i(K) = K\widetilde{V}_t^i\big(\alpha_i^{-1}K^{-1} - \alpha_i^{-1}\big)$$

$$= \alpha_i D_{tS_i} K \mathbb{E}_{\mathbb{F}^i}\Big[\big(X_{T_i}^i - (\alpha_i K)^{-1}\big)_+ \,\Big|\, \mathcal{F}_t\Big],$$

where $X_t^i = L_t^i + \alpha_i^{-1}$ and $\{\mathcal{F}_t\}$ is the augmented natural filtration generated by W. (Note that here $\{\mathcal{F}_t^A\} = \{\mathcal{F}_t\}$.)

For the Hull–White model in (i) we have

$$dX_t^i = \sigma_t^i X_t^i d\widetilde{W}_t^i,$$

where

$$\sigma_t^i = \sigma e^{at}\left(\frac{e^{-aT_i} - e^{-aS_i}}{a}\right)$$

and $\widetilde{W}$ is a Brownian motion under $\mathbb{F}^i$. Thus

$$X_t^i = X_0^i \exp\left(\int_0^t \sigma_u^i d\widetilde{W}_u^i - \tfrac{1}{2}\int_0^t (\sigma_u^i)^2 du\right)$$

and so, for $t < T_i$,

$$\log X_{T_i}^i = \log X_t^i - \tfrac{1}{2}\int_t^{T_i} (\sigma_u^i)^2 du + \int_t^{T_i} \sigma_u^i d\widetilde{W}_u^i.$$

Set

$$\Sigma_t := \left(\int_t^{T_i} (\sigma_u^i)^2 du\right)^{1/2} = \frac{1 - e^{-a(S_i - T_i)}}{a}\sqrt{\frac{\sigma^2}{2a}\big(1 - e^{-2a(T_i - t)}\big)}.$$

Noting that, conditional on $\mathcal{F}_t$, $\log X_{T_i}^i$ has a $N(\log X_t^i - \tfrac{1}{2}\Sigma_t^2, \Sigma_t^2)$ distribution, it follows by explicit calculation that

$$V_t^i(K) = D_{tT_i}KN(d_1) - D_{tS_i}N(d_2),$$

where

$$d_1 = \frac{\log(D_{tT_i}K/D_{tS_i})}{\Sigma_t} + \tfrac{1}{2}\Sigma_t$$

and

$$d_2 = \frac{\log(D_{tT_i}K/D_{tS_i})}{\Sigma_t} - \tfrac{1}{2}\Sigma_t.$$

Solution 17 (i) Recall from Chapter 11 that the time-zero value of a payers swaption with strike K is given by

$$V_0(K) = P_0 \mathbb{E}_{\mathbb{S}}\big[(y_T - K)_+\big],$$

where $\mathbb{S}$ denotes an EMM corresponding to numeraire P.

Differentiating both sides with respect to K and appealing to Fubini's theorem yields

$$\frac{\partial V_0(K)}{\partial K} = P_0 \frac{\partial}{\partial K} \mathbb{E}_{\mathbb{S}}\big[(y_T - K)_+\big] = P_0 \mathbb{E}_{\mathbb{S}}\big[-\mathbb{1}_{\{y_T > K\}}\big] = -P_0 \mathbb{S}(y_T > K).$$

Assuming $V_0(K)$ is given by Black's formula, we also have

$$\frac{\partial V_0(K)}{\partial K} = P_0 \left(\frac{-y_0}{\tilde{\sigma}_0 \sqrt{T} K} \phi(d_1) - N(d_2) + \frac{1}{\tilde{\sigma}_0 \sqrt{T}} \phi(d_2) \right)$$

$$= -P_0 N(d_2),$$

where $\phi(x)$ denotes the standard normal density.

Thus $\mathbb{S}(y_T > K) = N(d_2)$ and so

$$\mathbb{S}(y_T \leq K) = N(-d_2) = N\left(\frac{\log(K/y_0)}{\tilde{\sigma}_0 \sqrt{T}} + \tfrac{1}{2} \tilde{\sigma}_0 \sqrt{T} \right).$$

Hence, under $\mathbb{S}$, $\log y_T \sim N(\log y_0 - \tfrac{1}{2} \tilde{\sigma}_0^2 T, \, \tilde{\sigma}_0^2 T)$.

(ii) Let $\mathbb{S}$ denote the EMM corresponding to the numeraire P. Recall that

$$y_t = \frac{D_{tT} - D_{tS_n}}{P_t}$$

and so y_t must be a martingale under $\mathbb{S}$. We can write

$$dy_t = \sigma_t y_t d\widetilde{W}_t, \qquad (20.21)$$

where

$$\widetilde{W}_t = W_t - \int_0^t \frac{\mu}{\sigma_u} \, du.$$

Let $\{\mathcal{F}_t\}$ denote the augmented natural filtration generated by W. In order for $\widetilde{W}$ to be a Brownian motion under $\mathbb{S}$, Girsanov's theorem suggests we take

$$\left. \frac{d\mathbb{S}}{d\mathbb{P}} \right|_{\mathcal{F}_t} = \exp\left(\int_0^t \frac{\mu}{\sigma_u} dW_u - \tfrac{1}{2} \int_0^t \frac{\mu^2}{\sigma_u^2} \, du \right). \qquad (20.22)$$

This is an $(\{\mathcal{F}_t\}, \mathbb{P})$ martingale by Novikov since, for all $t \geq 0$,

$$\mathbb{E}\left[\exp\left(\tfrac{1}{2}\int_0^t \frac{\mu^2}{\sigma_u^2}du\right)\right] < \infty,$$

and so, by Girsanov's theorem, equation (20.22) defines a measure with the required properties.

The SDE (20.21) has a unique solution

$$y_t = y_0 \exp\left(\int_0^t \sigma_u d\widetilde{W}_u - \tfrac{1}{2}\int_0^t \sigma_u^2 du\right).$$

To establish that the model is consistent with Black's formula given in (i), it is sufficient to prove that

$$\int_0^T \sigma_u d\widetilde{W}_u \sim N(0, \tilde{\sigma}_0^2 T).$$

This follows immediately by Exercise 2 and we are done.

Solution 18   (i) Let $\mathbb{F}^i$ denote an EMM corresponding to numeraire $D_{\cdot S_i}$. Then the value of the ith caplet having strike K at time zero is given by

$$V_0^i(K) = D_{0S_i}\mathbb{E}_{\mathbb{F}^i}\left[\alpha_i(L_{T_i}^i - K)_+\right].$$

Differentiating both sides with respect to K yields

$$\frac{\partial V_0^i}{\partial K} = \alpha_i D_{0S_i}\mathbb{E}_{\mathbb{F}^i}\left[-\mathbf{1}_{\{L_{T_i}^i > K\}}\right] = -\alpha_i D_{0S_i}\mathbb{F}^i(L_{T_i}^i > K).$$

Assuming $V_0^i(K)$ is given by Black's formula, as given in the question, we obtain

$$\frac{\partial V_0^i}{\partial K} = -\alpha_i D_{0S_i} N\left(\frac{\log(L_0^i/K)}{\sigma^i\sqrt{T_i}} - \tfrac{1}{2}\sigma^i\sqrt{T_i}\right).$$

Equating terms yields

$$\mathbb{F}^i(L_{T_i}^i \leq K) = N\left(\frac{\log(K/L_0^i)}{\sigma^i\sqrt{T_i}} - \tfrac{1}{2}\sigma^i\sqrt{T_i}\right)$$

and thus, under $\mathbb{F}^i$,

$$\log L_{T_i}^i \sim N\left(\log L_0^i - \tfrac{1}{2}(\sigma^i)^2 T_i, (\sigma^i)^2 T_i\right).$$

(ii) We know from Section 18.2 that the μ_t^i have been chosen so that the model is arbitrage-free. If $\mathbb{F}^i$ denotes an EMM corresponding to numeraire $D_{\cdot S_i}$ then, recalling

$$L_t^i = \frac{D_{tT_i} - D_{tS_i}}{\alpha_i D_{tS_i}},$$

we see that L^i must be a martingale under $\mathbb{F}^i$. Proceeding as in Exercise 15(ii), we can show that

$$dL_t^i = \sigma^i L_t^i d\widetilde{W}_t,$$

where $\widetilde{W}_t$ is a Brownian motion under $\mathbb{F}^i$. The solution to this SDE is

$$L_t^i = L_0^i \exp\left(\sigma^i \widetilde{W}_t - \tfrac{1}{2}(\sigma^i)^2 t\right).$$

Thus, for each i, under $\mathbb{F}^i$ the random variable $L_{T_i}^i$ has the same log-normal distribution as found in (i) and so the model is consistent with Black's formula.

Remark 1: The LIBOR market model in (ii) above will of course be consistent with Black's formula for all $0 < t < S_n$.

Remark 2: Setting $\hat{D}_t^i := D_{tT_{i+1}}/D_{tS_n}$, the Radon–Nikodým derivative $\left.\frac{d\mathbb{F}^i}{d\mathbb{F}}\right|_{\mathcal{F}_t}$ is given by

$$\left.\frac{d\mathbb{F}^i}{d\mathbb{F}}\right|_{\mathcal{F}_t} = \frac{\hat{D}_t^i}{\hat{D}_0^i} = \mathcal{E}\left(\int_0^T \sum_{j=i+1}^n \frac{\alpha_j L_u^j}{1 + \alpha_j L_u^j} \sigma^j dW_u\right),$$

where $\{\mathcal{F}_t\}$ denotes the augmented natural filtration generated by W.

Solution 19   Observe that, for $i = 1, \ldots, n$,

$$\hat{D}_t^{(i-1)} := \frac{D_{tT_i}}{D_{tS_n}}$$

$$= \prod_{j=i}^n (1 + \alpha_j L_t^j)$$

$$= \hat{D}_t^i + \alpha_i L_t^i \hat{D}_t^i.$$

Taking expectations and using the model assumptions, we have

$$\hat{D}_0^i + \alpha_i L_0^i \hat{D}_0^i = \hat{D}_0^{i-1}$$

$$= \mathbb{E}_{\mathbb{F}}\left[\hat{D}_t^{(i-1)}\right]$$

$$= \mathbb{E}_{\mathbb{F}}\left[\hat{D}_t^i + \alpha_i L_t^i \hat{D}_t^i\right]$$

$$= \hat{D}_0^i + \alpha_i \mathbb{E}_{\mathbb{F}}\left[L_t^i \hat{D}_t^i\right].$$

For this to hold we require

$$\mathbb{E}_{\mathbb{F}}\left[L_t^i \hat{D}_t^i\right] = L_0^i \hat{D}_0^i,$$

which is equivalent to

$$\exp(\eta_t^i)\mathbb{E}_{\mathbb{F}}\left[\exp(\sigma^i W_t)\hat{D}_t^i\right] = \hat{D}_0^i.$$

Taking logarithms yields the required result.

For the model to be arbitrage-free we require $\hat{D}^{(i-1)}$ to be a martingale under $\mathbb{F}$. Noting that, under $\mathbb{F}$,

$$dL_t^i = \sigma^i L_t^i dW_t + L_t^i(d\eta_t^i + \tfrac{1}{2}(\sigma^i)^2 dt),$$

we would need

$$L_t^i\left(d\eta_t^i + \tfrac{1}{2}(\sigma^i)^2 dt\right) = \mu_t^i dt,$$

where μ_t^i is as in Exercise 18(ii). But then η_t^i would not be deterministic. Thus the model described is not arbitrage-free.

The model is easy to implement in practice: write

$$v^i = \sigma^i/\sigma^n$$

and observe that

$$L_t^i = (L_t^n)^{v^i}\beta_t^i,$$

where

$$\beta_t^i = \frac{L_0^i \exp(\eta_t^i)}{(L_0^n \exp(\eta_t^n))^{v^i}}.$$

In practice we need to find $L_{T_k}^i$, $1 \le k \le i \le n$. Observe that

$$\eta_t^n = -\tfrac{1}{2}(\sigma^n)^2 t.$$

For the other η's, work back from $i = n$ inductively: suppose we know $\eta_{T_k}^j$, $j > i$, then we know $\hat{D}_{T_k}^{(j-1)}$, $j > i$, so we can find $\eta_{T_k}^i$.

Solution 20   (i) Under $\mathbb{F}^k$ each $\frac{D_{\cdot T_k}}{D_{\cdot S_k}}$ is a martingale. Noting that

$$L_t^k = \frac{D_{tT_k} - D_{tS_k}}{\alpha_k D_{tS_k}},$$

we see that L^k is a martingale under $\mathbb{F}^k$ and so the drift in the SDE for L^k must be zero, i.e. $\mu^k \equiv 0$.

(ii) For $1 \leq i < k$,

$$\hat{D}_t^{(i-1)} := \frac{D_{tT_i}}{D_{tS_k}} = \prod_{j=i}^{k}(1 + \alpha_j L_t^j) \tag{20.23}$$

and so

$$\hat{D}_t^{(i-1)} = \hat{D}_t^i + \alpha_i L_t^i \hat{D}_t^i.$$

Since $\hat{D}^{(i-1)}$ and $\hat{D}^i$ are both martingales, so is $L^i \hat{D}^i$.

Now

$$d(L_t^i \hat{D}_t^i) = L_t^i d\hat{D}_t^i + \hat{D}_t^i dL_t^i + d\hat{D}_t^i dL_t^i$$

and equating finite variation terms to zero yields

$$\hat{D}_t^i \mu_t^i dt + \sigma^i L_t^i d\hat{D}_t^i dW_t^i = 0. \tag{20.24}$$

From (20.23), ignoring finite variation terms which must combine to give zero,

$$d\hat{D}_t^i = \sum_{j=i+1}^{k} \frac{\partial \hat{D}_t^i}{\partial L_t^j} \sigma^j L_t^j dW_t^j$$

$$= \hat{D}_t^i \sum_{j=i+1}^{k} \frac{\alpha_j L_t^j}{1 + \alpha_j L_t^j} \sigma^j dW_t^j.$$

Substituting into (20.24) yields

$$\hat{D}_t^i \mu_t^i dt + \hat{D}_t^i \sum_{j=i+1}^{k} \frac{\alpha_j L_t^j}{1 + \alpha_j L_t^j} \sigma^j \sigma^i L_t^i \rho_{ij} dt = 0$$

and thus the formula for μ_t^i, for $1 \leq i < k$, follows as required.

(iii) For $i = k + 2, \ldots, n + 1$,

$$\hat{D}_t^{(i-1)} := \frac{D_{tT_i}}{D_{tS_k}} = \prod_{j=k+1}^{i-1} \frac{D_{tS_j}}{D_{tT_j}} = \prod_{j=k+1}^{i-1} (1 + \alpha_j L_t^j)^{-1} \tag{20.25}$$

and so

$$\hat{D}_t^{(i-1)} = \prod_{j=k+1}^{i} (1 + \alpha_j L_t^j)^{-1} (1 + \alpha_i L_t^i),$$

i.e. $\hat{D}_t^{(i-1)} = \hat{D}_t^i + \alpha_i L_t^i \hat{D}_t^i$ as before.

Equation (20.24) follows as before for $i = k + 1, \ldots, n$, but now, from (20.25), we have

$$
d\hat{D}_t^i = \sum_{j=k+1}^{i-1} \frac{\partial \hat{D}_t^i}{\partial L_t^j} \sigma^j L_t^j dW_t^j
$$

$$
= -\hat{D}_t^i \sum_{j=k+1}^{i} \frac{\alpha_j L_t^j}{1 + \alpha_j L_t^j} \sigma^j dW_t^j.
$$

The formula for μ_t^i, $i = k + 1, \ldots, n$, follows by substituting into (20.24).

Solution 21   Calibrating to vanilla cap prices is equivalent to calibrating to the inferred prices of digital caplets. For the Hull–White model specified in Exercise 16, the value of the ith digital caplet of strike K is given, after a little calculation, by

$$
\tilde{V}_0^i(K) = D_{0S_i} \mathbb{E}_{\mathbb{S}^i} \left[\mathbf{1}_{\{L_{T_i}^i > K\}} \right]
$$

$$
= D_{0S_i} \Phi(d_2)
$$

where Φ denotes the standard cumulative normal distribution,

$$
d_2 = \frac{\log \left(\dfrac{D_{0T_i}}{D_{0S_i}} (1 + \alpha_i K)^{-1} \right)}{\Sigma_0} - \tfrac{1}{2} \Sigma_0
$$

and

$$
\Sigma_0 = \left(\frac{e^{-aT_i} - e^{-aS_i}}{a} \right) \sqrt{\mathrm{var}\, x_{T_i}}.
$$

Consider the problem of how to specify the Markov-functional model which calibrates to these digital caplet prices. The boundary curve for this problem is exactly that of (19.1). The only functional forms needed on the boundary are $D_{T_iT_i}(x_{T_i})$, $i = 1, 2, \ldots, n$, which are trivially the unit map, and $D_{T_nS_n}(x_{T_n})$. To complete the specification of the Markov-functional model it is sufficient to find the functional forms for the numeraire bond $D_{T_iS_n}(x_{T_i})$, $i = 1, \ldots, n - 1$. (Compare (P.ii′) and (P.iii′) in Section 19.3.)

Thus to establish that the Markov-functional model coincides with the Hull–White model on the grid it is sufficient to show that the functional forms $D_{T_iS_n}(x_{T_i})$, $i = 1, \ldots, n$, agree for the two models.

For the Hull–White model, it follows from equation (17.13) that, for $i = 1, \ldots, n$,

$$
D_{T_iS_n}(x_{T_i}) = \frac{D_{0S_n}}{D_{0T_i}} \exp\left(- c_i x_{T_i} + \tfrac{1}{2} c_i^2 \,\mathrm{var}(x_{T_i}) \right), \tag{20.26}
$$

where

$$c_i := \frac{e^{-aT_i} - e^{-aS_n}}{a}.$$

We now show that the same expression is obtained for the Markov-functional model.

The form $D_{T_n S_n}$ can be recovered directly from the closed-form expression available for $L_{T_n}^n$ for the Hull–White model. From the working in Solution 16 it follows that

$$L_{T_n}^n = (L_0^n + \alpha_n^{-1}) \exp\left(c_n x_{T_n} - \tfrac{1}{2} c_n^2 \operatorname{var}(x_{T_n})\right) - \alpha_n^{-1}$$

and so

$$D_{T_n S_n} = \frac{1}{1 + \alpha_n L_{T_n}^n} = \frac{D_{0 S_n}}{D_{0 T_n}} \exp\left(-c_n x_{T_n} + \tfrac{1}{2} c_n^2 \operatorname{var}(x_{T_n})\right)$$

as required.

To determine the remaining functional forms we work back iteratively from the terminal time T_n. Consider the ith step in this procedure. Assume that $D_{T_j S_n}$, $j = i+1, \ldots, n$, have already been determined. We can also assume that

$$\frac{D_{T_i S_i}(x_{T_i})}{D_{T_i S_n}(x_{T_i})} = \frac{D_{0 S_i}}{D_{0 S_n}} \exp\left(c_{i+1} x_{T_i} - \tfrac{1}{2} c_{i+1}^2 \operatorname{var}(x_{T_i})\right),$$

having been determined via (19.2) and the known (conditional) distribution of $x_{T_{i+1}}$ given x_{T_i}.

For some $x^* \in \mathbb{R}$, define

$$
\begin{aligned}
J_0^i(x^*) &:= D_{0 S_n} \mathbb{E}_{\mathbb{S}^n}\left[\frac{D_{T_i S_i}(x_{T_i})}{D_{T_i S_n}(x_{T_i})} \mathbb{1}_{\{x_{T_i} > x^*\}}\right] \\
&= D_{0 S_i} \exp\left(\tfrac{1}{2} c_{i+1}^2 \operatorname{var}(x_{T_i})\right) \mathbb{E}_{\mathbb{S}^n}\left[\exp(c_{i+1} x_{T_i}) \mathbb{1}_{\{x_{T_i} > x^*\}}\right] \\
&= D_{0 S_i} \Phi\left(\frac{c_{i+1} \operatorname{var}(x_{T_i}) - x^*}{\sqrt{\operatorname{var} x_{T_i}}}\right).
\end{aligned}
$$

From (19.5) we have

$$D_{T_i S_n} = \left(\frac{D_{T_i S_i}}{D_{T_i S_n}}\right)^{-1} (1 + \alpha_i L_{T_i}^i)^{-1},$$

and so to determine $D_{T_i S_n}$ it is sufficient to find $L_{T_i}^i(x^*)$. From (19.6) and (19.8),

$$L_{T_i}^i(x^*) = K^i(x^*)$$

where $K^i(x^*)$ solves

$$J_0^i(x^*) = \widetilde{V}_0^i(K^i(x^*)).$$

It follows that

$$\left(1 + \alpha_i L_{T_i}^i(x^*)\right)^{-1} = \frac{D_{0S_i}}{D_{0T_i}} \exp\left(\Sigma_0\left(\frac{c_{i+1}\,\mathrm{var}(x_{T_i}) - x^*}{\sqrt{\mathrm{var}(x_{T_i})}}\right) + \tfrac{1}{2}\Sigma_0^2\right)$$

and thus, after some algebra, we obtain equation (20.26) and we are done.

Appendix 1

The Usual Conditions

When working with continuous time processes, it is common to work on a filtered probability space $(\Omega, \{\mathcal{F}_t\}, \mathcal{F}, \mathbb{P})$ which satisfies *the usual conditions*. We do this throughout most of this book and it is needed, for example, to ensure that the stochastic integral (defined in Chapter 4) is adapted. Here we briefly describe these conditions and how to *augment* a probability space to make them hold.

A filtered probability space $(\Omega, \{\mathcal{F}_t\}, \mathcal{F}, \mathbb{P})$ is said to satisfy the usual conditions if the following three conditions hold:

(i) $\mathcal{F}$ is $\mathbb{P}$-complete: if $B \subset A \in \mathcal{F}$ and $\mathbb{P}(A) = 0$ then $B \in \mathcal{F}$.

(ii) $\mathcal{F}_0$ contains all $\mathbb{P}$-null sets.

(iii) $\{\mathcal{F}_t : t \geq 0\}$ is right-continuous: for all $t \geq 0$,

$$\mathcal{F}_t = \mathcal{F}_{t+} := \bigcap_{s>t} \mathcal{F}_s.$$

Let $(\Omega, \mathcal{F}^o, \mathbb{P})$ be a probability triple. The $\mathbb{P}$-*completion* $(\Omega, \mathcal{F}, \mathbb{P})$ of $(\Omega, \mathcal{F}^o, \mathbb{P})$ is the probability triple defined by

$$\mathcal{F} = \sigma(\mathcal{F}^o \cup \mathcal{N}),$$

where

$$\mathcal{N} := \{A \subset \Omega : A \subseteq B \text{ for some } B \in \mathcal{F} \text{ with } \mathbb{P}(B) = 0\}.$$

The *usual* $\mathbb{P}$-*augmentation* $(\Omega, \{\mathcal{F}_t\}, \mathcal{F}, \mathbb{P})$ of the filtered probability space $(\Omega, \{\mathcal{F}_t^o\}, \mathcal{F}^o, \mathbb{P})$ is produced by taking $\mathcal{F}$ to be the $\mathbb{P}$-completion of $\mathcal{F}^o$ and by defining

$$\mathcal{F}_t := \sigma(\mathcal{F}_{t+}^o \cup \mathcal{N}) = \bigcap_{s>t} \sigma(\mathcal{F}_s^o \cup \mathcal{N})$$

for all $t \geq 0$. Clearly the usual $\mathbb{P}$-augmentation satisfies the usual conditions.

Financial Derivatives in Theory and Practice Revised Edition. P. J. Hunt and J. E. Kennedy
© 2004 John Wiley & Sons, Ltd   ISBNs: 0-470-86358-7 (HB); 0-470-86359-5 (PB)

Appendix 2

L^2 Spaces

The study of L^2 spaces, and more generally L^p spaces for $p \geq 1$, is a standard part of functional analysis covered by many books. See, for example, the treatments of Priestley (1997) and Rudin (1987). Here we present a few of the basic facts used in this book. We omit all proofs, which can be found in the aforementioned references.

Definition A.1 *Let (S, Σ) be a measurable space and let μ be a finite measure on (S, Σ). Then, for $p \geq 1$, $\mathcal{L}^p(S, \Sigma, \mu)$ (or just $\mathcal{L}^p$ when the context is clear) is the space of Σ-measurable functions $f : S \to \mathbb{R}$ such that*

$$\int_S |f| \, d\mu < \infty.$$

Define an equivalence relation on $\mathcal{L}^p$ by

$$f \sim g \quad \text{if and only if } f = g, \ \mu \text{ almost everywhere,}$$

and denote by $[f]$ the equivalence class containing the function f. The space $L^p(S, \Sigma, \mu)$ (abbreviated to L^p) is defined to be the space of equivalence classes of $\mathcal{L}^p$,

$$L^p(S, \Sigma, \mu) = \{[f] : f \in \mathcal{L}^p(S, \Sigma, \mu)\}.$$

Notationally it is common practice to drop the $[\cdot]$ notation to denote an equivalence class and write, for example, $f \in L^p$ rather than $[f] \in L^p$. We shall adopt this convention.

That the relation $\sim$ is an equivalence relation is easily verified, as are the following simple lemmas.

Lemma A.2 *For a finite measure μ and any $p \geq q$,*

$$\mathcal{L}^p \subseteq \mathcal{L}^q, \quad L^p \subseteq L^q.$$

Financial Derivatives in Theory and Practice Revised Edition. P. J. Hunt and J. E. Kennedy
© 2004 John Wiley & Sons, Ltd ISBNs: 0-470-86358-7 (HB); 0-470-86359-5 (PB)

Lemma A.3   *For any f in either $\mathcal{L}^p$ or L^p, define*

$$\|f\|_p := \left(\int_S |f|^p d\mu \right)^{1/p}.$$

Then $\| \cdot \|_p$ defines a semi-norm on the space $\mathcal{L}^p$ and a norm on the space L^p.

It is the identification of functions which are almost everywhere equal that turns $\| \cdot \|_p$ from a semi-norm into a norm: for all $f, g \in \mathcal{L}^p$,

$$f \sim g \quad \Leftrightarrow \quad \|f - g\|_p = 0.$$

The normed space $(L^p, \| \cdot \|_p)$ has one very important property, completeness.

Theorem A.4   *For $p \geq 1$, the space L^p is a complete normed space, i.e. a Banach space.*

This is all we shall say about general L^p spaces.

The space L^2 has extra structure not present in a general L^p space, namely it supports an inner product. The existence of an inner product means we can talk about the idea of orthogonality, and this is crucial for the development of the stochastic integral.

Theorem A.5   *For $f, g \in L^2(S, \sum, \mu)$, let*

$$\langle f, g \rangle := \left(\int_s fg d\mu \right).$$

Then $\langle \cdot, \cdot \rangle$ defines an inner product on $L^2(S, \sum, \mu)$ (with corresponding norm $\|f\|_2^2 = \langle f, f \rangle$). Thus $(L^2, \langle \cdot, \cdot \rangle)$ is a Hilbert space (i.e. an inner product space which is also a Banach space).

Remark A.6: Note that in the main text we denote the L^2 norm described above by $\| \cdot \|$ and reserve the notation $\| \cdot \|_2$ for the corresponding (semi-)norm on the space of square-integrable martingales.

Appendix 3

Gaussian Calculations

Here we gather together some simple Gaussian integrals that arise frequently in applications. Throughout $X \sim N(\mu, \sigma^2)$, $\phi(x, \mu, \sigma^2)$ is the corresponding Gaussian density, $\phi(x)$ is the standard Gaussian density function and Φ is the corresponding distribution function.

Polynomial integrals

We derive expressions for

$$I_n(l, h, \mu, \sigma^2) = \mathbb{E}[X^n \mathbf{1}_{\{l \leq X \leq h\}}]$$

$$= \int_l^h x^n \phi(x, \mu, \sigma^2) \, dx.$$

We will abbreviate this to $I_n(l, h)$ when no confusion can arise.
For $m > 0$,

$$\frac{d}{dx}(x^m \phi(x, \mu, \sigma^2)) = m x^{m-1} \phi(x, \mu, \sigma^2) - \frac{x - \mu}{\sigma^2} x^m \phi(x, \mu, \sigma^2).$$

Integrating this over the interval $[l, h]$ gives

$$h^m \phi(h, \mu, \sigma^2) - l^m \phi(l, \mu, \sigma^2) = m I_{m-1}(l, h) + \frac{\mu}{\sigma^2} I_m(l, h) - \frac{1}{\sigma^2} I_{m+1}(l, h).$$

Rearranging and substituting $m = n - 1$ yields

$$I_n(l, h) = \sigma^2 (n - 1) I_{n-2}(l, h) + \mu I_{n-1}(l, h)$$

$$+ \sigma^2 (l^{n-1} \phi(l, \mu, \sigma^2) - h^{n-1} \phi(h, \mu, \sigma^2))$$

$$= \sigma^2 (n - 1) I_{n-2}(l, h) + \mu I_{n-1}(l, h)$$

$$+ \sigma \left(l^{n-1} \phi\left(\frac{l - \mu}{\sigma}\right) - h^{n-1} \phi\left(\frac{h - \mu}{\sigma}\right) \right).$$

Financial Derivatives in Theory and Practice Revised Edition. P. J. Hunt and J. E. Kennedy
© 2004 John Wiley & Sons, Ltd ISBNs: 0-470-86358-7 (HB); 0-470-86359-5 (PB)

It remains to calculate I_0 and I_1. I_0 is obtained by direct evaluation:

$$I_0(l, h) = \frac{1}{\sigma\sqrt{2\pi}} \int_l^h \exp\left(-\frac{1}{2}\left(\frac{x-\mu}{\sigma}\right)^2\right) dx$$

$$= \frac{1}{\sqrt{2\pi}} \int_{\frac{l-\mu}{\sigma}}^{\frac{h-\mu}{\sigma}} \exp\left(-\frac{1}{2}y^2\right) dy$$

$$= \Phi\left(\frac{h-\mu}{\sigma}\right) - \Phi\left(\frac{l-\mu}{\sigma}\right).$$

To calculate $I_1(l, h)$ note that

$$\frac{d}{dx}\phi(x, \mu, \sigma^2) = -\frac{x-\mu}{\sigma^2}\phi(x, \mu, \sigma^2)$$

and so

$$I_1(l, h) = \mu I_0(l, h) + \sigma^2(\phi(l, \mu, \sigma^2) - \phi(h, \mu, \sigma^2))$$

$$= \mu I_0(l, h) + \sigma\left(\phi\left(\frac{l-\mu}{\sigma}\right) - \phi\left(\frac{h-\mu}{\sigma}\right)\right)$$

which gives

$$I_1(l, h) = \mu\left(\Phi\left(\frac{h-\mu}{\sigma}\right) - \Phi\left(\frac{l-\mu}{\sigma}\right)\right) + \sigma^2(\phi(l, \mu, \sigma^2) - \phi(h, \mu, \sigma^2))$$

$$= \mu\left(\Phi\left(\frac{h-\mu}{\sigma}\right) - \Phi\left(\frac{l-\mu}{\sigma}\right)\right) + \sigma\left(\phi\left(\frac{l-\mu}{\sigma}\right) - \phi\left(\frac{h-\mu}{\sigma}\right)\right)$$

Exponential integrals

These are commonly used in Black-Scholes-type calculations.

$$\mathbb{E}[e^{\alpha X}\mathbf{1}_{\{l \le X \le h\}}] = \int_l^h \exp\left(\alpha x - \frac{1}{2}\left(\frac{x-\mu}{\sigma}\right)^2\right) dx$$

$$= e^{\alpha\mu + \frac{1}{2}\alpha^2\sigma^2} \int_l^h \exp\left(-\frac{1}{2}\left(\frac{x-(\mu+\alpha\sigma^2)}{\sigma}\right)^2\right) dx$$

$$= e^{\alpha\mu + \frac{1}{2}\alpha^2\sigma^2} I_0(l, h, \mu, +\alpha\sigma^2, \sigma^2)$$

$$= e^{\alpha\mu + \frac{1}{2}\alpha^2\sigma^2}\left(\Phi\left(\frac{h-(\mu+\alpha\sigma^2)}{\sigma}\right) - \Phi\left(\frac{l-(\mu+\alpha\sigma^2)}{\sigma}\right)\right).$$

References

Andersen L. and Andreasen J. (2000) Volatility skews and extensions of the Libor market model. *Applied Mathematical Finance*, 7(1), 1–32.

Babbs S.H. (1990) The term structure of interest rates: stochastic processes and contingent claims. PhD thesis, Imperial College, London.

Babbs S.H. and Selby M.J.P. (1998) Pricing by arbitrage under arbitrary information. *Mathematical Finance*, 8(2), 163–168.

Balland P. and Hughston L.P. (2000) Markov market model consistent with caplet smile. *International Journal of Theoretical and Applied Finance*, 3(2), 161–181

Baxter M.W. (1997) General interest rate models and the universality of HJM. In S. Pliska and M. Dempster (eds), *Mathematics of Derivative Securities*. Cambridge University Press, Cambridge.

Baxter M.W. and Rennie A. (1996) *Financial Calculus*. Cambridge University Press, Cambridge.

Bennett M.N. and Kennedy, J.E. (2004) A comparison of Markov-functional and market models: the one-dimensional case. Preprint, University of Warwick.

Billingsley P. (1986) *Probability and Measure*, 2nd edition. John Wiley & Sons, Chichester.

Björk T., Di Masi G., Kabanov Y. and Runggaldier W. (1997a) Towards a general theory of bond markets. *Finance and Stochastics*, 1, 141–174.

Björk T., Kabanov Y. and Runggaldier W. (1997b) Bond market structure in the presence of marked point processes. *Mathematical Finance*, 7, 211–239.

Black F. (1976) The pricing of commodity contracts. *Journal of Financial Economics*, 3, 167–179.

Black F., Derman E. and Toy W. (1990) A one-factor model of interest rates and its application to Treasury bond options. *Financial Analysts Journal*, Jan.-Feb., 33–39.

Black F. and Karasinski P. (1991) Bond and option pricing when short rates are lognormal. *Financial Analysts Journal*, July-Aug., 52–59.

Black F. and Scholes M. (1973) The pricing of options and corporate liabilities. *Journal of Political Economy*, 81, 637–659.

Brace A., Gatarek D. and Musiela M. (1997) The market model of interest rate dynamics. *Mathematical Finance*, 7, 127–154.

Breeden M.J. and Litzenberger R. (1978) Prices of state-contingent claims implicit in option prices. *Journal of Business*, 51, 621–651.

Breiman L. (1992) *Probability*, classic edition. Society of Industrial and Applied Mathematics, Philadelphia.

Financial Derivatives in Theory and Practice Revised Edition. P. J. Hunt and J. E. Kennedy
© 2004 John Wiley & Sons, Ltd   ISBNs: 0-470-86358-7 (HB); 0-470-86359-5 (PB)

Chung K.L. (1974) *A Course in Probability Theory*, 2nd edition. Academic Press, New York.

Chung K.L. and Williams R.J. (1990) *Introduction to Stochastic Integration*. Birkhäuser, Boston.

Coleman T.S. (1995) Convexity adjustment for constant maturity swaps and Libor-in-arrears basis swaps. *Derivatives Quarterly*, Winter, 19–27.

Cox J.C., Ingersoll J.E. and Ross, S.A. (1985) A theory of the term structure of interest rates. *Econometrica*, 53, 385–408.

Delbaen F. and Schachermayer W. (1999) Non-arbitrage and the fundamental theorem of asset pricing: Summary of main results. In D.C. Heath and G. Swindle (eds), *Introduction to Mathematical Finance*. Proceedings of Symposia in Applied Mathematics of the AMS, 57, 49–58.

Dellacherie C. and Meyer P.-A. (1980) *Probabilité et Potentiel. Théorie des Martingales*. Hermann, Paris.

Doust P. (1995) Relative pricing techniques in the swaps and options markets. *Journal of Financial Engineering*, 4(1), 11–46.

Duffie D. and Stanton R. (1992) Pricing continuously resettled contingent claims. *J. Economic Dynamics and Control*, 16, 561–573.

Duffie D. (1996) *Dynamic Asset Pricing Theory*, 2nd edition. Princeton University Press, Princeton, NJ.

Duffie D. and Singleton K.J. (1999) Modeling term structures of defaultable bonds. *Review of Financial Studies*, 12(4), 687–720.

Dupire B. (1994) Pricing with a smile. *RISK Magazine*, 7(1), 18–20.

Durrett R. (1984) *Brownian Motion and Martingales in Analysis*. Wadsworth, Belmont, CA.

Durrett R. (1996) *Stochastic Calculus: A Practical Introduction*. CRC Press, Boca Raton, FL.

Elliott R.J. (1982) *Stochastic Calculus and Applications*. Springer-Verlag, Berlin, Heidelberg and New York.

Flesaker B. and Hughston L.P. (1996) Positive interest. *RISK Magazine*, 9(1), 46–49.

Flesaker B. and Hughston L.P. (1997) International models for interest rates and foreign exchange. *Net Exposure*, 3, 55–79.

Gandhi S.K. and Hunt P.J. (1997) Numerical option pricing using conditioned diffusions. In S. Pliska and M. Dempster (eds), *Mathematics of Derivative Securities*. Cambridge University Press, Cambridge.

Geman H., El Karoui N. and Rochet J. (1995) Changes of numéraire, changes of probability measure and option pricing. *Journal of Applied Probability*, 32, 443–458.

Glasserman P. and Zhao X. (2000) Arbitrage-free discretization of lognormal forward Libor and swap rate models. *Finance and Stochastics*, 4(1).

Harrison J.M. and Kreps D. (1979) Martingales and arbitrage in multiperiod securities markets. *Journal of Economic Theory*, 20, 381–408.

Harrison, J.M. and Pliska S. (1981) Martingales and stochastic integrals in the theory of continuous trading. *Stochastic Processes and Their Applications*, 11, 215–260.

Heath D., Jarrow R. and Morton A. (1992) Bond pricing and the term structure of interest rates: a new methodology for contingent claims valuation. *Econometrica*, 61(1), 77–105.

Hogan M. and Weintraub K. (1993) The log-normal interest rate model and Eurodollar futures. Working Paper, Citibank, New York.

Hull J. and White A. (1990) Pricing interest rate derivative securities. *Review of Financial Studies*, 3(4), 573–592.

Hull J. and White A. (1994) Numerical procedures for implementing term structure models I: Single-factor models. *Journal of Derivatives*, Fall, 7–16.

Hunt P.J. and Kennedy J.E. (1996) On multi-currency interest rate models. Working Paper, University of Warwick.

Hunt P.J. and Kennedy J.E. (1997) On convexity corrections. Working Paper, University of Warwick.

Hunt P.J. and Kennedy J.E. (1998) Implied interest rate pricing models. *Finance and Stochastics*, 3, 275–293.

Hunt P.J., Kennedy J.E. and Pelsser A.A.J. (2000) Markov-functional interest rate models. *Finance and Stochastics*, 4(4).

Hunt P.J., Kennedy J.E. and Scott E.M. (1996) Terminal swap-rate models. Working Paper, University of Warwick.

Hunt P.J. and Pelsser A.A.J. (1996) Arbitrage-free pricing of quanto-swaptions. *Journal of Financial Engineering*, 7(1), 25–33.

Jacka S.D., Hamza K. and Klebaner F. (1998) Arbitrage-free term structure models. Research Report No. 303, University of Warwick Statistics Department.

Jacod J. (1979) *Calcul Stochastique et Problèmes de Martingales, Lecture Notes in Math. 714*. Springer-Verlag, Berlin, Heidelberg and New York.

Jamshidian F. (1989) An exact bond option formula. *Journal of Finance*, 44(1), 205–209.

Jamshidian F. (1991) Forward induction and construction of yield curve diffusion models. *Journal of Fixed Income*, 1, 62–74.

Jamshidian F. (1996a) Sorting out swaptions. *RISK Magazine*, 9(3), 59–60.

Jamshidian F. (1996b) Libor and swap market models and measures, II. Working Paper, Sakura Global Capital, London.

Jamshidian F. (1997) Libor and swap market models and measures. *Finance and Stochastics*, 1, 261–291.

Jarrow R.A. and Madan D. (1995) Option pricing using the term structure of interest rates to hedge systematic discontinuities in asset returns. *Mathematical Finance*, 5, 311–336.

Jarrow R.A. and Turnbull S.M. (1995) Pricing derivatives on financial securities subject to credit risk. *Journal of Finance*, 50, 53–85.

Jin Y. and Glasserman P. (2001) Equilibrium positive interest rates: a unified view. *Review of Financial Studies*, 14, 187–214.

Karatzas I. (1993) IMA Tutorial Lectures 1–3: Minneapolis. Department of Statistics, Columbia University.

Karatzas I. and Shreve S.E. (1991) *Brownian Motion and Stochastic Calculus*, 2nd edition. Springer-Verlag, Berlin, Heidelberg and New York.

Karatzas I. and Shreve S.E. (1998) *Methods of Mathematical Finance*. Springer-Verlag, New York.

Kennedy D.P. (1994) The term structure of interest rates as a Gaussian random field. *Mathematical Finance*, 4, 247–258.

Kunita H. and Watanabe S. (1967) On square integrable martingales. *Nagoya Mathematical Journal*, 30, 209–245.

Lamberton D. and Lapeyre B. (1996) *Introduction to Stochastic Calculus Applied to Finance*, English edition, translated by N. Rabeau and F. Mantion. Chapman & Hall, London.

Margrabe W. (1978) The value of an option to exchange one asset for another. *Journal of Finance*, 33, 177–186.

Merton R. (1973) The theory of rational option pricing. *Bell Journal of Economics and Management Science*, 4, 141–183.

Miltersen K., Sandmann K. and Sondermann D. (1997) Closed form solutions for term structure derivatives with log-normal interest rates. *Journal of Finance*, 52, 409–430.

Musiela M. and Rutkowski M. (1997) *Martingale Methods in Financial Modelling*. Springer-Verlag, Berlin, Heidelberg and New York.

Neuberger A. (1990) Pricing swap options using the forward swap market. Working Paper, London Business School.

Oksendal B. (1989) *Stochastic Differential Equations*, 2nd edition. Springer-Verlag, Berlin, Heidelberg and New York.

Press W.H., Teukolsky S.A., Vetterling W.T. and Flannery B.P. (1988) *Numerical Recipes in C*, 2nd edition. Cambridge University Press, Cambridge.

Priestley H.A. (1997) *Introduction to Integration*. Clarendon Press, Oxford.

Protter P. (1990) *Stochastic Integration and Differential Equations*. Springer-Verlag, Berlin, Heidelberg and New York.

Rebonato, R. (2002) *Modern Pricing of Interest-Rate Derivatives*. Princeton University Press, Princeton, NJ.

Revuz D. and Yor M. (1991) *Continuous Martingales and Brownian Motion*. Springer-Verlag, Berlin, Heidelberg and New York.

Ritchken P. and Sankarasubramanian L. (1995) Volatility structures of forward rates and the dynamics of the term structure. *Mathematical Finance*, 5, 55–72.

Rogers L.C.G. and Williams D. (1987) *Diffusions, Markov Processes, and Martingales. Vol. 2: Itô Calculus*. John Wiley & Sons, Chichester.

Rogers L.C.G. and Williams D. (1994) *Diffusions, Markov Processes, and Martingales. Vol. 1: Foundations*, 2nd edition. John Wiley & Sons, Chichester.

Rudin W. (1987) *Real and Complex Analysis*, 3rd edition. McGraw-Hill, Singapore.

Rutkowski M. (1997) A note on the Flesaker–Hughston model of term structure of interest rates. *Applied Mathematical Finance*, 4(3), 151–163.

Rutkowski M. (2001) Modelling of forward LIBOR and swap rates. In E. Jouini, J. Cvitanic and M. Musiela (eds), *Option Pricing, Interest Rates and Risk Management*, 336–395. Cambridge University Press, Cambridge.

Stroock D.W. and Varadhan S.R.S. (1979) *Multidimensional Diffusion Processes*. Springer-Verlag, Berlin, Heidelberg and New York.

Vasicek O.A. (1977) An equilibrium characterisation of the term structure. *Journal of Financial Economics*, 5, 177–188.

Williams D. (1991) *Probability with Martingales*. Cambridge University Press, Cambridge.

Willmot P., Dewynne J.N. and Howison S. (1993) *Option Pricing: Numerical Models and Computation*. Oxford Financial Press, Oxford.

Yamada T. and Watanabe S. (1971) On the uniqueness of solutions of stochastic differential equations. *Journal of Mathematics of Kyoto University*, 11, 155–167.

Index

Financial Derivatives in Theory and Practice Revised Edition. P. J. Hunt and J. E. Kennedy
© 2004 John Wiley & Sons, Ltd ISBNs: 0-470-86358-7 (HB); 0-470-86359-5 (PB)

WILEY SERIES IN PROBABILITY AND STATISTICS

ESTABLISHED BY WALTER A. SHEWHART AND SAMUEL S. WILKS

Editors: *David J. Balding, Noel A. C. Cressie, Nicholas I. Fisher,*
Iain M. Johnstone, J. B. Kadane, Geert Molenberghs, Louise M. Ryan,
David W. Scott, Adrian F. M. Smith, Jozef L. Teugels
Editors Emeriti: *Vic Barnett, J. Stuart Hunter David G. Kendall*

The *Wiley Series in Probability and Statistics* is well established and authoritative. It covers many topics of current research interest in both pure and applied statistics and probability theory. Written by leading statisticians and institutions, the titles span both state-of-the-art developments in the field and classical methods.

Reflecting the wide range of current research in statistics, the series encompasses applied, methodological and theoretical statistics, ranging from applications and new techniques made possible by advances in computerized practice to rigorous treatment of theoretical approaches.

This series provides essential and invaluable reading for all statisticians, whether in academia, industry, government, or research.

ABRAHAM and LEDOLTER · Statistical Methods for Forecasting
AGRESTI · Analysis of Ordinal Categorical Data
AGRESTI · An Introduction to Categorical Data Analysis
AGRESTI · Categorical Data Analysis, *Second Edition*
ALTMAN, GILL, and McDONALD · Numerical Issues in Statistical Computing
   for the Social Scientist
AMARATUNGA and CABRERA · Exploration and Analysis of DNA Microarray and Protein
   Array Data
ANDĚL · Mathematics of Chance
ANDERSON · An Introduction to Multivariate Statistical Analysis, *Third Edition*
*ANDERSON · The Statistical Analysis of Time Series
ANDERSON, AUQUIER, HAUCK, OAKES, VANDAELE, and WEISBERG ·
   Statistical Methods for Comparative Studies
ANDERSON and LOYNES · The Teaching of Practical Statistics
ARMITAGE and DAVID (editors) · Advances in Biometry
ARNOLD, BALAKRISHNAN, and NAGARAJA · Records
*ARTHANARI and DODGE · Mathematical Programming in Statistics
*BAILEY · The Elements of Stochastic Processes with Applications to the Natural
   Sciences
BALAKRISHNAN and KOUTRAS · Runs and Scans with Applications
BARNETT · Comparative Statistical Inference, *Third Edition*
BARNETT · Environmental Statistics: Methods & Applications
BARNETT and LEWIS · Outliers in Statistical Data, *Third Edition*
BARTOSZYNSKI and NIEWIADOMSKA-BUGAJ · Probability and Statistical Inference
BASILEVSKY · Statistical Factor Analysis and Related Methods: Theory and
   Applications
BASU and RIGDON · Statistical Methods for the Reliability of Repairable Systems
BATES and WATTS · Nonlinear Regression Analysis and Its Applications
BECHHOFER, SANTNER, and GOLDSMAN · Design and Analysis of Experiments for
   Statistical Selection, Screening, and Multiple Comparisons
BELSLEY · Conditioning Diagnostics: Collinearity and Weak Data in Regression

*Now available in a lower priced paperback edition in the Wiley Classics Library.

BELSLEY, KUH, and WELSCH · Regression Diagnostics: Identifying Influential
Data and Sources of Collinearity
BENDAT and PIERSOL · Random Data: Analysis and Measurement Procedures,
Third Edition
BERRY, CHALONER, and GEWEKE · Bayesian Analysis in Statistics and
Econometrics: Essays in Honor of Arnold Zellner
BERNARDO and SMITH · Bayesian Theory
BHAT and MILLER · Elements of Applied Stochastic Processes, *Third Edition*
BHATTACHARYA and JOHNSON · Statistical Concepts and Methods
BHATTACHARYA and WAYMIRE · Stochastic Processes with Applications
BILLINGSLEY · Convergence of Probability Measures, *Second Edition*
BILLINGSLEY · Probability and Measure, *Third Edition*
BIRKES and DODGE · Alternative Methods of Regression
BLISCHKE AND MURTHY (editors) · Case Studies in Reliability and Maintenance
BLISCHKE AND MURTHY · Reliability: Modeling, Prediction, and Optimization
BLOOMFIELD · Fourier Analysis of Time Series: An Introduction, *Second Edition*
BOLLEN · Structural Equations with Latent Variables
BOROVKOV · Ergodicity and Stability of Stochastic Processes
BOULEAU · Numerical Methods for Stochastic Processes
BOX · Bayesian Inference in Statistical Analysis
BOX · R. A. Fisher, the Life of a Scientist
BOX and DRAPER · Empirical Model-Building and Response Surfaces
*BOX and DRAPER · Evolutionary Operation: A Statistical Method for Process
Improvement
BOX, HUNTER, and HUNTER · Statistics for Experimenters: An Introduction to
Design, Data Analysis, and Model Building
BOX and LUCEÑO · Statistical Control by Monitoring and Feedback Adjustment
BRANDIMARTE · Numerical Methods in Finance: A MATLAB-Based Introduction
BROWN and HOLLANDER · Statistics: A Biomedical Introduction
BRUNNER, DOMHOF, and LANGER · Nonparametric Analysis of Longitudinal
Data in Factorial Experiments
BUCKLEW · Large Deviation Techniques in Decision, Simulation, and Estimation
CAIROLI and DALANG · Sequential Stochastic Optimization
CHAN · Time Series: Applications to Finance
CHATTERJEE and HADI · Sensitivity Analysis in Linear Regression
CHATTERJEE and PRICE · Regression Analysis by Example, *Third Edition*
CHERNICK · Bootstrap Methods: A Practitioner's Guide
CHERNICK and FRIIS · Introductory Biostatistics for the Health Sciences
CHILÈS and DELFINER · Geostatistics: Modeling Spatial Uncertainty
CHOW and LIU · Design and Analysis of Clinical Trials: Concepts and Methodologies, *Second
Edition*
CLARKE and DISNEY · Probability and Random Processes: A First Course with
Applications, *Second Edition*
*COCHRAN and COX · Experimental Designs, *Second Edition*
CONGDON · Applied Bayesian Modelling
CONGDON · Bayesian Statistical Modelling
CONOVER · Practical Nonparametric Statistics, *Second Edition*
COOK · Regression Graphics
COOK and WEISBERG · Applied Regression Including Computing and Graphics
COOK and WEISBERG · An Introduction to Regression Graphics
CORNELL · Experiments with Mixtures, Designs, Models, and the Analysis of Mixture
Data, *Third Edition*
COVER and THOMAS · Elements of Information Theory

*Now available in a lower priced paperback edition in the Wiley Classics Library.

COX · A Handbook of Introductory Statistical Methods
*COX · Planning of Experiments
CRESSIE · Statistics for Spatial Data, *Revised Edition*
CSÖRGŐ and HORVÁTH · Limit Theorems in Change Point Analysis
DANIEL · Applications of Statistics to Industrial Experimentation
DANIEL · Biostatistics: A Foundation for Analysis in the Health Sciences, *Sixth Edition*
*DANIEL · Fitting Equations to Data: Computer Analysis of Multifactor Data,
   Second Edition
DASU and JOHNSON · Exploratory Data Mining and Data Cleaning
DAVID and NAGARAJA · Order Statistics, *Third Edition*
*DEGROOT, FIENBERG, and KADANE · Statistics and the Law
DEL CASTILLO · Statistical Process Adjustment for Quality Control
DENISON, HOLMES, MALLICK and SMITH · Bayesian Methods for Nonlinear Classification
   and Regression
DETTE and STUDDEN · The Theory of Canonical Moments with Applications in
   Statistics, Probability, and Analysis
DEY and MUKERJEE · Fractional Factorial Plans
DILLON and GOLDSTEIN · Multivariate Analysis: Methods and Applications
DODGE · Alternative Methods of Regression
*DODGE and ROMIG · Sampling Inspection Tables, *Second Edition*
*DOOB · Stochastic Processes
DOWDY, WEARDEN and CHILKO · Statistics for Research, *Third Edition*
DRAPER and SMITH · Applied Regression Analysis, *Third Edition*
DRYDEN and MARDIA · Statistical Shape Analysis
DUDEWICZ and MISHRA · Modern Mathematical Statistics
DUNN and CLARK · Applied Statistics: Analysis of Variance and Regression, *Second Edition*
DUNN and CLARK · Basic Statistics: A Primer for the Biomedical Sciences,
   Third Edition
DUPUIS and ELLIS · A Weak Convergence Approach to the Theory of Large Deviations
*ELANDT-JOHNSON and JOHNSON · Survival Models and Data Analysis
ENDERS · Applied Econometric Time Series
ETHIER and KURTZ · Markov Processes: Characterization and Convergence
EVANS, HASTINGS, and PEACOCK · Statistical Distributions, *Third Edition*
FELLER · An Introduction to Probability Theory and Its Applications, Volume I,
   Third Edition, Revised; Volume II, *Second Edition*
FISHER and VAN BELLE · Biostatistics: A Methodology for the Health Sciences
*FLEISS · The Design and Analysis of Clinical Experiments
FLEISS · Statistical Methods for Rates and Proportions, *Third Edition*
FLEMING and HARRINGTON · Counting Processes and Survival Analysis
FULLER · Introduction to Statistical Time Series, *Second Edition*
FULLER · Measurement Error Models
GALLANT · Nonlinear Statistical Models
GIESBRECHT and GUMPERTZ · Planning, Construction, and Statistical Analysis of Compar-
   ative Experiments
GIFI · Nonlinear Multivariate Analysis
GHOSH, MUKHOPADHYAY, and SEN · Sequential Estimation
GLASSERMAN and YAO · Monotone Structure in Discrete-Event Systems
GNANADESIKAN · Methods for Statistical Data Analysis of Multivariate Observations,
   Second Edition
GOLDSTEIN and LEWIS · Assessment: Problems, Development, and Statistical Issues
GREENWOOD and NIKULIN · A Guide to Chi-Squared Testing
GROSS and HARRIS · Fundamentals of Queueing Theory, *Third Edition*
*HAHN and SHAPIRO · Statistical Models in Engineering

*Now available in a lower priced paperback edition in the Wiley Classics Library.

HAHN and MEEKER · Statistical Intervals: A Guide for Practitioners

HALD · A History of Probability and Statistics and their Applications Before 1750

HALD · A History of Mathematical Statistics from 1750 to 1930

HAMPEL · Robust Statistics: The Approach Based on Influence Functions

HANNAN and DEISTLER · The Statistical Theory of Linear Systems

HEIBERGER · Computation for the Analysis of Designed Experiments

HEDAYAT and SINHA · Design and Inference in Finite Population Sampling

HELLER · MACSYMA for Statisticians

HINKELMAN and KEMPTHORNE: · Design and Analysis of Experiments, Volume 1: Introduction to Experimental Design

HOAGLIN, MOSTELLER, and TUKEY · Exploratory Approach to Analysis of Variance

HOAGLIN, MOSTELLER, and TUKEY · Exploring Data Tables, Trends and Shapes

*HOAGLIN, MOSTELLER, and TUKEY · Understanding Robust and Exploratory Data Analysis

HOCHBERG and TAMHANE · Multiple Comparison Procedures

HOCKING · Methods and Applications of Linear Models: Regression and the Analysis of Variance, Second Edition

HOEL · Introduction to Mathematical Statistics, Fifth Edition

HOGG and KLUGMAN · Loss Distributions

HOLLANDER and WOLFE · Nonparametric Statistical Methods, Second Edition

HOSMER and LEMESHOW · Applied Logistic Regression, Second Edition

HOSMER and LEMESHOW · Applied Survival Analysis: Regression Modeling of Time to Event Data

HØYLAND and RAUSAND · System Reliability Theory: Models and Statistical Methods

HUBER · Robust Statistics

HUBERTY · Applied Discriminant Analysis

HUNT and KENNEDY · Financial Derivatives in Theory and Practice, Revised Edition

HUSKOVA, BERAN, and DUPAC · Collected Works of Jaroslav Hajek – with Commentary

IMAN and CONOVER · A Modern Approach to Statistics

JACKSON · A User's Guide to Principle Components

JOHN · Statistical Methods in Engineering and Quality Assurance

JOHNSON · Multivariate Statistical Simulation

JOHNSON and BALAKRISHNAN · Advances in the Theory and Practice of Statistics: A Volume in Honor of Samuel Kotz

JUDGE, GRIFFITHS, HILL, LÜTKEPOHL, and LEE · The Theory and Practice of Econometrics, Second Edition

JOHNSON and KOTZ · Distributions in Statistics

JOHNSON and KOTZ (editors) · Leading Personalities in Statistical Sciences: From the Seventeenth Century to the Present

JOHNSON, KOTZ, and BALAKRISHNAN · Continuous Univariate Distributions, Volume 1, Second Edition

JOHNSON, KOTZ, and BALAKRISHNAN · Continuous Univariate Distributions, Volume 2, Second Edition

JOHNSON, KOTZ, and BALAKRISHNAN · Discrete Multivariate Distributions

JOHNSON, KOTZ, and KEMP · Univariate Discrete Distributions, Second Edition

JUREČKOVÁ and SEN · Robust Statistical Procedures: Asymptotics and Interrelations

JUREK and MASON · Operator-Limit Distributions in Probability Theory

KADANE · Bayesian Methods and Ethics in a Clinical Trial Design

KADANE and SCHUM · A Probabilistic Analysis of the Sacco and Vanzetti Evidence

KALBFLEISCH and PRENTICE · The Statistical Analysis of Failure Time Data Second Edition

KARIYA and KURATA · Generalized Least Squares

KASS and VOS · Geometrical Foundations of Asymptotic Inference

*Now available in a lower priced paperback edition in the Wiley Classics Library.

KAUFMAN and ROUSSEEUW · Finding Groups in Data: An Introduction to Cluster Analysis

KEDEM and FOKIANOS · Regression Models for Time Series Analysis

KENDALL, BARDEN, CARNE, and LE · Shape and Shape Theory

KHURI · Advanced Calculus with Applications in Statistics, *Second Edition*

KHURI, MATHEW, and SINHA · Statistical Tests for Mixed Linear Models

KLEIBER and KOTZ · Statistical Size Distributions in Economics and Actuarial Sciences

KLUGMAN, PANJER, and WILLMOT · Loss Models: From Data to Decisions

KLUGMAN, PANJER, and WILLMOT · Solutions Manual to Accompany Loss Models: From Data to Decisions

KOTZ, BALAKRISHNAN, and JOHNSON · Continuous Multivariate Distributions, Volume 1, *Second Edition*

KOTZ and JOHNSON (editors) · Encyclopedia of Statistical Sciences: Volumes 1 to 9 with Index

KOTZ and JOHNSON (editors) · Encyclopedia of Statistical Sciences: Supplement Volume

KOTZ, READ, and BANKS (editors) · Encyclopedia of Statistical Sciences: Update Volume 1

KOTZ, READ, and BANKS (editors) · Encyclopedia of Statistical Sciences: Update Volume 2

KOVALENKO, KUZNETZOV, and PEGG · Mathematical Theory of Reliability of Time-Dependent Systems with Practical Applications

LACHIN · Biostatistical Methods: The Assessment of Relative Risks

LAD · Operational Subjective Statistical Methods: A Mathematical, Philosophical, and Historical Introduction

LAMPERTI · Probability: A Survey of the Mathematical Theory, *Second Edition*

LANGE, RYAN, BILLARD, BRILLINGER, CONQUEST, and GREENHOUSE · Case Studies in Biometry

LARSON · Introduction to Probability Theory and Statistical Inference, *Third Edition*

LAWLESS · Statistical Models and Methods for Lifetime Data, *Second Edition*

LAWSON · Statistical Methods in Spatial Epidemiology

LE · Applied Categorical Data Analysis

LE · Applied Survival Analysis

LEE and WANG · Statistical Methods for Survival Data Analysis, *Third Edition*

LePAGE and BILLARD · Exploring the Limits of Bootstrap

LEYLAND and GOLDSTEIN (editors) · Multilevel Modelling of Health Statistics

LIAO · Statistical Group Comparison

LINDVALL · Lectures on the Coupling Method

LINHART and ZUCCHINI · Model Selection

LITTLE and RUBIN · Statistical Analysis with Missing Data, *Second Edition*

LLOYD · The Statistical Analysis of Categorical Data

MAGNUS and NEUDECKER · Matrix Differential Calculus with Applications in Statistics and Econometrics, *Revised Edition*

MALLER and ZHOU · Survival Analysis with Long Term Survivors

MALLOWS · Design, Data, and Analysis by Some Friends of Cuthbert Daniel

MANN, SCHAFER, and SINGPURWALLA · Methods for Statistical Analysis of Reliability and Life Data

MANTON, WOODBURY, and TOLLEY · Statistical Applications Using Fuzzy Sets

MARDIA and JUPP · Directional Statistics

MASON, GUNST, and HESS · Statistical Design and Analysis of Experiments with Applications to Engineering and Science, *Second Edition*

McCULLOCH and SEARLE · Generalized, Linear, and Mixed Models

McFADDEN · Management of Data in Clinical Trials

*Now available in a lower priced paperback edition in the Wiley Classics Library.

McLACHLAN · Discriminant Analysis and Statistical Pattern Recognition

McLACHLAN and KRISHNAN · The EM Algorithm and Extensions

McLACHLAN and PEEL · Finite Mixture Models

McNEIL · Epidemiological Research Methods

MEEKER and ESCOBAR · Statistical Methods for Reliability Data

MEERSCHAERT and SCHEFFLER · Limit Distributions for Sums of Independent Random Vectors: Heavy Tails in Theory and Practice

*MILLER · Survival Analysis, *Second Edition*

MONTGOMERY, PECK, and VINING · Introduction to Linear Regression Analysis, *Third Edition*

MORGENTHALER and TUKEY · Configural Polysampling: A Route to Practical Robustness

MUIRHEAD · Aspects of Multivariate Statistical Theory

MURRAY · X-STAT 2.0 Statistical Experimentation, Design Data Analysis, and Nonlinear Optimization

MURTHY, XIE, and JIANG · Weibull Models

MYERS and MONTGOMERY · Response Surface Methodology: Process and Product Optimization Using Designed Experiments, *Second Edition*

MYERS, MONTGOMERY, and VINING · Generalized Linear Models. With Applications in Engineering and the Sciences

NELSON · Accelerated Testing, Statistical Models, Test Plans, and Data Analyses

NELSON · Applied Life Data Analysis

NEWMAN · Biostatistical Methods in Epidemiology

OCHI · Applied Probability and Stochastic Processes in Engineering and Physical Sciences

OKABE, BOOTS, SUGIHARA, and CHIU · Spatial Tesselations: Concepts and Applications of Voronoi Diagrams, *Second Edition*

OLIVER and SMITH · Influence Diagrams, Belief Nets and Decision Analysis

PALTA · Quantitative Methods in Population Health: Extensions of Ordinary Regressions

PANKRATZ · Forecasting with Dynamic Regression Models

PANKRATZ · Forecasting with Univariate Box-Jenkins Models: Concepts and Cases

*PARZEN · Modern Probability Theory and It's Applications

PEÑA, TIAO, and TSAY · A Course in Time Series Analysis

PIANTADOSI · Clinical Trials: A Methodologic Perspective

PORT · Theoretical Probability for Applications

POURAHMADI · Foundations of Time Series Analysis and Prediction Theory

PRESS · Bayesian Statistics: Principles, Models, and Applications

PRESS · Subjective and Objective Bayesian Statistics, *Second Edition*

PRESS and TANUR · The Subjectivity of Scientists and the Bayesian Approach

PUKELSHEIM · Optimal Experimental Design

PURI, VILAPLANA, and WERTZ · New Perspectives in Theoretical and Applied Statistics

PUTERMAN · Markov Decision Processes: Discrete Stochastic Dynamic Programming

*RAO · Linear Statistical Inference and Its Applications, *Second Edition*

RENCHER · Linear Models in Statistics

RENCHER · Methods of Multivariate Analysis, *Second Edition*

RENCHER · Multivariate Statistical Inference with Applications

RIPLEY · Spatial Statistics

RIPLEY · Stochastic Simulation

ROBINSON · Practical Strategies for Experimenting

ROHATGI and SALEH · An Introduction to Probability and Statistics, *Second Edition*

*Now available in a lower priced paperback edition in the Wiley Classics Library.

ROLSKI, SCHMIDLI, SCHMIDT, and TEUGELS · Stochastic Processes for Insurance and Finance

ROSENBERGER and LACHIN · Randomization in Clinical Trials: Theory and Practice

ROSS · Introduction to Probability and Statistics for Engineers and Scientists

ROUSSEEUW and LEROY · Robust Regression and Outlier Detection

RUBIN · Multiple Imputation for Nonresponse in Surveys

RUBINSTEIN · Simulation and the Monte Carlo Method

RUBINSTEIN and MELAMED · Modern Simulation and Modeling

RYAN · Modern Regression Methods

RYAN · Statistical Methods for Quality Improvement, Second Edition

SALTELLI, CHAN, and SCOTT (editors) · Sensitivity Analysis

*SCHEFFE · The Analysis of Variance

SCHIMEK · Smoothing and Regression: Approaches, Computation, and Application

SCHOTT · Matrix Analysis for Statistics

SCHOUTENS · Levy Processes in Finance: Pricing Financial Derivatives

SCHUSS · Theory and Applications of Stochastic Differential Equations

SCOTT · Multivariate Density Estimation: Theory, Practice, and Visualization

*SEARLE · Linear Models

SEARLE · Linear Models for Unbalanced Data

SEARLE · Matrix Algebra Useful for Statistics

SEARLE, CASELLA, and McCULLOCH · Variance Components

SEARLE and WILLETT · Matrix Algebra for Applied Economics

SEBER and LEE · Linear Regression Analysis, Second Edition

SEBER · Multivariate Observations

SEBER and WILD · Nonlinear Regression

SENNOTT · Stochastic Dynamic Programming and the Control of Queueing Systems

*SERFLING · Approximation Theorems of Mathematical Statistics

SHAFER and VOVK · Probability and Finance: Its Only a Game!

SMALL and McLEISH · Hilbert Space Methods in Probability and Statistical Inference

SRIVASTAVA · Methods of Multivariate Statistics

STAPLETON · Linear Statistical Models

STAUDTE and SHEATHER · Robust Estimation and Testing

STOYAN, KENDALL, and MECKE · Stochastic Geometry and Its Applications, Second Edition

STOYAN and STOYAN · Fractals, Random Shapes and Point Fields: Methods of Geometrical Statistics

STYAN · The Collected Papers of T. W. Anderson: 1943–1985

SUTTON, ABRAMS, JONES, SHELDON, and SONG · Methods for Meta-Analysis in Medical Research

TANAKA · Time Series Analysis: Nonstationary and Noninvertible Distribution Theory

THOMPSON · Empirical Model Building

THOMPSON · Sampling, Second Edition

THOMPSON · Simulation: A Modeler's Approach

THOMPSON and SEBER · Adaptive Sampling

THOMPSON, WILLIAMS, and FINDLAY · Models for Investors in Real World Markets

TIAO, BISGAARD, HILL, PEÑA, and STIGLER (editors) · Box on Quality and Discovery: with Design, Control, and Robustness

TIERNEY · LISP-STAT: An Object-Oriented Environment for Statistical Computing and Dynamic Graphics

TSAY · Analysis of Financial Time Series

UPTON and FINGLETON · Spatial Data Analysis by Example, Volume II: Categorical and Directional Data

*Now available in a lower priced paperback edition in the Wiley Classics Library.

*Now available in a lower priced paperback edition in the Wiley Classics Library.